I0072252

Advanced Molecular Biology

Volume I

Advanced Molecular Biology
Volume I

Edited by **Gildroy Swan**

R Callisto Reference

New York

Published by Callisto Reference,
106 Park Avenue, Suite 200,
New York, NY 10016, USA
www.callistoreference.com

Advanced Molecular Biology: Volume I
Edited by Gildroy Swan

© 2015 Callisto Reference

International Standard Book Number: 978-1-63239-019-6 (Hardback)

This book contains information obtained from authentic and highly regarded sources. Copyright for all individual chapters remain with the respective authors as indicated. A wide variety of references are listed. Permission and sources are indicated; for detailed attributions, please refer to the permissions page. Reasonable efforts have been made to publish reliable data and information, but the authors, editors and publisher cannot assume any responsibility for the validity of all materials or the consequences of their use.

The publisher's policy is to use permanent paper from mills that operate a sustainable forestry policy. Furthermore, the publisher ensures that the text paper and cover boards used have met acceptable environmental accreditation standards.

Trademark Notice: Registered trademark of products or corporate names are used only for explanation and identification without intent to infringe.

Printed in the United States of America.

Contents

Preface

The quest to know about the origins of life led to the birth of various branches of science. It opened up new avenues of study and research. One of them is molecular biology. Molecular biology's history goes back to 18th century when understanding the functioning of living organisms at the microscopic level led to numerous questions and subsequent answers. The development of technologies like x-ray diffraction, electron microscopy, ultracentrifugization, and electrophoresis by physicists and chemists led to learning the structure and function of the macromolecules.

Oxford dictionaries define molecular biology as the branch of biology that deals with the structure and function of the macromolecules (e.g. proteins and nucleic acids) essential to life. The proteins are active agents of living organisms, while deoxyribonucleic acid (DNA) is the most famous nucleic acid and the constituent of genes. The techniques of this subject are combined with skills of genetics and biochemistry for further study. Moreover, studies are also combined in bioinformatics and computational biology.

Since the late 1950s and early 1960s, molecular biologists have learned to characterize, isolate, and manipulate the molecular components of cells and organisms. These components include DNA, the repository of genetic information; RNA, a close relative of DNA whose functions range from serving as a temporary working copy of DNA to actual structural and enzymatic functions as well as a functional and structural part of the translational apparatus; and proteins the major structural and enzymatic type of molecules in cells. Also, molecular biology has extensive use in medicine, diagnosis of disease, etc.

I wish to thank all the contributing authors who have shared their knowledge with us in this book. Not only did they just contribute their studies, but they also helped me with the required editing as and when required. I also wish to thank my publisher for giving me this incredible opportunity of reviewing the work of these renowned international experts.

Editor

RASSF Signalling and DNA Damage: Monitoring the Integrity of the Genome?

Simon F. Scrace and Eric O'Neill

Department of Oncology, The Gray Institute, University of Oxford, Roosevelt Drive, Oxford OX3 7DQ, UK

Correspondence should be addressed to Eric O'Neill, eric.oneill@oncology.ox.ac.uk

Academic Editor: Geoffrey J. Clark

The RASSF family of proteins has been extensively studied in terms of their genetics, structure and function. One of the functions that has been increasingly studied is the role of the RASSF proteins in the DNA damage response. Surprisingly, this research, which encompasses both the classical and N-terminal RASSF proteins, has revealed an involvement of the RASSFs in oncogenic pathways as well as the more familiar tumour suppressor pathways usually associated with the RASSF family members. The most studied protein with respect to DNA damage is RASSF1A, which has been shown, not only to be activated by ATM, a major regulator of the DNA damage response, but also to bind to and activate a number of different pathways which all lead to and feedback from the guardian of the genome, p53. In this review we discuss the latest research linking the RASSF proteins to DNA damage signalling and maintenance of genomic integrity and look at how this knowledge is being utilised in the clinic to enhance the effectiveness of traditional cancer therapies such as radiotherapy.

1. Introduction

RASSF proteins were originally designated on the basis of sequence homology to domains that associate with Ras-like small GTP-binding proteins. These domains are known as Ras association (RA) domains [RalGDS (Ral guanine nucleotide dissociation stimulator)/AF6 (ALL-1 fusion partner from chromosome 6)] and are distinct from Ras-binding domains (RBD) which bind an alternative set of Ras effectors [1, 2]. Ras belongs to a family of small G-proteins that are ubiquitously expressed and oscillate between an inactive, GDP-bound state, and an active, GTP-bound state, in response to diverse cellular signals. Various GTP-bound Ras-like proteins bind effector proteins to mediate distinct biological responses. There are 150 Ras-like proteins encoded in the human genome which can be grouped by homology or functionality, as being similar to Ras, Rho, Rab, Arf (ADP-ribosylation factor), or Ran. While originally suggested to associate with Ras [3], the RASSF family has a differential affinity for Ras-like GTPases, with NORE1 (RAPL/RASSF5) displaying a much greater affinity for the closely related Ras homolog, Rap1B, than H-Ras itself [4]. The RA domain of RASSF1 associates with K-Ras, rather than H-Ras or N-Ras and is also described to associate with Ran [5, 6]. There are now 10 members in the RASSF family (RASSF1-10) subdivided into two distinct subgroups, the classical RASSF proteins (RASSF1-6) and the N-terminal RASSF proteins (RASSF7-10) based on the location of the RA domain [7]. Little is known about the GTP-binding proteins that may interact with the majority of the RASSF family or how they function but the potential exists for a greater number of signalling connections. In addition to an RA domain, the classical RASSF proteins also have a protein-protein interaction motif known as the SARAH domain that is responsible for scaffolding and regulatory interactions [8]. This domain is a short coiled-coil region and so named due to its location in the extreme C-terminus of genetically linked *Drosophila* proteins; Salvador (hSav1/WW45), dRASSF and Hippo (hMST1/2) (SARAH: SAlvador, RAssf, Hippo) which can form both homo- and heterodimers [9]. The N-terminal RASSFs lack an identifiable SARAH domain, although the SMART database predicts that RASSF7, 8 and 10 contain extensive coiled-coil regions, which can dimerise [10].

RASSF1A and RASSF5A [also known as NORE1A (Novel Ras Effector 1 isoform A)] also contain an N-terminal atypical diacylglycerol/phorbol ester-binding (DAG) domain also known as the protein kinase C conserved region (C1) domain that contains a central zinc finger (Zinc-binding domain) [11]. The Zinc finger in the RASSF family members is denoted "atypical" because it lacks critical residues required for binding of phorbol esters or DNA and therefore probably mediates protein-protein interactions. Indeed, structural analysis indicates that the C1 domain of NORE1A associates with the RA domain to occlude RAS association [12]. As none of the family members have any known enzymatic activity they are thought to be scaffold/adaptor proteins using these binding domains to bring target proteins together to impart their functions.

There are a number of reviews that introduce the RASSF family and the pathways within which they function; however, this paper will focus on the emerging roles of the RASSF family and their effectors in the response to DNA damage. The best described protein in this family with respect to DNA damage is RASSF1 thus the review will concentrate on this protein with particular reference to a recently elucidated signalling network from RASSF1A and the potential clinical significance of targeting this pathway [13].

2. RASSF1

It had long been suspected that the 3p21.3 region of the human genome harboured one or more important tumour suppressors because loss of heterozygosity (LOH) was found at this locus in lung, breast, and kidney tumours and genetic instability in this region is the earliest most frequently detected deficiency in lung cancers [14–20]. This 120 kb region contains 8 genes namely *CACNA2D2, PL6, 101F6, NPRL2/ G21, ZMYND10/BLU, RASSF1/123F2, FUS1,* and *MYAL2.* However, none of these candidate genes are frequently mutated in cancers [16, 21]. At the same time as these LOH studies, Dammann et al. identified RASSF1 as an interacting partner of the DNA damage repair protein xeroderma pigmentosum complementation group A (XPA) [22]. While the role of RASSF1 in nucleotide excision repair could not be verified, it may yet prove to be significant given the emerging role of RASSF1 in the DNA damage response. The *RASSF1* gene consists of 8 exons spanning a region of about 11 kb. The C-terminal showed high-sequence homology with NORE1, containing an RA domain and thus the gene was named *RASSF1* for Ras association domain family member 1 [22]. Alternative splicing generates 8 isoforms A–H from promoters held within 2 CpG islands. The first CpG island encompasses the promoter regions for RASSF1A, D, E, F, and G. Epigenetic inactivation by DNA methylation at this CpG island is one of the most common events in human cancers (reviewed in [23–25]). This methylation has recently been attributed to HOXB3 driven overexpression of the DNA methyltransferase, *DNMT3B* [26]. RASSF1B, C, and H are generated from a promoter located within the larger 3′ CpG island [27]. This commonly remains unmethylated in cancers and consequently cells retain expression of these isoforms [23]. RASSF1A and RASSF1C are the major transcripts of the *RASSF1* gene and are expressed ubiquitously in normal tissues [28].

3. RASSF1A

Exogenous expression of RASSF1A reduces colony formation in soft agar and reduced tumourigenicity in nude mice [22, 29–31]. Similarly, reexpression of RASSF1A using demethyltransferase inhibitors such as zebularine and 5-aza-2′-deoxycytidine caused significant growth arrest in ovarian cancer cell lines [32]. Reciprocally, RASSF1A knockout mice develop spontaneous tumours, particularly when combined with a knockout of p53, highlighting the significance of RASSF1A in tumour development [33–35]. In addition these RASSF1A$^{-/-}$, p53$^{-/-}$ mice showed high levels of aneuploidy/ tetraploidy suggesting an important role for RASSF1A in maintaining genomic integrity. RASSF1A has been shown to have many roles in cell cycle control and microtubule organisation [23, 27], the response to DNA damage is, however, only beginning to be elucidated. It is therefore timely to present these pathways and highlight their importance to the DNA damage response, genomic integrity, and cell survival during cancer development.

4. RASSF1 Phosphorylation

The majority of the phosphorylation of RASSF1A has being attributed to the phosphorylation of Serine 202/203. These sites have been demonstrated to be targeted by a number of kinases including, both CDK (Cyclin-Dependent Kinase) and Aurora kinases [36–40]. These phosphorylation events prevent the association of RASSF1A with microtubules during prometaphase. The phosphorylation of RASSF1A on these sites also coordinates the regulation of mitosis by controlling activation of the anaphase-promoting complex/ cyclosome (APC/C), and regulation of syntaxin16 to promote cytokinesis [39, 40]. Loss of phosphorylation at these sites leads to defects in mitosis resulting in aneuploidy and genomic instability.

In the DNA damage response phosphorylation of RASSF1A serine 131 (S131) is emerging as an important phosphorylation site. The initial kinases that respond to breaks in DNA are the phosphatidyl-inositol 3-kinase like kinases ATM (Ataxia Telangiectasia Mutated), ATR (ATM- and Rad3-Related), and DNA-PK$_{cs}$ (DNA-dependent protein kinase catalytic subunit) [41]. RASSF1A has a consensus site for ATM phosphorylation on serine 131 that is conserved in vertebrates and unique amongst family members and has recently been confirmed as a bone-fide target for ATM [13, 42]. Serine 131 phosphorylation appears important for RASSF1A activation and inactivating mutations of this site have been identified in human cancers [43]. Indeed Shivakumar et al., showed that mutation of the predicted phosphorylation site, S131F, removed the ability to induce cell cycle arrest and block cell proliferation [43]. ATM-dependent phosphorylation at the 131 site is also restricted by S131F and disables the ability of RASSF1A to respond to various DNA-damaging agents [13]. Mutations near the ATM site are hypothesised to function by inactivating ATM phosphorylation. One of these is a nonsynonymous single nucleotide polymorphism (SNP) at p.RASSF1A-A133S (rs2073498),

ATM phosphorylation site

Mutant serine	F**Q**A
Polymorphism	S**Q**S
WT sequence	ETPDL**SQA**E

126   134

Amino acid 1   51   101   195   288 290   340

RASSF1A   C1   RA   SARAH

MOAP-1 and   Ras/Ran   MST1/2 and
MDM2 binding   binding.   Salvador binding.

Amino acid   1   56–64 125   218 220   270

RASSF1C   RA   SARAH

DAXX binding

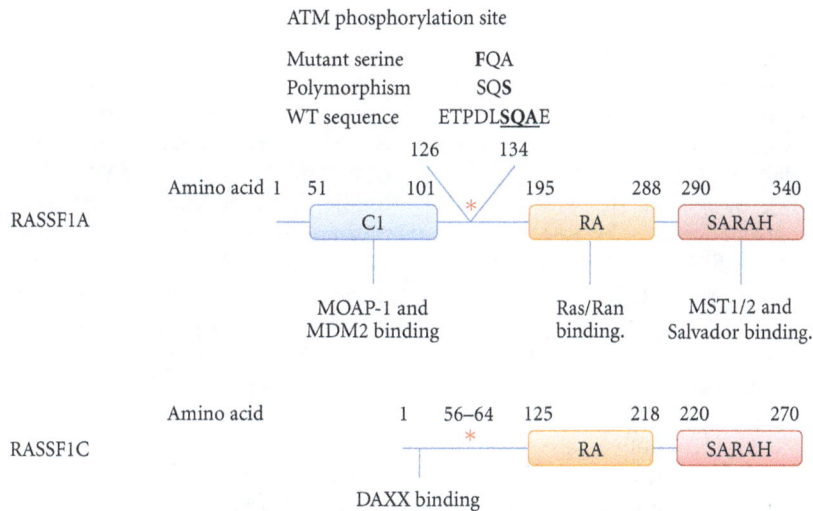

FIGURE 1: Cartoon depicting the interactions of RASSF1A and RASSF1C. RASSF1A and RASSF1C share a common C-terminal aminoacid sequence, which includes the ATM phosphorylation site (red asterisk), the RA domain, and the SARAH domain but differs at the N-terminal. RASSF1A has a C1 domain which interacts with MOAP-1 and MDM2. RASSF1C lacks the C1 domain but has an alternative DAXX interaction domain. Serine 131 of RASSF1A has been shown to be mutated from serine (S) to phenylalanine (F). An alanine (A) to Serine (S) polymorphism also exists at the 133 site.

which has significant allele frequencies in human populations (http://hapmap.ncbi.nlm.nih.gov/). The minor allele of the SNP encodes a serine (A133S) and decreases the ability of RASSF1A to become phosphorylated which, like S131F, results in a defective G1 arrest [43]. This suggests that sequence changes to the ATM consensus sequence (aminoacids 125–138) may severely inhibit the function of RASSF1A by disrupting the phosphorylation of S131 and preventing the activation of RASSF1A (Figure 1).

RASSF1A association with Ran directs the formation of a Ran-GTP gradient between the spindle poles and the metaphase plate which is important for the formation of mitotic spindle and for successful completion of mitosis. RASSF1A targets MST1/2 kinase activity towards the RanGEF (GTP exchange factor) RCC1, which inhibits its function and results in elevated Ran-GTP near the metaphase plate. Taken together, these studies indicate that RASSF1A is important for the maintenance of genomic stability by acting as an integrity checkpoint factor. Loss of RASSF1A is likely to weaken the prometaphase checkpoint and increase the potential to create genomic instability and DNA damage leading to cancer development. Indeed, the restriction of RASSF1A activity by modulation of the ATM site may be linked to numerous observations regarding the early onset of tumours in individuals carrying one minor allele of the p.RASSF1A-A133S polymorphism [44, 45]. This has been controversially linked to the exacerbation of a BRCA1/2 genomic instability phenotype; however, the inconsistency may be due to confounding factors other than BRCA2 and may be due to genomic instability via defects in RASSF1A itself [46]. All this may indicate that DNA damage activation of RASSF1A may provide an extra level of regulatory response, whereby the prometaphase checkpoint senses cells entering into mitosis with DNA damage.

5. Regulation by Domain Interaction

As a scaffold, RASSF1A must exert its tumour suppressor function through its interaction domains. The two most important domains in the context of DNA damage are the C1 domain and the SARAH domain. The most significant binding partners identified to interact with the C1 domain are the TNF-R1/TRAIL-R1—Modulator of Apoptosis-1 (MOAP-1) complexes and the MDM2/DAXX/HAUSP/p53 complex [47, 48] (Figure 1). MOAP-1 and RASSF1A are recruited to either TNF-R1 or TRAIL-R1 in response to TNFα stimulation. RASSF1A binds MOAP-1 causing an activating conformational change to the structure of MOAP-1. The active structure can bind to the proapoptotic Bcl-2 family member BAX which creates a pore in the outer mitochondrial membrane leading to the release of cytochrome C and induction of caspase-dependent apoptotic signalling pathways [47, 49]. BAX and the associated negative regulator BAK tightly regulate the cell's response to apoptotic signals and are often coordinated with other apoptotic signals such as DNA damage. It is reasonable to assume that RASSF1A-MOAP-1 may be affected by DNA damage but whether this contributes to the regulation of BAX/BAK at the mitochondria remains uncertain.

The response of tumour suppressor p53 to DNA damage results in a variety of outcomes including cell cycle arrest, apoptosis, and senescence, combining to protect the integrity of the genome [50, 51]. In unstressed cells p53 levels are low, being controlled by the RING domain-containing E3 ubiquitin ligase MDM2 (Mouse Double Minute 2) [52, 53]. Induction of DNA damage results in phosphorylation of p53 by the DNA damage checkpoint proteins ATM (on serine 15) and CHK2 (Checkpoint Kinase 2) (on serine 20) [54, 55]. These phosphorylation events combine with an ATM-mediated restriction of MDM2 activity to stabilize p53. Song et al. have recently shown that the C1 domain of RASSF1A can bind and sequester MDM2 in an ATM-dependent manner [48]. They describe a complex consisting of MDM2, DAXX (death-domain-associated protein), and HAUSP1 (a deubiquitinating enzyme). HAUSP1 removes ubiquitin molecules from MDM2 and increases its stability. Upon DNA damage, ATM activates RASSF1A driving its association with MDM2, potentially through phosphorylation at S131. RASSF1A disrupts the MDM2-DAXX-HAUSP1 complex, sequestering MDM2 away from p53, and preventing HAUSP1-regulated deubiquitination of MDM2 promoting its degradation. Release of DAXX from the complex is thought to allow DAXX relocation to the plasma membrane where it can bind the death receptor Fas and activate c-Jun NH_2-terminal kinase (JNK) [56]. Activated p53 exerts its tumour suppressor function by acting as a transcription factor. It has recently been shown that the *RASSF1* promoter is a target for p53 [57]. Interestingly, p53 appears to downregulate the transcription of RASSF1A hinting at a second mechanism through which p53 can negatively regulate itself in addition to the upregulation of MDM2.

RASSF1A makes two significant interactions through its SARAH domain; the first with mammalian sterile 20-like kinases 1 and 2 (MST1/2) and the second to the scaffold protein Salvador (Figure 1). The RASSF1A interaction with MST1/2 leads to an increase in the local concentration of MST molecules allowing them to undergo transphosphorylation and autoactivation [58]. The interaction further stabilises the MST1/2 kinase activity by preventing dephosphorylation of MST1/2 [59]. MST1/2 were initially cloned from lymphoid cDNA library when looking for human relatives of *Saccharomyces cerevisiae* protein Ste20 and subsequently shown to be activated by a wide variety of cellular stresses [60–63]. Of note is that both *Drosophila* dMST (Hippo) and MST2 are activated in response to DNA damage. In mammals, DNA damage induction of MST2 requires direct binding of RASSF1A- and ATM-mediated phosphorylation of S131 [13, 64, 65]. Interestingly MST1 was shown to be able to activate p53 in response to cisplatin-induced DNA damage by phosphorylating and inactivating Sirt1, a deacetylase that inactivates p53 [66]. Additional substrates of MST kinases that may prove subject to DNA damage are the histones H2B and H2AX, JNK and FOXO transcription factors [67–70]. However, a clear example of signalling through RASSF1A-MST after DNA damage is the recruitment and activation of the large tumour suppressor kinases 1 and 2 (LATS1/2) [71, 72].

Studies on the *Drosophila* homolog of MST1/2, Hippo have discovered that the pathway through Warts (LATS1) is responsible for controlling proliferation and apoptosis and is conserved in both vertebrates and invertebrates. Mutations in pathway member's Hippo (MST1/2), Warts (LATS1/2), Salvador (WW45), or Mats (Mob1 as a tumour suppressor) result in vast tissue overgrowth. The pathway generates a signal to inhibit Yorkie (YAP). Yorkie mutants therefore inevitably show a reduced tissue growth phenotype (reviewed in [73]). Yorkie is a non-DNA binding transcriptional coactivator that binds Scalloped (TEAD1-4) leading to the upregulation of proteins such as Cyclin E and Diap-1 to promote cell division and inhibit apoptosis. In this case Warts phosphorylates Yorkie creating a site for 14-3-3 binding. This sequesters Yorkie in the cytoplasm inhibiting its oncogenic activity [74]. In mammals, in the presence of RASSF1A and a DNA damage signal, LATS1 phosphorylation of YAP maintains a pool of YAP in the nucleus which switches binding partner from the antiapoptotic, YAP-TEAD complex to a proapoptotic YAP-p73 complex [75]. The interaction between YAP1 and p73 stabilises p73 by preventing its nuclear export and subsequent degradation [76–78]. YAP1 functions as a coactivator of p73 and this complex upregulates p73 responsive genes such as the proapoptotic BH3 only Bcl-2 family member, PUMA [79, 80]. This idea is in agreement with the finding that both LATS1 and LATS2 mediate apoptosis through p53. In certain cases LATS2-mediated apoptosis is p53 independent, potentially indicating a switch to YAP1 and p73 [13, 81, 82].

LATS2 has been shown to activate p53 both directly, by binding to and inhibiting MDM2 and indirectly by driving the nuclear accumulation of ASPP1 (apoptosis-stimulating protein of p53) [83, 84]. Interestingly, cytoplasmic ASPP1 appears to behave in an opposite manner and inactivates the ability of LATS1 to interact with YAP1 [85]. As RASSF1A activates LATS1/2 in response to DNA damage this could potentially drive ASPP1 activation of p53 and contribute to the overall p53 response. Interestingly the *Drosophila* ASPP protein (dASPP) has also been shown to interact with dRASSF8 to regulate C-terminal Src kinase (dCsk) and adherens junctions [86], a site key to the regulation of the core hippo pathway [87].

LATS2 has been implicated in the G1 tetraploidy checkpoint, a process that is thought to be driven by LATS2 activation by ATR and leads to direct stabilisation of p53 [83, 88]. Active p53 then creates a positive feedback loop with LATS2 by upregulating its activity further [88]. In response to UV radiation CHK1 activation by ATR has been shown to activate LATS2 [89].

Although not addressed in a RASSF1A-dependent manner, YAP forms an additional DNA damage promoted complex with the transcription factor early growth response 1 (EGR1) [90]. The interaction promotes enhanced transcriptional activity of EGR1 for the Bcl-2-associated X (BAX) promoter. Thus YAP can act as an oncogene and a tumour suppressor in a RASSF1A-context-dependent manner.

In *Drosophila* dRASSF and Salvador are known to compete for MST binding. Here Salvador acts as an adaptor to bring Hippo and Warts together to activate the hippo pathway, which is antagonised by dRASSF [91]. In mammals, however, RASSF1A can bind both MST1/2 and Salvador at the same time using different regions with the SARAH domain. Using an L308P mutant of RASSF1A that cannot bind MST but remains bound to Salvador, Donninger et al. have shown that the RASSF1A Salvador interaction can activate p73 in an MST-independent manner [92].

6. RASSF1C

RASSF1C is the second ubiquitously expressed isoform of the *RASSF1* gene. Like RASSF1A, RASSF1C contains the ATM consensus sequence (Figure 1). This site, at Serine 61, has not yet been confirmed but the sequence is identical between RASSF1A and RASSF1C at this site so it is plausible to suggest that RASSF1C is also phosphorylated and activated by ATM. Indeed, the Serine 61 to phenylalanine (S61F) mutant of RASSF1C was unable to block the genomic destabilising effects of Ras which can be ablated by overexpression of wild-type RASSF1C in the embryonic kidney cell line 293T and human lung tumour cell line NCI-H1299 [93]. This suggests that DNA damage activation of RASSF1C may require phosphorylation of Serine 61 (RASSF1A-131) site. Further to this, RASSF1C has recently been implicated in a DNA damage response pathway involving DAXX (which also binds to RASSF1A) and JNK [94] (Figure 1). In unstressed conditions RASSF1C is shown to be in a complex with DAXX in the nucleus, recently resolved by NMR [95]. Upon ultraviolet radiation or MMS-induced DNA damage this interaction is lost allowing RASSF1C to move to the cytoplasm where it aids the activation of SAPK/JNK signalling [94]. DAXX, however, remains in the nucleus concentrating at PML bodies. The signal that leads to release of RASSF1C from DAXX is unknown; however, it would be interesting to see if the signal relies upon the ATM phosphorylation site. Conversely, another study has identified that RASSF1C, far from being activated by DNA damage, is targeted for degradation under stress conditions. Exposure to UV radiation or treatment of cells with doxorubicin leads to RASSF1C phosphorylation by GSK3β creating a phosphodegron at S19/23 which is bound to by SCF$^{\beta\text{-TrCP}}$ targeting RASSF1C for degradation [96]. This GSK3β-dependent degradation was shown to be inhibited by the PI3-K/AKT pathway. Since AKT activity can lead to RASSF1C upregulation it suggests that RASSF1C could function as an oncogene. This is in keeping with several recent reports showing that RASSF1C increased cell proliferation in lung cancer cells and migration in breast cancer cell lines [97, 98].

7. Therapeutic Implications of RASSF1A Loss

One of the most common and widespread events to occur during cancer development is the loss of RASSF1A expression. This loss is due to methylation of the upstream CpG islands in the *RASSF1* gene [22, 29]. The frequency of epigenetically driven loss of RASSF1A correlates well with the increasing grade of the tumour. Methylation has been reported in over 37 tumour types (comprehensively reviewed in [24, 99]) and is thought to be an early event in breast and thyroid tumourigenesis, childhood neoplasia, and endometrial carcinogenesis [27].

RASSF1A methylation correlates with a decreased responsiveness to DNA-damaging therapies [100–102]. The DNA methyltransferase (DNMT) inhibitor zebularine has been used to effectively reexpress RASSF1A and show an increase in cancer cell sensitivity to radiation-induced damage *in vitro* and *in vivo* [101] as well as to cisplatin [32]. Dote et al. showed that 48 h treatment with zebularine, which corresponded to the maximum reexpression of RASSF1A increased the radiosensitivity of PaCa, DU145, and U251 cancer cell lines

by 1.5 times and caused an increased tumour delay in U251 xenograph models in mice [101]. A 48 h treatment with zebularine also increased cancer cell sensitivity to DNA damage and a 16-fold reduction in IC$_{50}$ of cisplatin in resistant ovarian cancer cell lines [32]. Sensitivity of testicular germ cell tumours to cisplatin could also be enhanced by another DNMT inhibitor that is in clinical trials, 5-aza-2'-deoxycytidine [103]. Interestingly, they noted that effectiveness of the 5-aza-2'-deoxycytidine treatment was dependent on the level of DNMT3B levels. The higher the DNMT3B level the greater the effect. The most significant target gene for DNMT3B was shown to be RASSF1A (as mentioned above) and thus it can be extrapolated that the increase in sensitivity to cisplatin is due to the reexpression of RASSF1A. Reexpression of RASSF1A using 5-aza-2'-deoxycytidine or reintroduction of RASSF1A into the hepatocellular carcinoma cell line, SMMC-7721, was also shown to increase sensitivity to chemotherapeutics such as fluorouracil, mitomycin, and cisplatin [104]. Together these results support a clinically relevant role for RASSF1A in the DNA damage response that is backed up by phase I and II clinical trials in myelodysplasia and leukaemia patients where 5-aza-2'-deoxycytidine has shown efficacy both alone and in combination with the histone deacetylase (HDAC) inhibitor valproic acid [105, 106]. Therapeutic failure upon RASSF1A loss can also be counteracted by targeting the downstream DNA damage responsive signalling pathway. Direct activation of BAX via the BH3 mimetic ABT-737 has recently put forward as a potential treatment for RASSF1A methylated medulloblastoma [107]. The role of RASSF1A in checkpoint activation and maintenance of genomic integrity is highlighted in a study by Zhang et al. which showed a significant increase in DNA damage caused by aflatoxin B$_1$ in tumour tissues where RASSF1A has been lost due to DNA methylation [108].

8. Other RASSFs and DNA Damage

This paper has concentrated primarily upon the role of RASSF1 in DNA damage; however, it is worth noting that other RASSF proteins have also been linked to DNA damage pathways. The RASSF2 gene resides on chromosome 20. The gene can be spliced into two very similar proteins RASSF2A and RASSF2C both of which contain the RA domain and the SARAH domain. They show 28% identity to RASSF1A and like RASSF1A, the promoter has been shown to be inactivated by hypermethylation in primary tumours [109–114]. RASSF2 has been reported to be upregulated in lymphocytes from individuals exposed to ionising radiation [115]. RASSF2 has also been shown to associate with, and is phosphorylated by, MST2 leading to stabilisation of MST2 and the generation of proapoptotic signals [116].

The *RASSF6* gene is located on chromosome 4. While the expression of RASSF6 is lost in cancer, *in silico* analysis did not find any CpG islands located near the promoter; therefore, it is assumed that this loss is not due to DNA methylation [117, 118]. RASSF6 is known to activate apoptosis in both caspase-dependent and -independent mechanisms in response to TNFα; however, it is unknown whether it is also

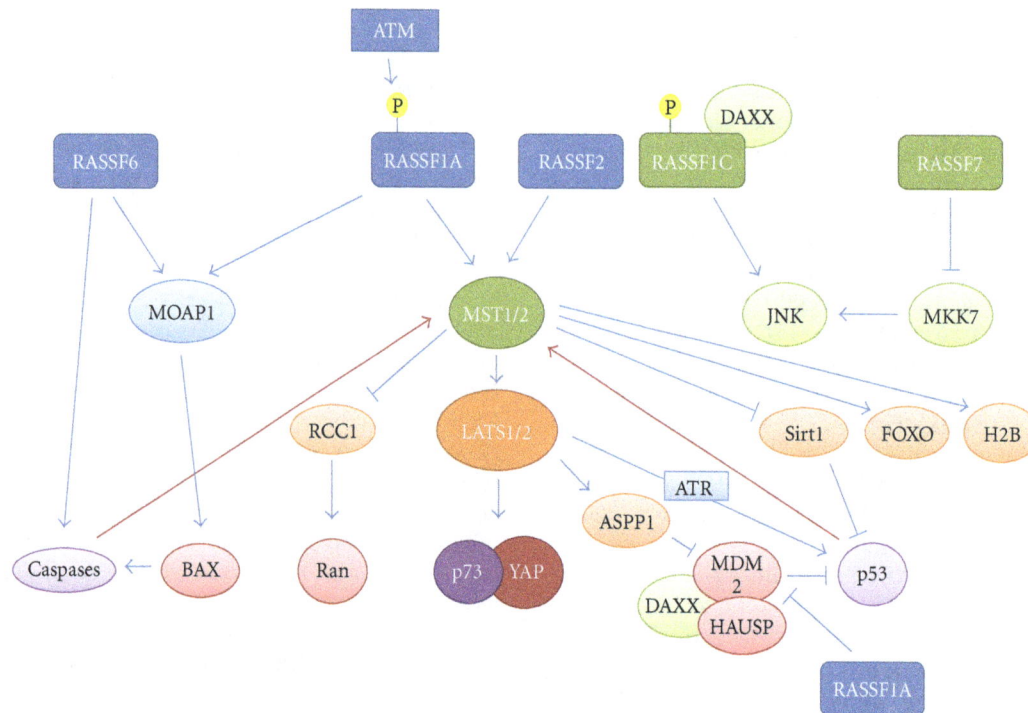

FIGURE 2: Cartoon depicting DNA damage activated pathways downstream of RASSF family members. RASSF family members, activated by DNA damage, signal through various intermediates (primary interaction: light blue and green [involving RASSF1C or RASSF7]; secondary: orange and tertiary: red) to activate p53, p73, and caspases (purple) to control apoptosis, genome stability, and senescence. Feedback loops exist from caspases and p53 that further activate the pathways and amplify the signal. RASSF1A can also directly sequester MDM2 leading to p53 activation. RASSF1C can transfer DNA damage signals from the nucleus to the cytoplasm by activating JNK signalling. RASSF7 acts as an oncogene inhibiting the activation of JNK.

activated by DNA damage signals [118]. RASSF6 contains both the RA domain and SARAH domain and like RASSF1A it has been shown to bind to MOAP-1 [117], which could be responsible for its induction of apoptosis in response to TNFα. Unlike other family members, RASSF6 contains a number of ATM consensus sites (SQ/TQ) upstream of the RA domain; however it is not clear if these are functional.

RASSF family members efficiently form heterodimers [119]. This provides a potential mechanism through which additional RASSF proteins could be involved in DNA damage signalling. A heterodimer between RASSF1A and RASSF5A has been suggested to be important for the interaction of RASSF1A with Ras [120]. Given that each of the RASSF proteins above is thought to impart its tumour suppressor function through the MST kinases we could propose that heterodimeric interactions between RASSF family members may be important for their DNA damage-induced apoptotic signalling.

RASSF7 is the best studied N-terminal RASSF protein and the first to be shown to be linked to the DNA damage response. Located on chromosome 11 close to the H-Ras gene (HRAS1), it forms part of a microsatellite that is associated with increased cancer risk [121–123]. Unlike the majority of the RASSF family members that are silenced in cancer, RASSF7 has been shown to be upregulated in a number of cancers including pancreatic, endometrial, and ovarian [124–128]. The upregulation of RASSF7 in cancers suggests

an oncogenic function, the mechanism of which has only just started to be explored. RASSF7, in concert with N-Ras, is thought to suppress the activation of JNK in response to low doses of UV radiation by binding and inhibiting MKK7, preventing its interaction with JNK. At higher doses of UV, RASSF7, like RASSF1C, is targeted for degradation through an ubiquitin-dependent mechanism. This frees MKK7 to activate a stress response through JNK [129].

9. Conclusion

Ras-association domain containing family members are important tumour suppressors involved in linking cellular stresses to cell cycle arrest and apoptosis (Figure 2). RASSF1A is an adaptor protein with three major interaction domains through which it imparts its functions. Each of these domains is involved in binding different effector proteins in response to DNA damage. The C1 domain binds MDM2 to stabilise p53 and the RA and SARAH domains are required to activate the mammalian Hippo pathway. The mammalian homolog of Hippo, MST1/2, can activate apoptosis in response to cellular stresses either directly, in the case of FOXO1 and histone H2B or via LATS1/2. RASSF2, RASSF5, and RASSF6 which share the RA and SARAH domains with RASSF1A have also been shown to active MST1/2 to induce apoptosis as well as being able to induce apoptosis independently of the Hippo pathway. LATS1 and 2 have been implicated in apoptosis by stabilising both p53, either directly through an interaction with MDM2 or indirectly via ASPP1 and stabilising p73 via YAP, in response to DNA damage. RASSF1C has been shown to be released from DAXX and p53 upon DNA damage where it can go and transmit the damage signal from the nucleus to the cytoplasm by activating JNK signalling. Each of these proteins appears to act both upstream and downstream of the "guardian of the genome" p53 to create a network which feeds back upon itself to enhance the DNA damage signaling within the cell. Greater than 50% of human tumours has either lost or mutated p53. Disruption of these networks will inactivate p53 and may contribute to tumourigenesis in a number of the cases where wild-type p53 is retained. Although not correlated with p53 loss or mutation, RASSF proteins are epigenetically lost in human cancers by DNA methylation. It has been shown that, as with p53, loss of *RASSF1* expression is associated with more aggressive tumours and increased resistance to radiation-induced DNA damage and platinum-based drugs. DMNT inhibitors such as zebularine have been shown to reexpress RASSF1A and increase the radiosensitivity of these cancers suggesting that reexpression of RASSF1A and other silenced RASSFs maybe a path through which chemoradioresistant tumours can be combated.

References

[1] C. P. Ponting and D. R. Benjamin, "A novel family of Ras-binding domains," *Trends in Biochemical Sciences*, vol. 21, no. 11, pp. 422–425, 1996.

[2] T. Yamamoto, S. Taya, and K. Kaibuchi, "Ras-induced transformation and signaling pathway," *Journal of Biochemistry*, vol. 126, no. 5, pp. 799–803, 1999.

[3] D. Vavvas, X. Li, J. Avruch, and X. F. Zhang, "Identification of Nore1 as a potential Ras effector," *Journal of Biological Chemistry*, vol. 273, no. 10, pp. 5439–5442, 1998.

[4] S. Wohlgemuth, C. Kiel, A. Krämer, L. Serrano, F. Wittinghofer, and C. Herrmann, "Recognizing and defining true ras binding domains I: biochemical analysis," *Journal of Molecular Biology*, vol. 348, no. 3, pp. 741–758, 2005.

[5] A. Dallol, L. B. Hesson, D. Matallanas et al., "RAN GTPase is a RASSF1A effector involved in controlling microtubule organization," *Current Biology*, vol. 19, no. 14, pp. 1227–1232, 2009.

[6] D. Matallanas, D. Romano, F. Al-Mulla et al., "Mutant K-Ras activation of the proapoptotic MST2 pathway is antagonized by wild-type K-Ras," *Molecular Cell*, vol. 44, no. 6, pp. 893–906, 2011.

[7] V. Sherwood, R. Manbodh, C. Sheppard, and A. D. Chalmers, "RASSF7 is a member of a new family of RAS association domain-containing proteins and is required for completing mitosis," *Molecular Biology of the Cell*, vol. 19, no. 4, pp. 1772–1782, 2008.

[8] E. Hwang, K. S. Ryu, K. Pääkkönen et al., "Structural insight into dimeric interaction of the SARAH domains from Mst1 and RASSF family proteins in the apoptosis pathway," *Proceedings of the National Academy of Sciences of the United States of America*, vol. 104, no. 22, pp. 9236–9241, 2007.

[9] H. Scheel and K. Hofmann, "A novel interaction motif, SARAH, connects three classes of tumor suppressor," *Current Biology*, vol. 13, no. 23, pp. R899–R900, 2003.

[10] G. Grigoryan and A. E. Keating, "Structural specificity in coiled-coil interactions," *Current Opinion in Structural Biology*, vol. 18, no. 4, pp. 477–483, 2008.

[11] A. C. Newton, "Protein kinase C: seeing two domains," *Current Biology*, vol. 5, no. 9, pp. 973–976, 1995.

[12] E. Harjes, S. Harjes, S. Wohlgemuth et al., "GTP-Ras disrupts the intramolecular complex of C1 and RA domains of nore1," *Structure*, vol. 14, no. 5, pp. 881–888, 2006.

[13] G. Hamilton, K. S. Yee, S. Scrace, and E. O'Neill, "ATM regulates a RASSF1A-dependent DNA damage response," *Current Biology*, vol. 19, no. 23, pp. 2020–2025, 2009.

[14] J. Hung, Y. Kishimoto, K. Sugio et al., "Allele-specific chromosome 3p deletions occur at an early stage in the pathogenesis of lung carcinoma," *Journal of the American Medical Association*, vol. 273, no. 24, p. 1908, 1995.

[15] K. Kok, S. L. Naylor, and C. H. C. M. Buys, "Deletions of the short arm of chromosome 3 in solid tumors and the search for suppressor genes," *Advances in Cancer Research*, vol. 71, pp. 27–92, 1997.

[16] Y. Sekido, M. Ahmadian, I. I. Wistuba et al., "Cloning of a breast cancer homozygous deletion junction narrows the region of search for a 3p21.3 tumor suppressor gene," *Oncogene*, vol. 16, no. 24, pp. 3151–3157, 1998.

[17] V. Sundaresan, S. P. Ganly, P. Hasleton et al., "p53 and chromosome 3 abnormalities, characteristic of malignant lung tumours, are detectable in preinvasive lesions of the bronchus," *Oncogene*, vol. 7, no. 10, pp. 1989–1997, 1992.

[18] I. I. Wistuba, C. Behrens, S. Milchgrub et al., "Sequential molecular abnormalities are involved in the multistage development of squamous cell lung carcinoma," *Oncogene*, vol. 18, no. 3, pp. 643–650, 1999.

[19] I. I. Wistuba, C. Behrens, A. K. Virmani et al., "High resolution chromosome 3p allelotyping of human lung cancer and preneoplastic/preinvasive bronchial epithelium reveals multiple, discontinuous sites of 3p allele loss and three regions of frequent breakpoints," *Cancer Research*, vol. 60, no. 7, pp. 1949–1960, 2000.

[20] Y. Sekido, K. M. Fong, and J. D. Minna, "Progress in understanding the molecular pathogenesis of human lung cancer," *Biochimica et Biophysica Acta*, vol. 1378, no. 1, pp. F21–F59, 1998.

[21] M. I. Lerman and J. D. Minna, "The 630-kb lung cancer homozygous deletion region on human chromosome 3p21.3: identification and evaluation of the resident candidate tumor suppressor genes," *Cancer Research*, vol. 60, no. 21, pp. 6116–6133, 2000.

[22] R. Dammann, C. Li, J. H. Yoon, P. L. Chin, S. Bates, and G. P. Pfeifer, "Epigenetic inactivation of a RAS association domain family protein from the lung tumour suppressor locus 3p21.3," *Nature Genetics*, vol. 25, no. 3, pp. 315–319, 2000.

[23] H. Donninger, M. D. Vos, and G. J. Clark, "The RASSF1A tumor suppressor," *Journal of Cell Science*, vol. 120, no. 18, pp. 3163–3172, 2007.

[24] L. B. Hesson, W. N. Cooper, and F. Latif, "The role of RASSF1A methylation in cancer," *Disease Markers*, vol. 23, no. 1-2, pp. 73–87, 2007.

[25] P. A. Jones and S. B. Baylin, "The epigenomics of cancer," *Cell*, vol. 128, no. 4, pp. 683–692, 2007.

[26] R. K. Palakurthy, N. Wajapeyee, M. K. Santra et al., "Epigenetic silencing of the RASSF1A tumor suppressor gene

through HOXB3-mediated induction of DNMT3B expression," *Molecular Cell*, vol. 36, no. 2, pp. 219–230, 2009.

[27] L. van der Weyden and D. J. Adams, "The Ras-association domain family (RASSF) members and their role in human tumourigenesis," *Biochimica et Biophysica Acta*, vol. 1776, no. 1, pp. 58–85, 2007.

[28] A. M. Richter, G. P. Pfeifer, and R. H. Dammann, "The RASSF proteins in cancer; from epigenetic silencing to functional characterization," *Biochimica et Biophysica Acta*, vol. 1796, no. 2, pp. 114–128, 2009.

[29] D. G. Burbee, E. Forgacs, S. Zöchbauer-Müller et al., "Epigenetic inactivation of RASSF1A in lung and breast cancers and malignant phenotype suppression," *Journal of the National Cancer Institute*, vol. 93, no. 9, pp. 691–699, 2001.

[30] K. Dreijerink, E. Braga, I. Kuzmin et al., "The candidate tumor suppressor gene, RASSF1A, from human chromosome 3p21.3 is involved in kidney tumorigenesis," *Proceedings of the National Academy of Sciences of the United States of America*, vol. 98, no. 13, pp. 7504–7509, 2001.

[31] I. Kuzmin, J. W. Gillespie, A. Protopopov et al., "The RASSF1A tumor suppressor gene is inactivated in prostate tumors and suppresses growth of prostate carcinoma cells," *Cancer Research*, vol. 62, no. 12, pp. 3498–3502, 2002.

[32] C. Balch, P. Yan, T. Craft et al., "Antimitogenic and chemosensitizing effects of the methylation inhibitor zebularine in ovarian cancer," *Molecular Cancer Therapeutics*, vol. 4, no. 10, pp. 1505–1514, 2005.

[33] S. Tommasi, R. Dammann, Z. Zhang et al., "Tumor susceptibility of Rassf1a knockout mice," *Cancer Research*, vol. 65, no. 1, pp. 92–98, 2005.

[34] L. Van Der Weyden, K. K. Tachibana, M. A. Gonzalez et al., "The RASSF1A isoform of RASSF1 promotes microtubule stability and suppresses tumorigenesis," *Molecular and Cellular Biology*, vol. 25, no. 18, pp. 8356–8367, 2005.

[35] S. Tommasi, A. Besaratinia, S. P. Wilczynski, and G. P. Pfeifer, "Loss of Rassf1a enhances p53-mediated tumor predisposition and accelerates progression to aneuploidy," *Oncogene*, vol. 30, no. 6, pp. 690–700, 2011.

[36] L. Liu, C. Guo, R. Dammann, S. Tommasi, and G. P. Pfeifer, "RASSF1A interacts with and activates the mitotic kinase Aurora-A," *Oncogene*, vol. 27, no. 47, pp. 6175–6186, 2008.

[37] R. Rong, L. Y. Jiang, M. S. Sheikh, and Y. Huang, "Mitotic kinase Aurora-A phosphorylates RASSF1A and modulates RASSF1A-mediated microtubule interaction and M-phase cell cycle regulation," *Oncogene*, vol. 26, no. 55, pp. 7700–7708, 2007.

[38] M. S. Song, S. J. Song, S. J. Kim, K. Nakayama, K. I. Nakayama, and D. S. Lim, "Skp2 regulates the antiproliferative function of the tumor suppressor RASSF1A via ubiquitin-mediated degradation at the G1-S transition," *Oncogene*, vol. 27, no. 22, pp. 3176–3185, 2008.

[39] S. J. Song, S. J. Kim, M. S. Song, and D. S. Lim, "Aurora B-mediated phosphorylation of RASSF1A maintains proper cytokinesis by recruiting syntaxin16 to the midzone and midbody," *Cancer Research*, vol. 69, no. 22, pp. 8540–8544, 2009.

[40] S. J. Song, M. S. Song, S. J. Kim et al., "Aurora a regulates prometaphase progression by inhibiting the ability of RASSF1A to suppress APC-Cdc20 activity," *Cancer Research*, vol. 69, no. 6, pp. 2314–2323, 2009.

[41] D. Durocher and S. P. Jackson, "DNA-PK, ATM and ATR as sensors of DNA damage: variations on a theme?" *Current Opinion in Cell Biology*, vol. 13, no. 2, pp. 225–231, 2001.

[42] S. T. Kim, D. S. Lim, C. E. Canman, and M. B. Kastan, "Substrate specificities and identification of putative substrates of ATM kinase family members," *Journal of Biological Chemistry*, vol. 274, no. 53, pp. 37538–37543, 1999.

[43] L. Shivakumar, J. Minna, T. Sakamaki, R. Pestell, and M. A. White, "The RASSF1A tumor suppressor blocks cell cycle progression and inhibits cyclin D1 accumulation," *Molecular and Cellular Biology*, vol. 22, no. 12, pp. 4309–4318, 2002.

[44] H. Endoh, Y. Yatabe, S. Shimizu et al., "RASSF1A gene inactivation in non-small cell lung cancer and its clinical implication," *International Journal of Cancer*, vol. 106, no. 1, pp. 45–51, 2003.

[45] B. Gao, X. J. Xie, C. Huang et al., "RASSF1A polymorphism A133S is associated with early onset breast cancer in BRCA1/2 mutation carriers," *Cancer Research*, vol. 68, no. 1, pp. 22–25, 2008.

[46] J. Bergqvist, A. Latif, S. A. Roberts et al., "RASSF1A polymorphism in familial breast cancer," *Familial Cancer*, vol. 9, no. 3, pp. 263–265, 2010.

[47] C. J. Foley, H. Freedman, S. L. Choo et al., "Dynamics of RASSF1A/MOAP-1 association with death receptors," *Molecular and Cellular Biology*, vol. 28, no. 14, pp. 4520–4535, 2008.

[48] M. S. Song, S. J. Song, S. Y. Kim, H. J. Oh, and D. S. Lim, "The tumour suppressor RASSF1A promotes MDM2 self-ubiquitination by disrupting the MDM2-DAXX-HAUSP complex," *EMBO Journal*, vol. 27, no. 13, pp. 1863–1874, 2008.

[49] S. Baksh, S. Tommasi, S. Fenton et al., "The tumor suppressor RASSF1A and MAP-1 link death receptor signaling to bax conformational change and cell death," *Molecular Cell*, vol. 18, no. 6, pp. 637–650, 2005.

[50] D. Michael and M. Oren, "The p53-Mdm2 module and the ubiquitin system," *Seminars in Cancer Biology*, vol. 13, no. 1, pp. 49–58, 2003.

[51] B. Vogelstein, D. Lane, and A. J. Levine, "Surfing the p53 network," *Nature*, vol. 408, no. 6810, pp. 307–310, 2000.

[52] Y. Haupt, R. Maya, A. Kazaz, and M. Oren, "Mdm2 promotes the rapid degradation of p53," *Nature*, vol. 387, no. 6630, pp. 296–299, 1997.

[53] M. H. G. Kubbutat, S. N. Jones, and K. H. Vousden, "Regulation of p53 stability by Mdm2," *Nature*, vol. 387, no. 6630, pp. 299–303, 1997.

[54] A. M. Bode and Z. Dong, "Post-translational modification of p53 in tumorigenesis," *Nature Reviews Cancer*, vol. 4, no. 10, pp. 793–805, 2004.

[55] R. Maya, M. Balass, S. T. Kim et al., "ATM-dependent phosphorylation of Mdm2 on serine 395: role in p53 activation by DNA damage," *Genes and Development*, vol. 15, no. 9, pp. 1067–1077, 2001.

[56] X. Yang, R. Khosravi-Far, H. Y. Chang, and D. Baltimore, "Daxx, a novel fas-binding protein that activates JNK and apoptosis," *Cell*, vol. 89, no. 7, pp. 1067–1076, 1997.

[57] Y. Tian et al., "Tumor suppressor RASSF1A promoter: p53 binding and methylation," *PLoS One*, vol. 6, no. 3, article e17017, 2011.

[58] H. Glantschnig, G. A. Rodan, and A. A. Reszka, "Mapping of MST1 kinase sites of phosphorylation: activation and auto-phosphorylation," *Journal of Biological Chemistry*, vol. 277, no. 45, pp. 42987–42996, 2002.

[59] C. Guo, X. Zhang, and G. P. Pfeifer, "The tumor suppressor RASSF1A prevents dephosphorylation of the mammalian STE20-like kinases MST1 and MST2," *Journal of Biological Chemistry*, vol. 286, no. 8, pp. 6253–6261, 2011.

[60] C. L. Creasy, D. M. Ambrose, and J. Chernoff, "The Ste20-like protein kinase, Mst1, dimerizes and contains an inhibitory

domain," *Journal of Biological Chemistry*, vol. 271, no. 35, pp. 21049–21053, 1996.

[61] J. Chernoff, "Cloning and characterization of a member of the MST subfamily of Ste20-like kinases," *Gene*, vol. 167, no. 1-2, pp. 303–306, 1995.

[62] C. L. Creasy and J. Chernoff, "Cloning and characterization of a human protein kinase with homology to Ste20," *Journal of Biological Chemistry*, vol. 270, no. 37, pp. 21695–21700, 1995.

[63] L. K. Taylor, H. C. R. Wang, and R. L. Erikson, "Newly identified stress-responsive protein kinases, Krs-1 and Krs-2," *Proceedings of the National Academy of Sciences of the United States of America*, vol. 93, no. 19, pp. 10099–10104, 1996.

[64] J. Colombani, C. Polesello, F. Josué, and N. Tapon, "Dmp53 activates the hippo pathway to promote cell death in response to DNA damage," *Current Biology*, vol. 16, no. 14, pp. 1453–1458, 2006.

[65] W. Wen, F. Zhu, J. Zhang et al., "MST1 promotes apoptosis through phosphorylation of histone H2AX," *Journal of Biological Chemistry*, vol. 285, no. 50, pp. 39108–39116, 2010.

[66] F. Yuan, Q. Xie, J. Wu et al., "MST1 promotes apoptosis through regulating Sirt1-dependent p53 deacetylation," *Journal of Biological Chemistry*, vol. 286, no. 9, pp. 6940–6945, 2011.

[67] W. L. Cheung, K. Ajiro, K. Samejima et al., "Apoptotic phosphorylation of histone H2B is mediated by mammalian sterile twenty kinase," *Cell*, vol. 113, no. 4, pp. 507–517, 2003.

[68] M. K. Lehtinen, Z. Yuan, P. R. Boag et al., "A conserved MST-FOXO signaling pathway mediates oxidative-stress responses and extends life span," *Cell*, vol. 125, no. 5, pp. 987–1001, 2006.

[69] S. Ura, H. Nishina, Y. Gotoh, and T. Katada, "Activation of the c-Jun N-terminal kinase pathway by MST1 is essential and sufficient for the induction of chromatin condensation during apoptosis," *Molecular and Cellular Biology*, vol. 27, no. 15, pp. 5514–5522, 2007.

[70] M. Praskova, F. Xia, and J. Avruch, "MOBKL1A/MOBKL1B phosphorylation by MST1 and MST2 inhibits cell proliferation," *Current Biology*, vol. 18, no. 5, pp. 311–321, 2008.

[71] C. Guo, S. Tommasi, L. Liu, J. K. Yee, R. Dammann, and G. Pfeifer, "RASSF1A is part of a complex similar to the drosophila hippo/salvador/lats tumor-suppressor network," *Current Biology*, vol. 17, no. 8, pp. 700–705, 2007.

[72] E. E. O'Neill, D. Matallanas, and W. Kolch, "Mammalian sterile 20-like kinases in tumor suppression: an emerging pathway," *Cancer Research*, vol. 65, no. 13, pp. 5485–5487, 2005.

[73] B. V. V. G. Reddy and K. D. Irvine, "The fat and warts signaling pathways: new insights into their regulation, mechanism and conservation," *Development*, vol. 135, no. 17, pp. 2827–2838, 2008.

[74] B. Zhao, X. Wei, W. Li et al., "Inactivation of YAP oncoprotein by the Hippo pathway is involved in cell contact inhibition and tissue growth control," *Genes and Development*, vol. 21, no. 21, pp. 2747–2761, 2007.

[75] K. S. Yee and E. O'Neill, "YAP1: friend and foe," *Cell Cycle*, vol. 9, no. 8, pp. 1447–1448, 2010.

[76] S. Basu, N. F. Totty, M. S. Irwin, M. Sudol, and J. Downward, "Akt phosphorylates the Yes-associated protein, YAP, to induce interaction with 14-3-3 and attenuation of p73-mediated apoptosis," *Molecular Cell*, vol. 11, no. 1, pp. 11–23, 2003.

[77] M. Dobbelstein, S. Strano, J. Roth, and G. Blandino, "p73-induced apoptosis: a question of compartments and cooperation," *Biochemical and Biophysical Research Communications*, vol. 331, no. 3, pp. 688–693, 2005.

[78] G. Melino, F. Bernassola, M. Ranalli et al., "p75 induces apoptosis via PUMA transactivation and bax mitochondrial translocation," *Journal of Biological Chemistry*, vol. 279, no. 9, pp. 8076–8083, 2004.

[79] D. Matallanas, D. Romano, K. Yee et al., "RASSF1A elicits apoptosis through an MST2 pathway directing proapoptotic transcription by the p73 tumor suppressor protein," *Molecular Cell*, vol. 27, no. 6, pp. 962–975, 2007.

[80] S. Strano, O. Monti, N. Pediconi et al., "The transcriptional coactivator yes-associated protein drives p73 gene-target specificity in response to DNA damage," *Molecular Cell*, vol. 18, no. 4, pp. 447–459, 2005.

[81] H. Ke, J. Pei, Z. Ni et al., "Putative tumor suppressor Lats2 induces apoptosis through downregulation of Bcl-2 and Bcl-xL," *Experimental Cell Research*, vol. 298, no. 2, pp. 329–338, 2004.

[82] M. Kawahara, T. Hori, K. Chonabayashi, T. Oka, M. Sudol, and T. Uchiyama, "Kpm/Lats2 is linked to chemosensitivity of leukemic cells through the stabilization of p73," *Blood*, vol. 112, no. 9, pp. 3856–3866, 2008.

[83] Y. Aylon, N. Yabuta, H. Besserglick et al., "Silencing of the lats2 tumor suppressor overrides a p53-dependent oncogenic stress checkpoint and enables mutant H-Ras-driven cell transformation," *Oncogene*, vol. 28, no. 50, pp. 4469–4479, 2009.

[84] Y. Aylon, Y. Ofir-Rosenfeld, N. Yabuta et al., "The Lats2 tumor suppressor augments p53-mediated apoptosis by promoting the nuclear proapoptotic function of ASPP1," *Genes and Development*, vol. 24, no. 21, pp. 2420–2429, 2010.

[85] A. M. Vigneron, R. L. Ludwig, and K. H. Vousden, "Cytoplasmic ASPP1 inhibits apoptosis through the control of YAP," *Genes and Development*, vol. 24, no. 21, pp. 2430–2439, 2010.

[86] P. F. Langton, J. Colombani, E. H. Y. Chan, A. Wepf, M. Gstaiger, and N. Tapon, "The dASPP-dRASSF8 complex regulates cell-cell adhesion during drosophila retinal morphogenesis," *Current Biology*, vol. 19, no. 23, pp. 1969–1978, 2009.

[87] D. Pan, "The hippo signaling pathway in development and cancer," *Developmental Cell*, vol. 19, no. 4, pp. 491–505, 2010.

[88] Y. Aylon, D. Michael, A. Shmueli, N. Yabuta, H. Nojima, and M. Oren, "A positive feedback loop between the p53 and Lats2 tumor suppressors prevents tetraploidization," *Genes and Development*, vol. 20, no. 19, pp. 2687–2700, 2006.

[89] N. Okada, N. Yabuta, H. Suzuki, Y. Aylon, M. Oren, and H. Nojima, "A novel Chk1/2-Lats2-14-3-3 signaling pathway regulates P-body formation in response to UV damage," *Journal of Cell Science*, vol. 124, no. 1, pp. 57–67, 2011.

[90] M. Zagurovskaya, M. M. Shareef, A. Das et al., "EGR-1 forms a complex with YAP-1 and upregulates Bax expression in irradiated prostate carcinoma cells," *Oncogene*, vol. 28, no. 8, pp. 1121–1131, 2009.

[91] C. Polesello, S. Huelsmann, N. Brown, and N. Tapon, "The drosophila RASSF homolog antagonizes the hippo pathway," *Current Biology*, vol. 16, no. 24, pp. 2459–2465, 2006.

[92] H. Donninger, N. Allen, A. Henson et al., "Salvador protein is a tumor suppressor effector of RASSF1A with hippo pathway-independent functions," *Journal of Biological Chemistry*, vol. 286, no. 21, pp. 18483–18491, 2011.

[93] M. D. Vos, A. Martinez, C. Elam et al., "A role for the RASSF1A tumor suppressor in the regulation of tubulin polymerization and genomic stability," *Cancer Research*, vol. 64, no. 12, pp. 4244–4250, 2004.

[94] D. Kitagawa, H. Kajiho, T. Negishi et al., "Release of RASSF1C from the nucleus by Daxx degradation links DNA damage

and SAPK/JNK activation," *EMBO Journal*, vol. 25, no. 14, pp. 3286–3297, 2006.

[95] E. Escobar-Cabrera, D. K. W. Lau, S. Giovinazzi, A. M. Ishov, and L. P. McIntosh, "Structural characterization of the DAXX N-terminal helical bundle domain and its complex with Rassf1C," *Structure*, vol. 18, no. 12, pp. 1642–1653, 2010.

[96] X. Zhou, T.-T. Li, X. Feng et al., "Targeted polyubiquitylation of RASSF1C by the Mule and SCF β-TrCP ligases in response to DNA damage," *Biochemical Journal*, vol. 441, no. 1, pp. 227–236, 2012.

[97] Y. G. Amaar, M. G. Minera, L. K. Hatran, D. D. Strong, S. Mohan, and M. E. Reeves, "Ras association domain family 1C protein stimulates human lung cancer cell proliferation," *American Journal of Physiology*, vol. 291, no. 6, pp. L1185–L1190, 2006.

[98] M. E. Reeves, S. W. Baldwin, M. L. Baldwin et al., "Ras-association domain family 1C protein promotes breast cancer cell migration and attenuates apoptosis," *BMC Cancer*, vol. 10, article 562, 2010.

[99] R. Dammann, U. Schagdarsurengin, C. Seidel et al., "The tumor suppressor RASSF1A in human carcinogenesis: an update," *Histology and Histopathology*, vol. 20, no. 2, pp. 645–663, 2005.

[100] J. W. F. Catto, A. R. Azzouzi, I. Rehman et al., "Promoter hypermethylation is associated with tumor location, stage, and subsequent progression in transitional cell carcinoma," *Journal of Clinical Oncology*, vol. 23, no. 13, pp. 2903–2910, 2005.

[101] H. Dote, D. Cerna, W. E. Burgan et al., "Enhancement of in vitro and in vivo tumor cell radiosensitivity by the DNA methylation inhibitor zebularine," *Clinical Cancer Research*, vol. 11, no. 12, pp. 4571–4579, 2005.

[102] S. Honda, M. Haruta, W. Sugawara et al., "The methylation status of RASSF1A promoter predicts responsiveness to chemotherapy and eventual cure in hepatoblastoma patients," *International Journal of Cancer*, vol. 123, no. 5, pp. 1117–1125, 2008.

[103] M. J. Beyrouthy, K. M. Garner, M. P. Hever et al., "High DNA methyltransferase 3B expression mediates 5-aza-deoxycytidine hypersensitivity in testicular germ cell tumors," *Cancer Research*, vol. 69, no. 24, pp. 9360–9366, 2009.

[104] Q.-Z. Zhao and K.-F. Dou, "Methylation of Ras association domain family protein 1, isoform a correlated with proliferation and drug resistance in hepatocellular carcinoma cell line SMMC-7721," *Journal of Gastroenterology and Hepatology*, vol. 22, no. 5, pp. 683–689, 2007.

[105] G. Garcia-Manero, H. M. Kantarjian, B. Sanchez-Gonzalez et al., "Phase 1/2 study of the combination of 5-aza-2′-deoxycytidine with valproic acid in patients with leukemia," *Blood*, vol. 108, no. 10, pp. 3271–3279, 2006.

[106] B. Rüter, P. W. Wijermans, and M. Lübbert, "Superiority of prolonged low-dose azanucleoside administration? Results of 5-aza-2′-deoxycytidine retreatment in high-risk myelodysplasia patients," *Cancer*, vol. 106, no. 8, pp. 1744–1750, 2006.

[107] J. Levesley et al., "RASSF1A and the BH3-only mimetic ABT-737 promote apoptosis in pediatric medulloblastoma cell lines," *Neuro-Oncology*, vol. 13, no. 12, pp. 1265–1276, 2011.

[108] Y. J. Zhang, H. Ahsan, Y. Chen et al., "High frequency of promoter hypermethylation of RASSF1A and p16 and its relationship to aflatoxin B1-DNA adduct levels in human hepatocellular carcinoma," *Molecular Carcinogenesis*, vol. 35, no. 2, pp. 85–92, 2002.

[109] K. Akino, M. Toyota, H. Suzuki et al., "The Ras effector RASSF2 is a novel tumor-suppressor gene in human colorectal cancer," *Gastroenterology*, vol. 129, no. 1, pp. 156–169, 2005.

[110] M. Endoh, G. Tamura, T. Honda et al., "RASSF2, a potential tumour suppressor, is silenced by CpG island hypermethylation in gastric cancer," *British Journal of Cancer*, vol. 93, no. 12, pp. 1395–1399, 2005.

[111] L. B. Hesson, R. Wilson, D. Morton et al., "CpG island promoter hypermethylation of a novel Ras-effector gene RASSF2A is an early event in colon carcinogenesis and correlates inversely with K-ras mutations," *Oncogene*, vol. 24, no. 24, pp. 3987–3994, 2005.

[112] H. W. Park, C. K. Hio, I. J. Kim et al., "Correlation between hypermethylation of the RASSF2A promoter and K-ras/BRAF mutations in microsatellite-stable colorectal cancers," *International Journal of Cancer*, vol. 120, no. 1, pp. 7–12, 2007.

[113] M. D. Vos, C. A. Ellis, C. Elam, A. S. Ülkü, B. J. Taylor, and G. J. Clark, "RASSF2 is a novel K-Ras-specific effector and potential tumor suppressor," *Journal of Biological Chemistry*, vol. 278, no. 30, pp. 28045–28051, 2003.

[114] Z. Zhang, D. Sun, N. V. Do, A. Tang, L. Hu, and G. Huang, "Inactivation of RASSF2A by promoter methylation correlates with lymph node metastasis in nasopharyngeal carcinoma," *International Journal of Cancer*, vol. 120, no. 1, pp. 32–38, 2007.

[115] E. T. Sakamoto-Hojo, S. S. Mello, E. Pereira et al., "Gene expression profiles in human cells submitted to genotoxic stress," *Mutation Research - Reviews in Mutation Research*, vol. 544, no. 2-3, pp. 403–413, 2003.

[116] W. N. Cooper, L. B. Hesson, D. Matallanas et al., "RASSF2 associates with and stabilizes the proapoptotic kinase MST2," *Oncogene*, vol. 28, no. 33, pp. 2988–2998, 2009.

[117] N. P. C. Allen, H. Donninger, M. D. Vos et al., "RASSF6 is a novel member of the RASSF family of tumor suppressors," *Oncogene*, vol. 26, no. 42, pp. 6203–6211, 2007.

[118] M. Ikeda, S. Hirabayashi, N. Fujiwara et al., "Ras-association domain family protein 6 induces apoptosis via both caspase-dependent and caspase-independent pathways," *Experimental Cell Research*, vol. 313, no. 7, pp. 1484–1495, 2007.

[119] M. Ikeda, A. Kawata, M. Nishikawa et al., "Hippo pathway-dependent and -independent roles of RASSF6," *Science Signaling*, vol. 2, no. 90, article ra59, 2009.

[120] S. Ortiz-Vega, A. Khokhlatchev, M. Nedwidek et al., "The putative tumor suppressor RASSF1A homodimerizes and heterodimerizes with the Ras-GTP binding protein Nore1," *Oncogene*, vol. 21, no. 9, pp. 1381–1390, 2002.

[121] J. N. Weitzel, A. Kasperczyk, C. Mohan, and T. G. Krontiris, "The HRAS1 gene cluster: two upstream regions recognizing transcripts and a third encoding a gene with a leucine zipper domain," *Genomics*, vol. 14, no. 2, pp. 309–319, 1992.

[122] T. G. Krontiris, B. Devlin, D. D. Karp, N. J. Robert, and N. Risch, "An association between the risk of cancer and mutations in the HRAS1 minisatellite locus," *New England Journal of Medicine*, vol. 329, no. 8, pp. 517–523, 1993.

[123] T. G. Krontiris, N. A. DiMartino, M. Colb, and D. R. Parkinson, "Unique allelic restriction fragments of the human Ha-ras locus in leukocyte and tumour DNAs of cancer patients," *Nature*, vol. 313, no. 6001, pp. 369–374, 1985.

[124] H. Friess, J. Ding, J. Kleeff et al., "Microarray-based identification of differentially expressed growth- and metastasis-associated genes in pancreatic cancer," *Cellular and Molecular Life Sciences*, vol. 60, no. 6, pp. 1180–1199, 2003.

RASSF Signalling and DNA Damage: Monitoring the Integrity of the Genome?

11

[125] C. D. Logsdon, D. M. Simeone, C. Binkley et al., "Molecular profiling of pancreatic adenocarcinoma and chronic pancreatitis identifies multiple genes differentially regulated in pancreatic cancer," *Cancer Research*, vol. 63, no. 10, pp. 2649–2657, 2003.

[126] A. W. Lowe, M. Olsen, Y. Hao et al., "Gene expression patterns in pancreatic tumors, cells and tissues," *PLoS One*, vol. 2, no. 3, article e323, 2007.

[127] G. L. Mutter, J. P.A. Baak, J. T. Fitzgerald et al., "Global expression changes of constitutive and hormonally regulated genes during endometrial neoplastic transformation," *Gynecologic Oncology*, vol. 83, no. 2, pp. 177–185, 2001.

[128] D. S. P. Tan, M. B. K. Lambros, S. Rayter et al., "PPM1D is a potential therapeutic target in ovarian clear cell carcinomas," *Clinical Cancer Research*, vol. 15, no. 7, pp. 2269–2280, 2009.

[129] S. Takahashi, A. Ebihara, H. Kajiho, K. Kontani, H. Nishina, and T. Katada, "RASSF7 negatively regulates pro-apoptotic JNK signaling by inhibiting the activity of phosphorylated-MKK7," *Cell Death and Differentiation*, vol. 18, no. 4, pp. 645–655, 2011.

2

TRIM22: A Diverse and Dynamic Antiviral Protein

Clayton J. Hattlmann, Jenna N. Kelly, and Stephen D. Barr

Department of Microbiology and Immunology, Center for Human Immunology, The University of Western Ontario, London, ON, Canada N6A 5C1

Correspondence should be addressed to Stephen D. Barr, stephen.barr@uwo.ca

Academic Editor: Abraham Brass

The tripartite motif (TRIM) family of proteins is an evolutionarily ancient group of proteins with homologues identified in both invertebrate and vertebrate species. Human TRIM22 is one such protein that has a dynamic evolutionary history that includes gene expansion, gene loss, and strong signatures of positive selection. To date, TRIM22 has been shown to restrict the replication of a number of viruses, including encephalomyocarditis virus (EMCV), hepatitis B virus (HBV), and human immunodeficiency virus type 1 (HIV-1). In addition, TRIM22 has also been implicated in cellular differentiation and proliferation and may play a role in certain cancers and autoimmune diseases. This comprehensive paper summarizes our current understanding of TRIM22 structure and function.

1. Introduction

The TRIM gene family encodes a diverse group of proteins that are involved in many biological and antiviral processes. There are currently 100 known TRIM genes in the human genome and many of these genes are upregulated by multiple, distinct stimuli [1–3]. Historically, TRIM genes have been researched mainly for their antiviral properties; however this paradigm is changing. Two recent reports discussing the role of TRIM genes in autoimmunity and cancer highlight the importance of the TRIM family in the development of nonviral diseases [4, 5]. Many TRIM genes also have a dynamic evolutionary history and the TRIM family has been shown to undergo extensive gene duplication in both primates and teleost fish [1, 6]. In addition, several TRIM genes have experienced strong positive selection in primates [7]. Although the forces behind TRIM evolution remain unclear, it is possible that the TRIM family has evolved and continues to evolve, in response to new viral pathogens or endogenous danger signals. This paper provides an overview of the TRIM22 gene and summarizes its structure, evolution, expression, and antiviral activities.

2. Structure

TRIM proteins typically contain a conserved RBCC motif, which consists of an amino-terminal RING domain, one or two B-box domains, and a predicted coiled-coil region. Approximately 60% of TRIM proteins, including TRIM22, also contain a carboxyl-terminal domain B30.2 domain (Figure 1) [8, 9]. The RING domain of TRIM22 has homology with E3 ligases and has been shown to possess E3 ubiquitin ligase activity [9, 10]. The catalytic cysteine residues Cys15 and Cys18 are essential for this activity and mediate the transfer of ubiquitin to target proteins (Figure 1) [11, 12]. TRIM22 can also modify itself with ubiquitin which leads to proteasomal degradation [10, 11]. Interestingly, the *TRIM* family represents one of the largest groups of E3 ubiquitin ligases and E3 ligase activity seems to be crucial for *TRIM*-mediated carcinogenesis [4]. In addition, E3 ligase activity is important for many *TRIM*-mediated antiviral activities and for TRIM22, it is required for the inhibition of EMCV, HBV, and HIV-1 [11, 13, 14].

TRIM proteins typically contain one or two B-box domains, although B-box 1 is never present without B-box 2, and the two domains have different consensus sequences

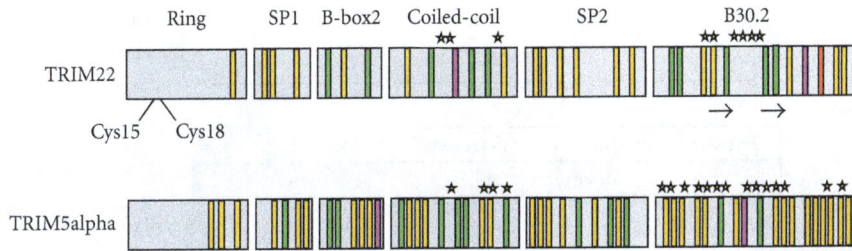

FIGURE 1: Structure and variability of TRIM22 and TRIM5α protein domains. TRIM22 contains an amino-terminal RING domain, one B-box domain (B-box 2), a coiled-coil region, and a carboxyl-terminal B30.2 domain (SP1 = Spacer 1 and SP2 = Spacer 2). Two cysteine residues (Cys15 and Cys18) in the RING domain are required for the E3 ligase activity of TRIM22, and a number of positively selected amino acids are found in the coiled-coil and B30.2 domains. The location and spacing of positively selected amino acids in TRIM22 are similar to those found in TRIM5α, which may reflect species-specific pathogenic pressures. The approximate location of positively selected amino acids in TRIM22 and TRIM5α is denoted with a star, and the location of the $\beta2$-$\beta3$ surface loop of TRIM22 is also indicated (arrows). Single nucleotide polymorphisms (SNPs) in the coding regions of TRIM22 and TRIM5α are shown as vertical bars, along with the type of mutation that each SNP can generate (green: nonsynonymous mutations; yellow: missense mutations; pink: frameshift mutations; red: nonsense mutations).

[15, 16]. TRIM22 contains one B-box domain (B-box 2), of which no clear function has been assigned (Figure 1). Certain B-box 2 mutations have been shown to affect viral recognition by other TRIM proteins, such as TRIM5α. Similar to TRIM22, TRIM5α has been shown to inhibit HIV-1 replication albeit at an earlier stage in the viral lifecycle. Interestingly, the human orthologue of TRIM5α only modestly inhibits HIV-1 replication whereas the rhesus orthologue of TRIM5α (rhTRIM5α) has potent anti-HIV-1 activity [17]. Several mechanisms of rhTRIM5α-mediated HIV-1 inhibition have been proposed; however, the favoured mechanism involves rhTRIM5α binding to the HIV-1 core and disruption of the normal uncoating process (reviewed in [18, 19]). For rhTRIM5α, the RING and B-box 2 domains promote its dimerization and higher-order self-association on the HIV-1 capsid [17]. It is unknown whether the B-box 2 domain of TRIM22 is required for higher-order self-association; however, it has been shown to play a role in the nuclear localization of TRIM22 [20].

The coiled-coil domain contains multiple predicted hypersecondary structures and intertwined α-helices [21]. In TRIM proteins, the coiled-coil domain is thought to promote homo-oligomerization, as its deletion prevents TRIM protein self-association [22]. Homo-oligomerization can be important for the formation of higher-molecular-weight complexes that define specific subcellular structures, such as nuclear bodies [21, 22]. Although the role of the coiled-coil region of TRIM22 remains unclear, self-association is a function of the coiled-coil region in other TRIM proteins. For example, the coiled-coil region of rhTRIM5α is required for rhTRIM5α trimerization and may be involved in the formation of cytoplasmic bodies. Importantly, rhTRIM5α trimerization is thought to drive its interaction with the HIV-1 capsid and the coiled-coil region is required for rhTRIM5α-mediated HIV-1 restriction [17, 23]. TRIM22 has also been shown to form trimers and to restrict HIV-1 replication but it is unknown whether the coiled-coil domain is required for these processes [14, 22, 24–26].

The B30.2 domain of TRIM proteins consists of two separate domains called the PRY and SPRY domains that form a putative protein-protein interaction site [27, 28]. This interaction site is likely important for the antiviral activities of TRIM22 and other TRIM proteins. Indeed, the B30.2 domain of rhTRIM5α is required for trimerization and HIV-1 restriction [23]. Three hyper-variable regions in the B30.2 domain of rhTRIM5α are thought to form the binding surface for the HIV-1 capsid protein [29]. In addition, these hypervariable regions confer the virus specificity of rhTRIM5α. The B30.2 domain of TRIM22 also contains these three hypervariable regions but their role in HIV-1 restriction has not yet been established. Similar to rhTRIM5α, the hypervariable regions in TRIM22 are highly polymorphic and contain a large number of positively selected amino acids (Figure 1) [7]. It will be interesting to learn if the B30.2 domain of TRIM22 confers specificity for targets such as viral pathogens. Notably, the B30.2 domain is required for the formation of nuclear bodies [13, 20, 30].

3. Evolution of TRIM22

The human *TRIM22* gene is located on chromosome 11, immediately adjacent to the *TRIM5*, *TRIM6*, and *TRIM34* genes [7, 31]. The origins of *TRIM22*, and the entire *TRIM5/6/22/34* gene cluster, can be traced back to the Cretaceous period, sometime after the divergence of Metatherian (marsupial) and Eutherian (placental) mammals (Figure 2). Previous studies have shown that the *TRIM5/6/22/34* locus is absent in Metatherian mammals such as opossum and chicken but presents in the major Eutherian groups containing cow, dog, and human [7]. Thus, this gene cluster must have emerged after the Metatherian-Eutherian division but before the separation of the major Eutherian groups. Taken together, this dates the birth of *TRIM22* (along with *TRIM5*, *TRIM6* and *TRIM34*) to approximately 90–180 million years ago (Figure 2) [7].

The *TRIM5/6/22/34* gene cluster likely arose through tandem gene duplication, as these four *TRIM* genes are close human paralogs and because major gene rearrangements have been documented in this chromosomal region

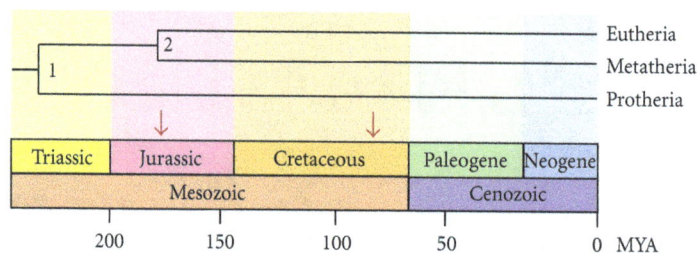

FIGURE 2: Timeline of Metatherian and Eutherian mammalian evolution showing the emergence of *TRIM22*. The divergence of Metatherian (marsupial) and Eutherian (placental) mammals occurred approximately 180 million years ago in the Jurassic period of the Mesozoic era. The *TRIM22* gene emerged sometime after this division, as it is absent in Metatherian mammals but present in all major Eutherian groups. In addition, since *TRIM22* is present in all Eutherian mammals, it must have emerged before further Eutherian division occurred (approximately 90 million years ago). Taken together, this dates the birth of *TRIM22* to approximately 90–180 million years ago. The predicted window of time for *TRIM22* emergence in Eutherian mammals is demarcated with two red arrows. MYA: millions of years.

[1, 7, 32]. Gene duplication plays a major role in evolution and *TRIM* genes have been shown to undergo extensive gene duplication in both primates and teleost fish [1, 6]. One of the most important outcomes of gene duplication is neofunctionalization, whereby one copy of the duplicate gene acquires a novel, beneficial function, and the other copy of the gene retains its original function [33–35]. This type of gene manipulation is a potent driver of evolution because it allows an organism to create new, potentially advantageous genes without disrupting the integrity of the original gene.

Recently, a genomic analysis of a different branch of the *TRIM* gene family identified several *TRIM* genes on chromosome 11 that have given rise to multiple *TRIM* paralogs in humans and African apes [1]. A group of 7 *TRIM* genes that are present in all Eutherian mammals (*TRIM43*, *TRIM48*, *TRIM49*, *TRIM51*, *TRIM53*, *TRIM64*, and *TRIM77*) were shown to spawn 11 new *TRIM* genes in certain primates and 6 new *TRIM* genes in humans, primarily through segmental duplications [1]. These new *TRIM* genes have presumably evolved and adapted to react against more recently emerged pathogenic threats. In addition, a Han Chinese woman with 12 new *TRIM* genes was identified, documenting for the first time *TRIM* gene copy number variation in humans [1]. Given its role in antiviral immunity, *TRIM22* probably emerged in a similar manner as a means of counteracting new viral pathogens; however the exact selective pressures giving rise to the *TRIM22* gene remain unclear.

According to a recent study, *TRIM* genes can be divided into two main groups based on their structural similarities and evolutionary properties [36]. Group 1 members have two B-box domains, have variable C-terminal domains, and are represented in both vertebrate and invertebrate species. In contrast, Group 2 members have only one B-box domain (B-box 2), are characterized by a C-terminal SPRY domain, and are found only in vertebrates. In addition, Group 2 genes are younger and smaller and evolve more rapidly than Group 1 genes [36]. Compared to some other *TRIM* genes, *TRIM22* is young and has evolved under strong positive selection, thus *TRIM22* (along with the *TRIM5/6/22/34* gene cluster) is classified as a Group 2 gene. Interestingly, the authors suggest that Group 2 genes may act as *TRIM* gene reservoirs, spawning new genes to respond to species-specific changes

at the host-pathogen interface. Consistent with this interpretation, there are a number of positively selected amino acids in TRIM22 which all cluster at predicted virus interaction sites in the coiled-coil and B30.2 domains (Figure 1) [7, 36].

Within the *TRIM5/6/22/34* gene cluster, *TRIM22* and *TRIM5* have a unique evolutionary relationship. In some Eutherian groups, such as cow, there are multiple copies of the *TRIM5* gene and no *TRIM22* gene. However in others such as dog, the *TRIM22* gene is present and the *TRIM5* gene is absent [7]. In addition, the strong positive selection that each of these two genes has experienced over millions of years has occurred in a mutually exclusive manner. This type of anticorrelative pattern is probably due to genetic linkage between the two genes, whereby positive selection of an advantageous mutation in one gene indirectly leads to the selection of a linked mutation in the other [7]. The location and spacing of positively selected amino acids in TRIM22 is very similar to those found in TRIM5α (Figure 1). In both proteins, the positively selected amino acids are located in the coiled-coil and B30.2 domains, which is interesting because their amino acid sequences are actually the least similar in these regions. The majority of positively selected amino acids in TRIM22 are found within the $\beta2$-$\beta3$ surface loop of the B30.2 domain, an area that is important for HIV-1 recognition in TRIM5α (Figure 1) [7, 37, 38]. It is possible that TRIM22 and TRIM5α once possessed a similar antiretroviral mechanism, and that they evolved over time to respond to species-specific pathogenic pressures. Indeed, many studies have shown that rhesus TRIM5α, but not human TRIM5α, can potently inhibit HIV-1 replication [18, 39]. In contrast, human TRIM22 can inhibit HIV-1 replication and thus may have evolved to compensate for the loss of TRIM5α's anti-HIV function.

The *TRIM22* gene has a dynamic evolutionary history that includes gene expansion, gene loss, and strong signatures of positive selection in primates [1, 6, 7, 36]. The high number of nonsynonymous mutations found in *TRIM22*, along with its classification as a Group 2 *TRIM* gene, suggests that this gene continues to evolve at a rapid pace. Given the volatile state of other *TRIM* genes in chromosome 11, it is possible that the *TRIM5/6/22/34* gene cluster takes part in

TABLE 1: Summary of the localization patterns observed for TRIM22.

Localization	Pattern	Cell Type	Epitope Tag	Reference
Cytoplasm	Diffuse	293T	GFP or V5/His	[40]
	Diffuse	COS7	GFP or V5/His	[40]
	Diffuse	HeLa	Endogenous	[40]
	Diffuse with speckles/bodies	HeLa	GFP	[22]
	Diffuse	HeLa	GFP or V5/His	[40]
	Diffuse	PBMCs	Endogenous	[40]
	Diffuse with speckles/bodies	U2OS	GFP	[22]
Cytoplasm & Nucleus	Nucleoplasmic, with nuclear bodies[1]	ABC28	Endogenous	[30]
	Diffuse throughout, or nuclear bodies[2]	HeLa	EGFP	[30]
	Nucleoplasmic, with nuclear bodies	HeLa	Endogenous	[30]
	Diffuse, with cytoplasmic bodies[3]	HeLa	FLAG	[43]
	Nucleoplasmic with NB[4]	MCF7	EGFP, EYFP, or FLAG	[30]
	Nucleoplasmic, with nuclear bodies	MCF7	Endogenous	[30]
	Nucleoplasmic and cytoplasmic	T47D	Endogenous	[30]
	Diffuse with speckles[5-7]	U2OS	Endogenous	[42]
Nucleus	Aggregates/bodies	293	Myc	[41]
	Aggregates/bodies	COS7	Myc	[10]
	Diffuse with speckles/bodies	HepG2	Endogenous	[13]
	Diffuse with speckles/bodies	HepG2	Myc	[13]
	Diffuse with bodies	MCF7	FLAG	[20]

[1] Some colocalization with fibrillarin (Nucleoli).
[2] Pattern changes with cell cycle phase: (G0/G1: nuclear bodies; S-phase: nuclear speckles and cytoplasmic; mitosis: diffuse throughout cell).
[3] TRIM22 plasmid was coexpressed with Rhesus TRIM5α.
[4] Partial colocalization with Cajal bodies.
[5] Potential colocalization with calnexin (Endoplasmic reticulum).
[6] Localization was primarily cytoplasmic when cells were fixed with paraformaldehyde, or both cytoplasmic and nuclear when fixed with ice-cold methanol.
[7] Partial colocalization with the centrosome.

gene and/or segmental duplication in humans. Presumably, individuals with an increased number of these *TRIM* genes may have an augmented antiviral response and could be particularly adept at controlling retroviral infections. Similar to copy number variations, a number of single nucleotide polymorphisms (SNPs) exist in *TRIM22* that may influence its antiviral capacity or biological function for that matter. For instance, there are two documented frameshift mutations and one documented nonsense mutation in the National Center for Biotechnology Information SNP database for the *TRIM22* gene (Figure 1). If present, these SNPs would generate different truncated versions of the TRIM22 protein, which may alter its structure, E3 ubiquitin ligase activity and/or antiviral function. There are also twenty documented missense mutations in the *TRIM22* gene, the majority of which are found in its B30.2 domain (Figure 1). Many of these SNPs have the potential to impact TRIM22 function and their presence or absence may contribute to individual differences in TRIM22-mediated activities.

4. Biological Functions of TRIM22

4.1. TRIM22 Localization. There are several contradictory reports detailing the subcellular localization of TRIM22. Some reports have observed that TRIM22 localizes predominantly to the cytoplasm [22, 40] or to the nucleus [10, 13, 20, 41], whereas other reports have observed that TRIM22 can localize to both the cytoplasm and the nucleus (Table 1) [30, 42, 43]. The pattern of localization also varied between diffuse, speckled, and aggregated. A number

of explanations have been given in the literature for the differences in localization, including whether the expression was endogenous (e.g., IFN-treatment) or exogenous (e.g., overexpression). In addition, the method of fixation and the type of epitope tag used for detection have also been reported to affect the localization pattern. Given the diverse range of cell lines used in these studies, it is also possible that cell type-specific factors influence the localization of TRIM22.

A number of determinants affecting TRIM22 localization have been identified. A bipartite nuclear localization signal (NLS) located in the Spacer 2 domain of TRIM22 was shown to be necessary, but not sufficient, for nuclear localization [20]. Although there are no known NLSs present in the B30.2/SPRY domain, several groups have shown that this domain is required for nuclear localization [13, 20, 40, 41]. More specifically, Val 493 and Cys 494 of the B30.2 domain were shown to be critical for nuclear localization and the formation of nuclear bodies [20]. In an independent study, amino acids Ser 395, Lys 396, and Ser 400 located in variable loops 1 and 3 of the B30.2 domain were shown to be important for certain localization patterns of TRIM22 [40].

In some cell types, TRIM22 localizes in the nucleus as punctate bodies, which have been shown to partially colocalize with Cajal bodies [20]. Cajal bodies play important roles in RNA processing and modification as well as in cell cycle progression [44]. TRIM22 also interacts with p80-coilin, which is a major component of Cajal bodies. Similar to Cajal bodies, TRIM22 localization has been shown to change during the cell cycle. In G0/G1 TRIM22 localizes in nuclear bodies, in S-phase it localizes in a more diffuse and speckled pattern throughout the nucleus, and during mitosis it assumes a diffuse pattern in both the nucleus and cytoplasm [30]. In an independent study, TRIM22 was shown to colocalize with the centrosome independently of the cell cycle and also with vimentin-containing aggresome-like structures next to the endoplasmic reticulum [42]. From these data, it appears that multiple factors influence the localization of TRIM22, possibly indicating that TRIM22 has several biological roles.

4.2. Antiviral Function of TRIM22. Several reports including published transcriptional profiling datasets (e.g., GDS1096, GDS3113, and GDS596) deposited in the Gene Expression Omnibus database repository (http://www.ncbi.nlm.nih.gov/gds) show that TRIM22 is ubiquitously expressed in several human tissues and is highly upregulated in response to Type I and II interferons (Table 2) [7, 13, 14, 24, 25, 45–50]. Interestingly, the 5′ flanking region of the *TRIM22* gene contains two regions matching the consensus sequence for an IFN-stimulating response element (ISRE) and a third region matching that for an IFN-γ activation site (GAS); however ISRE1 or GAS is not required for IFN-γ induction of TRIM22. In contrast, the ISRE2 plus six upstream nucleotides (extended ISRE) is capable of binding IFN regulatory factor 1 (IRF1) in a manner dependent on the chromatin remodelling enzyme Brahma-related gene 1 (BRG1) [48, 49]. Furthermore, this extended ISRE appears to be important for both stimulation by IFN-α and IFN-γ as well as for basal TRIM22 expression [48]. The significant upregulation of TRIM22 in response to IFNs, together with the finding that TRIM22 has evolved under strong positive selection for millions of years, suggests that TRIM22 plays an important fundamental role in cell biology. To date, several lines of evidence suggest that this role is as an antiviral factor.

Human TRIM22 was first discovered by Tissot and Mechti in 1995 during a search for IFN-induced genes in Daudi cells, where exogenous expression of TRIM22 was shown to downregulate transcription from the HIV-1 LTR [45]. Although this was performed using a luciferase reporter gene under the transcriptional control of the HIV-1 LTR and not in the context of the entire HIV-1 proviral genome, it provided the first evidence suggesting that TRIM22 blocks HIV-1 transcription and replication. In 2006, Bouazzaoui et al. showed that TRIM22 was highly upregulated in primary monocyte-derived macrophages (MDMs) in response to HIV-1 infection, IFNα treatment, or stimulation with lipopolysaccharide (LPS). They provided the first evidence that TRIM22 can restrict HIV-1 replication *in vitro* by showing that exogenous expression of TRIM22 inhibited HIV-1 infection by 50–90% in 293T cells modified to express the CD4 and CCR5 receptors and in primary MDM. Furthermore, cotransfection of TRIM22 with a three-plasmid system for replication-defective HIV-1 resulted in reduced infectious titres of pseudotyped virus, suggesting that TRIM22 inhibited a late stage of HIV-1 pseudoparticle production and/or subsequent infection with the pseudo-typed virus [24].

In 2008, Barr et al. showed that TRIM22 was an integral part of the Type I interferon-induced inhibition of HIV-1 replication and provided the first mechanistic data for the inhibition of HIV-1 replication by TRIM22. TRIM22 expression in several human cell lines potently inhibited HIV-1 replication, and interestingly, analysis of Gag production in those cells revealed that TRIM22 may inhibit HIV-1 replication by two different mechanisms. In the HOS and HeLa cell lines, TRIM22 inhibited HIV-1 particle production by interfering with the trafficking of the Gag polyprotein to the plasma membrane. Since TRIM22 and Gag proteins interact, and that the E3 ligase activity of TRIM22 is required for this restriction [14], it is possible that TRIM22 posttranslationally modifies Gag, resulting in altered Gag trafficking to the plasma membrane. In the U2OS and 143B cell lines, TRIM22 inhibited HIV-1 particle production by inhibiting the accumulation of intracellular Gag protein [14]. Although no mechanism of restriction was identified in U2OS or 143B cells, several possibilities could explain the decrease in intracellular Gag protein levels, including inhibition of transcription from the LTR as previously suggested [25, 45], or degradation of the Gag RNA and/or polyprotein. Given that TRIM22 exhibits cell type-specific differences in localization (as discussed earlier), it is likely that the mechanism of TRIM22-induced restriction of HIV-1 particle production is cell type-specific and/or dependent on the subcellular localization of TRIM22. Future experiments are required to further elucidate the mechanism of TRIM22-induced inhibition of HIV-1 particle production (Figure 3).

TRIM22 was also independently identified and shown to inhibit HIV-1 replication by several laboratories [25]. Following observations made by Franzoso et al. in 1994 that clones of the U937 promonocytic cell line were either permissive or nonpermissive to HIV-1 replication, Kajaste-Rudnitski et al. (2011) identified *TRIM22* as the only known restriction factor that was expressed in the nonpermissive and absent from the permissive U937 cells. Using a luciferase reporter plasmid under the control of the HIV-1 LTR, they showed that LTR-mediated transcription was decreased 7–10-fold in nonpermissive clones. They also showed that by knocking down *TRIM22* expression in nonpermissive cells, the levels of transcription from the LTR approached those observed in permissive cells. Exogenous expression of TRIM22 in permissive clones also decreased LTR transcription to levels comparable to those observed in nonpermissive clones. Further investigation revealed that TRIM22 inhibited basal and phorbol myristate acetate-ionomycin-induced HIV-1 transcription. These effects were independent of NFκB, HIV-1 Tat and the E3 ubiquitin ligase activity of TRIM22 [25]. It is important to note that all direct evidence showing that TRIM22 inhibits HIV-1 transcription has been through the use of LTR-driven reporter constructs. It will be important to test the effects of TRIM22 on HIV-1 LTR transcription in the context of full-length replication-competent HIV-1.

In 2011, Singh et al. provided the first clinically relevant evidence supporting a role for TRIM22 as an anti-HIV-1 effector *in vivo*. They showed that expression of

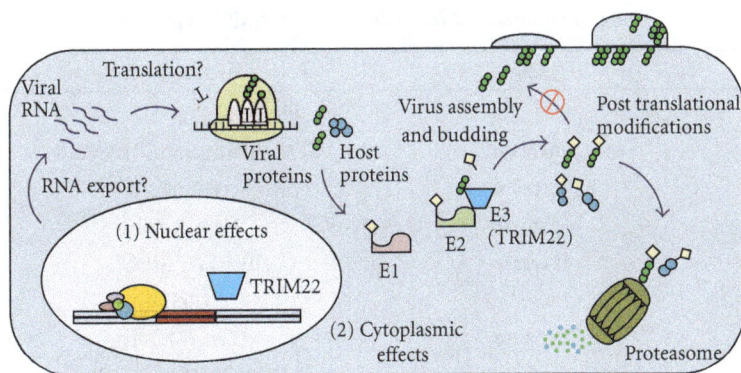

FIGURE 3: Possible mechanisms of TRIM22 antiviral functions. Based on current reports, TRIM22 can inhibit viral replication through nuclear-associated effects such as inhibiting viral transcription. Although not investigated to date, RNA export and translation are also potential targets of TRIM22. Given its E3 ligase activity, TRIM22 may posttranslationally modify host or viral proteins that are required for viral assembly and/or budding. Posttranslational modifications occur when an E1 activating enzyme (E1), E2 conjugating enzyme (E2), and E3 ligase protein (E3) work together to transfer ubiquitin or ubiquitin-like molecules to a target protein. These modifications could target the protein for proteasomal degradation or alter its subcellular localization or ability to interact with other proteins or DNA.

TRIM22 in peripheral blood mononuclear cells (PBMCs) of HIV-1-infected individuals was significantly increased in patients after HIV-1 infection. Importantly, infected patients expressing higher *TRIM22* levels exhibited significantly lower viral loads and significantly higher CD4+ T-cell counts [26]. These findings are quite significant, as this suggests that TRIM22 has a potential effect on the severity and/or progression of HIV-1 infection. Additional research on the role of TRIM22 during primary infection will be important to provide a greater understanding of the effects TRIM22 may have on HIV-1 replication *in vivo*.

The antiviral activities of TRIM22 are not limited to HIV-1. In 2009, Eldin et al. identified TRIM22 as a potent inhibitor of encephalomyocarditis virus (EMCV) replication. TRIM22 was shown to interact with the EMCV 3C protease via the C-terminal domain of TRIM22, and expression of TRIM22 corresponded with increased ubiquitination of the 3C protease (Figure 3). 3C protease is essential for successful viral replication and has several roles, including processing of the viral polyprotein and inhibition of the host immune defences [11]. There are also reports that TRIM22 may play an important role in protecting the liver from viral pathogens. In 2009, Gao et al. reported that TRIM22 is highly upregulated in response to type I or II IFN in the hepatocellular carcinoma cell line HepG2. Cotransfection of plasmids encoding TRIM22 and replication-competent hepatitis B virus (HBV) inhibited the accumulation of HBV antigens in the supernatants of cells and significantly reduced levels of intracellular HBV RNA and DNA replication intermediates. Similar results were observed in the sera of mice during codelivery of plasmids to mouse livers, showing that TRIM22 can restrict HBV infection in an *in vivo* system. Using a luciferase reporter plasmid, they showed that TRIM22 downregulates expression from the HBV core promoter (Figure 3). This mechanism of action was dependent on the nuclear localization of TRIM22 and its E3 ubiquitin ligase activity [13]. Although there is no direct evidence for a protective role of TRIM22

against HBV in primates, *TRIM22* expression is significantly upregulated during clearance of HBV in chimpanzees [51]. Moreover, TRIM22 expression is significantly upregulated during clearance of hepatitis C virus (HCV) in chimpanzees [52]. These findings are paralleled in human infections, as *TRIM22* is significantly upregulated in cirrhotic liver from HCV patients and patients with mild chronic HCV infection and no fibrosis [53]. Further research is needed to assess the role of TRIM22 in inhibiting HBV and HCV *in vivo*.

In further support of the notion that TRIM22 is involved in the host antiviral response, TRIM22 expression is modulated in response to several other viruses and viral antigens (Table 2). TRIM22 expression is upregulated in response to infection with rubella virus [54] and Epstein-Barr virus (EBV) [55] and downregulated during infection with human papillomavirus type 31 [56]. A couple intriguing reports elude to the possibility that TRIM22 may also contribute to viral latency. Exogenous expression of TRIM22 significantly upregulates expression of the EBV latent membrane protein 1 (LMP-1) [55]. LMP-1 is required for latency during EBV infection and appears to induce an antiviral state by upregulating expression of several ISGs via an IFN- and STAT1-independent mechanism. The Kaposi's sarcoma-associated herpesvirus (KSHV) latency-associated nuclear antigen (LANA) also activates several ISGs including TRIM22, which was shown to be upregulated by LANA both in culture and in tissues from KSHV lesions. LANA also represses transcription from the HIV-1 LTR, an NFκB consensus sequence, and the SV40 promoter [57]. Furthermore, TRIM22 is expressed in resting T cells, which are known reservoirs of latent HIV-1, and is strongly repressed during T-cell activation [47]. Although much more research is needed to directly implicate TRIM22 in viral latency, it is tempting to hypothesize that TRIM22 contributes to viral latency.

4.3. Other Functions of TRIM22. Several reports in the literature suggest that TRIM22 may have a role in other biological processes, such as cell differentiation and proliferation. One

TABLE 2: Summary of factors that alter TRIM22 expression.

Stimulation	Change	Tissue	Reference
Cytokines			
IFN-α	increase	CEM, Jurkat, and THP-1 cells	[26]
IFN-α	increase	H9 cells	[47]
IFN-α	increase	HepG2 cells	[13]
IFN-α	increase	Primary MDM	[24]
IFN-α	increase	U937	[25]
IFN-α	increase	U-937-4 cells	[46]
IFN-α/β	increase	Daudi, and HeLa cells	[45]
IFN-β	increase	HOS cells	[14]
IFN-γ	increase	HeLa cells	[30, 45]
IFN-γ	increase	HepG2 cells	[13, 48, 49]
IFN-γ	increase	MCF7 cells	[30]
IL-1-β	increase	Coronary artery endothelium	[58]
IL-2	increase	CD4+, CD8+, NK cells	[50]
IL-15	increase	CD4+, CD8+, NK cells	[50]
Progesterone	increase	ABC28, and T47D cells	[30]
TNF-α	increase	Coronary artery endothelium	[58]
Antigens/Infections			
EBV infection[1]	increase	BL41-EBV cells[1]	[55]
EBV LMP-1	increase	DG75 cells	[55]
Hepatitis B virus infection[2]	increase	Liver tissue[2]	[51]
Hepatitis C virus infection[2]	increase	Liver tissue[2]	[52]
Hepatitis C virus infection	increase	Liver tissue	[53]
HIV-1 infection	increase	Immature DC	[55]
HIV-1 infection	increase	Primary MDM	[24]
HIV-1 infection	increase	Primary PBMCs	[26]
HIV-1 Tat	increase	Immature DC	[55]
HPV infection	decrease	Human keratinocytes	[56]
KSHV infection	increase	KSHV lesion	[57]
KSHV LANA	increase	BJAB cells	[57]
LPS	increase	Primary MDM	[24]
Rubella virus infection	increase	ECV304 cells	[54]
Activation/Differentiation/Cell Cycle			
1α,25-dihydroxyvitamine D3[3]	increase	Primary MDM	[24]
Anti-CD2	increase	Primary T cells	[47]
Anti-CD2/CD28	decrease	Primary T cells	[47]
Anti-CD2/CD28/CD3	decrease	CD4+, CD8+, NK cells[4]	[50]
All-trans retinoic acid	increase	HL60 and NB4 cells	[46]
All-trans retinoic acid	increase	Primary MDM	[24]
p53	increase	K562 and U-937-4 cells[5]	[46]
p73	increase	U-937-4 cells	[46]
Pioglitazone	increase	Primary MDM	[24]
UV-irradiation[6]	increase	MCF-7 cells	[46]
Disease			
SLE	decrease	CD4+ T cells from SLE patient	[59]

TABLE 2: Continued.

Stimulation	Change	Tissue	Reference
Wilms tumor	decrease	Tumor tissue	[60, 61]

[1] BL41 cells that are latently infected with EBV.
[2] From infected chimpanzees.
[3] Hormonally active form of Vitamin D.
[4] Only reached significance in CD8+ cells.
[5] Cells lack endogenous p53 but stably express a plasmid encoding p53 under control of a temperature-sensitive promoter. Cells were grown at the permissive temperature (32°C) to induce p53 expression.
[6] UV-irradiation induces p53 expression.

group showed that the expression of TRIM22 is directly activated by p53 in myeloid cells via a functional p53-response element in intron 1 of the *TRIM22* gene [46]. They also showed that the p53-family member p73 can bind to this response element and activate *TRIM22* expression [46]. Since p73 has been linked to the differentiation of leukemic cells [62], the authors speculated that TRIM22 may be involved in cell differentiation. Another group reported that TRIM22 expression is significantly upregulated during differentiation of the promyelocytic cell line NB4 [63]. They also showed that TRIM22 expression is high in monocytes and early granulocytes but decreases in the lymphocyte and late granulocyte populations and is undetectable in erythroid cells [63]. Obad et al. (2004) provided the first direct evidence supporting an antiproliferative role for TRIM22 by showing that overexpression of TRIM22 in the promonocytic cell line U937 resulted in decreased clonogenic growth [46]. An inverse correlation between TRIM22 expression and cell differentiation has also been reported, showing that TRIM22 is highly expressed in human immature CD34+ bone marrow progenitor cells, but declines in mature populations [63]. Despite the correlations of TRIM22 expression levels with cell differentiation and proliferation, the evidence lacks key experiments such as loss-of-function studies (i.e., TRIM22 knockdown) to conclusively implicate TRIM22 as a key player in any of these processes.

A couple of reports have associated TRIM22 with human disease. Downregulation of TRIM22 expression is associated with progression, relapse and increased mortality in cases of Wilms tumor [60, 61]. Although TRIM22 is a p53-responsive gene and may promote cell-cycle arrest [46], its role in tumour development and progression, including Wilms tumor, is yet to be determined. The involvement of TRIM proteins in cancer is not unprecedented. TRIM13, 24, and 29, which are also involved in p53 regulation, have also been implicated as important regulators for carcinogenesis. Moreover, TRIM19/PML may act as a tumour suppressor protein (reviewed in [4]). TRIM22 expression is also downregulated in CD4+ T cells from patients with active systemic lupus erythematosus (SLE) [59]. Although it is also unclear what role TRIM22 plays in this disease, it is notable that several other TRIM proteins, including TRIM 21, 25, 56, and 68, have been linked to SLE and other autoimmune diseases [5]. It will be interesting to learn more about the role (if any) TRIM22 plays in these and other human diseases.

Although it is clear that TRIM22 is an exciting and dynamic protein, it appears that we have only begun to understand its role in cellular biology and antiviral immunity. A rich evolutionary history, together with its potential involvement in numerous biological processes, suggests that TRIM22 is an important and multifarious protein. Despite its importance, the function of TRIM22 remains poorly understood, and a number of issues will need to be addressed in future research. One discrepancy that needs clarification is the disparate observations and contradictory reports surrounding TRIM22 subcellular localization. In particular, we need to understand why TRIM22 localization is so heterogeneous, as this may provide useful insight into its biological function. Another priority will be to consolidate previous reports on the antiviral mechanism of TRIM22. In the case of HIV-1, it will be important to determine the stage(s) of the virus lifecycle that TRIM22 targets. In this regard, future studies that identify the host and/or virus targets of TRIM22 will be extremely useful. In addition, it will be interesting to discover if TRIM22 has antiviral activity against additional viruses, and to determine the role it plays in other nonviral diseases. Overall, its breadth of involvement in antiviral immunity, combined with the range of possible mechanisms by which TRIM22 acts, presents a number of exciting research opportunities. Future work on TRIM22 will help us understand this important player in the host antiviral response and contribute to our knowledge of host-pathogen interactions.

Acknowledgments

S. D. Barr is supported by funds from the Department of Microbiology and Immunology (The University of Western Ontario) and a Scholarship Award from The Ontario HIV Treatment Network (OHTN). J. N. Kelly is supported by an Ontario Graduate Scholarship.

References

[1] K. Han, D. I. Lou, and S. L. Sawyer, "Identification of a genomic reservoir for new TRIM genes in primate genomes," *PLoS Genetics*, vol. 7, no. 12, Article ID e1002388, 2011.

[2] R. Rajsbaum, J. P. Stoye, and A. O'Garra, "Type I interferon-dependent and -independent expression of tripartite motif proteins in immune cells," *European Journal of Immunology*, vol. 38, no. 3, pp. 619–630, 2008.

[3] L. Carthagena, A. Bergamaschi, J. M. Luna et al., "Human TRIM gene expression in response to interferons," *Plos one*, vol. 4, no. 3, Article ID e4894, 2009.

[4] S. Hatakeyama, "TRIM proteins and cancer," *Nature Reviews Cancer*, vol. 11, pp. 792–804, 2011.

[5] C. Jefferies, C. Wynne, and R. Higgs, "Antiviral TRIMs: friend or foe in autoimmune and autoinflammatory disease?" *Nature Reviews Immunology*, vol. 11, no. 9, pp. 617–625, 2011.

[6] L. M. van der Aa, J. P. Levraud, M. Yahmi et al., "A large new subset of TRIM genes highly diversified by duplication and positive selection in teleost fish," *BMC Biology*, vol. 7, article 7, 2009.

[7] S. L. Sawyer, M. Emerman, and H. S. Malik, "Discordant evolution of the adjacent antiretroviral genes TRIM22 and TRIM5 in mammals," *PLoS Pathogens*, vol. 3, no. 12, article e197, 2007.

[8] S. Nisole, J. P. Stoye, and A. Saïb, "TRIM family proteins: retroviral restriction and antiviral defence," *Nature Reviews Microbiology*, vol. 3, no. 10, pp. 799–808, 2005.

[9] G. Meroni and G. Diez-Roux, "TRIM/RBCC, a novel class of "single protein RING finger" E3 ubiquitin ligases," *Bioessays*, vol. 27, no. 11, pp. 1147–1157, 2005.

[10] Z. Duan, B. Gao, W. Xu, and S. Xiong, "Identification of TRIM22 as a RING finger E3 ubiquitin ligase," *Biochemical and Biophysical Research Communications*, vol. 374, no. 3, pp. 502–506, 2008.

[11] P. Eldin, L. Papon, A. Oteiza, E. Brocchi, T. G. Lawson, and N. Mechti, "TRIM22 E3 ubiquitin ligase activity is required to mediate antiviral activity against encephalomyocarditis virus," *Journal of General Virology*, vol. 90, no. 3, pp. 536–545, 2009.

[12] K. L. Lorick, J. P. Jensen, S. Fang, A. M. Ong, S. Hatakeyama, and A. M. Weissman, "RING fingers mediate ubiquitin-conjugating enzyme (E2)-dependent ubiquitination," *Proceedings of the National Academy of Sciences of the United States of America*, vol. 96, no. 20, pp. 11364–11369, 1999.

[13] B. Gao, Z. Duan, W. Xu, and S. Xiong, "Tripartite motif-containing 22 inhibits the activity of hepatitis b virus core promoter, which is dependent on nuclear-located RING domain," *Hepatology*, vol. 50, no. 2, pp. 424–433, 2009.

[14] S. D. Barr, J. R. Smiley, and F. D. Bushman, "The interferon response inhibits hiv particle production by induction of TRIM22," *Plos Pathogens*, vol. 4, no. 2, Article ID e1000007, 2008.

[15] M. Torok and L. D. Etkin, "Two B or not two B? overview of the rapidly expanding b-box family of proteins," *Differentiation*, vol. 67, no. 3, pp. 63–71, 2001.

[16] K. Ozato, D. M. Shin, T. H. Chang, and H. C. Morse, "TRIM family proteins and their emerging roles in innate immunity," *Nature Reviews Immunology*, vol. 8, no. 11, pp. 849–860, 2008.

[17] X. Li, D. F. Yeung, A. M. Fiegen, and J. Sodroski, "Determinants of the higher order association of the restriction factor TRIM5α and other tripartite motif (TRIM) proteins," *Journal of Biological Chemistry*, vol. 286, no. 32, pp. 27959–27970, 2011.

[18] E. E. Nakayama and T. Shioda, "Anti-retroviral activity of TRIM5α," *Reviews in Medical Virology*, vol. 20, no. 2, pp. 77–92, 2010.

[19] Z. Lukic and E. M. Campbell, "The cell biology of TRIM5α," *Current HIV/AIDS Reports*, vol. 9, pp. 73–80, 2012.

[20] G. Sivaramakrishnan, Y. Sun, R. Rajmohan, and V. C. L. Lin, "B30.2/SPRY domain in tripartite motif-containing 22 is essential for the formation of distinct nuclear bodies," *FEBS Letters*, vol. 583, no. 12, pp. 2093–2099, 2009.

[21] D. A. D. Parry, R. D. B. Fraser, and J. M. Squire, "Fifty years of coiled-coils and α-helical bundles: a close relationship between sequence and structure," *Journal of Structural Biology*, vol. 163, no. 3, pp. 258–269, 2008.

[22] A. Reymond, G. Meroni, A. Fantozzi et al., "The tripartite motif family identifies cell compartments," *EMBO Journal*, vol. 20, no. 9, pp. 2140–2151, 2001.

[23] C. C. Mische, H. Javanbakht, B. Song et al., "Retroviral restriction factor TRIM5α is a TRIMer," *Journal of Virology*, vol. 79, no. 22, pp. 14446–14450, 2005.

[24] A. Bouazzaoui, M. Kreutz, V. Eisert et al., "Stimulated trans-acting factor of 50 kDa (Staf50) inhibits HIV-1 replication in human monocyte-derived macrophages," *Virology*, vol. 356, no. 1-2, pp. 79–94, 2006.

[25] A. Kajaste-Rudnitski, S. S. Marelli, C. Pultrone et al., "TRIM22 inhibits HIV-1 transcription independently of its E3 ubiquitin ligase activity, Tat, and NF-κB-responsive long terminal repeat elements," *Journal of Virology*, vol. 85, no. 10, pp. 5183–5196, 2011.

[26] R. Singh, G. Gaiha, L. Werner et al., "Association of TRIM22 with the type 1 interferon response and viral control duRING primary HIV-1 infection," *Journal of Virology*, vol. 85, no. 1, pp. 208–216, 2011.

[27] D. A. Rhodes, B. De Bono, and J. Trowsdale, "Relationship between spry and b30.2 protein domains. evolution of a component of immune defence," *Immunology*, vol. 116, no. 4, pp. 411–417, 2005.

[28] C. Grütter, C. Briand, G. Capitani et al., "Structure of the pryspry-domain: implications for autoinflammatory diseases," *FEBS Letters*, vol. 580, no. 1, pp. 99–106, 2006.

[29] S. Ohkura, M. W. Yap, T. Sheldon, and J. P. Stoye, "All three variable regions of the TRIM5α B30.2 domain can contribute to the specificity of retrovirus restriction," *Journal of Virology*, vol. 80, no. 17, pp. 8554–8565, 2006.

[30] G. Sivaramakrishnan, Y. Sun, S. K. Tan, and V. C. L. Lin, "Dynamic localization of tripartite motif-containing 22 in nuclear and nucleolar bodies," *Experimental Cell Research*, vol. 315, no. 8, pp. 1521–1532, 2009.

[31] C. Tissot, S. A. Taviaux, S. Diriong, and N. Mechti, "Localization of Staf50, a member of the RING finger family, to 11p15 by ruorescence in situ hybridization," *Genomics*, vol. 34, no. 1, pp. 151–153, 1996.

[32] J. Zhang, S. Qin, S. N. J. Sait et al., "The pericentromeric region of human chromosome 11: evidence for a chromosome-specific duplication," *Cytogenetics and Cell Genetics*, vol. 94, no. 3-4, pp. 137–141, 2001.

[33] S. P. Otto and P. Yong, "The evolution of gene duplicates," *Advances in Genetics*, vol. 46, pp. 451–483, 2002.

[34] R. C. Moore and M. D. Purugganan, "The early stages of duplicate gene evolution," *Proceedings of the National Academy of Sciences of the United States of America*, vol. 100, no. 26, pp. 15682–15687, 2003.

[35] B. Conrad and S. E. Antonarakis, "Gene duplication: a drive for phenotypic diversity and cause of human disease," *Annual Review of Genomics and Human Genetics*, vol. 8, pp. 17–35, 2007.

[36] M. Sardiello, S. Cairo, B. Fontanella, A. Ballabio, and G. Meroni, "Genomic analysis of the TRIM family reveals two groups of genes with distinct evolutionary properties," *BMC Evolutionary Biology*, vol. 8, no. 1, article 225, 2008.

[37] S. L. Sawyer, L. I. Wu, M. Emerman, and H. S. Malik, "Positive selection of primate TRIM5α identifies a critical species-specific retroviral restriction domain," *Proceedings of the National Academy of Sciences of the United States of America*, vol. 102, no. 8, pp. 2832–2837, 2005.

[38] M. W. Yap, S. Nisole, and J. P. Stoye, "A single amino acid change in the spry domain of human TRIM5α leads to HIV-1 restriction," *Current Biology*, vol. 15, no. 1, pp. 73–78, 2005.

[39] M. Stremlau, M. Perron, S. Welikala, and J. Sodroski, "Species-specific variation in the B30.2(SPRY) domain of TRIM5α determines the potency of human immunodeficiency virus restriction," *Journal of Virology*, vol. 79, no. 5, pp. 3139–3145, 2005.

[40] A. M. Herr, R. Dressel, and L. Walter, "Different subcellular localisations of TRIM22 suggest species-specific function," *Immunogenetics*, vol. 61, no. 4, pp. 271–280, 2009.

[41] S. Yu, B. Gao, Z. Duan, W. Xu, and S. Xiong, "Identification of tripartite motif-containing 22 (TRIM22) as a novel NF-κB activator," *Biochemical and Biophysical Research Communications*, vol. 410, no. 2, pp. 247–251, 2011.

[42] J. Petersson, P. Lönnbro, A. M. Herr, M. Mörgelin, U. Gullberg, and K. Drott, "The human IFN-inducible p53 target gene TRIM22 colocalizes with the centrosome independently of cell cycle phase," *Experimental Cell Research*, vol. 316, no. 4, pp. 568–579, 2010.

[43] X. Li, B. Gold, C. O'hUigin et al., "Unique features of TRIM5α among closely related human TRIM family members," *Virology*, vol. 360, no. 2, pp. 419–433, 2007.

[44] M. Cioce and A. I. Lamond, "Cajal bodies: a long history of discovery," *Annual Review of Cell and Developmental Biology*, vol. 21, pp. 105–131, 2005.

[45] C. Tissot and N. Mechti, "Molecular cloning of a new interferon-induced factor that represses human immunodeficiency virus type i long terminal repeat expression," *Journal of Biological Chemistry*, vol. 270, no. 25, pp. 14891–14898, 1995.

[46] S. Obad, H. Brunnström, J. Vallon-Christersson, A. Borg, K. Drott, and U. Gullberg, "Staf50 is a novel p53 target gene conferRING reduced clonogenic growth of leukemic U-937 cells," *Oncogene*, vol. 23, no. 23, pp. 4050–4059, 2004.

[47] C. Gongora, C. Tissot, C. Cerdan, and N. Mechti, "The interferon-inducible Staf50 gene is downregulated during t cell costimulation by CD2 and CD28," *Journal of Interferon and Cytokine Research*, vol. 20, no. 11, pp. 955–961, 2000.

[48] B. Gao, Y. Wang, W. Xu, Z. Duan, and S. Xiong, "A 55′ extended IFN-stimulating response element is crucial for IFN-γ-induced tripartite motif 22 expression via interaction with ifn regulatory factor-1," *Journal of Immunology*, vol. 185, no. 4, pp. 2314–2323, 2010.

[49] Y. Wang, B. Gao, W. Xu, and S. Xiong, "Brg1 is indispensable for IFN-γ-induced TRIM22 expression, which is dependent on the recruitment of IRF-1," *Biochemical and Biophysical Research Communications*, vol. 410, no. 3, pp. 549–554, 2011.

[50] S. Obad, T. Olofsson, N. Mechti, U. Gullberg, and K. Drott, "Regulation of the interferon-inducible p53 target gene TRIM22 (Staf50) in human t lymphocyte activation," *Journal of Interferon and Cytokine Research*, vol. 27, no. 10, pp. 857–864, 2007.

[51] S. Wieland, R. Thimme, R. H. Purcell, and F. V. Chisari, "Genomic analysis of the host response to hepatitis b virus infection," *Proceedings of the National Academy of Sciences of the United States of America*, vol. 101, no. 17, pp. 6669–6674, 2004.

[52] A. I. Su, J. P. Pezacki, L. Wodicka et al., "Genomic analysis of the host response to hepatitis C virus infection," *Proceedings of the National Academy of Sciences of the United States of America*, vol. 99, no. 24, pp. 15669–15674, 2002.

[53] M. E. Folkers, D. A. Delker, C. I. Maxwell et al., "Encode tiling array analysis identifies differentially expressed annotated and novel 5′ capped RNAs in hepatitis C infected liver," *Plos ONE*, vol. 6, no. 2, Article ID e14697, 2011.

[54] X. Y. Mo, W. Ma, Y. Zhang et al., "Microarray analyses of differentially expressed human genes and biological processes in ECV304 cells infected with rubella virus," *Journal of Medical Virology*, vol. 79, no. 11, pp. 1783–1791, 2007.

[55] E. Izmailova, F. M. N. Bertley, Q. Huang et al., "HIV-1 Tat reprograms immature dendritic cells to express chemoattractants for activated T cells and macrophages," *Nature Medicine*, vol. 9, no. 2, pp. 191–197, 2003.

[56] Y. E. Chang and L. A. Laimins, "Microarray analysis identifies interferon-inducible genes and Stat-1 as major transcriptional targets of human papillomavirus type 31," *Journal of Virology*, vol. 74, no. 9, pp. 4174–4182, 2000.

[57] Y. Wang, H. Li, Q. Tang, G. G. Maul, and Y. Yuan, "Kaposi's sarcoma-associated herpesvirus ori-lyt-dependent DNA replication: involvement of host cellular factors," *Journal of Virology*, vol. 82, no. 6, pp. 2867–2882, 2008.

[58] O. Bandman, R. T. Coleman, J. F. LoRING, J. J. Seilhamer, and B. G. Cocks, "Complexity of inflammatory responses in endothelial cells and vascular smooth muscle cells determined by microarray analysis," *Annals of the New York Academy of Sciences*, vol. 975, pp. 77–90, 2002.

[59] Y. J. Deng, Z. X. Huang, C. J. Zhou et al., "Gene profiling involved in immature CD4$^+$ T lymphocyte responsible for systemic lupus erythematosus," *Molecular Immunology*, vol. 43, no. 9, pp. 1497–1507, 2006.

[60] B. Zirn, O. Hartmann, B. Samans et al., "Expression profiling of wilms tumors reveals new candidate genes for different clinical parameters," *International Journal of Cancer*, vol. 118, no. 8, pp. 1954–1962, 2006.

[61] S. Wittmann, C. Wunder, B. Zirn et al., "New prognostic markers revealed by evaluation of genes correlated with clinical parameters in wilms tumors," *Genes Chromosomes and Cancer*, vol. 47, no. 5, pp. 386–395, 2008.

[62] T. X. Liu, J. W. Zhang, J. Tao et al., "Gene expression networks underlying retinoic acid-induced differentiation of acute promyelocytic leukemia cells," *Blood*, vol. 96, no. 4, pp. 1496–1504, 2000.

[63] S. Obad, T. Olofsson, N. Mechti, U. Gullberg, and K. Drott, "Expression of the IFN-inducible p53-target gene TRIM22 is down-regulated duRING erythroid differentiation of human bone marrow," *Leukemia Research*, vol. 31, no. 7, pp. 995–1001, 2007.

3

RASSF1A Signaling in the Heart: Novel Functions beyond Tumor Suppression

Dominic P. Del Re and Junichi Sadoshima

Cardiovascular Research Institute, Department of Cell Biology and Molecular Medicine, UMDNJ-New Jersey Medical School,
185 South Orange Avenue, MSB G-609, Newark, NJ 07103-2714, USA

Correspondence should be addressed to Junichi Sadoshima, sadoshju@umdnj.edu

Academic Editor: Shairaz Baksh

The RASSF proteins are a family of polypeptides, each containing a conserved Ras association domain, suggesting that these scaffold proteins may be effectors of activated Ras or Ras-related small GTPases. RASSF proteins are characterized by their ability to inhibit cell growth and proliferation while promoting cell death. RASSF1 isoform A is an established tumor suppressor and is frequently silenced in a variety of tumors and human cancer cell lines. However, our understanding of its function in terminally differentiated cell types, such as cardiac myocytes, is relatively nascent. Herein, we review the role of RASSF1A in cardiac physiology and disease and highlight signaling pathways that mediate its function.

1. Introduction

The Ras association domain family (RASSF) consists of 10 members: RASSF1-10. Additionally, splice variants of RASSF1, 5 and 6 have been identified [1]. Importantly, all isoforms contain a Ras association (RA) domain either in their C-terminal (RASSF1-6) or N-terminal (RASSF7-10) regions [2]. To date, no known catalytic activity has been described for this family, and the general consensus supposes that RASSF proteins function as scaffolds to localize signaling in the cell. Accordingly, protein-protein interactions are critical in mediating their biological functions. RASSF1 isoform A (RASSF1A) is the most characterized member of the RASSF family. This paper will focus primarily on RASSF1A and its role in cardiovascular biology.

2. RASSF1A

RASSF1A was first identified and described by Dammann et al. in 2000 [3]. The *RASSF1* gene encodes multiple splice variants, including the two predominant isoforms, RASSF1A and C. The RASSF1A isoform is the longest variant of the *RASSF1* gene. Structurally, RASSF1A is a product of exons 1α, $2\alpha/\beta$, 3, 4, 5, and 6, while RASSF1C consists of exons 2γ, 3, 4, 5, and 6. Both isoforms contain a C-terminal RA domain; however, RASSF1A has an additional C1 domain that is not present in RASSF1C.

The *RASSF1* gene is located on Chr3p21.3 [3]. This short arm of chromosome 3 is known to exhibit loss of heterozygosity in many tumor models and is thought to harbor tumor suppressor genes. As the literature has shown, RASSF1A fits this description. The *RASSF1A* promoter contains a CpG island that shows a high frequency of hypermethylation in tumors, thereby silencing RASSF1A expression in many human cancers including lung, breast, ovarian, renal, and bladder [4–7]. RASSF1A expression is also lost in numerous cancer cell lines, while RASSF1C expression is seemingly unaffected [4]. Interestingly, recent work suggests that RASSF1C may actually promote tumor progression [8, 9], further distinguishing these two splice variants.

All RASSF proteins have an RA domain, which is thought to necessitate their binding to activated, GTP-bound Ras proteins. While RASSF5 (Nore1) is thought to bind Ras directly, whether RASSF1A is able to associate with Ras is less clear. It has been shown that RASSF1A binds K-Ras *in vitro* [10], and an interaction between ectopically expressed RASSF1A and activated K-Ras has been observed in HEK293 cells [11, 12]. However, other work has found that this interaction only occurs in the presence of Nore1, arguing for an indirect association [13]. Importantly, to our knowledge, there are no reports demonstrating the interaction of endogenous RASSF1A and Ras proteins.

RASSF1A has several key biological functions typical of tumor suppressor proteins. It has been implicated in the negative regulation of cell cycle progression, cell proliferation, and cell survival [2]. RASSF1A has been shown to localize to microtubules of proliferating cells, increasing microtubule stability and inhibiting cell division [14, 15]. This may be mediated through direct binding or though interaction with microtubule-associated proteins such as C19ORF5 [16]. RASSF1A has also been shown to inhibit proliferation by inhibiting the accumulation of cyclin D1 and arresting cell division [17, 18].

RASSF1A also promotes apoptosis, which can reportedly occur through multiple mechanisms and is likely cell-type dependent. One mechanism that mediates the apoptotic function of RASSF1A involves protein interaction with modulator of apoptosis-1 (MOAP-1 or MAP-1) [19]. MOAP-1 is normally sequestered in an inactive form in healthy cells. Upon death receptor stimulation, RASSF1A binds MOAP-1, causing its activation and subsequent association with Bax, which leads to apoptosis [19]. Previous work has also demonstrated enhancement of RASSF1A/Mst-mediated cell death by the scaffold CNK1 [20].

2.1. RASSF1A and Hippo Signaling. RASSF1A can also elicit inhibitory effects on growth and survival through engagement of the Hippo pathway. The Hippo signaling pathway is a highly conserved kinase cascade that was originally discovered in Drosophila and has been shown to be a critical regulator of cell proliferation, survival, and organ growth [21]. Three members of this pathway, dRASSF, Salvador and Hippo, contain the SARAH (Salvador-RASSF-Hippo) domain, which is conserved in its mammalian counterparts RASSF1-6, WW45, and Mst1/2, respectively [22]. The SARAH domain is critical for homo- and heterodimerization between components [23–27]. While the Drosophila ortholog dRASSF is known to antagonize Hippo activation in the fly [28], it has been demonstrated that RASSF1A promotes phosphorylation and activation of Mst 1/2 by inhibiting the phosphatase PP2A in mammalian systems [29, 30].

The biological relevance of RASSF1A-mediated activation of Hippo signaling has also been investigated. Matallanas et al. reported a RASSF1A-Mst2-Yap-p73-PUMA signaling axis that promotes apoptosis in mammalian cells [31]. Hippo signaling is also important for maintaining intestinal homeostasis and tissue regeneration in response to injury. Mouse models with conditional disruption of either Mst1/2 or Sav1 in the intestinal epithelium displayed hyperactivation of Yes-associated protein (Yap), increased intestinal stem cell

(ISC) proliferation, and increased polyp formation following dextran sodium sulfate (DSS) treatment [32, 33]. Similarly, loss-of-function mutations of Hippo components in the fly midgut caused increased ISC proliferation [34]. These findings suggest that perhaps Hippo signaling serves a more global role in regulating organ integrity, structure, and response to injury, and that perturbation of this pathway can lead to aberrant growth and dysfunction.

3. Cardiovascular Function of RASSF1A

In 2005, two independent groups generated and published findings regarding the systemic deletion of the *Rassf1a* gene variant in mice [35, 36]. Both described similar phenotypes involving the spontaneous generation of tumors, particularly in aged mice, thus further supporting the notion that RASSF1A is a bona fide tumor suppressor [35, 36]. Not surprisingly, nearly all studies involving RASSF1A to date are related to cancer biology with few reports related to the cardiovascular field.

RASSF1A is ubiquitously expressed and has been detected in heart tissue [3, 37, 38]. Initial investigation into the role of *Rassf1* gene products in a cardiac context came from the Neyses laboratory [39]. Their findings demonstrated that both RASSF1A and RASSF1C could associate with the sarcolemmal calcium pump, PMCA4b, in neonatal rat cardiac myocytes. This interaction was shown to mediate the inhibition of ERK, and subsequent Elk transcription and suggested the possibility that RASSF1A could modulate cardiac myocyte growth [39].

3.1. $Rassf1a^{-/-}$ Mice. Five years later, the same group demonstrated that RASSF1A does in fact negatively regulate cardiac hypertrophy *in vivo* using $Rassf1a^{-/-}$ mice [37]. Although these mice have increased susceptibility to spontaneous tumorigenesis [36], no apparent cardiovascular phenotype was observed under basal conditions, that is, no differences in heart size, morphology, or function compared to WT. However, when $Rassf1a^{-/-}$ mice were challenged with pressure overload, they responded with an exaggerated hypertrophic response, evidenced by significantly greater increases in heart weight/body weight and hypertrophic gene expression (ANP, BNP, β-MHC). Cardiac myocytes of $Rassf1a^{-/-}$ mice were significantly larger, which explains the augmented heart growth. Chamber dilation of $Rassf1a^{-/-}$ mouse hearts was observed by echocardiography, consistent with eccentric hypertrophic remodeling. Hemodynamic analysis of WT and $Rassf1a^{-/-}$ mice showed a rightward shift in PV loops following pressure overload in $Rassf1a^{-/-}$ hearts, yet dP/dt_{max}, dP/dt_{min}, and fractional shortening were not altered in $Rassf1a^{-/-}$ mice compared to WT.

To examine RASSF1A function in cardiac myocytes, Oceandy et al. utilized a neonatal rat cardiac myocyte (NRCM) culture and the forced expression of RASSF1A through adenoviral gene transfer [37]. Increased RASSF1A expression inhibited phenylephrine-(PE-) induced cardiac myocyte growth and suppressed Raf-1 and ERK1/2 activation by PE treatment. Conversely, both Raf-1 and ERK1/2 phosphorylation were increased in $Rassf1a^{-/-}$ hearts following pressure overload, suggesting negative regulation of MAPK signaling by RASSF1A. Deletion mutants of RASSF1A revealed an important function of the N-terminus of RASSF1A that disrupts the binding of active Ras and Raf-1, thus preventing ERK activation and cardiac myocyte growth.

3.2. *Cardiac Myocyte-Specific Rassf1a Deletion.* To better understand the function of RASSF1A in cardiac myocytes *in vivo*, we crossed genetically altered mice harboring a floxed *Rassf1a* allele [35] with mice harboring the Cre recombinase transgene driven by the α-MHC promoter. This strategy disrupted endogenous *Rassf1a* gene expression and ensured cardiac myocyte specificity [38, 40]. Similar to the $Rassf1a^{-/-}$ mice, $Rassf1a^{F/F}$-*Cre* mice had no obvious baseline cardiac phenotype. Although we also found exaggerated heart growth in the $Rassf1a^{-/-}$ mice in response to pressure overload, the $Rassf1a^{F/F}$-*Cre* mice unexpectedly had attenuated hypertrophy, that is, smaller hearts and cardiac myocytes, compared to $Rassf1a^{F/F}$ and α-MHC-Cre controls [38]. Furthermore, $Rassf1a^{F/F}$-*Cre* mice had significantly less fibrosis and myocyte apoptosis, and better cardiac function following pressure overload. This was in stark contrast to the $Rassf1a^{-/-}$ mice, which presented significantly more fibrosis and a decline in cardiac function comparable to the levels found in WT mice.

As an alternative approach we also generated two different cardiac-specific transgenic mouse lines: the first expressing wild-type RASSF1A and the second expressing a RASSF1A SARAH domain point mutant (L308P) that renders it unable to bind Mst1 [41]. Interestingly, we found that increased RASSF1A expression in the heart caused increases in Mst1 activation, cardiac myocyte apoptosis, and fibrosis, and led to worsened function following pressure overload. Conversely, RASSF1A L308P TG mice had significant reductions in Mst1 activation, apoptosis and fibrosis, while cardiac function was preserved after stress [38]. These opposing phenotypes strongly implicate Mst1 as a critical effector of RASSF1A-mediated myocardial dysfunction.

In cultured NRCMs, increased RASSF1A expression elicited activation of Mst1 and caused Mst1-mediated apoptosis. However, in primary rat cardiac fibroblasts, RASSF1A had a more pronounced effect on inhibition of cell proliferation rather than survival. Indeed, we found that silencing of RASSF1A in fibroblasts caused increased cell proliferation. Additionally, RASSF1A depletion led to an upregulation of NF-κB-dependent TNF-α expression and secretion in cardiac fibroblasts, while no change in IL-1β, IL-6, or TGF-β1 was observed. Through conditioned medium transfer experiments, we demonstrated that TNF-α secretion from fibroblasts promotes cardiac myocyte growth. Furthermore, treatment of $Rassf1a^{-/-}$ mice with a neutralizing antibody against TNF-α was able to rescue the augmented heart growth and fibrosis observed following pressure overload [38]. These data strongly implicated TNF-α as a critical paracrine factor influencing the cardiac myocyte growth response to stress *in vivo*. This work also demonstrated

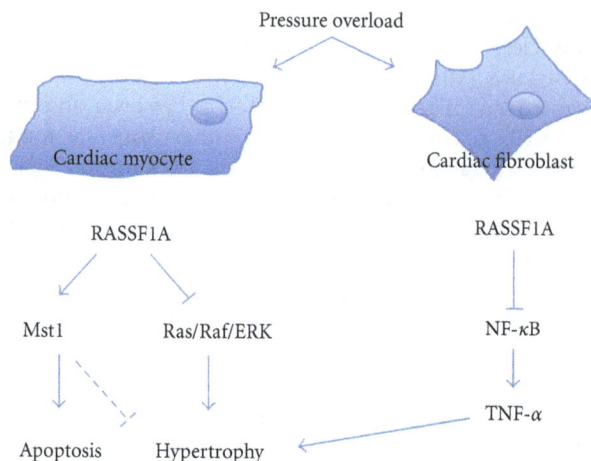

FIGURE 1: In cardiac myocytes, RASSF1A can prevent hypertrophy through disruption of Ras/Raf-1/ERK MAPK signaling. RASSF1A can also activate Mst1 to elicit apoptosis. In cardiac fibroblasts, RASSF1A represses NF-κB transcriptional activity and inhibits TNF-α production and secretion, thereby preventing paracrine-mediated hypertrophic signaling between fibroblast and myocyte.

the cell-type specificity of RASSF1A signaling in the heart and highlighted a novel signaling pathway downstream of RASSF1A/Mst1 that mediates a paracrine effect *in vivo* (see Figure 1). This mechanism involving multiple cell types, and paracrine signaling among them is rather unique and contrasts with more established signaling paradigms of cardiac hypertrophy including calcineurin/NFAT, HDAC/MEF2 and MEK/ERK pathways, which have been elucidated in the cardiac myocyte [42].

3.3. *Hippo Signaling in the Heart.* Our previous work has demonstrated the functional importance of Hippo signaling in the heart. Using genetically altered mouse models we showed that increased expression of Mst1, and subsequent activation of the Hippo pathway, caused increased apoptosis, dilated cardiomyopathy, and premature death [43]. Interestingly, expression of Mst1 also attenuated cardiac myocyte hypertrophy thereby impairing the heart's ability to appropriately respond to stress. In contrast, expression of a kinase-inactive Mst1 mutant (DN-Mst1) prevented cell death and protected the heart from insult [43]. Lats1/2 kinases (mammalian homologs of Warts) are targets of Mst1/2 that can phosphorylate and inactivate Yap, thereby inhibiting Yap-mediated gene transcription [44]. Similar to our findings related to Mst1, we demonstrated that transgenic expression of Lats2 in the heart led to inhibited growth and worsened function [45]. Conversely, kinase-inactive Lats2 (DN-Lats2) transgenic mice had larger hearts both at baseline and following pressure overload and displayed attenuated cardiac myocyte apoptosis in response to stress [45]. Taken together, these results provide further evidence that activation of Hippo signaling, via increased Mst1 or Lats2 expression, inhibits cardiac myocyte growth and promotes apoptosis in the adult heart. Furthermore, selective inhibition of Hippo signaling in the cardiac myocyte (DN-Mst1 or DN-Lats2

TG) confers protection against insult, similar to what we observed in the cardiac myocyte-specific RASSF1A deleted mice [38]. However, the hypertrophic response in these two models was opposite, which may result from a Hippo-independent pathway(s) downstream of RASSF1A. It should be pointed out that studies of adult mouse models using cardiac myocyte-restricted deletion of Mst1/2, Lats1/2 or Yap have not been published. Findings from these models should be helpful in further elucidating the role of Hippo signaling components in the adult murine heart.

Recent work from the Martin laboratory demonstrated the importance of mammalian Hippo signaling during cardiac development and cardiac myocyte proliferation [46]. Conditional deletion of Salvador (Sav1) in the embryonic heart, driven by Nkx2.5-Cre expression, caused increased myocyte proliferation and cardiac enlargement and was mediated by hyperactivation of Yap and subsequent Wnt/β-catenin-regulated gene expression. In a similar vein, direct targeting of Yap expression in the developing mouse heart further demonstrated its role in governing both myocyte proliferation and heart growth [47]. Interestingly, both reports described an interaction between Yap and Wnt signaling, highlighting additional Hippo signaling crosstalk in the heart.

4. Conclusion

Fueled by the initial reports described herein, investigation into the role of RASSF1A in cardiovascular biology has begun to accelerate. Yet many questions remain outstanding. Among them, what are the upstream inputs that regulate RASSF1A function? What is the mechanism responsible for RASSF1A cell-type-specific signaling? What are the molecular constituents of the RASSF1A complex? Does RASSF1A have additional Mst1-independent functions in the heart, as has been demonstrated in tumor cell lines [41]? Recent work identified activated K-Ras as a promoter of RASSF1A signaling in colorectal cancer cells [48]. This finding begs the question of whether K-Ras or additional Ras isoforms regulate RASSF1A in other systems and cell types. Based on our findings in *Rassf1a*-deleted mice [38], we speculate that the difference in proliferative capacity between cardiac myocytes and fibroblasts may explain the distinct effects of RASSF1A signaling in the heart. There may also be differences in the expression or localization of signaling components, thereby modulating their ability to effectively signal in certain cell types. Exposure to diverse signals and cues in the extracellular milieu may also contribute to varied outcomes downstream RASSF1A.

As we continue to elucidate the role of RASSF1A and Hippo signaling in the heart, its importance in cardiac development, physiology, and disease is becoming apparent. Of course, translating these findings into meaningful therapeutic strategies remains the greatest challenge. Our work has shed light on the importance of cell type specificity RASSF1A in determining pathological outcomes [38]. We also defined a paracrine mechanism functioning downstream of RASSF1A in response to cardiac stress [38].

It is likely that additional complexities remain to be uncovered and will ultimately influence possible interventions to manipulate RASSF1A and treat heart disease.

RASSF1A signaling is diverse and our knowledge regarding RASSF1A function is rapidly expanding. Given that a bridge from cancer to cardiovascular biology is in place, it is likely that as additional RASSF1A mechanisms of action are discovered, its impact on cardiac biology will continue to grow.

Acknowledgments

This work was supported in part by U.S. Public Health Service Grants HL59139, HL67724, HL69020, HL91469, HL102738, and AG27211. This work was also supported by an American Heart Association Scientist Development Grant (11SDG7240067) and Foundation Leducq Transatlantic Network of Excellence.

References

[1] M. Gordon and S. Baksh, "RASSF1A: not a prototypical Ras effector," *Small GTPases*, vol. 2, no. 3, pp. 148–157, 2011.

[2] J. Avruch, R. Xavier, N. Bardeesy et al., "Rassf family of tumor suppressor polypeptides," *Journal of Biological Chemistry*, vol. 284, no. 17, pp. 11001–11005, 2009.

[3] R. Dammann, C. Li, J. H. Yoon, P. L. Chin, S. Bates, and G. P. Pfeifer, "Epigenetic inactivation of a RAS association domain family protein from the lung tumour suppressor locus 3p21.3," *Nature Genetics*, vol. 25, no. 3, pp. 315–319, 2000.

[4] L. van der Weyden and D. J. Adams, "The Ras-association domain family (RASSF) members and their role in human tumourigenesis," *Biochimica et Biophysica Acta*, vol. 1776, no. 1, pp. 58–85, 2007.

[5] A. Agathanggelou, S. Honorio, D. P. Macartney et al., "Methylation associated inactivation of RASSF1A from region 3p21.3 in lung, breast and ovarian tumours," *Oncogene*, vol. 20, no. 12, pp. 1509–1518, 2001.

[6] K. Dreijerink, E. Braga, I. Kuzmin et al., "The candidate tumor suppressor gene, RASSF1A, from human chromosome 3p21.3 is involved in kidney tumorigenesis," *Proceedings of the National Academy of Sciences of the United States of America*, vol. 98, no. 13, pp. 7504–7509, 2001.

[7] R. Dammann, T. Takahashi, and G. P. Pfeifer, "The CpG island of the novel tumor suppressor gene RASSF1A is intensely methylated in primary small cell lung carcinomas," *Oncogene*, vol. 20, no. 27, pp. 3563–3567, 2001.

[8] Y. G. Amaar, M. G. Minera, L. K. Hatran, D. D. Strong, S. Mohan, and M. E. Reeves, "Ras association domain family 1C protein stimulates human lung cancer cell proliferation," *American Journal of Physiology*, vol. 291, no. 6, pp. L1185–L1190, 2006.

[9] E. Estrabaud, I. Lassot, G. Blot et al., "RASSF1C, an isoform of the tumor suppressor RASSF1A, promotes the accumulation of β-catenin by interacting with βTrCP," *Cancer Research*, vol. 67, no. 3, pp. 1054–1061, 2007.

[10] M. D. Vos, C. A. Ellis, A. Bell, M. J. Birrer, and G. J. Clark, "Ras uses the novel tumor suppressor RASSF1 as an effector to mediate apoptosis," *Journal of Biological Chemistry*, vol. 275, no. 46, pp. 35669–35672, 2000.

[11] P. Rodriguez-Viciana, C. Sabatier, and F. McCormick, "Signaling specificity by ras family GTPases is determined by the full

spectrum of effectors they regulate," *Molecular and Cellular Biology*, vol. 24, no. 11, pp. 4943–4954, 2004.

[12] H. Donninger, M. D. Vos, and G. J. Clark, "The RASSF1A tumor suppressor," *Journal of Cell Science*, vol. 120, pp. 3163–3172, 2007.

[13] S. Ortiz-Vega, A. Khokhlatchev, M. Nedwidek et al., "The putative tumor suppressor RASSF1A homodimerizes and heterodimerizes with the Ras-GTP binding protein Nore1," *Oncogene*, vol. 21, no. 9, pp. 1381–1390, 2002.

[14] L. Liu, S. Tommasi, D. H. Lee, R. Dammann, and G. P. Pfeifer, "Control of microtubule stability by the RASSF1A tumor suppressor," *Oncogene*, vol. 22, no. 50, pp. 8125–8136, 2003.

[15] A. Dallol, A. Agathanggelou, S. L. Fenton et al., "RASSF1A interacts with microtubule-associated proteins and modulates microtubule dynamics," *Cancer Research*, vol. 64, no. 12, pp. 4112–4116, 2004.

[16] L. Liu, A. Vo, and W. L. McKeehan, "Specificity of the methylation-suppressed A isoform of candidate tumor suppressor RASSF1 for microtubule hyperstabilization is determined by cell death inducer C19ORF5," *Cancer Research*, vol. 65, no. 5, pp. 1830–1838, 2005.

[17] L. Shivakumar, J. Minna, T. Sakamaki, R. Pestell, and M. A. White, "The RASSF1A tumor suppressor blocks cell cycle progression and inhibits cyclin D1 accumulation," *Molecular and Cellular Biology*, vol. 22, no. 12, pp. 4309–4318, 2002.

[18] A. Agathanggelou, I. Bieche, J. Ahmed-Choudhury et al., "Identification of novel gene expression targets for the ras association domain family 1 (RASSF1A) tumor suppressor gene in non-small cell lung cancer and neuroblastoma," *Cancer Research*, vol. 63, no. 17, pp. 5344–5351, 2003.

[19] S. Baksh, S. Tommasi, S. Fenton et al., "The tumor suppressor RASSF1A and MAP-1 link death receptor signaling to bax conformational change and cell death," *Molecular Cell*, vol. 18, no. 6, pp. 637–650, 2005.

[20] S. Rabizadeh, R. J. Xavier, K. Ishiguro et al., "The scaffold protein CNK1 interacts with the tumor suppressor RASSF1A and augments RASSF1A-induced cell death," *Journal of Biological Chemistry*, vol. 279, no. 28, pp. 29247–29254, 2004.

[21] D. Pan, "The hippo signaling pathway in development and cancer," *Developmental Cell*, vol. 19, no. 4, pp. 491–505, 2010.

[22] B. Zhao, K. Tumaneng, and K. L. Guan, "The Hippo pathway in organ size control, tissue regeneration and stem cell self-renewal," *Nature Cell Biology*, vol. 13, no. 8, pp. 877–883, 2011.

[23] A. Khokhlatchev, S. Rabizadeh, R. Xavier et al., "Identification of a novel Ras-regulated proapoptotic pathway," *Current Biology*, vol. 12, no. 4, pp. 253–265, 2002.

[24] H. Scheel and K. Hofmann, "A novel interaction motif, SARAH, connects three classes of tumor suppressor," *Current Biology*, vol. 13, no. 23, pp. R899–R900, 2003.

[25] M. Praskova, A. Khoklatchev, S. Ortiz-Vega, and J. Avruch, "Regulation of the MST1 kinase by autophosphorylation, by the growth inhibitory proteins, RASSF1 and NORE1, and by Ras," *Biochemical Journal*, vol. 381, pp. 453–462, 2004.

[26] E. Hwang, K. S. Ryu, K. Paakkonen et al., "Structural insight into dimeric interaction of the SARAH domains from Mst1 and RASSF family proteins in the apoptosis pathway," *Proceedings of the National Academy of Sciences of the United States of America*, vol. 104, no. 22, pp. 9236–9241, 2007.

[27] C. Guo, S. Tommasi, L. Liu, J. K. Yee, R. Dammann, and G. P. Pfeifer, "RASSF1A is part of a complex similar to the drosophila hippo/salvador/lats tumor-suppressor network," *Current Biology*, vol. 17, no. 8, pp. 700–705, 2007.

[28] C. Polesello, S. Huelsmann, N. H. Brown, and N. Tapon, "The Drosophila RASSF homolog antagonizes the hippo pathway," *Current Biology*, vol. 16, no. 24, pp. 2459–2465, 2006.

[29] H. J. Oh, K. K. Lee, S. J. Song et al., "Role of the tumor suppressor RASSF1A in Mst1-mediated apoptosis," *Cancer Research*, vol. 66, no. 5, pp. 2562–2569, 2006.

[30] C. Guo, X. Zhang, and G. P. Pfeifer, "The tumor suppressor RASSF1A prevents dephosphorylation of the mammalian STE20-like kinases MST1 and MST2," *Journal of Biological Chemistry*, vol. 286, no. 8, pp. 6253–6261, 2011.

[31] D. Matallanas, D. Romano, K. Yee et al., "RASSF1A elicits apoptosis through an MST2 pathway directing proapoptotic transcription by the p73 tumor suppressor protein," *Molecular Cell*, vol. 27, no. 6, pp. 962–975, 2007.

[32] J. Cai, N. Zhang, Y. Zheng, R. F. de Wilde, A. Maitra, and D. Pan, "The hippo signaling pathway restricts the oncogenic potential of an intestinal regeneration program," *Genes and Development*, vol. 24, no. 21, pp. 2383–2388, 2010.

[33] D. Zhou, Y. Zhang, H. Wu et al., "Mst1 and Mst2 protein kinases restrain intestinal stem cell proliferation and colonic tumorigenesis by inhibition of Yes-associated protein (Yap) overabundance," *Proceedings of the National Academy of Sciences of the United States of America*, vol. 108, no. 49, pp. E1312–E1320, 2011.

[34] F. Ren, B. Wang, T. Yue, E. Y. Yun, Y. T. Ip, and J. Jiang, "Hippo signaling regulates Drosophila intestine stem cell proliferation through multiple pathways," *Proceedings of the National Academy of Sciences of the United States of America*, vol. 107, no. 49, pp. 21064–21069, 2010.

[35] L. van der Weyden, K. K. Tachibana, M. A. Gonzalez et al., "The RASSF1A isoform of RASSF1 promotes microtubule stability and suppresses tumorigenesis," *Molecular and Cellular Biology*, vol. 25, no. 18, pp. 8356–8367, 2005.

[36] S. Tommasi, R. Dammann, Z. Zhang et al., "Tumor susceptibility of Rassf1a knockout mice," *Cancer Research*, vol. 65, no. 1, pp. 92–98, 2005.

[37] D. Oceandy, A. Pickard, S. Prehar et al., "Tumor suppressor ras-association domain family 1 Isoform A is a novel regulator of cardiac hypertrophy," *Circulation*, vol. 120, no. 7, pp. 607–616, 2009.

[38] D. P. Del Re, T. Matsuda, P. Zhai et al., "Proapoptotic Rassf1A/Mst1 signaling in cardiac fibroblasts is protective against pressure overload in mice," *Journal of Clinical Investigation*, vol. 120, no. 10, pp. 3555–3567, 2010.

[39] A. L. Armesilla, J. C. Williams, M. H. Buch et al., "Novel functional interaction between the plasma membrane Ca^{2+} pump 4b and the proapoptotic tumor suppressor Ras-associated factor 1 (RASSF1)," *Journal of Biological Chemistry*, vol. 279, no. 30, pp. 31318–31328, 2004.

[40] R. Agah, P. A. Frenkel, B. A. French, L. H. Michael, P. A. Overbeek, and M. D. Schneider, "Gene recombination in postmitotic cells: targeted expression of Cre recombinase provokes cardiac-restricted, site-specific rearrangement in adult ventricular muscle *in vivo*," *Journal of Clinical Investigation*, vol. 100, no. 1, pp. 169–179, 1997.

[41] H. Donninger, N. Allen, A. Henson et al., "Salvador protein is a tumor suppressor effector of RASSF1A with hippo pathway-independent functions," *Journal of Biological Chemistry*, vol. 286, no. 21, pp. 18483–18491, 2011.

[42] J. Heineke and J. D. Molkentin, "Regulation of cardiac hypertrophy by intracellular signalling pathways," *Nature Reviews Molecular Cell Biology*, vol. 7, no. 8, pp. 589–600, 2006.

[43] S. Yamamoto, G. Yang, D. Zablocki et al., "Activation of Mst1 causes dilated cardiomyopathy by stimulating apoptosis

without compensatory ventricular myocyte hypertrophy," *Journal of Clinical Investigation*, vol. 111, no. 10, pp. 1463–1474, 2003.

[44] Y. Hao, A. Chun, K. Cheung, B. Rashidi, and X. Yang, "Tumor suppressor LATS1 is a negative regulator of oncogene YAP," *Journal of Biological Chemistry*, vol. 283, no. 9, pp. 5496–5509, 2008.

[45] Y. Matsui, N. Nakano, D. Shao et al., "Lats2 is a negative regulator of myocyte size in the heart," *Circulation Research*, vol. 103, no. 11, pp. 1309–1318, 2008.

[46] T. Heallen, M. Zhang, J. Wang et al., "Hippo pathway inhibits wnt signaling to restrain cardiomyocyte proliferation and heart size," *Science*, vol. 332, no. 6028, pp. 458–461, 2011.

[47] M. Xin, Y. Kim, L. B. Sutherland et al., "Regulation of insulin-like growth factor signaling by Yap governs cardiomyocyte proliferation and embryonic heart size," *Science Signaling*, vol. 4, no. 196, p. ra70, 2011.

[48] D. Matallanas, D. Romano, F. Al-Mulla et al., "Mutant K-Ras activation of the proapoptotic MST2 pathway is antagonized by wild-type K-Ras," *Molecular Cell*, vol. 44, no. 6, pp. 893–906, 2011.

4

The SARAH Domain of RASSF1A and Its Tumor Suppressor Function

Claudia Dittfeld,[1,2] Antje M. Richter,[3] Katrin Steinmann,[1,3]
Antje Klagge-Ulonska,[1] and Reinhard H. Dammann[1,3]

[1] *AWG Tumor Genetics of the Medical Faculty, Martin-Luther-University Halle-Wittenberg, 06108 Halle, Germany*
[2] *OncoRay, National Center for Radiation Research in Oncology, Medical Faculty Carl Gustav Carus, University of Technology, 06108 Halle, Dresden, Germany*
[3] *Institute for Genetics, Justus-Liebig University Giessen, 35392 Giessen, Germany*

Correspondence should be addressed to Reinhard H. Dammann, reinhard.dammann@gen.bio.uni-giessen.de

Academic Editor: Geoffrey J. Clark

The Ras association domain family 1A (RASSF1A) tumor suppressor encodes a Sav-RASSF-Hpo domain (SARAH), which is an interaction domain characterized by hWW45 (dSAV) and MST1/2 (dHpo). In our study, the interaction between RASSF1A and RASSF1C with MST1 and MST2 was demonstrated and it was shown that this interaction depends on the SARAH domain. SARAH domain-deleted RASSF1A had a similar growth-reducing effect as full-length RASSF1A and inhibited anchorage independent growth of the lung cancer cell lines A549 significantly. In cancer cells expressing the SARAH deleted form of RASSF1A, reduced mitotic rates ($P = 0.001$) with abnormal metaphases ($P < 0.001$) were observed and a significantly increased rate of apoptosis was found ($P = 0.006$) compared to full-length RASSF1A. Although the association with microtubules and their stabilization was unaffected, mitotic spindle formation was altered by deletion of the SARAH domain of RASSF1A. In summary, our results suggest that the SARAH domain plays an important role in regulating the function of RASSF1A.

1. Introduction

The Ras association domain family 1 gene (*RASSF1*) was identified on chromosome 3p21.3, a region frequently deleted in cancer [1]. There are two major transcripts of *RASSF1*, termed *RASSF1A* and *RASSF1C*, which are transcribed from different CpG island promoters [1]. The promoter of *RASSF1A* is often hypermethylated in cancer, whereas the promoter region of *RASSF1C* is never methylated [2, 3]. Both isoforms encode a Ras association domain in the C-terminus, an ATM-kinase phosphorylation site, a SARAH protein interaction domain, and the N-terminal sequence of *RASSF1A* harbors a diacyl glycerol binding domain [1, 4]. It has been demonstrated that *RASSF1A* encodes a tumor suppressor gene, which reduces tumor growth *in vivo* and *in vitro* [1, 5–8]. Deletion of *Rassf1a* in mice significantly increased spontaneous and induced tumorigenesis [9–11]. It has been reported that RASSF1A binds to microtubules and protects cells from microtubule destabilizing agents [7, 12–15]. This interaction contributes to cell cycle regulation and mitotic progression.

RASSF1A is regulated by the binding of RAS and the novel Ras effector 1 (NORE1) and mediates proapoptotic signals through binding of the mammalian sterile 20-like kinase 1 and 2 (MST1 and MST2) [16–19]. Moreover, an association of RASSF1A with the BH3-like protein modulator of apoptosis was observed and this interaction regulates conformational change of BAX and apoptosis [20, 21]. RASSF1A promotes MDM2 self-ubiquitination and prevents p53 degradation [22]. Additionally, it was reported that RASSF1A inhibits the anaphase promoting complex (APC) through its binding to CDC20 and induces mitotic arrest by stabilizing mitotic cyclins [23] and it was further shown that the Aurora mitotic kinases are involved [24]. However, we were not able to verify the interaction between RASSF1A and CDC20 [25].

In the C-terminal part of RASSF1A and RASSF1C, a protein-protein interaction domain called SARAH (Sav/RASSF/Hpo) has been determined [26]. The SARAH domain is a key feature of the Hippo signaling pathway components, by which the interaction of Sav, Rassf, and Hpo is accomplished [26]. In the Drosophila Hippo pathway, Salvador (Sav, the human homologue is named WW45) acts as a scaffold protein that interacts with the proapoptotic kinase Hippo (Hpo, human homologue MST) [27–29]. Hpo is able to phosphorylate the kinase Warts (human homologue LATS), which in Drosophila leads to cell cycle arrest and apoptosis [28–30].

It was shown that the single Drosophila orthologue of the human RASSF proteins restricts Hpo activity by competing with Sav for binding to Hpo [31]. Praskova et al. previously showed that human RASSF1A interacts with MST1 through the C-terminus [16] and more precisely through the SARAH domain [32]. MST1 has two caspase 3 cleavage sites and both MST1 and MST2 play a role in processes of apoptosis both before and after caspase cleavage [33]. The cleaved form of MST1 translocates in the nucleus and phosphorylates histone H2B at Ser14 [34, 35]. H2B phosphorylation correlates with apoptotic chromatin condensation and nuclear fragmentation in mammalians and yeast [35, 36]. Following death receptor activation, MST1 (homologue of Hpo) is known to become activated through caspase-dependent cleavage [19]. The cleaved fragment then localizes from the cytoplasm to the nucleus, where it induces apoptosis [19] by chromatin condensation through activation of the c-Jun N-terminal kinase pathway [37].

Both RASSF1A and WW45 activate MST2 by promoting its autophosphorylation [38]. Moreover, RASSF1A stabilizes MST1/2 activation by preventing the dephosphorylation of these kinases [39]. Activated MST1/2 phosphorylates different targets including LATS kinases, which in turn activate the transcription coactivator YAP1 [40, 41]. Other MST1/2 targets are H2AX [42], FOXO [43], and troponin [44].

To gain new insights into the tumor suppressor function of RASSF1A, we deleted its SARAH domain and analyzed its altered function. Deletion of the SARAH domain resulted in a decreased colony formation of tumor cells. During mitosis, abnormal spindle formation was observed. We demonstrate that the interaction of RASSF1A and RASSF1C with MST1 and MST2 depends on the SARAH domain. Deregulation of the SARAH domain may contribute to altered proapoptotic and mitotic signaling of RASSF1A.

2. Materials and Methods

2.1. Tissues and Cell Lines.
The localization experiments and the protein expression experiments were performed in HEK293 and COS7 cells (ATCC, Manassas, Virginia, USA). For stable transfection, the lung cancer cell line A549 (ATCC) was used. A549 cells harbor epigenetic silenced RASSF1A, but express RASSF1C [1].

2.2. Interaction Studies Using the Yeast Two-Hybrid System.
The Matchmaker Two-hybrid system (Clontech, Mountain View, USA) was utilized. cDNAs of RASSF1A and RASSF1C were described previously [1]. The genes MST1 and MST2 were cloned after amplification of the fragments from EST-clones IRAKp961C0282Q and IRAKp961I0613Q (RZPD, Berlin, Germany), respectively. RASSF1AΔSARAH and RASSF1A were cloned into the vector pGADT7, RASSF1C and RASSF1CΔSARAH into the vector pAS2-1, and MST1 and MST2 into pGBKT7 [25, 45]. Mutant forms of RASSF1 were generated with the QuickChange XL Site-Directed Mutagenesis Kit (Stratagene, La Jolla, USA) and forward primer (5′-AGGAAAATGACTCTGGGCCCCTTGGGTGACCTCT) and the complementary reverse primer. All constructs were confirmed by sequencing. The yeast strain PJ69-4A was cotransformed with 0.1 μg of each plasmid using the PEG/LiAc method. The interaction analysis was carried out on SD minimal medium plates without adenine and histidine and the transformation efficiency was controlled on SD plates with adenine and histidine. The strength of interaction was investigated by quantification of the expression of the β-galactosidase reporter gene with o-nitrophenyl-β-D-galactopyranoside (ONPG) as substrate at 420 nm.

2.3. Interaction Studies by Coprecipitation.
MST1 and MST2 were cloned into the vector pCMV-Tag1 (Stratagene, La Jolla, USA) and/or in the vector pEBG. To investigate the interaction of specific RASSF1 forms, MST1 and MST2, cotransfections (Lipofectamine 2000, Invitrogen, Carlsbad, USA) were performed in HEK293 cells. Plasmids (pEBG and pCMV-Tag 1) were used, that express GST-Flag-RASSF1A, GST-Flag-RASSF1C, GST-Flag-RASSF1AΔSARAH, GST-Flag-RASSF1CΔSARAH, and GST or Flag-MST1, Flag-MST2, and Flag-WW45 [45]. Two days after transfection, total protein was extracted in RIPA buffer. The GST-fused proteins were precipitated with glutathione-sepharose (Amersham Biosciences, Freiburg, Germany). Samples were separated on a 10% PAGE gel and blotted. The interaction was determined with anti-Flag-antibodies (F3165, Sigma, Steinheim, Germany) and anti-GST antibodies (Santa Cruz, Santa Cruz, USA).

2.4. MST1 and MST2 Phosphorylation.
A549 were treated with 3 μM staurosporine for 3 h or transfected with 10 μg of constructs with Turbofect for 36 h (Fermentas, St. Leon-Rot, Germany). Total protein was isolated using Flag-lysis buffer, samples were denatured with Laemmli-buffer, separated in 10% SDS-PAGE, and blotted onto PVDF membranes. First antibodies are: anti-GAPDH (FL332 Santa Cruz, USA), anti-P-MST1 (Thr183)/MST2 (Thr180) (3681 Cell signaling, Frankfurt, Germany), anti-Flag (F3165 Sigma, Steinheim, Germany), and secondary antibodies are HRP coupled (sc2004/5 Santa Cruz, USA). ECL (WBKLS0100 Millipore, Schwalbach, Germany) was used for detection with Versadoc (Bio-Rad, München, Germany).

2.5. Generation of Stable Transfected Cell Lines. RASSF1A, RASSF1AΔSARAH, and RASSF1C were cloned into the vector pCMV-Tag1 (Stratagene, La Jolla, USA). The lung cancer cell line A549 was transfected using Lipofectamine 2000 (Invitrogen, Carlsbad, USA). Colonies were selected under 1 mg/mL Geneticin (Gibco, Karlsruhe, Germany) in DMEM and clones were picked after 4 weeks. Expression of Flag-RASSF1 was confirmed by RT-PCR using the FLAG-specific primer (5′-TGGATTACAAGGATGACGACG) and RASSF1-specific primer L27111 (5′-TCCTGCAAGGAGGGTGGCTTC). PCR products were analyzed on a 2% Tris-borate EDTA agarose gel.

2.6. Proliferation Analyses of Stable Transfected Cell Lines. Growth curves of stable transfected clones were analyzed by seeding 150,000 cells in triplicates in 6-well plates. Every 24 hours, cells were counted using a Neubauer counting chamber. In order to investigate the proliferation in soft agar, stable transfected cells were seeded in 0.3% agarose. Experiments were performed in duplicates with 5,000 cells per plate under selection with 1 mg/mL Geneticin. Colony size was measured after 4 weeks with a microscope (LEICA DMIRB, Wetzlar, Germany). Therefore, colonies were stained with 400 μL of 5 mg/mL INT and the size of 25 colonies was determined with MetaVue (Molecular Devices GmbH, Ismaning, München).

2.7. Localization Studies. RASSF1A and RASSF1C were cloned into the fluorescence vector pEYFP-C2 (Clontech, Mountain View, USA). The deletion of the SARAH domain of RASSF1A and RASSF1C was accomplished by the Quick-Change XL Site-Directed Mutagenesis Kit (Stratagene, La Jolla, USA) with the upper SARAH deletion primer 5′-AGGAAAATGACTCTGGGCCCCTTGGGTGACCTCT and the complementary lower primer. After transient transfection into HEK293 cells with Lipofectamine 2000 (Invitrogen, Carlsbad, USA), the localization of YFP-RASSF1A, YFP-RASSF1AΔS, YFP-RASSF1C, and the vector control were investigated with a fluorescence microscope. Cells were costained with anti-α-tubulin (Molecular Probes, Invitrogen, Carlsbad, USA) antibodies and Alexa Fluor goat anti-mouse (Molecular Probes, Invitrogen, Carlsbad, USA) antibodies to show the colocalization with the microtubules and spindle poles, respectively. Nuclei of the cells were visualized by staining with DAPI (0,1 μg/mL in PBS). Cells in mitoses were scored by microscopy (ZEISS Axioplan 2). Cells with highly condensed chromosomes and spindle structures were classified as mitotic cells. For microtubule stability analysis, the cells were treated one day after transfection for one hour with 20 μM nocodazole, fixed, and stained with DAPI and anti-α-tubulin antibodies. YFP constructs are shown in green color.

2.8. Apoptosis Analysis by TUNEL Staining. The lung cancer cell line A549 was transiently transfected with different RASSF1A constructs tagged with yellow fluorescence (pEYFP-C2) using Lipofectamine 2000 (Invitrogen, Carlsbad, USA). After two days, the transfected cells were harvested and

centrifuged on a slide. After fixation with formaldehyde, a TUNEL staining was performed with the *In Situ* Cell Death Detection Kit, TMR red (Roche, Mannheim, Germany). The nuclei were costained with DAPI (0,1 μg/mL) solution. Slides were quantified with a fluorescence microscope (ZEISS Axioplan 2). Yellow fluorescence expressing cells (500) were counted and the rate of apoptotic cells (red fluorescence) was calculated. All experiments were done in triplicates.

2.9. Statistical Analysis. All statistical evaluations were performed using the SPSS 12.0 Software (SPSS Science, Chicago, IL). A Probability of $P < 0.05$ was considered as significant.

3. Results

3.1. The RASSF1-SARAH Domain Binds MST1 and MST2. *In silico* analysis of RASSF1A (340 aa) and RASSF1C (270 aa) by PROSITE (www.expasy.org) revealed the presence of a Sav-RASSF-Hpo (SARAH) domain at their C-terminus (290 to 337 and 220 to 267, resp.) (Figure 1(a)). To gain insight into the function of the SARAH domain of RASSF1, the interaction of RASSF1A, RASSF1C, RASSF1AΔSARAH, and RASSF1CΔSARAH with MST1 and MST2 was investigated in the yeast two-hybrid system (Figures 1(b) and 1(c)). Both RASSF1 isoforms (RASSF1A and RASSF1C) interacted with MST1 and MST2, but when the SARAH domain was deleted the proteins were not able to interact anymore. These interactions were quantified in an o-nitrophenyl-β-D-galactopyranoside (ONPG) assay (Figure 1(c)). The interaction of RASSF1A and RASSF1C with MST1 and MST2 was verified by coprecipitation experiments (Figure 1(d) and data not shown). The interaction between RASSF1A and MST1 and MST2 was confirmed (Figure 1(d)). There was also an interaction between RASSF1C and MST1 and MST2 (data not shown). Interaction of RASSF1A and RASSF1C with MST1 and MST2 was abolished, when the SARAH domain was deleted (Figure 1(d) and data not shown).

3.2. RASSF1A Constructs with Deletion of the SARAH Domain Inhibit Cell Growth. Growth effects of the SARAH domain of RASSF1 were investigated in stable transfected lung cancer cells. The lung cancer cell line A549 was transfected with RASSF1A, RASSF1C, and RASSF1AΔS, and the control vector (pCMV-Tag1) and the growth of these cells were evaluated (Figure 2). A549 harbor epigenetically silenced RASSF1A but express RASSF1C. We picked stable transfected colonies and expression of *RASSF1*-specific forms was confirmed by RT-PCR (Figure 2(a)). Subsequently, the proliferation- and anchorage-independent growth of these clones was analyzed. Proliferation of RASSF1A expressing cells was significantly reduced at 96 h compared to RASSF1C expressing cells and control cells ($P = 0.022$ and $P = 0.007$, resp.). Two RASSF1AΔS expressing clones showed a similar growth to RASSF1A expressing cells. RASSF1AΔS clone1 and clone2 had a significant reduction of growth at 96 h compared to RASSF1C ($P = 0.027$ and 0.042, resp.) and controls ($P = 0.018$ and 0.029, resp.). Subsequently, the anchorage independent growth was determined in soft agar experiments

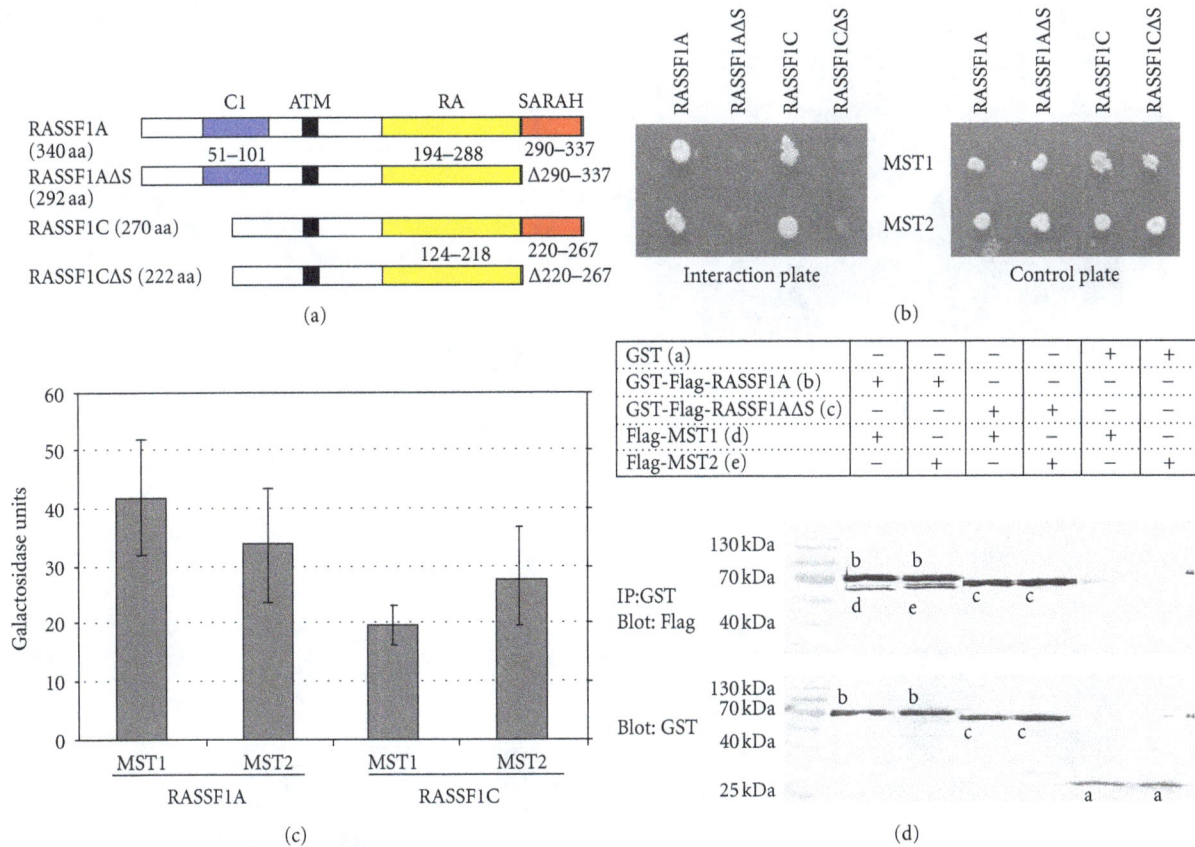

FIGURE 1: Binding studies of RASSF1, MST1, and MST2. (a) Characteristic domains of RASSF1 isoforms and SARAH deletion (ΔS) mutants are the protein kinase C conserved region (C1; blue), the ATM-kinase phosphorylation site (black), Ras-association (RalGDS/AF-6) domain (RA; yellow), and the Sav-RASSF-Hpo interaction site (SARAH; red). (b) Interaction analysis using the yeast two hybrid system. The indicated constructs were cotransformed into yeast strain PJ 69-4A. Interaction was evaluated on SD plates without alanine and histidine (interaction plate). Transformation was controlled on SD plates with alanine and histidine (control plate). (c) Quantitative interaction analysis using the ONPG assay. In three independent colonies, the activation of the β-galactosidase reporter gene was quantified with ONPG as substrate. The standard deviation is indicated. (d) Binding studies in coprecipitation. Constructs that express GST (a), GST-Flag-RASSF1A (b), GST-Flag-RASSF1AΔS (c), Flag-MST1 (d), or Flag-MST2 (e) were transfected into HEK293 cells. After two days, total protein was extracted and GST-tagged proteins were precipitated with glutathione sepharose. Samples were separated on a 10% PAGE gel and blotted. The precipitated and coprecipitated proteins were determined with anti-Flag-antibodies and anti-GST antibodies.

(Figures 2(c) and 2(d)). In these experiments, cells expressing RASSF1AΔS (clone1) exhibited a significantly reduced colony growth (average colony size: $22\,\mu$m) compared to RASSF1A (average: $36\,\mu$m; $P < 0.01$, Welch's test) and control (average: $66\,\mu$m; $P < 0.01$, Welch's test).

3.3. Expression of RASSF1A with a Deleted SARAH Domain Induces Aberrant Mitosis and Apoptosis. We transfected yellow fluorescent protein-tagged RASSF1A and RASSF1AΔSARAH (RASSF1AΔS) into HEK293 and COS7 cells and the localization was determined by fluorescence microscopy (Figure 3(a)). RASSF1A and RASSF1AΔS are both localized at the tubulin-containing cytoskeleton during interphase. In mitotic cells, RASSF1A and RASSF1AΔS were detected at spindles and centrosomes (Figure 3(a)). In the RASSF1AΔS expressing HEK293 cells, multipolar spindles and unequal alignment of the chromosomes between poles were observed (Figure 3(a)). In COS7 cells, overexpression of

RASSF1AΔS also induced monopolar spindles (Figure 3(a)). Interestingly, mitotic rate of RASSF1AΔS expressing cells was significantly ($P = 0.001$) reduced to 3.4% compared to 8.6% in RASSF1A transfected HEK293 cells (Figure 3(b)). The mitotic rate in vector transfected cells was 1.2% (data not shown). The majority (74.3%; $P < 0.001$) of mitoses in RASSF1AΔS expressing cells were abnormal (multi- or monopolar) compared to RASSF1A transfected cells, where only 0.6% of abnormal mitosis were counted (Figure 3(c)).

To determine if the aberrant spindle formation in RASSF1AΔS expressing cells is due to an altered microtubule stability of these cells, transiently transfected cells were treated with $20\,\mu$M nocodazole for one hour (Figure 4(a)). In YFP control cells, this treatment caused massive depolymerisation of the microtubules in interphase and during mitosis. RASSF1A, RASSF1AΔS, and RASSF1C overexpressing cells were able to stabilize microtubules from depolymerization by nocodazole (Figure 4(a) and data not shown).

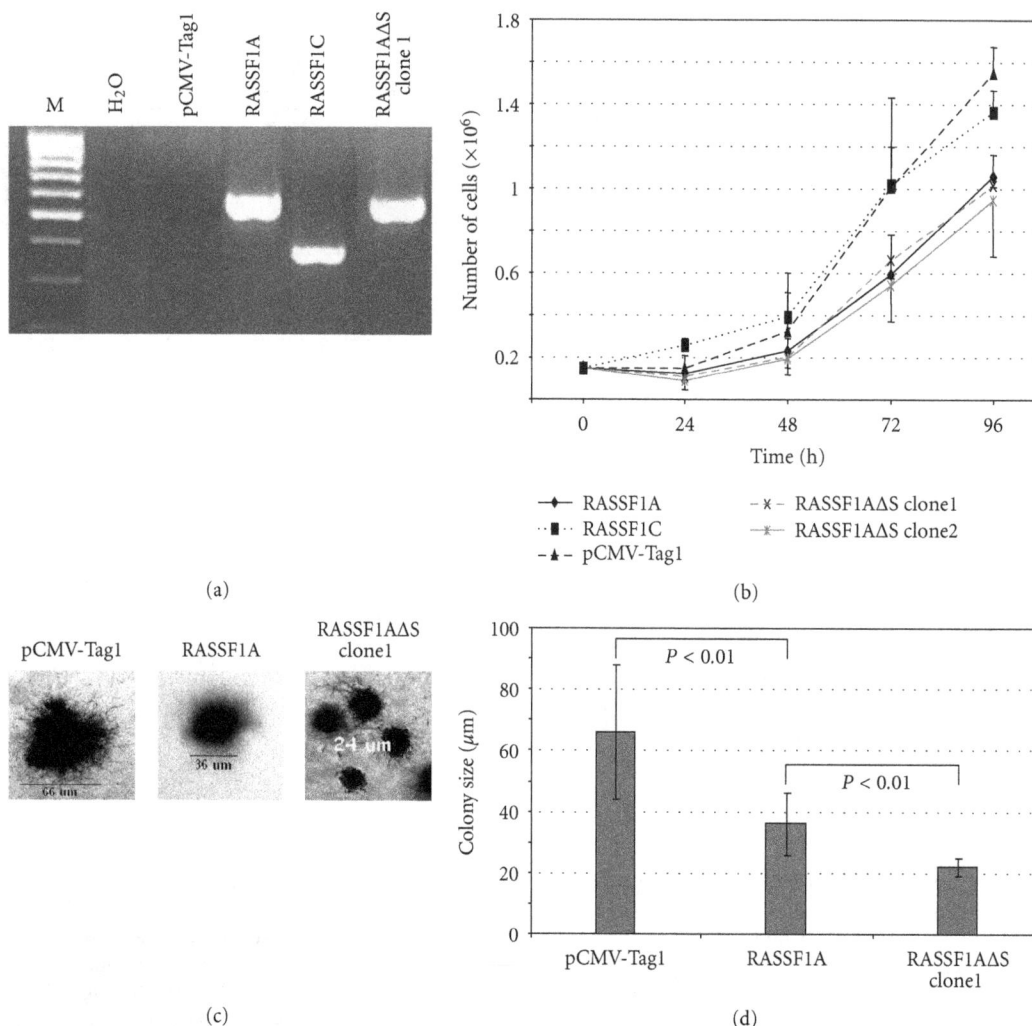

(a)

(b)

(c)

(d)

FIGURE 2: Proliferation analysis of RASSF1AΔSARAH expressing lung cancer cells. (a) A549 lung cancer cells were transfected with pCMV-Tag1, RASSF1A, RASSF1C, and RASSF1AΔSARAH (RASSF1AΔS) and stable clones were analyzed. RASSF1 expression of clones was confirmed by RT-PCR using a Flag-specific forward primer and primer L27111. Products (RASSF1A and RASSF1AΔS, 585 bp; RASSF1C, 374 bp) were analysed with controls (pCMV-Tag1 and H_2O) and a 100 bp ladder (M) on a 2% Tris-borate EDTA agarose gel. (b) Growth curve of A549 cells stably transfected with the indicated constructs. Clones were analyzed by seeding 1.5×10^5 cells in 6-well plates. Every 24 hours, cells were counted using a Neubauer chamber. Three independent experiments were performed and the mean and standard deviation is plotted. (c) Colony sizes in a soft agar experiment after 4 weeks. Examples of colonies expressing pCMV-Tag1, RASSF1A, and RASSF1AΔS are shown. (d) 25 colonies were measured and the average colony size was calculated. Statistical significance P-values are indicated.

To analyze the effect of the deletion of the SARAH domain on the proapoptotic function of RASSF1A, a transient transfection into the lung cancer cell line A549 was performed and the rate of transfected and apoptotic cells was calculated after 1 to 2 days (Figure 4(b)). Apoptotic cells were stained using red fluorescence TUNEL-Kit and YFP was used as a control. The rate of apoptosis was significantly ($P = 0.001$) higher, when the cells expressed RASSF1A (28%) in comparison to RASSF1C (16%) and the YFP control (14%, Figure 4(b)). The deletion of the SARAH domain (RASSF1AΔS) resulted in a significantly increased apoptotic rate of 39% compared to RASSF1C ($P < 0.001$) and RASSF1A ($P = 0.006$, Figure 4(b)). In summary, our results

show that expression of RASSF1A with a deletion of the SARAH domain deregulates normal mitotic progression and enhances apoptosis.

Subsequently, RASSF1A-induced autophosphorylation of MST1 and MST2 was analyzed in A549 lung cancer cells (Figure 5). For this propose, we have utilized an antibody that detects endogenous MST1 and MST2 only when phosphorylated at Thr183 and Thr180, respectively. Treatment of A549 cells with 3 μM staurosporine induced phosphorylation of MST1 and MST2 (Figure 5), as described previously [34, 46]. However, when A549 cells were transfected with RASSF1A and RASSF1AΔSARAH, phosphorylation of MST1 and MST2 was not detected (Figure 5). Similar results were

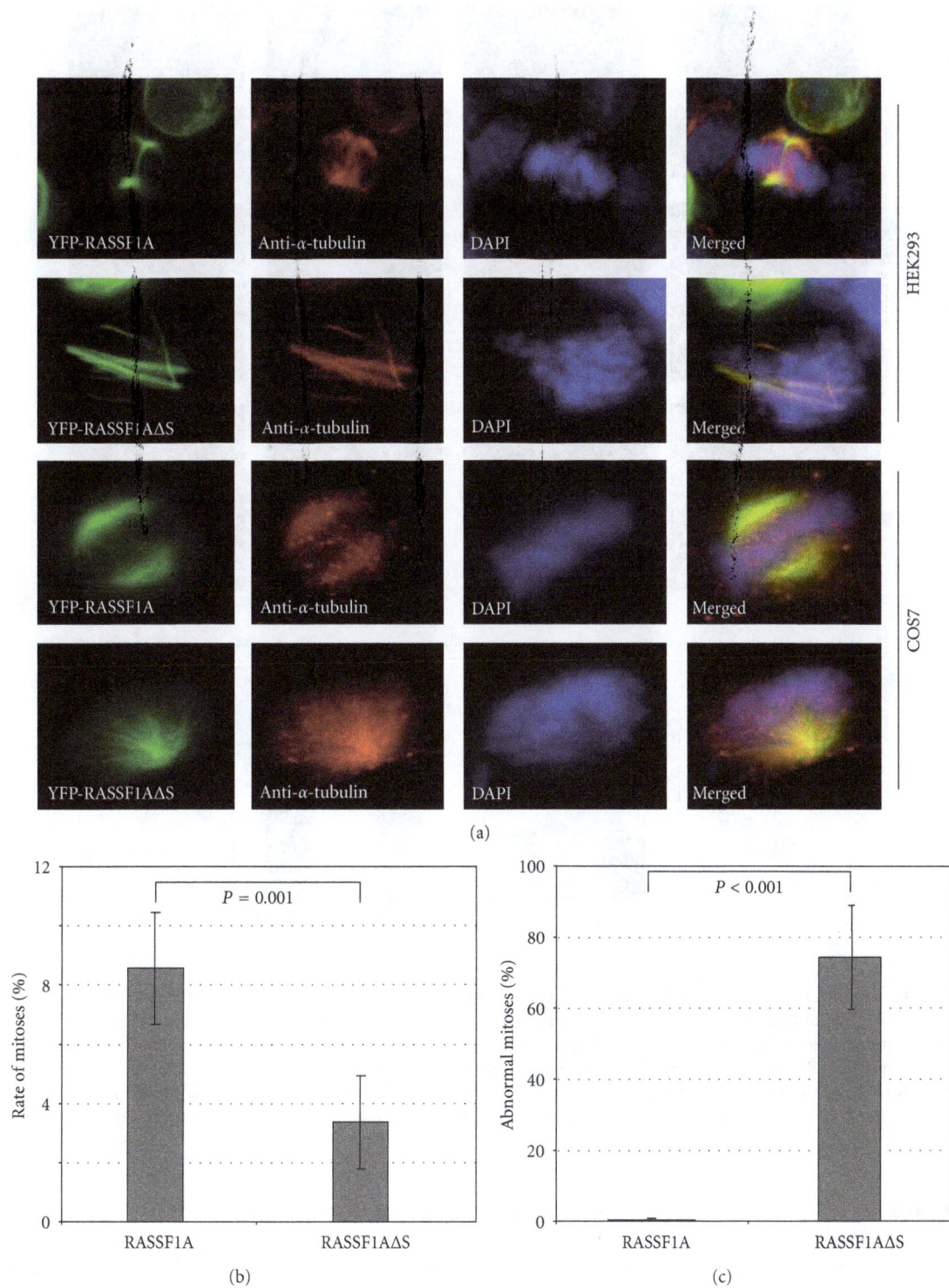

FIGURE 3: Effects of RASSF1AΔSARAH expression on mitosis. (a) HEK293 and COS7 cells were transfected with YFP-RASSF1A and YFP-RASSF1AΔSARAH (RASSF1AΔS) and stained using DAPI and anti-α-tubulin antibody. Yellow fluorescent is shown in green. (b) Rate of mitosis in HEK293 after transient expression. Cells with highly condensed chromosomes and spindle structures were classified as mitotic cells (c) Abnormal mitoses (monopolar, multipolar spindles and abnormal spindle fibers) were counted in HEK293 cells and the rate of abnormal mitosis is plotted. All experiments were done in triplicates and 500 cells each were evaluated. The mean and standard deviation were determined. Statistical significant P-values are indicated.

(a)

(b)

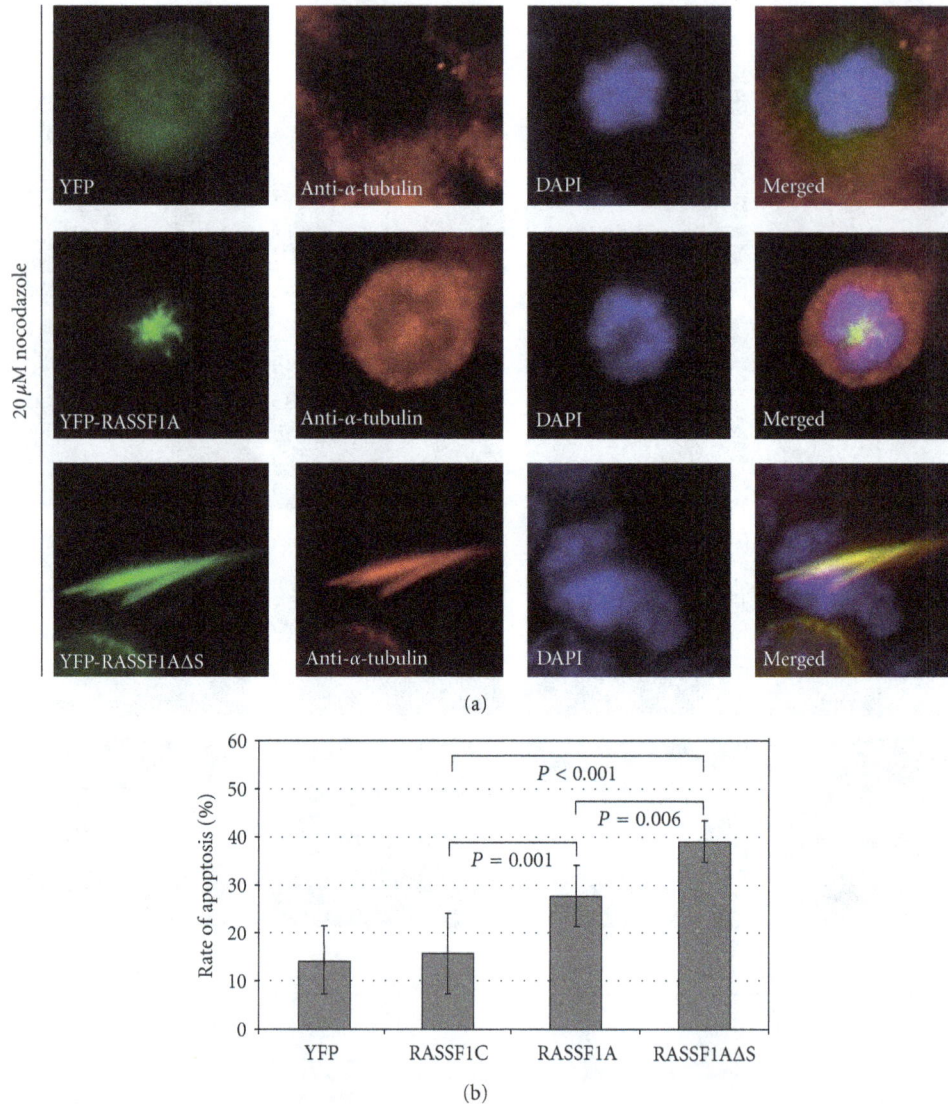

FIGURE 4: (a) Microtubule stability of RASSF1AΔSARAH expressing mitotic cells. HEK293 cells were transfected with plasmids YFP, YFP-RASSF1A or YFP-RASSF1AΔSARAH (YFP-RASSF1AΔS). One day after transfection, the cells were treated for one hour with 20 μM nocodazole, fixed and stained with DAPI and anti-α-tubulin antibodies. Yellow fluorescence is shown in green. (b) Induction of apoptosis after transient expression of RASSF1A and RASSF1AΔSARAH. Lung cancer cells A549 were transfected with YFP, YFP-RASSF1C (RASSF1C), YFP-RASSF1A (RASSF1A), and YFP-RASSF1AΔSARAH (RASSF1AΔS). TUNEL staining was utilized to determine the frequency of apoptotic cells in transfected cells. All experiments were done in triplicates and the mean and standard deviation were determined. Statistical significant P-values are indicated.

obtained in HEK293 cells (data not shown). Praskova et al. showed that the MST1 kinase autoactivation through phosphorylation is inhibited by coexpression of RASSF1A and RASSF1C [16].

4. Discussion

RASSF1A is a tumor suppressor gene, which is involved in several signaling pathway including apoptosis, microtubule stability, proliferation, and mitotic regulation [2, 3]. In our study, we have analyzed the function of the Sav-RASSF1-Hpo (SARAH) domain of RASSF1A. Here, we report that the SARAH domain regulates several pathways, which are frequently altered in tumors. The SARAH domain is involved

in apoptosis and growth-suppressing functions of RASSF1A like anchorage-independent proliferation. Moreover, the SARAH domain is important for mitotic progression and spindle formation. It has been previously reported that RASSF1 interacts through its C-terminal domain with MST1 and thereby regulates MST1-mediated apoptosis [16, 18, 19]. We demonstrate that RASSF1A and RASSF1C interact with both MST1 and MST2. This complex may regulate several pivotal signaling pathway including apoptosis and phosphorylation of Warts/LATS serine threonine kinases that regulate mitotic progression.

It has been reported that RASSF1A regulates a proapoptotic pathway through its interaction with the proto-oncogene Ras and the novel Ras effector 1 (Nore1) [18, 32].

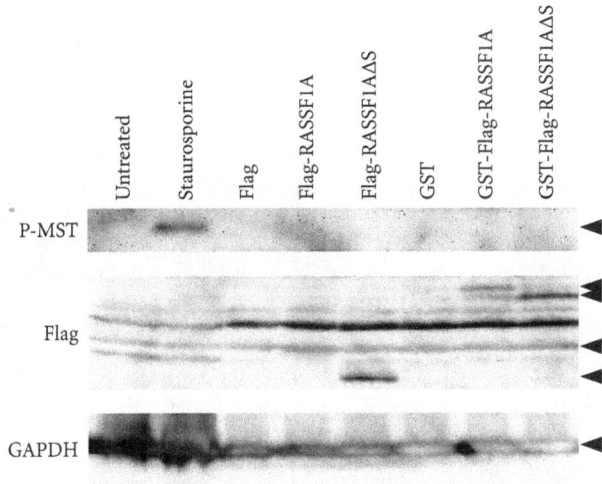

FIGURE 5: Phosphorylation of MST1/2 is induced by staurosporine. A549 were treated with $3\,\mu M$ Staurosporine for 3 h. For transfections Turbofect (Fermentas) was used with $10\,\mu g$ of indicated constructs for 36 h. Total protein was isolated using Flag-lysis buffer, samples were denatured with Laemmli-buffer, separated in 10% SDS-PAGE and blotted onto PVDF membrane. First antibodies are: α-GAPDH, α-P-MST1 and α-Flag and secondary antibody are HRP coupled (sc2004 and sc2005 Santa cruz). ECL (WBKLS0100 Millipore) was used for detection with Versadoc (Biorad). Arrowheads indicate top down: P-MST, GST-Flag-RASSF1A, GST-Flag-RASSF1AΔSARAH (ΔS), Flag-RASSF1A, Flag-RASSF1AΔSARAH (ΔS), and GAPDH.

RASSF1A and Nore1 interact with the proapoptotic Ste20 protein kinase MST1 via the C-terminus [16, 19] through the SARAH domain [32]. The homologue of RASSF1, NORE1 forms a complex with MST1 that mediates a proapoptotic pathway induced by Ras [18]. Early on, it was reported that MST1 is a serine/threonine protein kinase that could autophosphorylate itself [47] and later Praskova et al. demonstrated that MST1 phosphorylates and activates itself, whereas this autophosphorylation is inhibited when MST1 is bound to RASSF1A and RASSF1C [16]. We show that the interaction of RASSF1A and RASSF1C with MST1 and MST2 depends on the C-terminal SARAH domain. Since the SARAH domain binds the proapoptotic kinases MST1 and MST2, a deregulation of these kinases may contribute to the apoptotic rate in the cells with truncated RASSF1A. However, we could not observe an autophosphorylation of MST1/2 after transfection of RASSF1A and RASSF1AΔSARAH. In contrast, staurosporine induced strong phosphorylation of MST1/2. This indicates that RASSF1A-induced apoptosis observed in A549 cells occurs MST independent, possibly through the N-terminus of RASSF1A, that associates with MDM2 and death-domain-associated protein (DAXX) and contributes to p53 activation in response to DNA damage [22]. Alternatively, RASSF1A was further linked to apoptosis through interacting with the microtubule-associated protein C19ORF5 [48, 49]. Furthermore, Donninger et al. described RASSF1A to interact with the potential tumor suppressor Salvador to promote apoptosis independently of Hippo signaling by modulating p73 [50].

It was demonstrated that RASSF1A colocalizes with the microtubule network during interphase and is found at the spindles and centrosomes during mitosis [14, 23]. RASSF1A binds to tubulin [7], thereby stabilizing microtubules [7, 12–14] and regulating the mitotic progression. RASSF1A overexpression leads to a mitotic arrest at metaphase [14], to a G1 arrest [51], to a G2/M arrest [52], to a G1 and G2/M arrest [7], and a prometaphase arrest [23]. The domain required for both microtubule association and stabilization was mapped to an amino-acid fragment from 120 to 288 [14]. Thus, the microtubule binding site and the SARAH domain are different [12, 14, 53] and this is consistent with our observation. Rong et al. showed that the microtubule binding was lost upon mutation of the phosphorylation site 203 in RASSF1A [54]. However, we and others did not observe an altered microtubule binding using phosphomimicking or nonphosphorylatable mutants of RASSF1A [15, 24].

RASSF1A was also reported to interact with MAP1B (microtubule-associated protein 1B) and C19ORF5 (chromosome 19 open reading frames 5), both microtubule-associated proteins [13, 49, 53]. C19ORF5 is a hyperstabilized microtubule-specific binding protein of which accumulation causes mitochondrial aggregation and cell death [48]. Regarding C19ORF5, it was demonstrated that its knockdown led to mitotic abnormalities [55], that C19ORF5 localizes to centrosomes, and it was stated that C19ORF5 is required for the recruitment of RASSF1A to the spindle poles [53, 55]. Liu et al. reported that RASSF1A caused hyperstabilization of microtubules and the accumulation of C19ORF5 on them [48]. The complex LATS1/MST2/WW45 is found together with RASSF1A at the centrosome, and it was shown that defects in this pathway may lead to abnormal mitosis caused by cytokinesis failure. Thus, RASSF1A may mediate organization of mitotic spindle poles through the recruitment of MST and LATS to the centrosomes.

In summary, our data indicate that RASSF1A is important for several signals, which are frequently altered in tumorigenesis, including apoptosis, mitotic spindle organization, and proliferation. Our data suggest that other domains (e.g., microtubule association domain) than SARAH also significantly contribute to the proapoptotic and antiproliferative function of RASSF1A. Specific interaction of RASSF1A with MST/LATS and other binding partners (e.g., RAS, MDM2, DAXX, C19ORF5, and Salvador) might be important in the regulation of proliferation and apoptosis and in the formation of normal mitotic spindles and processes of dividing chromosomes by RASSF1A.

Abbreviations

RASSF1A:	Ras association domain family 1A
MST:	Mammalian STE20 like kinase
LATS:	Large tumor suppressor
WW45:	45 kDa WW domain protein
Sav:	Salvador
Hpo:	Hippo
SARAH domain:	Sav-RASSF-Hpo interaction domain
RT-PCR:	Reverse transcriptase PCR
YFP:	Yellow fluorescence protein
TUNEL:	Terminal transferase mediated dUTP nick end labeling.

Acknowledgment

This study was supported by Grants BMBF FKZ01ZZ0104, LOEWE UGMLC, Deutsche Krebshilfe and DFG DA552 to R. H. Dammann.

References

[1] R. Dammann, C. Li, J. H. Yoon, P. L. Chin, S. Bates, and G. P. Pfeifer, "Epigenetic inactivation of a RAS association domain family protein from the lung tumour suppressor locus 3p21.3," *Nature Genetics*, vol. 25, no. 3, pp. 315–319, 2000.

[2] R. Dammann, U. Schagdarsurengin, C. Seidel et al., "The tumor suppressor RASSF1A in human carcinogenesis: an update," *Histology and Histopathology*, vol. 20, no. 2, pp. 645–663, 2005.

[3] A. Agathanggelou, W. N. Cooper, and F. Latif, "Role of the Ras-association domain family 1 tumor suppressor gene in human cancers," *Cancer Research*, vol. 65, no. 9, pp. 3497–3508, 2005.

[4] A. M. Richter, G. P. Pfeifer, and R. H. Dammann, "The RASSF proteins in cancer; from epigenetic silencing to functional characterization," *Biochimica et Biophysica Acta*, vol. 1796, no. 2, pp. 114–128, 2009.

[5] D. G. Burbee, E. Forgacs, S. Zöchbauer-Müller et al., "Epigenetic inactivation of RASSF1A in lung and breast cancers and malignant phenotype suppression," *Journal of the National Cancer Institute*, vol. 93, no. 9, pp. 691–699, 2001.

[6] K. Dreijerink, E. Braga, I. Kuzmin et al., "The candidate tumor suppressor gene, RASSF1A, from human chromosome 3p21.3 is involved in kidney tumorigenesis," *Proceedings of the National Academy of Sciences of the United States of America*, vol. 98, no. 13, pp. 7504–7509, 2001.

[7] R. Rong, W. Jin, J. Zhang, M. S. Sheikh, and Y. Huang, "Tumor suppressor RASSF1A is a microtubule-binding protein that stabilizes microtubules and induces G2/M arrest," *Oncogene*, vol. 23, no. 50, pp. 8216–8230, 2004.

[8] I. Kuzmin, J. W. Gillespie, A. Protopopov et al., "The RASSF1A tumor suppressor gene is inactivated in prostate tumors and suppresses growth of prostate carcinoma cells," *Cancer Research*, vol. 62, no. 12, pp. 3498–3502, 2002.

[9] S. Tommasi, R. Dammann, Z. Zhang et al., "Tumor susceptibility of Rassf1a knockout mice," *Cancer Research*, vol. 65, no. 1, pp. 92–98, 2005.

[10] L. van der Weyden, K. K. Tachibana, M. A. Gonzalez et al., "The RASSF1A isoform of RASSF1 promotes microtubule stability and suppresses tumorigenesis," *Molecular and Cellular Biology*, vol. 25, no. 18, pp. 8356–8367, 2005.

[11] L. van der Weyden, M. J. Arends, O. M. Dovey et al., "Loss of Rassf1a cooperates with ApcMin to accelerate intestinal tumourigenesis," *Oncogene*, vol. 27, no. 32, pp. 4503–4508, 2008.

[12] M. D. Vos, A. Martinez, C. Elam et al., "A role for the RASSF1A tumor suppressor in the regulation of tubulin polymerization and genomic stability," *Cancer Research*, vol. 64, no. 12, pp. 4244–4250, 2004.

[13] A. Dallol, A. Agathanggelou, S. L. Fenton et al., "RASSF1A interacts with microtubule-associated proteins and modulates microtubule dynamics," *Cancer Research*, vol. 64, no. 12, pp. 4112–4116, 2004.

[14] L. Liu, S. Tommasi, D. H. Lee, R. Dammann, and G. P. Pfeifer, "Control of microtubule stability by the RASSF1A tumor suppressor," *Oncogene*, vol. 22, no. 50, pp. 8125–8136, 2003.

[15] A. M. Richter, U. Schagdarsurengin, M. Rastetter, K. Steinmann, and R. H. Dammann, "Protein kinase A-mediated phosphorylation of the RASSF1A tumour suppressor at Serine 203 and regulation of RASSF1A function," *European Journal of Cancer*, vol. 46, no. 16, pp. 2986–2995, 2010.

[16] M. Praskova, A. Khoklatchev, S. Ortiz-Vega, and J. Avruch, "Regulation of the MST1 kinase by autophosphorylation, by the growth inhibitory proteins, RASSF1 and NORE1, and by Ras," *Biochemical Journal*, vol. 381, no. 2, pp. 453–462, 2004.

[17] S. Ortiz-Vega, A. Khokhlatchev, M. Nedwidek et al., "The putative tumor suppressor RASSF1A homodimerizes and heterodimerizes with the Ras-GTP binding protein Nore1," *Oncogene*, vol. 21, no. 9, pp. 1381–1390, 2002.

[18] A. Khokhlatchev, S. Rabizadeh, R. Xavier et al., "Identification of a novel Ras-regulated proapoptotic pathway," *Current Biology*, vol. 12, no. 4, pp. 253–265, 2002.

[19] H. J. Oh, K. K. Lee, S. J. Song et al., "Role of the tumor suppressor RASSF1A in Mst1-mediated apoptosis," *Cancer Research*, vol. 66, no. 5, pp. 2562–2569, 2006.

[20] S. Baksh, S. Tommasi, S. Fenton et al., "The tumor suppressor RASSF1A and MAP-1 link death receptor signaling to bax conformational change and cell death," *Molecular Cell*, vol. 18, no. 6, pp. 637–650, 2005.

[21] M. D. Vos, A. Dallol, K. Eckfeld et al., "The RASSF1A tumor suppressor activates bax via MOAP-1," *Journal of Biological Chemistry*, vol. 281, no. 8, pp. 4557–4563, 2006.

[22] M. S. Song, S. J. Song, S. Y. Kim, H. J. Oh, and D. S. Lim, "The tumour suppressor RASSF1A promotes MDM2 self-ubiquitination by disrupting the MDM2-DAXX-HAUSP complex," *EMBO Journal*, vol. 27, no. 13, pp. 1863–1874, 2008.

[23] M. S. Song, S. J. Song, N. G. Ayad et al., "The tumour suppressor RASSF1A regulates mitosis by inhibiting the APC-Cdc20 complex," *Nature Cell Biology*, vol. 6, no. 2, pp. 129–137, 2004.

[24] S. J. Song, M. S. Song, S. J. Kim et al., "Aurora a regulates prometaphase progression by inhibiting the ability of RASSF1A to suppress APC-Cdc20 activity," *Cancer Research*, vol. 69, no. 6, pp. 2314–2323, 2009.

[25] L. Liu, K. Baier, R. Dammann, and G. P. Pfeifer, "The tumor suppressor RASSF1A does not interact with Cdc20, an activator of the anaphase-promoting complex," *Cell Cycle*, vol. 6, no. 13, pp. 1663–1665, 2007.

[26] H. Scheel and K. Hofmann, "A novel interaction motif, SARAH, connects three classes of tumor suppressor," *Current Biology*, vol. 13, no. 23, pp. R899–R900, 2003.

[27] N. Tapon, K. F. Harvey, D. W. Bell et al., "salvador promotes both cell cycle exit and apoptosis in Drosophila and is mutated in human cancer cell lines," *Cell*, vol. 110, no. 4, pp. 467–478, 2002.

[28] S. Wu, J. Huang, J. Dong, and D. Pan, "hippo encodes a Ste-20 family protein kinase that restricts cell proliferation and promotes apoptosis in conjunction with salvador and warts," *Cell*, vol. 114, no. 4, pp. 445–456, 2003.

[29] K. F. Harvey, C. M. Pfleger, and I. K. Hariharan, "The Drosophila Mst ortholog, hippo, restricts growth and cell proliferation and promotes apoptosis," *Cell*, vol. 114, no. 4, pp. 457–467, 2003.

[30] B. A. Hay and M. Guo, "Coupling cell growth, proliferation, and death: hippo weighs in," *Developmental Cell*, vol. 5, no. 3, pp. 361–363, 2003.

[31] C. Polesello, S. Huelsmann, N. Brown, and N. Tapon, "The drosophila RASSF homolog antagonizes the hippo pathway," *Current Biology*, vol. 16, no. 24, pp. 2459–2465, 2006.

[32] J. Avruch, M. Praskova, S. Ortiz-Vega, M. Liu, and X. F. Zhang, "Nore1 and RASSF1 regulation of cell proliferation and of the MST1/2 kinases," *Methods in Enzymology*, vol. 407, pp. 290–310, 2005.

[33] K. K. Lee, T. Ohyama, N. Yajima, S. Tsubuki, and S. Yonehara, "MST, a physiological caspase substrate, highly sensitizes apoptosis both upstream and downstream of caspase activation," *Journal of Biological Chemistry*, vol. 276, no. 22, pp. 19276–19285, 2001.

[34] S. Ura, N. Masuyama, J. D. Graves, and Y. Gotoh, "Caspase cleavage of MST1 promotes nuclear translocation and chromatin condensation," *Proceedings of the National Academy of Sciences of the United States of America*, vol. 98, no. 18, pp. 10148–10153, 2001.

[35] W. L. Cheung, K. Ajiro, K. Samejima et al., "Apoptotic phosphorylation of histone H2B is mediated by mammalian sterile twenty kinase," *Cell*, vol. 113, no. 4, pp. 507–517, 2003.

[36] S. H. Ahn, W. L. Cheung, J. Y. Hsu, R. L. Diaz, M. M. Smith, and C. D. Allis, "Sterile 20 kinase phosphorylates histone H2B at serine 10 during hydrogen peroxide-induced apoptosis in S. cerevisiae," *Cell*, vol. 120, no. 1, pp. 25–36, 2005.

[37] S. Ura, H. Nishina, Y. Gotoh, and T. Katada, "Activation of the c-Jun N-terminal kinase pathway by MST1 is essential and sufficient for the induction of chromatin condensation during apoptosis," *Molecular and Cellular Biology*, vol. 27, no. 15, pp. 5514–5522, 2007.

[38] C. Guo, S. Tommasi, L. Liu, J. K. Yee, R. Dammann, and G. Pfeifer, "RASSF1A is part of a complex similar to the drosophila hippo/salvador/lats tumor-suppressor network," *Current Biology*, vol. 17, no. 8, pp. 700–705, 2007.

[39] C. Guo, X. Zhang, and G. P. Pfeifer, "The tumor suppressor RASSF1A prevents dephosphorylation of the mammalian STE20-like kinases MST1 and MST2," *Journal of Biological Chemistry*, vol. 286, no. 8, pp. 6253–6261, 2011.

[40] D. Matallanas, D. Romano, K. Yee et al., "RASSF1A elicits apoptosis through an MST2 pathway directing proapoptotic transcription by the p73 tumor suppressor protein," *Molecular Cell*, vol. 27, no. 6, pp. 962–975, 2007.

[41] T. Oka, V. Mazack, and M. Sudol, "Mst2 and Lats kinases regulate apoptotic function of Yes kinase-associated protein (YAP)," *Journal of Biological Chemistry*, vol. 283, no. 41, pp. 27534–27546, 2008.

[42] W. Wen, F. Zhu, J. Zhang et al., "MST1 promotes apoptosis through phosphorylation of histone H2AX," *Journal of Biological Chemistry*, vol. 285, no. 50, pp. 39108–39116, 2010.

[43] M. K. Lehtinen, Z. Yuan, P. R. Boag et al., "A conserved MST-FOXO signaling pathway mediates oxidative-stress responses and extends life span," *Cell*, vol. 125, no. 5, pp. 987–1001, 2006.

[44] B. You, G. Yan, Z. Zhang et al., "Phosphorylation of cardiac troponin I by mammalian sterile 20-like kinase 1," *Biochemical Journal*, vol. 418, no. 1, pp. 93–101, 2009.

[45] U. Schagdarsurengin, A. M. Richter, J. Hornung, C. Lange, K. Steinmann, and R. H. Dammann, "Frequent epigenetic inactivation of RASSF2 in thyroid cancer and functional consequences," *Molecular Cancer*, vol. 9, article 264, 2010.

[46] J. D. Graves, Y. Gotoh, K. E. Draves et al., "Caspase-mediated activation and induction of apoptosis by the mammalian Ste20-like kinase Mst1," *EMBO Journal*, vol. 17, no. 8, pp. 2224–2234, 1998.

[47] C. L. Creasy and J. Chernoff, "Cloning and characterization of a human protein kinase with homology to Ste20," *Journal of Biological Chemistry*, vol. 270, no. 37, pp. 21695–21700, 1995.

[48] L. Liu, A. Vo, and W. L. McKeehan, "Specificity of the methylation-suppressed A isoform of candidate tumor suppressor RASSF1 for microtubule hyperstabilization is determined by cell death inducer C19ORF5," *Cancer Research*, vol. 65, no. 5, pp. 1830–1838, 2005.

[49] L. Liu, A. Vo, G. Liu, and W. L. McKeehan, "Novel complex integrating mitochondria and the microtubular cytoskeleton with chromosome remodeling and tumor suppressor RASSF1 deduced by in silico homology analysis, interaction cloning in yeast, and colocalization in cultured cells," *In Vitro Cellular and Developmental Biology*, vol. 38, no. 10, pp. 582–594, 2002.

[50] H. Donninger, N. Allen, A. Henson et al., "Salvador protein is a tumor suppressor effector of RASSF1A with hippo pathway-independent functions," *Journal of Biological Chemistry*, vol. 286, no. 21, pp. 18483–18491, 2011.

[51] A. Agathanggelou, I. Bieche, J. Ahmed-Choudhury et al., "Identification of novel gene expression targets for the ras association domain family 1 (RASSF1A) tumor suppressor gene in non-small cell lung cancer and neuroblastoma," *Cancer Research*, vol. 63, no. 17, pp. 5344–5351, 2003.

[52] A. Moshnikova, J. Frye, J. W. Shay, J. D. Minna, and A. V. Khokhlatchev, "The growth and tumor suppressor NORE1A is a cytoskeletal protein that suppresses growth by inhibition of the ERK pathway," *Journal of Biological Chemistry*, vol. 281, no. 12, pp. 8143–8152, 2006.

[53] S. S. Min, S. C. Jin, J. S. Su, T. H. Yang, H. Lee, and D. S. Lim, "The centrosomal protein RAS association domain family protein 1A (RASSF1A)-binding protein 1 regulates mitotic progression by recruiting RASSF1A to spindle poles," *Journal of Biological Chemistry*, vol. 280, no. 5, pp. 3920–3927, 2005.

[54] R. Rong, L. Y. Jiang, M. S. Sheikh, and Y. Huang, "Mitotic kinase Aurora-A phosphorylates RASSF1A and modulates RASSF1A-mediated microtubule interaction and M-phase cell cycle regulation," *Oncogene*, vol. 26, no. 55, pp. 7700–7708, 2007.

[55] A. Dallol, W. N. Cooper, F. Al-Mulla, A. Agathanggelou, E. R. Maher, and F. Latif, "Depletion of the Ras association domain family 1, isoform A-associated novel microtubule-associated protein, C19ORF5/MAP1S, causes mitotic abnormalities," *Cancer Research*, vol. 67, no. 2, pp. 492–500, 2007.

5

Interleukin-1 Two-Locus Haplotype Is Strongly Associated with Severe Chronic Periodontitis among Yemenis

Nezar Noor Al-hebshi,[1,2] Amat-alrahman Ahmed Shamsan,[1] and Mohammed Sultan Al-ak'hali[3]

[1] *Molecular Research Laboratory, Faculty of Medical Sciences, University of Science and Technology, Sana'a, Yemen*
[2] *Faculty of Dentistry, Jazan University, P.O. Box. 114, Jazan, Saudi Arabia*
[3] *Department of Periodontology, Faculty of Dentistry, University of Sana'a, Sana'a, Yemen*

Correspondence should be addressed to Nezar Noor Al-hebshi, nazhebshi@yahoo.com

Academic Editor: Joseph Rothnagel

Aim. To assess IL-1A C[−889]T and IL-1B C[3954]T genotypes as well as haplotypes in relation to sever chronic periodontitis (SCP) among Yemenis. *Materials and Methods.* 40 cases with SCP and 40 sex- and age-matched controls were included; all were nonsmokers and free of systemic diseases. Genotyping at each locus was performed using an established PCR-RFLP assay. The Haploview and SimHap software were used to assess data for Hardy-Weinberg's equilibrium (HWE) and linkage disequilibrium (LD) and to obtain subject-level haplotypes. Multiple logistic regression was used to seek for associations in dominant, additive, and recessive models. *Results.* Mean plaque index (MPI) showed the strongest association with SCP (OR = 16). A significant LD was observed in the cases ($D' = 0.80$ and $r^2 = 0.47$). The genotype at each locus showed significant association with SCP in the recessive model (TT versus TC + CC) even after adjustment for MPI (OR = 6.29 & 461, resp.). The C-T haplotype conferred protection against SCP in a dominant manner (OR = 0.16). On the other hand, the T-T haplotype in double dose (recessive model) showed strong association with CP (OR = 15.6). *Conclusions.* IL-1 two-locus haplotype is associated with SCP in Yemenis. Haplotype-based analysis may be more suited for use in genetic association studies of periodontitis.

1. Introduction

Chronic periodontitis is currently viewed as an immunologically mediated destruction of tooth supporting tissues, the periodontium, provoked by specific pathogenic bacterial consortia in subgingival biofilm [1]. However, the occurrence, extent, and severity of the destructive process are dependent on the individual's susceptibility to the disease, which is in turn influenced by risk factors independent of the microbial challenge [2]. Environmental factors, such as cigarette smoking, and systemic diseases, such as diabetes, are well-established, classical examples of such risk modifiers [3].

Lately, the role of genetics in defining host susceptibility to chronic periodontitis has drawn much attention. Initial evidence for a significant genetic element in the pathogenesis of periodontitis came from twin- and family studies, in which chronic periodontitis was shown to have around 50% heritability [4, 5]. Since then, many genes, such as the cytokine gene family, pattern-recognition receptor genes and the vitamin D receptor gene, have been explored for allelic variants that may be associated with chronic as well as aggressive periodontitis [6]. Because some cytokines, particularly the IL-1 and TNF-A proteins, are strongly implicated in the pathogenesis of chronic periodontitis [7], the genes encoding them have been more extensively investigated with the biological premise that certain polymorphisms result in hyper secretion of these proinflammatory molecules and thus in severe periodontal destruction [8].

Associations between polymorphisms of the IL-1, IL-4, IL-6, IL-10, and TNF-A genes and chronic periodontitis have been reported by several investigators; however, an equivalent number of studies failed to demonstrate such an association, making the overall evidence fragile [6]. However, the situation is somewhat more encouraging in the case of the IL-1 gene. In a recent meta-analysis of 53 publications, the IL-1A C[−889]T and IL-1B C[3954]T polymorphisms, and the so-called composite genotype [9] maintained significant associations with chronic periodontitis especially in its severe form, indicating that these polymorphisms should continue to be considered as potential "genuine" genetic determinants of chronic periodontitis [10]. However, it must be pointed out that assessment of association between the IL-1 gene cluster and chronic periodontitis has largely been based on genotype analysis only (individual SNPs or composite genotype); linkage disequilibrium or haplotype-based analyses have been nearly lacking. It is therefore recommended that these methods of analysis should be considered in future studies [10].

In the analysis referred to above, differences in effect and size between Caucasians and Asians were also noted, substantiating the view that genetic influence on chronic periodontitis may vary among ethnic groups. So far, studies on the association between the IL-1 gene polymorphisms and chronic periodontitis have mostly involved white Caucasian, Asian, and to a lesser extent, Hispanic study populations [6]. Information on such an association in other ethnic groups is sparse. The purpose of this study was, therefore, to assess the association between the IL-1A C[−889]T and IL-1B C[3954]T polymorphisms and sever chronic periodontitis among Yemenis based on haplotype analysis.

2. Material and Methods

2.1. Study Subjects.
The Quanto software (University of Southern California, USA) was used to calculate sample size, assuming a population risk of 10%, an allele T frequency of 20%, a log additive mode of inheritance, and an OR of 3 for the heterozygous genotype, with a 95% level of confidence and 80% power.

Accordingly, forty subjects, ≥ 35 years old, with advanced chronic periodontitis (having at least one tooth site with pocket depth ≥ 4 mm and clinical attachment loss ≥ 6 in each quadrant) and forty age- and sex-matched controls (having no site with clinical attachment loss ≥ 3) were recruited from among patients attending the dental clinics at the Faculty of Dentistry, Sana'a University, and also a number of private dental centers in Sana'a City. All subjects were required to have Yemeni parents. Smokers and those with systemic diseases such as diabetes mellitus were strictly excluded. All teeth, except third molars, were assessed for plaque index [11], pocket depth (PD), and clinical attachment loss (CAL) at four sites per tooth by a single examiner (AAS). To be involved in the study, every subject was required to sign a written consent. The clinical features of both study groups are shown in Table 1.

2.2. Sampling and DNA Extraction.
Oral epithelium was used as the source of genomic DNA. Each subject was instructed to rinse his/her mouth with water to remove any debris, rubber his/her cheeks against the molars and then swish with 10 mL saline for 30 seconds to obtain epithelial

TABLE 1: Clinical characteristics of the study groups.

Variable	Cases $n = 40$	Controls $n = 40$
No. of males (%)	21 (52.5%)	20 (50.0%)
Mean age ±SD	43.40 ± 7.06	42.95 ± 5.27
Mean clinical attachment loss ±SD*	05.53 ± 1.15	00.41 ± 0.94
Mean pocket depth ±SD*	03.06 ± 0.60	01.12 ± 0.32
Mean plaque index ±SD*	01.60 ± 0.49	00.97 ± 0.41

*Difference statistically significant; t-test.

cells. The samples, collected in sterile containers, were then stored at −20°C. In preparation for DNA extraction, epithelial cells were pelleted by centrifuging the samples at 3000 rpm for 3 minutes, and then resuspended in 400 mL phosphate buffer saline. DNA was extracted from 200 μL of the suspension using the Purelink Genomic DNA extraction kit (Invitrogen, USA) according to manufactures' instructions. Presence of valid DNA was confirmed using a human ATCB gene Taqman real-time PCR assay (Primerdesign, UK).

2.3. IL-1A C[−889]T and IL-1B C[3954]T Genotyping.
The regions of interest (99 and 194 bp for the IL-1Aand B genes, resp.) were amplified using sequence-specific primers (Table 2). The reaction mix for each locus consisted of 18 μL Platinum Blue PCR SuperMix (Invitrogen, USA), 1 μL primer mix (600 nM each), and 1 μL DNA sample. MgCl$_2$ concentration was adjusted to 2.5 mM for the IL-1A reaction. The following thermal cycling profile was used: an initial denaturation/enzyme activation cycle at 95°C for 10 min, 35 cycles (38 for IL-1A) of denaturation at 95°C for 15 s, annealing at 55°C (54°C for IL-1A) for 30 s, and extension at 72°C for 30 s, followed by final extension cycle at 72°C for 5 min. The PCR product of the IL-1A gene was digested using 2.5 units of Nco I enzyme (Invitrogen, USA) for 3 h at 37°C, while that of the IL-1B gene was digested with 4 units of Taq I enzyme (Invitrogen, USA) at 65°C for 3 h. The restriction fragments were resolved by electrophoresis in 5% agarose gel at 50 V for 75 minutes. The genotype at each locus was determined (homozygous allele C, heterozygous, or homozygous allele T) based on fragment patterns as previously described [9].

2.4. Linkage Disequilibrium and Haplotyping.
The genotyping data was checked for conformity with Hardy-Wienberg's equilibrium (HWE) as well as for linkage disequilibrium (LD) between the two loci in each study group using the Haploview software (Daly Lab at Broad Institute, USA). LD was described in terms of disequilibrium coefficient (D'), correlation coefficient (r^2), and log of the likelihood odds ratio (LOD), the later being a measure of significance (LOD > 2 is significant). Haplotyping at the subject level was performed using the SimHap software (Centre for Genetic Epidemiology and Biostatistics, University of Western Australia), which employs the estimation-maximization (EM) algorithm for the estimation of haplotypes [12].

TABLE 2: Primer sequences used in the study.

Locus	Primer sequence (5′-3′)	Reference
IL-1α	Forward: TGTTCTACCACCTGAACTAGGC *	
	Reverse: TTACATATGAGCCTTCCATG	[9]
IL-1β	Forward: CTCAGGTGTCCTCGAAGAAATCAAA	
	Reverse: GCTTTTTTGCTGTGAGTCCCG	

*There are additional 5 nucleotides (AAGCT) in the 5′ end of the primer according to the reference; since they were found to be noncomplementary to the target and change the amplicon size to 104 bp, they were omitted in this study.

TABLE 3: IL-1A C[−889]T allele frequency, genotype distribution, and HWE status in the cases and controls.

| Parameter | Overall $N = 80$ | Cases $n = 40$ | Controls $n = 40$ | OR¶ (95% CI) | | |
				Dominant model	Additive model	Recessive model
C allele	85 (53.1%)	36 (45.0%)	49 (61.2%)		—ᵃ	
T allele	75 (46.9%)	44 (55.0%)	31 (38.8%)		1.92 (0.86–4.27)ᵇ	
CC genotype	21 (26.3%)	08 (20.0%)	13 (32.5%)			
TC genotype	43 (53.7%)	20 (50.0%)	23 (57.5%)	1.43 (0.37–5.55)	2.18 (0.89–5.39)	6.29 (1.21–32.6)*
TT genotype	16 (20.0%)	12 (30.0%)	04 (10.0%)			
HWEᶜ	+	+	+		N/A	

¶OR: odds ratio in the cases compared to the controls; logistic regression adjusting for mean plaque index.
ᵃReference category.
ᵇModels are not applicable.
ᶜHWE: Hardy-Weinberg equilibrium; (+): consistent with; (−): significant deviation.
*$P \leq 0.05$.

2.5. Statistical Analysis. Association of allele frequencies, genotypes, or haplotypes with chronic periodontitis was assessed with logistic regression, adjusting for mean plaque index (MPI) and interaction terms if present. When zero frequencies were encountered, exact logistic regression was performed instead. For genotypes and haplotypes, the analysis was performed using the recessive, additive, and dominant models. The odds ratio and 95% confidence intervals were calculated for each model. A P-value ≤ 0.05 was considered significant. The LogXact software (Cytel Corporation, USA) was used for performing exact logistic regression; SPSS was used for all other analyses.

3. Results

3.1. The IL-1A C[−889]T Locus Analysis. Table 3 describes the IL-1A C[−889]T allele frequencies, genotype distribution, and HWE status in both study groups. The genotyping data conformed to HWE in the controls as well as the cases ($P = 0.35$ and 1.0, resp.). The T allele was observed at a higher frequency in the cases compared to the controls (55% versus 38.8%); however, the difference was statistically insignificant after adjustment for MPI. Genotype-wise, the T allele showed significant association with sever periodontitis only in the recessive model (TT versus TC + CC genotypes) with an odds ratio (OR) of 6.29.

3.2. The IL-1B C[3954]T Locus Analysis. The IL-1B C[3954]T allele frequencies, genotype distribution and HWE status in both study groups are shown in Table 4. There was a significant deviation of genotype distribution

from HWE in the controls but not in the cases ($P = 0.01$ and 0.7, resp.). The difference in the T allele frequencies between the cases (47.5%) and the controls (40%) was not statistically significant. Initially, genotype-based analysis did not reveal significant association with periodontitis in any of the models. However, a significant interaction between MPI and the IL-1B C[3954]T genotype was observed and after adjustment for this, the homozygous T allele genotype (recessive model) did show a significant association with the disease (OR = 461, 95% CI: 2.80–7.5E4).

3.3. The IL-1 Two-Locus Analysis. The IL-1A C[−889]T and IL-1B C[3954]T loci were found to be in significant LD in the cases ($D' = 0.80$ and $r^2 = 0.47$) but not in the controls as shown in Table 5. The frequencies of the C-C, C-T, T-C, and T-T haplotypes were 47.5%, 13.8%, 12.5%, and 26.2%, respectively, for the controls, and 41.3%, 3.7%, 11.2%, and 43.8%, respectively, for the cases. Haplotype-based analysis (Table 6) revealed that the C-T haplotype was significantly associated with being a control (protective effect) in the log additive as well as the dominant models (OR = 0.17 and 0.16, resp.). On the other hand, the T-T haplotype in double-dose (recessive model) showed very strong association with chronic periodontitis (OR = 15.6).

4. Discussion

The association between the IL-1 gene polymorphisms and chronic periodontitis was first reported in white Caucasians by Kornman et al [9]. Since then, many attempts have been made to reproduce the results in the same as well as other

TABLE 4: IL-1B C[3954]T allele frequency, genotype distribution, and HWE status in the cases and controls.

| Parameter | Overall $N = 80$ | Cases $n = 40$ | Controls $n = 40$ | OR[¶] (95% CI) | | |
				Dominant model	Additive model	Recessive model
C allele	90 (56.2%)	42 (52.5%)	48 (60.0%)		—[a]	
T allele	70 (43.8%)	38 (47.5%)	32 (40.0%)		4.20 (0.37–48.0)[b]	
CC genotype	22 (27.5%)	12 (30.0%)	10 (25.0%)			
TC genotype	46 (57.5%)	18 (45.0%)	28 (70.0%)	1.05 (0.02–60.0)	7.38 (0.45–121)	461 (2.80–7.5E4)*
TT genotype	12 (15.0%)	10 (25.0%)	02 (05.0%)			
HWE[c]	+	+	−	N/A		

[¶]OR: odds ratio in the cases compared to the controls; logistic regression adjusting for mean plaque index (MPI) and MPI-genotype interaction.
[a]Reference category.
[b]Models are not applicable.
[c]HWE: Hardy-Weinberg equilibrium; (+): consistent with; (−): significant deviation.
*$P \leq 0.05$.

TABLE 5: IL-1A C[−889]T and IL-1B C[3954]T linkage disequilibrium analysis.

Statistic	Overall	Cases $n = 40$	Controls $n = 40$
Disequilibrium coefficient (D')	0.62	0.80	0.44
Correlation coefficient (r^2)	0.34	0.47	0.18
Log of likelihood odds ratio (LOD)	>2	>2	<2

* Statistically significant when >2.

races/ethnic populations [6]. To the best of our knowledge, the current study is the first to involve a Yemeni population and the second to include Arabs, who are usually classified under the Caucasian race [13].

Because of time and financial constraints, it was necessary to keep the sample size to a minimum. However, the study was very carefully designed as to do so without compromising the statistical power. First, stringent diagnostic criteria for cases were used that only subjects with very sever chronic periodontitis were included (5.53 mean CAL), justifying the use of an OR of 3 for sample size calculation. In fact, this case-enrichment approach is recommended for minimizing phenotypic heterogeneity and improving statistical power [14]. Second, controls were matched for age and sex to avoid the need for statistically adjusting for their effects. This is actually an advantage over a number of previous studies in which controls were much younger than cases [15–17]. Finally, smokers and those with diabetes were excluded, eliminating the possibility of interactions and the need for data stratification.

Good quality DNA was obtained from the oral epithelium samples, providing further support for their reliable use as noninvasive alternative to blood samples. In fact, complete concordance in IL-1 genotyping results from oral epithelium and blood samples have been previously demonstrated [18].

The IL-1A T allele frequency in the controls (38.8%) is somewhat similar to that reported in Arabs from Kuwait (35%; [19]) and Jordon (33.7%; [16]) and falls within the range reported in the literature (19–57%) for Caucasians [20–25]. This is in contrast with the very low prevalence (5.6–11.5%) found in subjects of Asian origin [15, 26, 27]. On the other hand, the control IL-1B T allele frequency observed

in this study (40%) is considerably higher than the 7.5–29% range reported for Caucasians [20–24]; however, it is close to that reported for Jordanian Arabs (45%) which is the highest in the literature [16]. Again, a much lower prevalence (1.2–4%) has been found in Asians [15, 26, 28]. The IL-1B C[3954]T genotyping data in the controls significantly deviated from HWE; a nonsignificant deviation was also noted for the IL-1A locus. This may be explained, at least in part, by the fact that first-cousin marriage, and therefore non-random mating, is common in Arab societies [29]. However, such a deviation was not observed in previous studies on Arabs [16, 19].

In the current study, the IL-1A C[−889]T and IL-1B C[3954]T polymorphisms showed strong association with sever periodontitis in the TT versus CT + CC genotype contrast (recessive model). This is in line with results from a number of previous reports in which either or both loci showed an association with chronic periodontitis, mostly in its sever form, in the same contrast [24, 30, 31] or in the TT + CT versus CC contrast [25, 32–35]. As with the current study, the association was only observed in nonsmokers. On the other hand, however, several authors have been unable to demonstrate any such association [15, 17, 20, 23, 27, 36].

While this controversy may be correctly attributed to differences in the race/ethnicity of studied populations, especially when comparing Asians with Caucasians, other factors such as differences in study design and data analysis probably account for a greater part of the variation among the reported results. Lopez et al. [22], for example, showed no association between the IL-A C[−889]T polymorphism and chronic periodontitis in a Caucasian Chilean population; however, they were later [30] able to demonstrate a significant association in another, yet, Caucasian Chilean population. The major difference between the two studies was case definition: the latter study included older cases with more sever chronic periodontitis compared to the former (4.2 and 1.9 mm mean CAL, resp.); this substantiates the importance of recruiting extreme phenotypes to enhance statistical power in genetic association studies, especially when small sample size is used. Using a low cutoff for definition of cases may, therefore, explain failure to demonstrate significant associations in some studies [16, 20]. Matching controls for

TABLE 6: IL-1 two-locus haplotype analysis.

Haplotype	Bi-allelic	Cases $n = 40$	Controls $n = 40$	OR[¶] (95% CI)		
				Dominant model	Additive model	Recessive model
C-C	O/O	14 (35.0%)	08 (20.0%)			
	C-C/O	19 (47.5%)	26 (65.0%)	0.31 (0.08–1.17)	0.86 (0.36–2.09)	4.03 (0.77–21.2)
	C-C/C-C	07 (17.5%)	06 (15.0%)			
C-T	O/O	37 (92.5%)	30 (75.0%)			
	C-T/O	03 (07.5%)	09 (22.5%)	0.16 (0.03–0.85)*	0.17 (0.03–0.83)*	0.72 (0.00–28.0)
	C-T/C-T	00 (00.0%)	01 (02.5%)			
T-C	O/O	32 (80.0%)	30 (75.0%)			
	T-C/O	07 (17.5%)	10 (25.0%)	0.92 (0.24–3.50)	0.93 (0.25–3.50)	0.04 (0.001–INF)
	T-C/T-C	01 (02.5%)	00 (00.0%)			
T-T	O/O	13 (32.5%)	19 (47.5%)			
	T-T/O	19 (47.5%)	21 (52.5%)	1.73 (0.54–5.56)	2.55 (0.95–6.87)	15.6 (1.58–INF)*
	T-T/T-T	08 (20.0%)	00 (00.0%)			

O: other haplotype.
[¶]OR: odds ratio in the cases compared to the controls; logistic regression adjusting for mean plaque index.
*$P \leq 0.05$.

age is another very important aspect of study design that has been missed by a number of authors who failed to show an association between the IL-1 polymorphisms and chronic periodontitis [15–17]. Differences among studies in performing data analysis may be another source of variation. Moreira et al. [33, 34], for example, showed a significant association between the IL-1A C[−889]T and IL-1B C[3954]T polymorphisms, and chronic periodontitis in the TT + TC versus CC contrast (dominant model); however, they did not attempt to do a comparison in the recessive (TT versus TC + CC) or log additive models which would have probably resulted in reporting stronger association as the data suggests. In fact, the current study is the first to explicitly perform comparisons in the three models.

In addition to showing a significant association at the individual locus level, the current study, and for the very first time, shows a strong association between IL-1 polymorphisms and sever chronic periodontitis based on bi-locus haplotype analysis and LD contrast between cases and control; a properiodontitis (T-T) and an antiperiodontitis (C-T) haplotypes were identified, and a significant LD was observed in cases but not in controls. Karasneh et al. [16] and Trevilatto et al. [37] have very recently used a similar approach for data analysis, but they were unable to demonstrate any association, neither did they provide haplotyping or LD data. However, a number of previous studies did report a strong association between the unphased composite genotype (simultaneous carriage of the T allele at the IL-1A C[−889]T and IL-1B C[3954]T loci) and chronic periodontitis [9, 38–40], which is somewhat is in line with the current findings although haplotyping data cannot be directly compared with raw composite genotyping data.

In conclusion, findings from the current study provides further support for an association between the IL-1 polymorphisms and severity of chronic periodontitis, and a preliminary evidence for the usefulness of LD and haplotype-based analysis in exploring for genetic determinants of periodontitis. However, assessing more loci (extended haplotypes) in larger-scale studies is required to explore this further.

Conflict of Interests

There is no conflict of interests to declare.

References

[1] R. P. Darveau, "Periodontitis: a polymicrobial disruption of host homeostasis," *Nature Reviews Microbiology*, vol. 8, no. 7, pp. 481–490, 2010.

[2] K. S. Kornman, "Mapping the pathogenesis of periodontitis: a new look," *Journal of Periodontology*, vol. 79, no. 8, supplement, pp. 1560–1568, 2008.

[3] T. E. Van Dyke and D. Sheilesh, "Risk factors for periodontitis," *Journal of the International Academy of Periodontology*, vol. 7, no. 1, pp. 3–7, 2005.

[4] B. S. Michalowicz, D. Aeppli, J. G. Virag et al., "Periodontal findings in adult twins," *Journal of Periodontology*, vol. 62, no. 5, pp. 293–299, 1991.

[5] B. S. Michalowicz, S. R. Diehl, J. C. Gunsolley et al., "Evidence of a substantial genetic basis for risk of adult periodontitis," *Journal of Periodontology*, vol. 71, no. 11, pp. 1699–1707, 2000.

[6] M. L. Laine, B. G. Loos, and W. Crielaard, "Gene polymorphisms in chronic periodontitis," *International Journal of Dentistry*, vol. 2010, Article ID 324719, 22 pages, 2010.

[7] D. T. Graves and D. Cochran, "The contribution of interleukin-1 and tumor necrosis factor to periodontal tissue destruction," *Journal of Periodontology*, vol. 74, no. 3, pp. 391–401, 2003.

[8] J. J. Taylor, P. M. Preshaw, and P. T. Donaldson, "Cytokine gene polymorphism and immunoregulation in periodontal disease," *Periodontology 2000*, vol. 35, pp. 158–182, 2004.

[9] K. S. Kornman, A. Crane, H. Y. Wang et al., "The interleukin-1 genotype as a severity factor in adult periodontal disease," *Journal of Clinical Periodontology*, vol. 24, no. 1, pp. 72–77, 1997.

[10] G. K. Nikolopoulos, N. L. Dimou, S. J. Hamodrakas, and P. G. Bagos, "Cytokine gene polymorphisms in periodontal disease: a meta-analysis of 53 studies including 4178 cases and 4590 controls," *Journal of Clinical Periodontology*, vol. 35, no. 9, pp. 754–767, 2008.

[11] H. Loee and J. Silness, "Periodontal disease in pregnancy. I. Prevalence and severity," *Acta Odontologica Scandinavica*, vol. 21, pp. 533–551, 1963.

[12] L. Excoffier and M. Slatkin, "Maximum-likelihood estimation of molecular haplotype frequencies in a diploid population," *Molecular Biology and Evolution*, vol. 12, no. 5, pp. 921–927, 1995.

[13] N. Risch, E. Burchard, E. Ziv, and H. Tang, "Categorization of humans in biomedical research: genes, race and disease," *Genome Biology*, vol. 3, no. 7, article comment2007-comment2007.12, 2002.

[14] A. S. Schafer, S. Jepsen, and B. G. Loos, "Periodontal genetics: a decade of genetic association studies mandates better study designs," *Journal of Clinical Periodontology*, vol. 38, no. 2, pp. 103–107, 2011.

[15] O. Anusaksathien, A. Sukboon, P. Sitthiphong, and R. Teanpaisan, "Distribution of interleukin-1β+3954 and IL-1α-889 genetic variations in a Thai population group," *Journal of Periodontology*, vol. 74, no. 12, pp. 1796–1802, 2003.

[16] J. A. Karasneh, K. T. Ababneh, A. H. Taha, M. S. Al-Abbadi, and W. Ollier, "Investigation of the interleukin-1 gene cluster polymorphisms in Jordanian patients with chronic and aggressive periodontitis," *Archives of Oral Biology*, vol. 56, no. 3, pp. 269–276, 2011.

[17] D. Sakellari, S. Koukoudetsos, M. Arsenakis, and A. Konstantinidis, "Prevalence of IL-1A and IL-1B polymorphisms in a Greek population," *Journal of Clinical Periodontology*, vol. 30, no. 1, pp. 35–41, 2003.

[18] H. Duan, J. Zhang, P. Huang, and Y. Zhang, "Buccal swab: a convenient source of DNA for analysis of IL-1 gene polymorphisms," *Hua Xi Kou Qiang Yi Xue Za Zhi*, vol. 19, no. 1, pp. 11–13, 2001.

[19] H. P. Muller and K. M. Barrieshi-Nusair, "Site-specific gingival bleeding on probing in a steady-state plaque environment: influence of polymorphisms in the interleukin-1 gene cluster," *Journal of Periodontology*, vol. 81, no. 1, pp. 52–61, 2010.

[20] P. M. Brett, P. Zygogianni, G. S. Griffiths et al., "Functional gene polymorphisms in aggressive and chronic periodontitis," *Journal of Dental Research*, vol. 84, no. 12, pp. 1149–1153, 2005.

[21] E. A. Gore, J. J. Sanders, J. P. Pandey, Y. Palesch, and G. M. P. Galbraith, "Interleukin-1β+3953 allele 2: association with disease status in adult periodontitis," *Journal of Clinical Periodontology*, vol. 25, no. 10, pp. 781–785, 1998.

[22] N. J. Lopez, L. Jara, and C. Y. Valenzuela, "Association of interleukin-1 polymorphisms with periodontal disease," *Journal of Periodontology*, vol. 76, no. 2, pp. 234–243, 2005.

[23] D. Sakellari, V. Katsares, M. Georgiadou, A. Kouvatsi, M. Arsenakis, and A. Konstantinidis, "No correlation of five gene polymorphisms with periodontal conditions in a Greek population," *Journal of Clinical Periodontology*, vol. 33, no. 11, pp. 765–770, 2006.

[24] J. Wagner, W. E. Kaminski, C. Aslanidis et al., "Prevalence of OPG and IL-1 gene polymorphisms in chronic periodontitis," *Journal of Clinical Periodontology*, vol. 34, no. 10, pp. 823–827, 2007.

[25] S. Shiroddria, J. Smith, I. J. McKay, C. N. Kennett, and F. J. Hughes, "Polymorphisms in the IL-1A gene are correlated with levels of interleukin-1α protein in gingival crevicular fluid of teeth with severe periodontal disease," *Journal of Dental Research*, vol. 79, no. 11, pp. 1864–1869, 2000.

[26] G. C. Armitage, Y. Wu, H. Y. Wang, J. Sorrell, F. S. di Giovine, and G. W. Duff, "Low prevalence of a periodontitis-associated interleukin-1 composite genotype in individuals of Chinese heritage," *Journal of Periodontology*, vol. 71, no. 2, pp. 164–171, 2000.

[27] T. Kobayashi, S. Ito, T. Kuroda et al., "The interleukin-1 and Fcγ receptor gene polymorphisms in Japanese patients with rheumatoid arthritis and periodontitis," *Journal of Periodontology*, vol. 78, no. 12, pp. 2311–2318, 2007.

[28] Y. Soga, F. Nishimura, H. Ohyama, H. Maeda, S. Takashiba, and Y. Murayama, "Tumor necrosis factor-alpha gene (TNF-α) -1031/-863, -857 single-nucleotide polymorphisms (SNPs) are associated with severe adult periodontitis in Japanese," *Journal of Clinical Periodontology*, vol. 30, no. 6, pp. 524–531, 2003.

[29] Z. Radovanovic, N. Shah, and J. Behbehani, "Prevalence and social correlates of consanguinity in Kuwait," *Annals of Saudi Medicine*, vol. 19, no. 3, pp. 206–210, 1999.

[30] N. J. Lopez, C. Y. Valenzuela, and L. Jara, "Interleukin-1 gene cluster polymorphisms associated with periodontal disease in type 2 diabetes," *Journal of Periodontology*, vol. 80, no. 10, pp. 1590–1598, 2009.

[31] M. A. Rogers, L. Figliomeni, K. Baluchova et al., "Do interleukin-1 polymorphisms predict the development of periodontitis or the success of dental implants?" *Journal of Periodontal Research*, vol. 37, no. 1, pp. 37–41, 2002.

[32] K. Geismar, C. Enevold, L. K. Sorensen et al., "Involvement of interleukin-1 genotypes in the association of coronary heart disease with periodontitis," *Journal of Periodontology*, vol. 79, no. 12, pp. 2322–2330, 2008.

[33] P. R. Moreira, J. E. Costa, R. S. Gomez, K. J. Gollob, and W. O. Dutra, "The IL1A (-889) gene polymorphism is associated with chronic periodontal disease in a sample of Brazilian individuals," *Journal of Periodontal Research*, vol. 42, no. 1, pp. 23–30, 2007.

[34] P. R. Moreira, A. R. de Sa, G. M. Xavier et al., "A functional interleukin-1β gene polymorphism is associated with chronic periodontitis in a sample of Brazilian individuals," *Journal of Periodontal Research*, vol. 40, no. 4, pp. 306–311, 2005.

[35] P. S. G. Prakash and D. J. Victor, "Interleukin-1b gene polymorphism and its association with chronic periodontitis in South Indian population," *International Journal of Genetics and Molecular Biology*, vol. 2, no. 8, pp. 179–183, 2010.

[36] S. B. Ferreira Jr., A. P. F. Trombone, C. E. Repeke et al., "An interleukin-1β (IL-1β) single-nucleotide polymorphism at position 3954 and red complex periodontopathogens independently and additively modulate the levels of IL-1β in diseased periodontal tissues," *Infection and Immunity*, vol. 76, no. 8, pp. 3725–3734, 2008.

[37] P. C. Trevilatto, A. P. de Souza Pardo, R. M. Scarel-Caminaga et al., "Association of IL1 gene polymorphisms with chronic periodontitis in Brazilians," *Archives of Oral Biology*, vol. 56, no. 1, pp. 54–62, 2011.

[38] A. A. Agrawal, A. Kapley, R. K. Yeltiwar, and H. J. Purohit, "Assessment of single nucleotide polymorphism at IL-1A+4845 and IL-1B+3954 as genetic susceptibility test for chronic periodontitis in Maharashtrian ethnicity," *Journal of Periodontology*, vol. 77, no. 9, pp. 1515–1521, 2006.

[39] M. J. McDevitt, H. Y. Wang, C. Knobelman et al., "Interleukin-
      1 genetic association with periodontitis in clinical practice,"
      Journal of Periodontology, vol. 71, no. 2, pp. 156–163, 2000.

[40] P. Meisel, C. Schwahn, D. Gesch, O. Bernhardt, U. John, and T.
      Kocher, "Dose-effect relation of smoking and the interleukin-
      1 gene polymorphism in periodontal disease," *Journal of
      Periodontology*, vol. 75, no. 2, pp. 236–242, 2004.

6

The Impact of HIV Genetic Polymorphisms and Subtype Differences on the Occurrence of Resistance to Antiretroviral Drugs

Mark A. Wainberg and Bluma G. Brenner

Jewish General Hospital AIDS Centre, McGill University, 3755 Cote-Ste-Catherine Road, Montreal, QC, Canada H3T 1E2

Correspondence should be addressed to Mark A. Wainberg, mark.wainberg@mcgill.ca

Academic Editor: Gilda Tachedjian

The vast majority of reports on drug resistance deal with subtype B infections in developed countries, and this is largely due to historical delays in access to antiretroviral therapy (ART) on a worldwide basis. This notwithstanding the concept that naturally occurring polymorphisms among different non-B subtypes can affect HIV-1 susceptibility to antiretroviral drugs (ARVs) is supported by both enzymatic and virological data. These findings suggest that such polymorphisms can affect both the magnitude of resistance conferred by some major mutations as well as the propensity to acquire certain resistance mutations, even though such differences are sometimes difficult to demonstrate in phenotypic assays. It is mandatory that tools are optimized to assure accurate measurements of drug susceptibility in non-B subtypes and to recognize that each subtype may have a distinct resistance profile and that differences in resistance pathways may also impact on cross-resistance and the choice of regimens to be used in second-line therapy. Although responsiveness to first-line therapy should not theoretically be affected by considerations of viral subtype and drug resistance, well-designed long-term longitudinal studies involving patients infected by viruses of different subtypes should be carried out.

1. Introduction

Nonsubtype B infections are responsible for most HIV cases worldwide [1]. HIV-1 group M has been classified into subtypes, circulating and unique recombinant forms (CRF and URF, resp.), due to its significant natural genetic variation; this includes subtypes A–D, F–H, and J–K and many CRFs and URFs. Although subtype B is the most prevalent in the Western World (Western Europe, the Americas, Japan, and Australia), non-B subtypes predominate in the rest of the world: that is, subtype C in sub-Saharan Africa, India, and Brazil, CRF01_AE in South East Asia, CRF02_AG in West Africa, and subtype A in Eastern Europe and Northern Asia [1–3]. The proportion of non-B subtypes in North and South America and Western Europe is increasing [4–7]. Combination antiretroviral therapy (ART) is now used in many areas of the world, and HIV resistance to antiretroviral drugs (ARVs) has widely emerged. Thus, non-B subtypes

will presumably become even more common in western countries.

Reduced sensitivity to ARVs in non-B subtypes has been less well studied than in subtype B, mainly because of the predominance of subtype B in those countries in which ARVs first became available, coupled with the availability of genotypic and phenotypic antiretroviral drug resistance testing in such countries [8]. This notwithstanding there is a potential for genetic differences among subtypes to yield differential patterns of resistance-conferring mutations in response to ARVs and this possibility is supported by the fact that HIV-1 naturally varies in genetic content by as much as 35% among subtypes. Indeed, variation is higher in some areas of the genome (40% in the env gene) and lower in others (8–10% in the *pol*, *gag*, and *IN* genes) [8]. Since differences in codon sequences at positions associated with drug resistance mutations might predispose viral isolates from different subtypes to encode different amino acid substitutions, it is possible that HIV-1 genetic diversity may influence the types of resistance mutations that might eventually emerge upon drug exposure as well as the rate of emergence of such mutations and phenotypic resistance [8, 9]. Such diversity may also affect the degree of cross-resistance to ARVs of the same class, with the potential to impact on virologic failure, clinical outcomes, and preservation of immunological responsiveness [8].

For example, studies of single dose nevirapine (sdNVP) for prevention of mother-to-child transmission (PMTCT) showed a disparity in overall resistance among subtypes, with frequencies of 69, 36, 19, and 21% against NVP in women with subtypes C, D, A, and CRF02_AG infections, respectively. Often, this result occurred prior to treatment and despite the absence of resistance mutations [10–13]. Very sensitive PCR detection procedures, which reveal resistance due to minority species, have revealed a higher incidence of NVP resistance (K103N, Y181C) in 70–87% of individuals with subtype C compared with 42% of individuals with subtype A [14–16].

Evaluations of virological and biochemical data also suggest that natural amino acid background can affect the magnitude of resistance conferred by many mutations responsible for antiretroviral drug resistance [17], as is best illustrated by HIV-2 and group O viruses that show high-level innate resistance to nonnucleoside reverse transcriptase inhibitors (NNRTIs) through the presence of natural polymorphisms that can confer drug resistance (Table 1) [18, 19]. However, many studies on antiretroviral drug resistance in non-B subtypes exposed to chronic suppressive therapy have yielded less definitive results with respect to the importance of natural HIV-1 diversity as a factor leading to differences in types of drug resistance mutations and the propensity to develop drug resistance in the first place [8, 17].

Although genotypic ARV resistance testing is of proven benefit in deciding on best choice of ARVs for individual treatment and serves as a repository of information on HIV resistance mutations, several factors underscore the difficulties in defining intersubtype differences. For example, genotyping can classify the major viral subtypes, but significant proportions (~15%) of infections remain unassigned or differentially assigned using different subtyping algorithms [8, 20, 21]. Certainly, HIV resistance databases make efforts to incorporate newer subtype data into pools of data, but the availability of HIV genotypes from areas of the world with non-B subtype predominance is still comparatively low [22]. The factors responsible include lesser availability of ARV therapy, the high cost of drug resistance testing, and limited opportunities for research in resource-limited areas. In some cases, resistance tests may often be performed only on participants enrolled in study cohorts or trials but not in general practice.

2. Resistance to Nucleoside Reverse Transcriptase Inhibitors (NRTIs)

As an example of disparity, subtype C patients in Botswana treated with ZDV/ddI developed an atypical thymidine ana-

logue mutation (TAM) resistance pathway (67N/70R/215Y) compared to subtype B (the TAM 1 and TAM 2 pathways) [23]. This distinction was not observed in patients with subtype C in Malawi, India, or South Africa [24–27]. Results from Botswana also reported a high incidence of K65R (30%) in subtype C patients who received d4T/ddI plus NVP or efavirenz (EFV) [28]. A much larger study from Malawi detected K65R or K70E in 23% of patients failing first-line therapy with d4T/3TC/NVP [26], while K65R was detected in 7% and 15% of patients in South Africa failing first- or second-line regimens, respectively, whose nucleoside backbones included d4T/3TC or ddI/ZDV [29, 30]. A study from Israel also reported a high frequency of K65R in subtype C viruses from Ethiopian immigrants [31], and a report from India showed that K65R was present in about 10–12% of patients who had received d4T/3TC/NVP in first-line therapy [32]. Such differences in K65R and thymidine analogue mutations (TAMs) might be attributed to treatment regimen and disease stage [24–27].

Access to viral load testing lead in India was also associated with early detection of NRTI-treatment failure, leading to use of new, second-line regimens and preventing acquisition of TAMs and K65R [24]. Additional studies support regional differences among subtype C subepidemics from Ethiopia, Brazil, and sub-Saharan Africa, that impact on NRTI resistance rates as a result of different NRTI-based regimens [8, 33, 34].

Higher rates of the K65R mutation in subtype C [26, 28, 29] suggest that these viruses may have a particular predisposition toward acquiring this mutation [35]. A subtype C RNA template mechanism has been proposed to explain this phenomenon that involves higher rates of K65R mutagenesis in subtype C viruses than in other subtypes (Figure 1) [36, 37], and this mechanism seems to be template dependent and is independent of the source of the reverse transcriptase (RT) employed [36]. Subtype C viruses apparently have an intrinsic difficulty in synthesizing stretches of adenine homopolymeric runs that leads to template pausing at codon 65, facilitating the acquisition of K65R under selective drug pressure [37, 38], whereas the subtype B template favors pausing at codon 67 that may facilitate the generation of D67N and TAMs rather than K65R [37–39]. In addition, the introduction of codons from positions 64 and 65 in the RT of subtype C into a subtype B backbone was sufficient to lead to selection of K65R by multiple NRTIs [37–39]. Figure 1 provides a pictorial representation of the preferential development of K65R in subtype C viruses.

Ultrasensitive pyrosequencing has also been used to detect the spread of K65R as transmitted and/or minority species in treatment-naïve populations [40, 41]. Patients harboring subtype C infections showed a higher frequency of K65R than subtype B variants (1.04% versus 0.25%) by this method but these differences were not duplicated using limiting dilution clonal sequencing approaches [40]. While these findings are consistent with PCR-induced pausing, leading to low-level spontaneous generation of K65R in subtype C, they do not negate the higher rates of development of K65R in subtype C populations failing regimens containing d4T, ddI, or tenofovir (TDV) [32]. The occurrence of K65R in

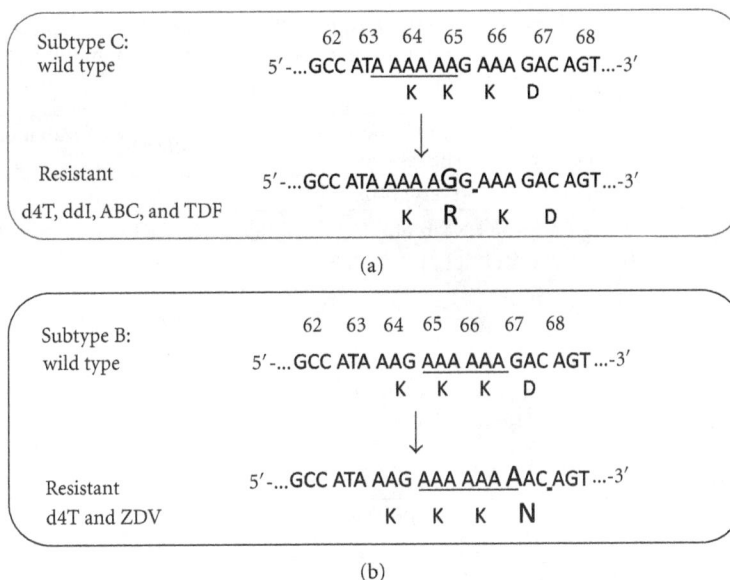

FIGURE 1: Subtype-specific poly-A nucleotide motifs lead to template pausing under pressure with thymidine analogues that favor K65R selection in subtype C and D67N selection in subtype B. Depiction of the template-based propensity of subtype C versus B viruses to develop the K65R mutation that is associated with broad cross-resistance among multiple members of the NRTI family of drugs. The codons located at positions 63, 64, and 65 in subtype C RT seem to be critically involved in the preferential development of K65R in subtype C. d4T: stavudine, ddI: didanosine, ABC: abacavir, TDF: tenofovir. It should be noted that the use of stavudine in particular has been shown to yeild K65R in subtype C infections with high frequency. Regimens that are based on the use of TDF and ABC, among other drugs, can help mitigate the development of the K65R mutation.

subtype C and CRF01_AE is also associated with the Y181C NVP mutation within the viral backbone [30, 42].

Subtype C selected the K65R mutation in drug resistance selection studies faster than subtype B under TFV pressure [35]. However, K65R may be less frequent in subtype A than other subtypes [43]. And a higher propensity to acquire TAMs was reported in patients carrying CRF_06 (AGK recombinants) as compared to patients carrying CRF02_AG from Burkina Faso [44].

The differential selection of K65R pathways in subtype C seems related to template differences, ddI and d4T-containing regimens, as well as to the presence of Y181C. Further genotypic studies will be required to ascertain subtype differences in acquisition of resistance to NRTIs.

3. Resistance to Nonnucleoside Reverse Transcriptase Inhibitors (NNRTIs)

Selection studies in culture have shown that a V106M mutation commonly develops in subtype C viruses following drug pressure with NVP or EFV, whereas a V106A mutation is more commonly selected in subtype B. This difference is due to a nucleotide polymorphism at codon 106 in RT [45, 46], and the clinical importance of V106M in non-B subtypes has been confirmed in multiple studies showing that V106M is frequently seen in non-B subtypes (C and CRF01_AE) after therapy with NVP or EFV [23, 25, 27, 47–50].

The G190A substitution was also relatively more frequent among subtype C infected patients failing NNRTI-based therapy in Israel and India, and G190A/S was seen in the Israeli study as a natural polymorphism in subtype C from Ethiopian immigrants [25, 49]. The frequencies of these mutations among treated patients in both studies were higher than in subtype B and C drug-naïve individuals.

Although the overall prevalence of V106M in subtype C is higher than in subtype B (12% versus, 0%) in individuals failing NNRTI-based regimens, K103N (29% versus 40%) and Y181C (12% versus 23%) remain important pathways for both subtype C and B, respectively (http://hivdb.stanford.edu/). Only minor differences in HIV resistance pathways seem to occur among subtypes A, B, and C with the second generation NNRTI etravirine (ETR) [50].

4. PR Mutations

The results of work with protease inhibitors PIs indicate that the D30N mutation was not observed in CRF02_AG and CRF02_AE isolates in patients failing nelfinavir (NFV) therapy but rather that the N88S mutation emerged after NFV use in CRF01_AE and after indinavir [51] use in subtype B [52, 53]. Although another study reported an absence of the D30N mutation in CRF01_AE, no information on the specific type of PIs received by the patients was available [54]. A lower frequency of D30N was seen in subtype C isolates from Ethiopian immigrants to Israel after NFV usage than in subtype C viruses from Botswana [55, 56], suggesting that subtype C viruses from Ethiopia (the origin of the samples

TABLE 1: Examples of polymorphisms and mutations in reverse transcriptase (RT), protease (PR), and integrase (IN) of different subtypes that may impact on emergent resistance to nucleoside and nonnucleoside reverse transcriptase inhibitors (NRTIs and NNRTIs), protease inhibitors (PIs), and integrase strand transfer inhibitors (INSTIs).

Drug class	Type/group/ subtype	Polymorphism or mutation associated with drug resistance	Drug(s) affected	Mutation(s) and their consequences	Reference
		Reverse transcriptase			
NRTI	C	64-65-66 KKK motif	ddI, d4T, TDF	K65R	[30]
NRTI	HIV-2	T69N, V75I, V118I, L210N, T215S, K219N	NRTIs	TAMs/K65R	[66]
NNRTI	C	V106V	EFV, NVP	V106M	[45]
NNRTI	G	A98S	NNRTIs		[66]
NNRTI	HIV-2	Y181I,Y188L, G190A K101A, V106I, V179I	All NNRTIs	Cross- NNRTI resistance	[68]
NNRTI	O	Y181C, A98S, K103R, V179E	All NNRTIs	Cross- NNRTI resistance	[18]
		Protease			
PI	Non-B	M36I	PIs		[59]
PI	G, AE	K20I	PIs		[63]
PI	G	V82I	PIs	I82M/T/S	[63]
PI	A, C, F, G, AE, AG	L89M	PIs	L89I	[71]
PI	HIV-2	L10I/V, K 20V, V32I, M36I, M46I, I47V, L63E/K, A71V, G73A, V77T, V82I/L,	PIs	APV and other PIs	[68]
		Integrase			
INSTIs	B	R263	MK-2048, DTG	R263K	[85]
	C	G118	MK-2048, DTG	G118R	[82]

ddI: didanosine; d4T: stavudine; TFV, tenofovir: EFV, efavirenz: NVP, nevirapine: DTG, dolutegravir.

identified in Israel) and southern Africa might behave in different fashion. M89I/V mutations were observed in F, G, and C subtypes but not in other subtypes [26], and the V82I natural polymorphism in subtype G led to the emergence of I82M/T/S in treatment failure [57]. The L90M mutation is rare in subtype F but common in subtype B from Brazil [58], and a recent paper suggests that polymorphisms at position 36 in PR may be important in determining the emergence of specific patterns of resistance mutations among viruses of different subtypes [59].

To gain an understanding of the underlying mechanisms leading to the overall higher preponderance of D30N in subtype B relative to other subtypes, molecular dynamic simulations were performed. D30N appeared to selectively confer resistance to NFV in subtype B by increasing the flexibility of the protease (PR) flap region and destabilizing the PR inhibitor complex [60]. In subtype C, D30N required the accessory N83T mutation to confer resistance and rescue fitness [61].

Two comprehensive surveys reported differences in natural protease polymorphisms among non-B subtypes [62, 63] and positions less frequently mutated in non-B subtypes than in subtype B after exposure to ARVs. Residues of importance in subtype A in PR were at positions 10, 20, and 63, whereas, in subtype C, they were at residues 20, 53, 63, 74, and 82. Other differences were at residues 13 and 20 in subtype D, residues 10, 14, 20, and 77 in subtype F, residues 20, 67,

73, 82, and 88 in subtype G, residues 20, 63, 82, and 89 in CRF01_AE, and residue 20 in CRF02_AG [63].

Higher rates of accumulation of NRTI and PI resistance mutations and equal rates of emergence of NNRTI mutations were also found in subtype B compared to C [64]. A study from southern Brazil also showed a lower frequency of primary resistance to PIs in subtype C compared to subtype B, suggesting that PI mutations may be less well tolerated at the structural level in subtype C [65].

However, HIV-1 subtype diversity has not limited the overall benefit of ART (Table 1). This notwithstanding there are subtype differences in the type and preference of pathways of resistance with some mutations emerging almost exclusively in some non-B subtypes, for example, the protease mutation 82 M in subtype G versus 82A/F/S in the others, 88D in subtype B versus 88S in subtypes C and CRF02_AG [66]. Furthermore, HIV-2 has major mutations in regard to NRTIs, NNRTIs, and PIs, which contribute to innate NNRTI resistance and rapid development of multiclass drug resistance (Table 1) [67, 68]. The V106M RT mutation in subtypes C and A versus V106A in subtype B is observed with resistance against NVP and EFV. Polymorphisms at RT residue 98, common in subtype G, are associated with NNRTI resistance in subtype B and may lower the resistance barrier and duration of efficacy of some NNRTIs [69]. The frequency of some resistance mutations shared by B and non-B subtypes can vary after failure of first-line therapeutic regimens, as in the case of the K65R mutation. Differences in type and frequency of resistance mutations should not be underestimated. However, the TAM pathway 67N/70R/215Y found in subtype C in Botswana will probably be adequately detected by most resistance algorithms, since it does not involve new mutations.

A lower risk for accumulation of major (primary) resistance mutations in subtype C than B has been reported [64]. The major mutations that emerged in both subtypes were the same. Since both subtypes B and C patients had similar profiles of virological failure after use of the same ART regimens, this rules out ancillary factors responsible for these differences. Minor mutations in subtype B PR may appear as frequent natural polymorphisms in several non-B subtypes (e.g., M36I, L89M) [58, 59]. The fact that the L89M polymorphism can lead to the M89I mutation that confers resistance to PIs suggests that there might be a lower accumulation of major mutations in C subtypes, if natural polymorphisms act similarly in subtype C as they do when present as secondary resistance mutations in subtype B.

The majority of non-B HIV-1 subtype isolates possess wild-type susceptibilities similar to those of subtype B wild-type isolates. Compared to B subtypes, diminished susceptibilities among wild-type isolates have been found for CRF02_AG recombinant viruses in three different studies in regard to ATV and NFV [63, 69, 70]. No study has yet assigned statistical significance of drug susceptibility levels due to polymorphisms and small sample size. One analysis performed molecular modeling and suggested that distortions in the K26 pocket of A/G proteases appear to be responsible for a lower binding energy of NFV and hence lower susceptibility of A/G viruses to this drug [70]. A/G isolates with lower susceptibilities to certain PIs (NFV and atazanavir (ATV)) have also been found. One study has detected an important proportion of WT isolates with lower susceptibilities to ATV [71]. In most cases, phenotypes have been determined by commercial or in-house assays that were developed primarily to measure B-subtype drug susceptibilities based on the laboratory adapted strains NL4-3 or HXB2, through use of a modified clone of a laboratory strain that lacks both the terminal part of Gag and most of Pol. It should be recognized that most commercial assays do not monitor polymorphisms, and indeed sequences that lie within particular regions, such as the substrates of PR within gag or the RNaseH and connection domains within pol, can influence drug resistance in both B and non-B subtypes but may not be easily recognized. Although some work has been carried out in this field, it is clear that other studies are required [72–75].

There are few data on the potential for cross-resistance to PIs among non-B subtypes in regard to NFV, although there is a tendency to select for the L90M pathway instead of D30N in subtype C. Competition fitness assays support the notion that subtype C viruses bearing D30N are impaired in replicative fitness, a finding that may explain the above results [61].

Thermodynamic studies performed on target-inhibitor interactions in PR have specifically described a lower affinity of non-B subtype proteases for PIs and amplification of

primary resistance mutations on the basis of polymorphisms that are present in background.

In addition to the foregoing, interesting results on polymorphisms that confer hypersusceptibility to some PIs have been recently reported [76]. Some of these polymorphisms can potentially delay acquisition of drug resistance and may therefore enhance the long-term effectiveness of relevant drugs.

5. Integrase Inhibitors and Drug Resistance

New data are emerging that subtype differences are also present in regard to integrase strand transfer inhibitors (INSTIs) despite the fact that HIV-1 subtype B and C wild-type integrase (IN) enzymes are similarly susceptible to clinically approved INSTIs [77–81]. This notwithstanding there are now data to indicate that the presence of resistance mutations may differentially affect susceptibility to specific INSTIs in viruses of different subtypes [77]. Moreover, such data have been obtained both in tissue culture using recombinant viruses of different subtypes that contain specific IN mutations as well as in biochemical integrase strand transfer and integrase 3′ synthetase assays, in which specific drug resistance mutations have been introduced into recombinant purified integrase enzymes derived from either subtype B or subtype C viruses [77].

Of particular interest may be that a novel next-generation INSTI termed MK-2078 with a higher genetic barrier for selection of resistance than either raltegravir (RAL) or elvitegravir (EGV) was able to differentially select for a novel G118R substitution in IN in subtype C compared with subtype B viruses [82]. This mutation conferred only slight resistance to MK-2048 but gave rise to 25-fold resistance against RAL when it was present together with a polymorphic substitution at position L74M in CRF02-AG cloned patient isolates [83]. It is also well known that INSTI Q148RHK resistance mutations that affect susceptibility to a novel INSTI, dolutegravir (DTG) in HIV-1 subtype B may not affect susceptibility of subtype C viruses or HIV-2 viruses and IN enzymes to the latter compound [84].

Finally, tissue culture selection with DTG has identified a novel R263K resistance mutation in subtype B but not subtype C viruses [85]. In contrast, the same series of selections with DTG in subtype C viruses yielded the same G118R mutation that had previously been obtained with MK-2048, also in subtype C. This raises the possibility that G118R may have the potential to be an important resistance mutation for next-generation INSTIs in subtype C viruses but that this role may be played by R263K in the context of subtype B viruses. Of course, definitive information on this topic may have to await the widespread clinical use of DTG and the characterization of mutations within IN that may arise in the event of rising viral loads and treatment failure.

6. Clinical Practice

HIV resistance in non-B subtypes has rarely been reported on the basis of single drugs or NRTI backbones but, rather, mutations have been reported for specific drug classes.

Cross-resistance can be estimated only for some NRTIs and NNRTIs but not for most PIs that are the only drugs eligible as part of second-line regimens in most regions of the world. The potential for cross-resistance to NFV in viruses of CRF01_AE and CRF02_AG origin could be higher than has been observed in subtype B, due to the preferential selection of the N88S and L90M substitutions, although such data are not yet available for most PIs in the context of non-subtype B viruses. NRTI backbones may also vary in the mutation profile they select for according to drug combinations that are used. Newer compounds (e.g., TFV and ATV/r) are now preferred both in resource-rich countries and non-B subtype prevalent areas. Although HIV resistance databases continue to enter HIV genotype data from nonB subtype variants, few data sets are available to date (stanford HIV resistance database, Agence Nationale pour la Recherche sur le SIDA-France (ANRS), etc.) for drugs that have become part of first-line therapy in developed countries, for example, TDF, ATV, darunavir, ETR, and RAL. In this context as well, it is relevant that some studies have attempted to address the clinical impact of HIV diversity on treatment response as well as the limitations of such approaches [86, 87].

7. Future Considerations

The preferential emergence of some mutations and changes in the frequency of these mutations in select non-B subtypes needs greater attention and research on the role of polymorphisms in nonsubtype B viruses that increase in frequency after drug exposure and that may contribute to drug resistance (e.g., A98G/S in RT and M36I and K20I in PR) [88] should be priorized, particularly in parts of Africa in which treatment failure has been reported in as many as 40% of patients after two years [89] and in India where resistance rates of 80% to two drug classes have been reported after failure of first-line regimens that employed various NRTI/NNRTI combinations [90]. To date, no study has tested the degree of resistance or cross-resistance that certain mutational combinations (67N/70R/215Y) may confer in tissue culture. Newer studies should assess pre- and posttreatment genotypes in order to determine associations of certain polymorphisms with drug resistance, including variations of polymorphisms in variants of the same subtype that are located in different geographical regions. This would improve the appropriateness of use of certain drugs over others in the context of second- or third-line therapeutic regimens.

The different studies conducted in populations affected by nonsubtype B viruses are too heterogeneous to permit pooling of data [8]. Such studies have addressed different research questions and used nonequivalent NRTI backbones (e.g., ZVD/ddI and ZDV/3TC) and have also grouped mutations by drug class without providing information on the nature of the regimen at virologic failure. Resistance has also been reported in different ways (e.g., different algorithms or resistance lists), making it difficult to relate resistance mutations to a specific drug or combination of drugs. More longitudinal studies on response to first-line ARV combinations are needed to better recognize intersubtype differences. Pre- and posttherapy genotype resistance testing is also desirable.

8. Conclusions

Virological and biochemical data provide compelling evidence on the differential effect of genetic background on both the type and degree of HIV-1 antiretroviral drug resistance. Genetic background can affect the degree of protein binding caused by primary mutations and restore the function of PR to a differential degree in different subtypes based on background polymorphisms, although this effect was not discernible in the absence of typical major resistance mutations but rather when particular backgrounds of combinations of major resistance mutations and background polymorphisms were represented. Clearly, some background polymorphisms can act as secondary resistance substitutions.

Phenotypic assays have failed to find differences of large magnitude in the susceptibilities of HIV B versus non-B subtypes, consistent with what has been learned at a molecular level. Unfortunately, only few datasets exist on relative susceptibility levels among subtypes carrying specific major resistance mutations, and more information is required, particularly because many polymorphisms in non-B viruses are considered to be secondary resistance mutations since they can emerge after drug exposure in subtype B viruses. The effect of such polymorphisms within different genetic backgrounds cannot always be extrapolated to non-B subtypes and might sometimes contribute to higher levels of resistance depending on genetic backbone. They could also have either a neutral effect or hypersensitize HIV to ARVs, and I93L is an example of a secondary resistance mutation in subtype B that in subtype C causes hypersusceptibility to PIs [61].

Novel NNRTI resistance mutations in subtype C were not recognized in subtype B. In tissue culture, subtype C can acquire a V106M mutation under NNRTI drug pressure compared to V106A in subtype B. V106M can confer broad cross-resistance to an extent that supersedes that conferred by V106A.

The acquisition of resistance could have important implications in regard to durability of therapy. In culture, the emergence of the K65R mutation is quicker in subtype C than in B [30, 35], and several biochemical mechanisms have been proposed to explain this observation, based on subtype C templates [36–38, 91]. K65R has been seen in approximately 70% of patients failing ddI-containing nucleoside backbones in Botswana [28] but does not appear to emerge frequently in subtype C patients who have received either TDF or TDF/FTC as part of triple therapy [30], a possible reflection of the use of well-tolerated effective drugs that have long mutually reinforcing intracellular half-lives that act in combination to suppress viral replication and prevent the emergence of resistance mutations. Higher numbers of patients and longer followup will be required to determine if there is a consistent impact of subtype C in the emergence of K65R in the clinic.

Multiple *in vitro* and clinical studies have confirmed that PR and Gag can act as a functional unit and coevolve when HIV is subject to drug pressure. Both genes can clearly mutate under PI pressure, and Gag mutations can act as compensatory substitutions that may increase levels of viral replication capacity and resistance. The recombinant phenotyping systems used for clinical samples do not now adequately monitor Gag. While differences among Gag may vary between only −2 to 2.5-fold between subtypes, different subtypes might develop compensatory Gag mutations at different rates, establishing a need to take Gag into account in determining a phenotype. One study reported that a recombinant construct included Gag of clinical origin but did not test the same subtypes as were reported in other work [57].

Although various mutations can impact on drug sensitivity to differential extent, such information cannot yet be generated with regard to non-B subtypes due to a paucity of paired phenotypic and genotypic data. Three studies analyzed genotypes and phenotypes of non-B subtypes in clinical trials: one on use of single dose NVP for prevention of mother-to-child transmission and two on double and triple NRTI combinations that are no longer used [8].

Cross-resistance acquires importance in settings with limited access to antiretroviral therapy, and few *in vitro* comparative data are available for PIs in non-B subtypes. However, such data may be crucial to understanding cross-resistance to specific drugs [58, 59], since some PIs may be the only potentially accessible option for drug sequencing in salvage therapy in many resource-limited settings. The fact that resistance to PIs commonly requires that large numbers of resistance mutations be present may yield a situation in which the individual contribution of any single mutation to drug resistance, with some exceptions, will be limited, a definite advantage of using drugs with a high genetic barrier toward the development of drug resistance. Thus, differences among subtypes with regard to development of drug resistance are more likely to be important for NRTIs and NNRTIs than for PIs. Clearly, large numbers of paired samples need to be systematically collected from naïve and treated patients infected with subtypes C, AE, AG, A, and G, in order for genotypic and phenotypic analysis to be conducted for both established drug classes as well as for newer classes of drugs such as inhibitors of integrase.

Finally, this paper has focused on classes of HIV drugs for which significant datasets are available in regard to subtypes and differential drug resistance. Limitations of both space and available datasets have precluded us from discussing the topic of entry inhibitors. However, most available data suggest that the only two approved entry inhibitors, that is, the fusion inhibitor, enfuvirtide, and the CCR5 entry antagonist, maraviroc, are both active against HIV isolates of multiple subtypes.

Acknowledgments

Work in our lab is supported by the Canadian Institutes of Health Research and by the Canadian Foundation for AIDS Research.

References

[1] K. K. Ariën, G. Vanham, and E. J. Arts, "Is HIV-1 evolving to a less virulent form in humans?" *Nature Reviews Microbiology*, vol. 5, no. 2, pp. 141–151, 2007.

[2] E. A. J. M. Soares, R. P. Santos, J. A. Pellegrini, E. Sprinz, A. Tanuri, and M. A. Soares, "Epidemiologic and molecular characterization of human immunodeficiency virus type 1 in southern Brazil," *Journal of Acquired Immune Deficiency Syndromes*, vol. 34, no. 5, pp. 520–526, 2003.

[3] E. A. J. M. Soares, A. M. B. Martínez, T. M. Souza et al., "HIV-1 subtype C dissemination in southern Brazil," *AIDS*, vol. 19, supplement 4, pp. S81–S86, 2005.

[4] C. A. Brennan, C. Brites, P. Bodelle et al., "HIV-1 strains identified in Brazilian blood donors: significant prevalence of B/F1 recombinants," *AIDS Research and Human Retroviruses*, vol. 23, no. 11, pp. 1434–1441, 2007.

[5] D. Locateli, P. H. Stoco, A. T. L. de Queiroz et al., "Molecular epidemiology of HIV-1 in Santa Catarina State confirms increases of subtype c in southern Brazil," *Journal of Medical Virology*, vol. 79, no. 10, pp. 1455–1463, 2007.

[6] A. Holguín, M. de Mulder, G. Yebra, M. López, and V. Soriano, "Increase of non-B subtypes and recombinants among newly diagnosed HIV-1 native spaniards and immigrants in Spain," *Current HIV Research*, vol. 6, no. 4, pp. 327–334, 2008.

[7] D. Descamps, M. L. Chaix, P. André et al., "French national sentinel survey of antiretroviral drug resistance in patients with HIV-1 primary infection and in antiretroviral-naive chronically infected patients in 2001-2002," *Journal of Acquired Immune Deficiency Syndromes*, vol. 38, no. 5, pp. 545–552, 2005.

[8] B. G. Brenner, "Resistance and viral subtypes: how important are the differences and why do they occur?" *Current Opinion in HIV and AIDS*, vol. 2, no. 2, pp. 94–102, 2007.

[9] R. Kantor, "Impact of HIV-1 pol diversity on drug resistance and its clinical implications," *Current Opinion in Infectious Diseases*, vol. 19, no. 6, pp. 594–606, 2006.

[10] T. D. Toni, B. Masquelier, E. Lazaro et al., "Characterization of nevirapine (NVP) resistance mutations and HIV type 1 subtype in women from Abidjan (Cote d'Ivoire) after NVP single-dose prophylaxis of HIV type 1 mother-to-child transmission," *AIDS Research and Human Retroviruses*, vol. 21, no. 12, pp. 1031–1034, 2005.

[11] S. H. Eshleman, D. R. Hoover, S. Chen et al., "Nevirapine (NVP) resistance in women with HIV-1 subtype C, compared with subtypes A and D, after the administration of single-dose NVP," *Journal of Infectious Diseases*, vol. 192, no. 1, pp. 30–36, 2005.

[12] S. H. Eshleman, J. D. Church, S. Chen et al., "Comparison of HIV-1 mother-to-child transmission after single-dose nevirapine prophylaxis among African women with subtypes A, C, and D," *Journal of Acquired Immune Deficiency Syndromes*, vol. 42, no. 4, pp. 518–521, 2006.

[13] M. L. Chaix, D. K. Ekouevi, F. Rouet et al., "Low risk of nevirapine resistance mutations in the prevention of mother-to-child transmission of HIV-1: Agence Nationale de Recherches sur le SIDA Ditrame Plus, Abidjan, Cote d'Ivoire," *Journal of Infectious Diseases*, vol. 193, no. 4, pp. 482–487, 2006.

[14] J. A. Johnson, J. F. Li, L. Morris et al., "Emergence of drug-resistant HIV-1 after intrapartum administration of single-dose nevirapine is substantially underestimated," *Journal of Infectious Diseases*, vol. 192, no. 1, pp. 16–23, 2005.

[15] T. S. Flys, S. Chen, D. C. Jones et al., "Quantitative analysis of HIV-1 variants with the K103N resistance mutation after single-dose nevirapine in women with HIV-1 subtypes A, C, and D," *Journal of Acquired Immune Deficiency Syndromes*, vol. 42, no. 5, pp. 610–613, 2006.

[16] T. Flys, D. V. Nissley, C. W. Claasen et al., "Sensitive drug-resistance assays reveal long-term persistence of HIV-1 variants with the K103N nevirapine (NVP) resistance mutation in some women and infants after the administration of single-dose NVP: HIVNET 012," *Journal of Infectious Diseases*, vol. 192, no. 1, pp. 24–29, 2005.

[17] J. L. Martínez-Cajas, N. Pant-Pai, M. B. Klein, and M. A. Wainberg, "Role of genetic diversity amongst HIV-1 non-B subtypes in drug resistance: a systematic review of virologic and biochemical evidence," *AIDS Reviews*, vol. 10, no. 4, pp. 212–223, 2008.

[18] D. Descamps, G. Collin, F. Letourneur et al., "Susceptibility of human immunodeficiency virus type 1 group O isolates to antiretroviral agents: in vitro phenotypic and genotypic analyses," *Journal of Virology*, vol. 71, no. 11, pp. 8893–8898, 1997.

[19] E. Tuaillon, M. Gueudin, V. Lemée et al., "Phenotypic susceptibility to nonnucleoside inhibitors of virion-associated reverse transcriptase from different HIV types and groups," *Journal of Acquired Immune Deficiency Syndromes*, vol. 37, no. 5, pp. 1543–1549, 2004.

[20] L. Vergne, J. Snoeck, A. Aghokeng et al., "Genotypic drug resistance interpretation algorithms display high levels of discordance when applied to non-B strains from HIV-1 naive and treated patients," *FEMS Immunology and Medical Microbiology*, vol. 46, no. 1, pp. 53–62, 2006.

[21] R. Gifford, T. de Oliveira, A. Rambaut et al., "Assessment of automated genotyping protocols as tools for surveillance of HIV-1 genetic diversity," *AIDS*, vol. 20, no. 11, pp. 1521–1529, 2006.

[22] S. Y. Rhee, R. Kantor, D. A. Katzenstein et al., "HIV-1 pol mutation frequency by subtype and treatment experience: extension of the HIVseq program to seven non-B subtypes," *AIDS*, vol. 20, no. 5, pp. 643–651, 2006.

[23] V. Novitsky, C. W. Wester, V. DeGruttola et al., "The reverse transcriptase 67N 70R 215Y genotype is the predominant TAM pathway associated with virologic failure among HIV type 1C-infected adults treated with ZDV/ddI-containing HAART in Southern Africa," *AIDS Research and Human Retroviruses*, vol. 23, no. 7, pp. 868–878, 2007.

[24] R. E. Barth, A. M. Wensing, H. A. Tempelman, R. Moraba, R. Schuurman, and A. I. Hoepelman, "Rapid accumulation of nonnucleoside reverse transcriptase inhibitor-associated resistance: evidence of transmitted resistance in rural South Africa," *AIDS*, vol. 22, no. 16, pp. 2210–2212, 2008.

[25] A. Deshpande, V. Jauvin, N. Magnin et al., "Resistance mutations in subtype C HIV type 1 isolates from Indian patients of Mumbai receiving NRTIs plus NNRTIs and experiencing a treatment failure: resistance to AR," *AIDS Research and Human Retroviruses*, vol. 23, no. 2, pp. 335–340, 2007.

[26] M. C. Hosseinipour, J. J. G. van Oosterhout, R. Weigel et al., "The public health approach to identify antiretroviral therapy failure: high-level nucleoside reverse transcriptase inhibitor resistance among Malawians failing first-line antiretroviral therapy," *AIDS*, vol. 23, no. 9, pp. 1127–1134, 2009.

[27] V. C. Marconi, H. Sunpath, Z. Lu et al., "Prevalence of HIV-1 drug resistance after failure of a first highly active antiretroviral therapy regimen in KwaZulu Natal, South Africa," *Clinical Infectious Diseases*, vol. 46, no. 10, pp. 1589–1597, 2008.

[28] F. Doualla-Bell, A. Avalos, B. Brenner et al., "High prevalence of the K65R mutation in human immunodeficiency virus type 1 subtype C isolates from infected patients in Botswana treated with didanosine-based regimens," *Antimicrobial Agents and Chemotherapy*, vol. 50, no. 12, pp. 4182–4185, 2006.

[29] C. Orrell, R. P. Walensky, E. Losina, J. Pitt, K. A. Freedberg, and R. Wood, "HIV type-1 clade C resistance genotypes in treatment-naive patients and after first virological failure in a large community antiretroviral therapy programme," *Antiviral Therapy*, vol. 14, no. 4, pp. 523–531, 2009.

[30] B. G. Brenner and D. Coutsinos, "The K65R mutation in HIV-1 reverse transcriptase: genetic barriers, resistance profile and clinical implications," *HIV Therapy*, vol. 3, no. 6, pp. 583–594, 2009.

[31] D. Turner, E. Shahar, E. Katchman et al., "Prevalence of the K65R resistance reverse transcriptase mutation in different HIV-1 subtypes in Israel," *Journal of Medical Virology*, vol. 81, no. 9, pp. 1509–1512, 2009.

[32] A. Deshpande, A. C. Jeannot, M. H. Schrive, L. Wittkop, P. Pinson, and H. J. Fleury, "Analysis of RT sequences of subtype C HIV-type 1 isolates from indian patients at failure of a first-line treatment according to clinical and/or immunological WHO guidelines," *AIDS Research and Human Retroviruses*, vol. 26, no. 3, pp. 343–350, 2010.

[33] W. Ayele, Y. Mekonnen, T. Messele et al., "Differences in HIV type 1 RNA plasma load profile of closely related cocirculating ethiopian subtype C strains: C and C'," *AIDS Research and Human Retroviruses*, vol. 26, no. 7, pp. 805–813, 2010.

[34] R. Fontella, M. A. Soares, and C. G. Schrago, "On the origin of HIV-1 subtype C in South America," *AIDS*, vol. 22, no. 15, pp. 2001–2011, 2008.

[35] B. G. Brenner, M. Oliveira, F. Doualla-Bell et al., "HIV-1 subtype C viruses rapidly develop K65R resistance to tenofovir in cell culture," *AIDS*, vol. 20, no. 9, pp. F9–F13, 2006.

[36] C. F. Invernizzi, D. Coutsinos, M. Oliveira, D. Moisi, B. G. Brenner, and M. A. Wainberg, "Signature nucleotide polymorphisms at positions 64 and 65 in reverse transcriptase favor the selection of the K65R resistance mutation in HIV-1 subtype C," *Journal of Infectious Diseases*, vol. 200, no. 8, pp. 1202–1206, 2009.

[37] D. Coutsinos, C. F. Invernizzi, H. Xu et al., "Template usage is responsible for the preferential acquisition of the K65R reverse transcriptase mutation in subtype C variants of human immunodeficiency virus type 1," *Journal of Virology*, vol. 83, no. 4, pp. 2029–2033, 2009.

[38] D. Coutsinos, C. F. Invernizzi, H. Xu, B. G. Brenner, and M. A. Wainberg, "Factors affecting template usage in the development of K65R resistance in subtype C variants of HIV type-1," *Antiviral Chemistry and Chemotherapy*, vol. 20, no. 3, pp. 117–131, 2010.

[39] P. R. Harrigan, C. W. Sheen, V. S. Gill et al., "Silent mutations are selected in HIV-1 reverse transcriptase and affect enzymatic efficiency," *AIDS*, vol. 22, no. 18, pp. 2501–2508, 2008.

[40] V. Varghese, E. Wang, F. Babrzadeh et al., "Nucleic acid template and the risk of a PCR-induced HIV-1 drug resistance mutation," *PLoS ONE*, vol. 5, no. 6, Article ID e10992, 2010.

[41] R. T. D'Aquila, A. M. Geretti, J. H. Horton et al., "Tenofovir (TDF)-selected or abacavir (ABC)-selected low-frequency HIV type 1 subpopulations during failure with persistent viremia as detected by ultradeep pyrosequencing," *AIDS Research and Human Retroviruses*, vol. 27, no. 2, pp. 201–209, 2011.

[42] M. Zolfo, J. M. Schapiro, V. Phan et al., "Genotypic impact of prolonged detectable HIV type 1 RNA viral load after HAART failure in a CRF01-AE-infected cohort," *AIDS Research and Human Retroviruses*, vol. 27, no. 7, pp. 727–735, 2011.

[43] R. K. Gupta, I. L. Chrystie, S. O'Shea, J. E. Mullen, R. Kulasegaram, and C. Y. W. Tong, "K65R and Y181C are less prevalent in HAART-experienced HIV-1 subtype A patients," *AIDS*, vol. 19, no. 16, pp. 1916–1919, 2005.

[44] D. M. Tebit, L. Sangaré, A. Makamtse et al., "HIV drug resistance pattern among HAART-exposed patients with sub-optimal virological response in Ouagadougou, Burkina Faso," *Journal of Acquired Immune Deficiency Syndromes*, vol. 49, no. 1, pp. 17–25, 2008.

[45] H. Loemba, B. Brenner, M. A. Parniak et al., "Genetic divergence of human immunodeficiency virus type 1 Ethiopian clade C reverse transcriptase (RT) and rapid development of resistance against nonnucleoside inhibitors of RT," *Antimicrobial Agents and Chemotherapy*, vol. 46, no. 7, pp. 2087–2094, 2002.

[46] B. Brenner, D. Turner, M. Oliveira et al., "A V106M mutation in HIV-1 clade C viruses exposed to efavirenz confers cross-resistance to non-nucleoside reverse transcriptase inhibitors," *AIDS*, vol. 17, no. 1, pp. F1–F5, 2003.

[47] L. Y. Hsu, R. Subramaniam, L. Bacheler, and N. I. Paton, "Characterization of mutations in CRF01_AE virus isolates from antiretroviral treatment-naive and -experienced patients in Singapore," *Journal of Acquired Immune Deficiency Syndromes*, vol. 38, no. 1, pp. 5–13, 2005.

[48] L. Rajesh, R. Karunaianantham, P. R. Narayanan, and S. Swaminathan, "Antiretroviral drug-resistant mutations at baseline and at time of failure of antiretroviral therapy in HIV type 1-coinfected TB patients," *AIDS Research and Human Retroviruses*, vol. 25, no. 11, pp. 1179–1185, 2009.

[49] Z. Grossman, V. Istomin, D. Averbuch et al., "Genetic variation at NNRTI resistance-associated positions in patients infected with HIV-1 subtype C," *AIDS*, vol. 18, no. 6, pp. 909–915, 2004.

[50] M. T. Lai, M. Lu, P. J. Felock et al., "Distinct mutation pathways of non-subtype B HIV-1 during in vitro resistance selection with nonnucleoside reverse transcriptase inhibitors," *Antimicrobial Agents and Chemotherapy*, vol. 54, no. 11, pp. 4812–4824, 2010.

[51] G. Aad, B. Abbott, J. Abdallah et al., "Search for new phenomena in $t\bar{t}$ events with large missing transverse momentum in proton-proton collisions at $\sqrt{s}$ = 7 TeV with the ATLAS detector," *Physical Review Letters*, vol. 108, no. 4, Article ID 041805, 2012.

[52] K. Ariyoshi, M. Matsuda, H. Miura, S. Tateishi, K. Yamada, and W. Sugiura, "Patterns of point mutations associated with antiretroviral drug treatment failure in CRF01_AE (subtype E) infection differ from subtype B infection," *Journal of Acquired Immune Deficiency Syndromes*, vol. 33, no. 3, pp. 336–342, 2003.

[53] M. L. Chaix, F. Rouet, K. A. Kouakoussui et al., "Genotypic human immunodeficiency virus type 1 drug resistance in highly active antiretroviral therapy-treated children in Abidjan, Cote d'Ivoire," *Pediatric Infectious Disease Journal*, vol. 24, no. 12, pp. 1072–1076, 2005.

[54] C. Sukasem, V. Churdboonchart, W. Sukeepaisarncharoen et al., "Genotypic resistance profiles in antiretroviral-naive HIV-1 infections before and after initiation of first-line HAART: impact of polymorphism on resistance to therapy," *International Journal of Antimicrobial Agents*, vol. 31, no. 3, pp. 277–281, 2008.

[55] Z. Grossman, E. E. Paxinos, D. Averbuch et al., "Mutation D30N is not preferentially selected by human immunodeficiency virus type 1 subtype C in the development of resistance to nelfinavir," *Antimicrobial Agents and Chemotherapy*, vol. 48, no. 6, pp. 2159–2165, 2004.

[56] F. Doualla-Bell, A. Avalos, T. Gaolathe et al., "Impact of human immunodeficiency virus type 1 subtype C on drug resistance mutations in patients from Botswana failing a nelfinavir-containing regimen," *Antimicrobial Agents and Chemotherapy*, vol. 50, no. 6, pp. 2210–2213, 2006.

[57] A. T. Dumans, M. A. Soares, E. S. Machado et al., "Synonymous genetic polymorphisms within Brazilian human immunodefidency virus type 1 subtypes may influence mutational routes to drug resistance," *Journal of Infectious Diseases*, vol. 189, no. 7, pp. 1232–1238, 2004.

[58] A. Calazans, R. Brindeiro, P. Brindeiro et al., "Low accumulation of L90M in protease from subtype F HIV-1 with resistance to protease inhibitors is caused by the L89M polymorphism," *Journal of Infectious Diseases*, vol. 191, no. 11, pp. 1961–1970, 2005.

[59] I. Lisovsky, S. M. Schader, J. L. Martinez-Cajas, M. Oliveira, D. Moisi, and M. A. Wainberg, "HIV-1 protease codon 36 polymorphisms and differential development of resistance to nelfinavir, lopinavir, and atazanavir in different HIV-1 subtypes," *Antimicrobial Agents and Chemotherapy*, vol. 54, no. 7, pp. 2878–2885, 2010.

[60] R. O. Soares, P. R. Batista, M. G. S. Costa, L. E. Dardenne, P. G. Pascutti, and M. A. Soares, "Understanding the HIV-1 protease nelfinavir resistance mutation D30N in subtypes B and C through molecular dynamics simulations," *Journal of Molecular Graphics and Modelling*, vol. 29, no. 2, pp. 137–147, 2010.

[61] L. M. F. Gonzalez, R. M. Brindeiro, R. S. Aguiar et al., "Impact of nelfinavir resistance mutations on in vitro phenotype, fitness, and replication capacity of human immunodeficiency virus type 1 with subtype B and C proteases," *Antimicrobial Agents and Chemotherapy*, vol. 48, no. 9, pp. 3552–3555, 2004.

[62] R. Kantor and D. Katzenstein, "Polymorphism in HIV-1 non-subtype b protease and reverse transcriptase and its potential impact on drug susceptibility and drug resistance evolution," *AIDS Reviews*, vol. 5, no. 1, pp. 25–35, 2003.

[63] R. Kantor, D. A. Katzenstein, B. Efron et al., "Impact of HIV-1 subtype and antiretroviral therapy on protease and reverse transcriptase genotype: results of a global collaboration," *PLoS Medicine*, vol. 2, Article ID e112, 2005.

[64] E. A. Soares, A. F. Santos, T. M. Sousa et al., "Differential drug resistance acquisition in HIV-1 of subtypes B and C," *PLoS ONE*, vol. 2, no. 1, article e730, 2007.

[65] E. Sprinz, E. M. Netto, M. Patelli et al., "Primary antiretroviral drug resistance among HIV type 1-infected individuals in Brazil," *AIDS Research and Human Retroviruses*, vol. 25, no. 9, pp. 861–867, 2009.

[66] R. Kantor, R. W. Shafer, and D. Katzenstein, "The HIV-1 Non-subtype B workgroup: an international collaboration for the collection and analysis of HIV-1 non-subtype B data," *MedGenMed*, vol. 7, no. 1, article 71, 2005.

[67] G. S. Gottlieb, N. M. D. Badiane, S. E. Hawes et al., "Emergence of multiclass drug-resistance in HIV-2 in antiretroviral-treated individuals in Senegal: implications for HIV-2 treatment in resouce-limited West Africa," *Clinical Infectious Diseases*, vol. 48, no. 4, pp. 476–483, 2009.

[68] M. L. Ntemgwa, T. D. Toni, B. G. Brenner, R. J. Camacho, and M. A. Wainberg, "Antiretroviral drug resistance in human

immunodeficiency virus type 2," *Antimicrobial Agents and Chemotherapy*, vol. 53, no. 9, pp. 3611–3619, 2009.

[69] M. Sylla, A. Chamberland, C. Boileau et al., "Characterization of drug resistance in antiretroviral-treated patients infected with HIV-1 CRF02_AG and AGK subtypes in Mali and Burkina Faso," *Antiviral Therapy*, vol. 13, no. 1, pp. 141–148, 2008.

[70] M. Kinomoto, R. Appiah-Opong, J. A. M. Brandful et al., "HIV-1 proteases from drug-naive West African patients are differentially less susceptible to protease inhibitors," *Clinical Infectious Diseases*, vol. 41, no. 2, pp. 243–251, 2005.

[71] H. J. Fleury, T. Toni, N. T. H. Lan et al., "Susceptibility to antiretroviral drugs of CRF01_AE, CRF02_AG, and subtype C viruses from untreated patients of Africa and Asia: comparative genotypic and phenotypic data," *AIDS Research and Human Retroviruses*, vol. 22, no. 4, pp. 357–366, 2006.

[72] K. A. Delviks-Frankenberry, G. N. Nikolenko, F. Maldarelli, S. Hase, Y. Takebe, and V. K. Pathak, "Subtype-specific differences in the human immunodeficiency virus type 1 reverse transcriptase connection subdomain of CRF01-AE are associated with higher levels of resistance to 3′azido-3′-deoxythymidine," *Journal of Virology*, vol. 83, no. 17, pp. 8502–8513, 2009.

[73] K. A. Delviks-Frankenberry, G. N. Nikolenko, R. Barr, and V. K. Pathak, "Mutations in human immunodeficiency virus type 1 RNase H primer grip enhance 3′-azido-3′-deoxythymidine resistance," *Journal of Virology*, vol. 81, no. 13, pp. 6837–6845, 2007.

[74] S. H. Yap, C. W. Sheen, J. Fahey et al., "N348I in the connection domain of HIV-1 reverse transcriptase confers zidovudine and nevirapine resistance," *PLoS Medicine*, vol. 4, no. 12, Article ID e335, 2007.

[75] R. K. Gupta, A. Kohli, A. L. McCormick, G. J. Towers, D. Pillay, and C. M. Parry, "Full-length HIV-1 gag determines protease inhibitor susceptibility within in-vitro assays," *AIDS*, vol. 24, no. 11, pp. 1651–1655, 2010.

[76] A. F. Santos, D. M. Tebit, M. S. Lalonde et al., "The role of natural polymorphisms in HIV-1 CRF02_AG protease on protease inhibitor hypersusceptibility," *Antimicrobial Agents and Chemotherapy*, vol. 56, no. 5, pp. 2719–2725, 2012.

[77] T. Bar-Magen, D. A. Donahue, E. I. McDonough et al., "HIV-1 subtype B and C integrase enzymes exhibit differential patterns of resistance to integrase inhibitors in biochemical assays," *AIDS*, vol. 24, no. 14, pp. 2171–2179, 2010.

[78] Y. Goldgur, R. Craigie, G. H. Cohen et al., "Structure of the HIV-1 integrase catalytic domain complexed with an inhibitor: a platform for antiviral drug design," *Proceedings of the National Academy of Sciences of the United States of America*, vol. 96, no. 23, pp. 13040–13043, 1999.

[79] T. Bar-Magen, R. D. Sloan, V. H. Faltenbacher et al., "Comparative biochemical analysis of HIV-1 subtype B and C integrase enzymes," *Retrovirology*, vol. 6, article 103, 2009.

[80] B. G. Brenner, M. Lowe, D. Moisi et al., "Subtype diversity associated with the development of HIV-1 resistance to integrase inhibitors," *Journal of Medical Virology*, vol. 83, no. 5, pp. 751–759, 2011.

[81] E. Z. Loizidou, I. Kousiappa, C. D. Zeinalipour-Yazdi, D. A. M. C. Van de Vijver, and L. G. Kostrikis, "Implications of HIV-1 M group polymorphisms on integrase inhibitor efficacy and resistance: genetic and structural in silico analyses," *Biochemistry*, vol. 48, no. 1, pp. 4–6, 2009.

[82] T. Bar-Magen, R. D. Sloan, D. A. Donahue et al., "Identification of novel mutations responsible for resistance to

MK-2048, a second-generation HIV-1 integrase inhibitor," *Journal of Virology*, vol. 84, no. 18, pp. 9210–9216, 2010.

[83] I. Malet, V. Fourati, C. Charpentier et al., "The HIV-1 integrase G118R mutation confers raltegravir resistance to the CRF02_AG HIV-1 subtype," *Journal of Antimicrobial Chemotherapy*, vol. 66, pp. 2827–2830, 2011.

[84] K. E. Hightower, R. Wang, F. Deanda et al., "Dolutegravir (S/GSK1349572) exhibits significantly slower dissociation than raltegravir and elvitegravir from wild-type and integrase inhibitor-resistant HIV-1 integrase-DNA complexes," *Antimicrobial Agents and Chemotherapy*, vol. 55, pp. 4552–4559, 2011.

[85] P. K. Quashie, T. Mesplede, Y. S. Han et al., "Characterization of the R263K mutation in HIV-1 integrase that confers low-level resistance to the second-generation integrase strand transfer inhibitor dolutegravir," *Journal of Virology*, vol. 86, no. 5, pp. 2696–2705, 2012.

[86] A. U. Scherrer, B. Ledergerber, V. von Wyl et al., "Improved virological outcome in White patients infected with HIV-1 non-B subtypes compared to subtype B," *Clinical Infectious Diseases*, vol. 53, pp. 1143–1152, 2011.

[87] E. A. Soares, A. F. Santos, and M. A. Soares, "HIV-1 subtype and virological response to antiretroviral therapy: acquired drug resistance," *Clinical Infectious Diseases*, vol. 54, pp. 738–739, 2012.

[88] A. Velazquez-Campoy, S. Vega, and E. Freire, "Amplification of the effects of drug resistance mutations by background polymorphisms in HIV-1 protease from African subtypes," *Biochemistry*, vol. 41, no. 27, pp. 8613–8619, 2002.

[89] J. B. Nachega, M. Hislop, D. W. Dowdy, R. E. Chaisson, L. Regensberg, and G. Maartens, "Adherence to nonnucleoside reverse transcriptase inhibitor-based HIV therapy and virologic outcomes," *Annals of Internal Medicine*, vol. 146, no. 8, pp. 564–573, 2007.

[90] N. Richard, M. Juntilla, A. Abraha et al., "High prevalence of antiretroviral resistance in treated Ugandans infected with non-subtype B human immunodeficiency virus type 1," *AIDS Research and Human Retroviruses*, vol. 20, no. 4, pp. 355–364, 2004.

[91] H. T. Xu, J. L. Martinez-Cajas, M. L. Ntemgwa et al., "Effects of the K65R and K65R/M184V reverse transcriptase mutations in subtype C HIV on enzyme function and drug resistance," *Retrovirology*, vol. 6, article 14, 2009.

7

RASSF1A and the Taxol Response in Ovarian Cancer

Susannah Kassler,[1] Howard Donninger,[1] Michael J. Birrer,[2] and Geoffrey J. Clark[1]

[1] *J.G. Brown Cancer Center, University of Louisville, 417 CTR Building, 505 S. Hancock Street, Louisville, KY 40202, USA*
[2] *Massachusetts General Hospital Cancer Center, Massachusetts General Hospital, Harvard Medical School, Boston, MA 02114, USA*

Correspondence should be addressed to Geoffrey J. Clark, geoff.clark@louisville.edu

Academic Editor: Farida Latif

The RASSF1A tumor suppressor gene is frequently inactivated by promoter methylation in human tumors. The RASSF1A protein forms an endogenous complex with tubulin and promotes the stabilization of microtubules. Loss of RASSF1A expression sensitizes cells to microtubule destabilizing stimuli. We have observed a strong correlation between the loss of RASSF1A expression and the development of Taxol resistance in primary ovarian cancer samples. Thus, we sought to determine if RASSF1A levels could dictate the response to Taxol and whether an epigenetic therapy approach might be able to reverse the Taxol resistant phenotype of RASSF1A negative ovarian tumor cells. We found that knocking down RASSF1A expression in an ovarian cancer cell line inhibited Taxol-mediated apoptosis and promoted cell survival during Taxol treatment. Moreover, using a combination of small molecule inhibitors of DNA Methyl Transferase enzymes, we were able restore RASSF1A expression and Taxol sensitivity. This identifies a role for RASSF1A in modulating the tumor response to Taxol and provides proof of principal for the use of epigenetic therapy to overcome Taxol resistance.

1. Introduction

RASSF1A is a poorly understood tumor suppressor that can modulate the cell cycle, tubulin dynamics and apoptosis [1–3]. It is subjected to epigenetic inactivation at high frequency in a broad range of human tumors, including approximately 50% of ovarian tumors [1, 4, 5]. Overexpression of RASSF1A promotes hyperstabilization of microtubules reminiscent of Taxol [6, 7], and previous investigations have shown that loss of RASSF1A sensitizes cells to microtubule destabilizing drugs such as nocodazole [7]. Thus, RASSF1A appears to play an important role in modulating microtubule stabilization. This implies that the RASSF1A levels in a tumor cell may impact how the cell responds to Taxol treatment. The development of resistance to Taxol remains a serious problem in the treatment of ovarian cancer.

The most frequent mechanism by which RASSF1A is inactivated in tumors is by hypermethylation promoter leading to transcriptional silencing [1, 4, 5]. Thus, the gene remains intact, just dormant. Over recent years, a series of small molecules have been identified that can inhibit the DNA methylation system and restore expression of genes that have suffered aberrant promoter methylation [8]. This has given rise to the concept of epigenetic therapy, whereby a tumor would be treated with drugs to restore the expression and function of RASSF1A or some other epigenetically inactivated target. If RASSF1A plays a key role in the response to Taxol, epigenetic therapy could be potentially serve as an approach to overcome the resistance.

In an attempt to address the issue of RASSF1A expression and Taxol resistance, we measured the expression levels of RASSF1A in a series of primary ovarian tumor samples that were characterized for resistance or sensitivity to Taxol. The results showed a very strong correlation between the reduced relative expression of RASSF1A and Taxol resistance in primary ovarian cancer. We then used an shRNA-based approach to generate a matched pair of ovarian tumor cell lines that were positive or negative for RASSF1A expression. In this system, loss of RASSF1A impaired the ability of Taxol to promote microtubule polymerization and rendered the cells resistant to the growth inhibitory effects of Taxol. Using an epigenetic therapy approach, we found that reactivating RASSF1A expression in a RASSF1A-negative ovarian tumor cell line enhanced the sensitivity of the cells to Taxol. Thus we confirm the hypothesis that RASSF1A plays a role in the cellular response to Taxol and provide proof of principal for the use of epigenetic therapy as strategy to address the problem of Taxol resistance ovarian cancer.

2. Materials and Methods

2.1. Tissue Culture. A547 and UCI-107 cells were grown in DMEM/10% FBS. Cells were transfected with shRNA constructs described previously [9] using lipofectamine 2000 (Invitrogen, Carlsbad, CA, USA) using the manufacturers protocol and selected in 1 μg/mL puromycin. Cells were treated with Taxol (Sigma, St. Louis, MI, USA) at the described doses for 48 hours prior to assay. Cell numbers were measured by trypsinization and counting in a haemocytometer. Cells were treated with Zebularine [10] and/or RG108 [11] dissolved in DMSO for 48 hours prior to assay. *t*-tests were used to determine statistical significance.

2.2. Quantitative Real-Time PCR. qRT-PCR analysis was used to evaluate the expression of RASSF1A in primary ovarian tumors essentially as described previously [12] using the following primers to RASSF1A: forward, 5′-GGACGAG-CCTGTGGAGTG-3′, and reverse, 5′-TGATGAAGCCTGT-GTAAGAACC-3′. β-actin was used as the reference gene. Sequences of the β-actin primers have been previously described [13].

2.3. Western Blotting. Cells were lysed in modified RIPA buffer as described previously [14], and subjected to Western analysis using an RASSF1A polyclonal antibody described previously [6]. Tubulin antibodies were purchased from Santa Cruz biochemical (Santa Cruz, CA, USA). Protein concentrations in lysates were measured prior to loading using the Bio-Rad Protein Assay (Bio-Rad, Hercules, CA, USA). Densitometry was performed using a Densitometer and Quantity One software. Values are expressed as adjusted volume Optical Density units/mm^2.

2.4. Caspase Assays. Cells were plated in 12-well plates at 30% confluency and treated with Taxol the next day. 22 hours later cells were lysed and assays with the Caspase-Glo kit (Promega, Madison, WI, USA) as described by the manufacturer.

3. Results

3.1. RASSF1A Downregulation Correlates with Acquisition of Taxol Resistance in Primary Ovarian Tumors. mRNA isolated from the tumors of patients with stage III or IV papillary serous ovarian cancer [12] whose tumors were either responsive or nonresponsive to Taxol were assayed by qRT-PCR for the levels of RASSF1A expression. Ten samples were used for each group and the data expressed as fold change relative to RASSF1A expression in the nonresponder group, after normalization to the expression of β-actin. Those tumors which responded to Taxol showed considerably higher levels of RASSF1A mRNA than those which were resistant (Figure 1).

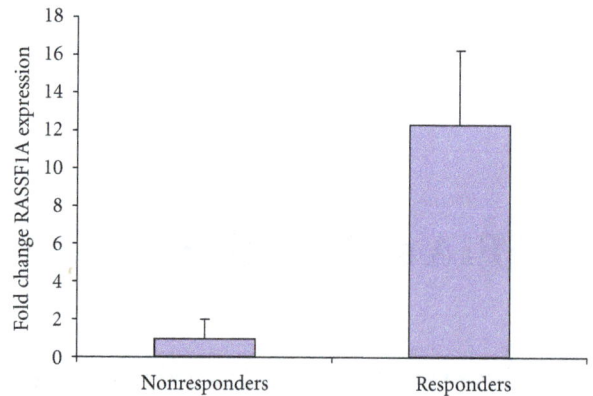

FIGURE 1: RASSF1A downregulation correlates with acquisition of Taxol resistance in primary ovarian tumors: qRT-PCR analysis of primary ovarian tumors correlates loss of RASSF1A expression with the development of Taxol resistance. Left column is relative expression of RASSF1A in Taxol-resistant patients; right column is relative expression in Taxol-sensitive patients. Data is expressed as fold change relative to the nonresponder group after normalization to β-actin expression. *t*-test was used to determine *P* was <.05.

3.2. RASSF1A Knockdown Induces Resistance to Taxol. UCI-107 cells are a Taxol-sensitive ovarian cancer cell line [15]. We transfected the cells with our validated RASSF1A shRNA [9] or the empty vector and generated a stable matched pair by selection in puromycin. The cells were then western blotted for RASSF1A using our polyclonal rabbit antibody [6]. Figure 2(a) shows that RASSF1A expression was effectively knocked down in the shRNA transfected cell line.

The matched pair system was then challenged with Taxol for 48 hours and cell survival measured. Loss of RASSF1A enhanced the survival of the treated cells (Figure 2(b)). RASSF1A is a proapoptotic protein and loss of RASSF1A expression may induce resistance to apoptosis [9]. To determine if that may be the case in ovarian cancer cells treated with Taxol, we then examined the effects of RASSF1A expression on apoptosis after Taxol treatment. The RASSF1A ± UCI-107 cells were treated with Taxol for 22 hours and then assayed for apoptosis using the Promega Caspase 3/7 kit, which is a fluorescent measure of caspase activation. Figure 2(c) shows that downregulation of RASSF1A promotes resistance to apoptosis induced by Taxol. We also observed a very slight reduction in the basal levels of caspase activation in the cells transfected with the RASSF1A shRNA.

3.3. Loss of RASSF1A Reduces the Ability of Taxol to Promote Microtubule Polymerization. RASSF1A binds microtubules and promotes their stabilization/polymerization [6, 7, 16]. Indeed, the effects of overexpressing RASSF1A in cells on tubulin is reminiscent of the effects of treating them with Taxol [6]. Moreover, downregulation of RASSF1A makes cells more sensitive to Nocodazole, a microtubule destabilizing drug [7]. Thus, we hypothesized that the presence of RASSF1A may be important to the ability of Taxol to induce microtubule polymerization. This would confirm RASSF1A loss as a component of the development of Taxol resistance in ovarian cancer and explain the results obtained in Figure 1.

(a)

(b)

(c)

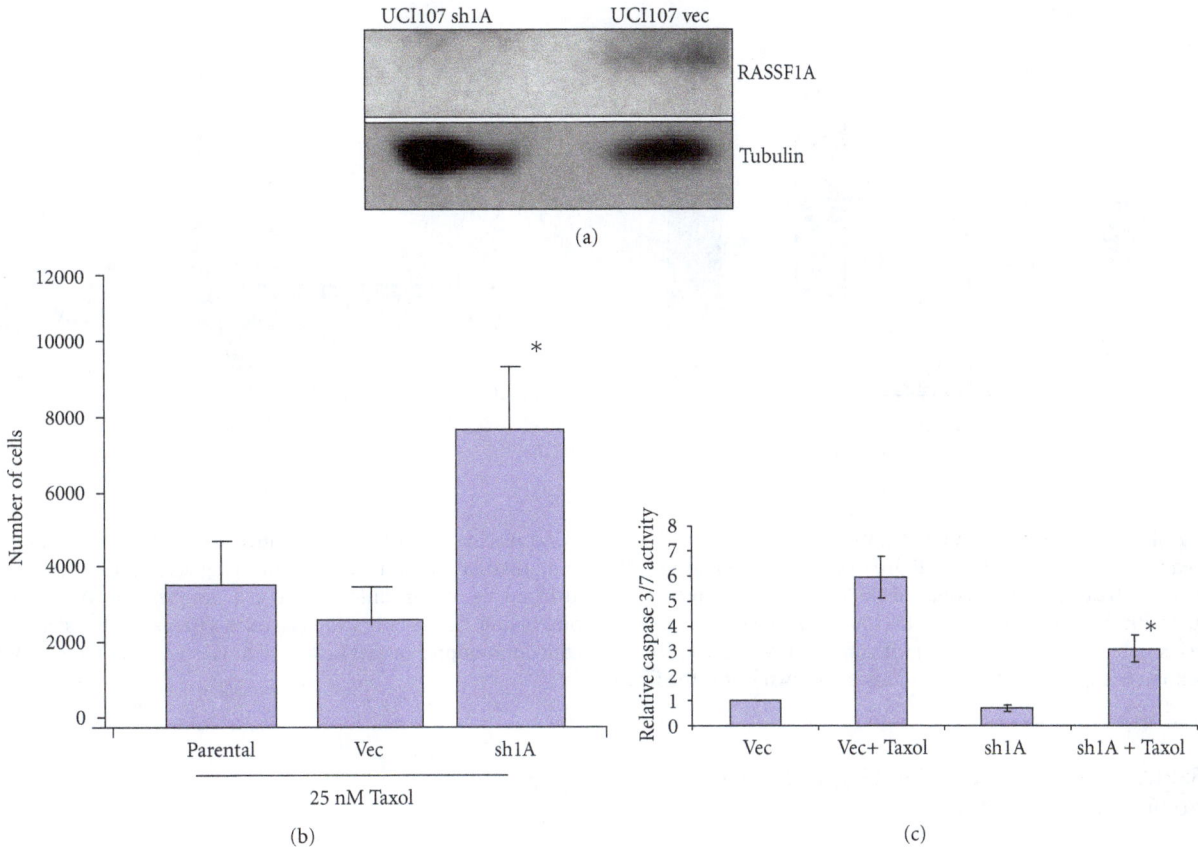

FIGURE 2: Loss of RASSF1A confers resistance to taxol-mediated apoptosis. A matched pair of RASSF1A ± cells was generated by stably knocking down RASSF1A expression in UCI-107 ovarian cancer cells using a RASSF1A-specific shRNA. Knockdown of RASSF1A was confirmed by western blotting. Tubulin served as a loading control (a). The UCI-107 RASSF1A ± cells were grown to 50% confluency and then treated with 25 nM Taxol or vehicle control 48 hours and cell number determined (b). Data represent an average of triplicate experiments, *$P < 0.1$ compared to parental or vector control cells. (c). The RASSF1A ± UCI-107 cells were treated with 25 nM Taxol for 22 hours and caspase activation measured as a readout for apoptosis using a luminescent caspase activation assay. Data represent the average of two assays performed in triplicate. *, statistically different from vector control cells treated with taxol, $P < 0.05$.

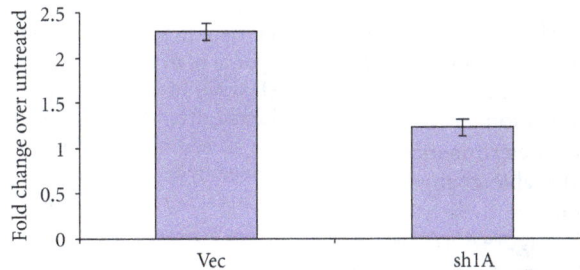

FIGURE 3: The ability of Taxol to promote tubulin acetylation is dependent on RASSF1A. The UCI-107 RASSF1A ± matched pair was treated with Taxol for 48 hours, cell lysates prepared and equal amounts of protein subjected to western blotting using antibodies specific for total or acetylated tubulin. The relevant bands from the western blot were quantified and average data from three experiments expressed as a ratio of acetylated tubulin to total tubulin to give a fold change. Knockdown of RASSF1A resulted in an approximately 50% reduction in the relative acetylation of tubulin, $P = 0.042275$.

When Taxol polymerizes, it becomes acetylated and this has been used as a marker for polymerization [17]. The UCI-107 RASSF1A ± matched pair of cell lines was treated with Taxol. After 48 hours the cells were lysed and equal quantities of protein subjected to Western analysis first for total tubulin and then for acetylated tubulin using an acetylated tubulin specific antibody. The ratio of acetylated tubulin to total tubulin was determined by densitometric scanning of the western blots to permit quantitative assessment of the effects of the presence of RASSF1A. Figure 3 shows that loss of

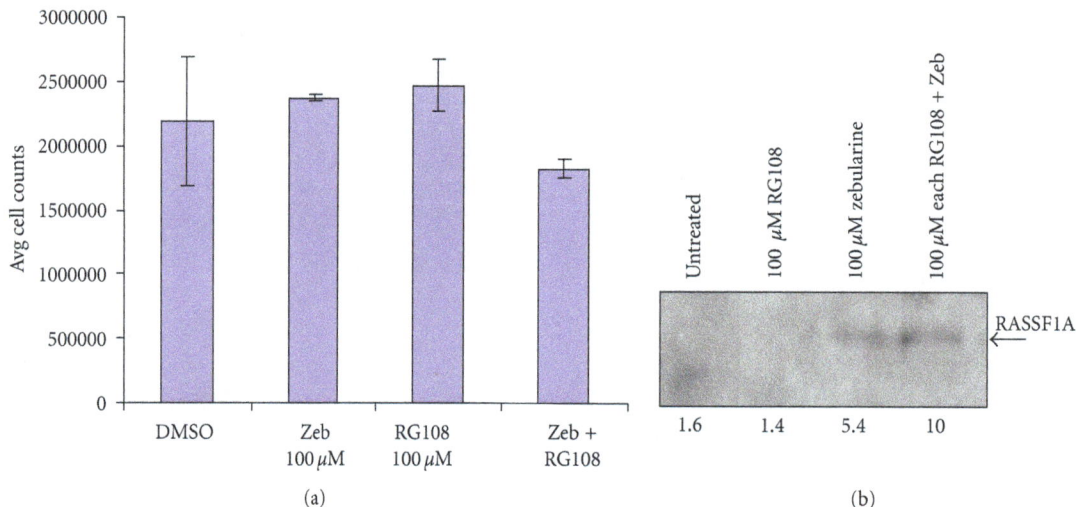

FIGURE 4: Synergistic reactivation of RASSF1A expression by RG108 and Zebularine. (a). RASSF1A negative A547 ovarian cancer cells were treated with DMSO, Zebularine, RG108 or Zebularine and RG108 in combination for 48 hours and surviving cells counted as a measure of toxicity. Treatment with either of the demethylating agents resulted in no significant difference in cell number. (b). A547 cells were treated with the indicated doses of RG108 and Zebularine alone or in combination for 48 hours and cell lysates prepared. Equal amounts of proteins were immunoprecipitated with an anti-RASSF1A antibody and the immunoprecipitates subjected to Western analysis for RASSF1A. Densitometric quantification of the bands is shown below the figure.

RASSF1A expression reduces the ability of Taxol to promote microtubule polymerization.

3.4. Synergistic Restoration of RASSF1A Expression with DNMT Inhibitors.

To examine the possibility that small molecule-induced restoration of RASSF1A expression might affect the cellular response to Taxol, we used the ovarian cancer cell line A547 that is negative for RASSF1A expression and exposed it to treatment with the DNA Methyl Transferase (DNMT) inhibitors Zebularine [10] and RG108 [11]. Zebularine has previously been shown to be active in restoring RASSF1A expression but is more specific and hence less toxic than the first generation DNMT inhibitor 5-AzaC [11, 18]. RG108 is a novel DNMT inhibitor that was designed to specifically inhibit the enzyme DNMT1 [19]. We also used the two in combination. Examination of the toxicity of RG108 and Zebularine allowed the determination of the minimal dose that provoked no detectable changes in cell growth or morphology. Combination of these two doses also resulted in no overt cell death (Figure 4(a)). Western analysis showed that Zebularine was more effective than RG108 at restoring RASSF1A expression but in combination their effects were greater than additive (Figure 4(b)).

3.5. Combined Epigenetic Therapy Restores Taxol Sensitivity.

Having determined that RG108 and Zebularine could act synergistically to restore RASSF1A expression at doses that were too low to induce cell toxicity, we examined the effect of the treatment on the Taxol response of the cells. Figure 5 shows that A547 cells pretreated with the Zebularine/RG108 epigenetic therapy regimen exhibited an enhanced sensitivity to Taxol.

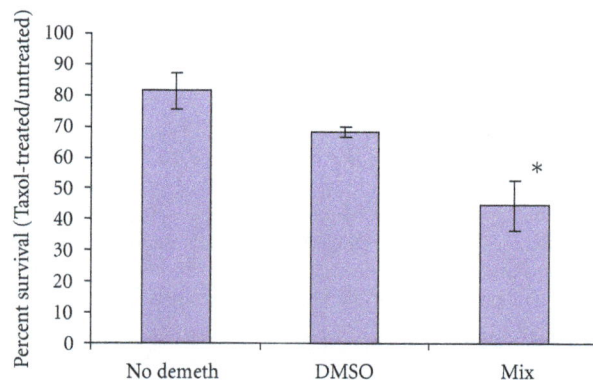

FIGURE 5: Synergistic epigenetic therapy enhances the taxol response of ovarian tumor cells. A547 cells were treated with carrier (DMSO) or a combination of RG108 and Zebularine (mix) for 48 hours, after which 400 nM Taxol was added and the cells incubated for an additional 48 hours. The number of viable cells was determined by trypan blue staining. Data are expressed as percent surviving cells relative to non-Taxol-treated cells for each condition.

4. Discussion

The RASSF1A tumor suppressor is frequently inactivated by an epigenetic process of aberrant promoter methylation in ovarian cancer [1]. RASSF1A complexes with microtubules and enhances their polymerization. Inactivation of RASSF1A results in an increased sensitivity to microtubule destabilizing drugs. Overall, the data suggests that RASSF1A plays an important role in the stabilization of microtubules. As the drug Taxol is thought to work in large part by stabilizing microtubules, we hypothesized that loss of RASSF1A expression might play a role in the development of resistance to Taxol. Our analysis of primary ovarian tumors showed that RASSF1A levels were much lower on average in Taxol resistant tumors. Based on this supporting evidence we proceeded to generate a matched pair of ovarian tumor cell lines that were identical other than for RASSF1A expression. Using this system, we showed that loss of RASSF1A expression caused a significant increase in the resistance of the cells to growth inhibition and apoptosis induction by Taxol.

These data supported the idea that if we could restore RASSF1A expression then we might be able to restore Taxol sensitivity to a tumor cell. Using a combination of demethylating drugs we were able to restore RASSF1A expression. These drugs, RG108 and Zebularine, appear much less toxic than the established demethylating drug 5-Aza-C, even when used in combination (unpublished observation, G. Clark). The cells with restored RASSF1A expression proved much more sensitive to Taxol. Thus, we provide proof of principle for the use of epigenetic therapy to overcome Taxol resistance in ovarian cancer. Moreover, the methylation of the RASSF1A promoter might serve as a predictive marker for the effectiveness of Taxol based therapy.

These studies focused on the role of RASSF1A in the Taxol response because of the apparent role of RASSF1A in supporting microtubule polymerization. However, RASSF1A has a general role in apoptosis and has now been shown to play a role in DNA repair. Thus, RASSF1A restoration might also be expected to enhance the effects of drugs which act by inducing apoptosis and DNA damage. Indeed, Zebularine has been shown to enhance the effects of Cisplatin in ovarian cancer models [20].

In these studies, we used Zebularine and RG108 as DNMT inhibitors. As they have different mechanisms of action, we hypothesized that they might have a synergistic activity. This would appear to be the case. As better agents arise that are more specific, for example Nanaomycin [21], the effectiveness and practicality of this strategy is likely to increase.

RASSF1A exhibits an SNP, which is present in excess of 20% of the Caucasian population. This SNP produces a variant protein where Alanine 133 is substituted for a serine. The A(133)S variant protein is defective for interacting with certain isoforms of tubulin [22] and is defective for binding the microtubule association protein MAP1a [23]. Mutations close to this SNP can impair the ability of RASSF1C to promote microtubule polymerization [6]. Thus, it may be interesting to determine if the presence of this SNP may also affect the response of an individual to Taxol treatment.

Acknowledgments

This work was supported by USAMRMC Grant W81XWH-07-OCRP-CA (G. Clark) and NCI intramural funds (M. Birrer).

References

[1] H. Donninger, M. D. Vos, and G. J. Clark, "The RASSF1A tumor suppressor," *Journal of Cell Science*, vol. 120, no. 18, pp. 3163–3172, 2007.

[2] R. Dammann, U. Schagdarsurengin, C. Seidel et al., "The tumor suppressor RASSF1A in human carcinogenesis: an update," *Histology and Histopathology*, vol. 20, no. 2, pp. 645–663, 2005.

[3] G. P. Pfeifer, R. Dammann, and S. Tommasi, "RASSF proteins," *Current Biology*, vol. 20, no. 8, pp. R344–R345, 2010.

[4] L. B. Hesson, W. N. Cooper, and F. Latif, "The role of RASSF1A methylation in cancer," *Disease Markers*, vol. 23, no. 1-2, pp. 73–87, 2007.

[5] A. Agathanggelou, W. N. Cooper, and F. Latif, "Role of the Ras-association domain family 1 tumor suppressor gene in human cancers," *Cancer Research*, vol. 65, no. 9, pp. 3497–3508, 2005.

[6] M. D. Vos, A. Martinez, C. Elam et al., "A role for the RASSF1A tumor suppressor in the regulation of tubulin polymerization and genomic stability," *Cancer Research*, vol. 64, no. 12, pp. 4244–4250, 2004.

[7] L. Liu, S. Tommasi, D. H. Lee, R. Dammann, and G. P. Pfeifer, "Control of microtubule stability by the RASSF1A tumor suppressor," *Oncogene*, vol. 22, no. 50, pp. 8125–8136, 2003.

[8] J. Ren, B. N. Singh, Q. Huang et al., "DNA hypermethylation as a chemotherapy target," *Cellular Signalling*, vol. 23, no. 7, pp. 1082–1093, 2011.

[9] M. D. Vos, A. Dallol, K. Eckfeld et al., "The RASSF1A tum-or suppressor activates bax via MOAP-1," *The Journal of Biological Chemistry*, vol. 281, no. 8, pp. 4557–4563, 2006.

[10] V. E. Marquez, J. A. Kelley, R. Agbaria et al., "Zebularine: a unique molecule for an epigenetically based strategy in cancer chemotherapy," *Annals of the New York Academy of Sciences*, vol. 1058, pp. 246–254, 2005.

[11] C. Stresemann, B. Brueckner, T. Musch, H. Stopper, and F. Lyko, "Functional diversity of DNA methyltransferase inhibitors in human cancer cell lines," *Cancer Research*, vol. 66, no. 5, pp. 2794–2800, 2006.

[12] H. Donninger, T. Bonome, M. Radonovich et al., "Whole genome expression profiling of advance stage papillary serous ovarian cancer reveals activated pathways," *Oncogene*, vol. 23, no. 49, pp. 8065–8077, 2004.

[13] K. A. Kreuzer, U. Lass, A. Bohn, O. Landt, and C. A. Schmidt, "Lightcycler technology for the quantitation of bcr/abl fusion transcripts," *Cancer Research*, vol. 59, no. 13, pp. 3171–3174, 1999.

[14] H. Donninger, N. P. Allen, A. Henson et al., "Salvador protein is a tumor suppressor effector of RASSF1A with hippo pathway-independent functions," *The Journal of Biological Chemistry*, vol. 286, no. 21, pp. 18483–18491, 2011.

[15] G. Gamboa, P. M. Carpenter, Y. D. Podnos et al., "Characterization and development of UCl 107, a primary human ovarian carcinoma cell line," *Gynecologic Oncology*, vol. 58, no. 3, pp. 336–343, 1995.

[16] A. Dallol, A. Agathanggelou, S. L. Fenton et al., "RASSF1A interacts with microtubule-associated proteins and modulates microtubule dynamics," *Cancer Research*, vol. 64, no. 12, pp. 4112–4116, 2004.

[17] A. Matsuyama, T. Shimazu, Y. Sumida et al., "In vivo destabilization of dynamic microtubules by HDAC6-mediated deacetylation," *The EMBO Journal*, vol. 21, no. 24, pp. 6820–6831, 2002.

[18] C. B. Yoo, J. C. Cheng, and P. A. Jones, "Zebularine: a new drug for epigenetic therapy," *Biochemical Society Transactions*, vol. 32, no. 6, pp. 910–912, 2004.

[19] B. Brueckner, R. G. Boy, P. Siedlecki et al., "Epigenetic reactivation of tumor suppressor genes by a novel small-molecule inhibitor of human DNA methyltransferases," *Cancer Research*, vol. 65, no. 14, pp. 6305–6311, 2005.

[20] C. Balch, P. Yan, T. Craft et al., "Antimitogenic and chemosensitizing effects of the methylation inhibitor zebularine in ovarian cancer," *Molecular Cancer Therapeutics*, vol. 4, no. 10, pp. 1505–1514, 2005.

[21] D. Kuck, T. Caulfield, F. Lyko, and J. L. Medina-Franco, "Nanaomycin A selectively inhibits DNMT3B and reactivates silenced tumor suppressor genes in human cancer cells," *Molecular Cancer Therapeutics*, vol. 9, no. 11, pp. 3015–3023, 2010.

[22] M. El-Kalla, C. Onyskiw, and S. Baksh, "Functional importance of RASSF1A microtubule localization and polymorphisms," *Oncogene*, vol. 29, no. 42, pp. 5729–5740, 2010.

[23] H. Donninger, T. Barnoud, N. Nelson et al., "RASSF1A and the rs2073498 cancer associated SNP," *Frontiers in Cancer Genetics*, vol. 1, article 54, 2011.

8

ASGR1 and *ASGR2*, the Genes that Encode the Asialoglycoprotein Receptor (Ashwell Receptor), Are Expressed in Peripheral Blood Monocytes and Show Interindividual Differences in Transcript Profile

Rebecca Louise Harris, Carmen Wilma van den Berg, and Derrick John Bowen

Institute of Molecular and Experimental Medicine, Cardiff University School of Medicine, Heath Park, Cardiff CF14 4XN, UK

Correspondence should be addressed to Derrick John Bowen, bowendj1@cardiff.ac.uk

Academic Editor: Mouldy Sioud

Background. The asialoglycoprotein receptor (ASGPR) is a hepatic receptor that mediates removal of potentially hazardous glycoconjugates from blood in health and disease. The receptor comprises two proteins, asialoglycoprotein receptor 1 and 2 (ASGR1 and ASGR2), encoded by the genes *ASGR1* and *ASGR2*. *Design and Methods*. Using reverse transcription amplification (RT-PCR), expression of *ASGR1* and *ASGR2* was investigated in human peripheral blood monocytes. *Results*. Monocytes were found to express *ASGR1* and *ASGR2* transcripts. Correctly spliced transcript variants encoding different isoforms of ASGR1 and ASGR2 were present in monocytes. The profile of transcript variants from both *ASGR1* and *ASGR2* differed among individuals. Transcript expression levels were compared with the hepatocyte cell line HepG2 which produces high levels of ASGPR. Monocyte transcripts were 4 to 6 orders of magnitude less than in HepG2 but nonetheless readily detectable using standard RT-PCR. The monocyte cell line THP1 gave similar results to monocytes harvested from peripheral blood, indicating it may provide a suitable model system for studying ASGPR function in this cell type. *Conclusions*. Monocytes transcribe and correctly process transcripts encoding the constituent proteins of the ASGPR. Monocytes may therefore represent a mobile pool of the receptor, capable of reaching sites remote from the liver.

1. Introduction

The asialoglycoprotein receptor (ASGPR) (also known as the Ashwell receptor) mediates the capture and endocytosis of galactose- (Gal) and N-acetylgalactosamine- (GalNAc) terminating glycoproteins. The relevance of this function has been the subject of much debate; the primary role may be the removal of potentially hazardous glycoconjugates arising from normal tissue turnover, tissue injury, disease, and other causes [1]. Studies using knock-out mice have provided evidence for direct involvement of the ASGPR in removal of abnormally sialylated plasma glycoproteins. Mice lacking the sialyltransferase ST3Gal4 showed prolonged bleeding which was attributed to ASGPR-mediated clearance of at least one plasma hemostatic component, von Willebrand factor

(VWF), that showed decreased sialylation [2]. Findings in mice lacking the ASGPR demonstrated that, during sepsis, the receptor removed components of hemostasis (VWF and platelets) that had been desialylated by bacterial neuraminidase and thereby allowed hemostatic adaptation that moderated disseminated intravascular coagulation and improved host survival [3]. The ASGPR may therefore be poised for rapid clearance of plasma glycoproteins that, for whatever reason, show decreased or abnormal sialylation.

The highest level of expression of the ASGPR is in the liver, in which it is located on the sinusoidal face of hepatocytes [4]. Low-level expression has been shown in various other cell types, such as, peritoneal macrophages in rat [5], human intestinal epithelial cells [6], mouse testis [7], and the rat thyroid gland [8]. The role of the receptor at these extrahepatic sites is uncertain. The galactosyl homeostasis theory proposes that the balance of both Gal and GalNAc glycoconjugates is important for normal tissue physiology [1, 9]; whether or not expression of the ASGPR at these other sites represents a need for localized constitutive control of such glycoconjugates is not known.

A major area of interest focussing on the ASGPR is the targeted gene transfer/delivery of drugs to the liver [10, 11]. For example, using GalNAc as a ligand on a siRNA-containing complex, the latter was successfully targeted to hepatocytes after simple intravenous injection into the tail vein in mice [11]. Moreover, siRNA-mediated knock-down of targeted genes was demonstrated within the hepatocytes. Besides its potential importance in allowing liver-specific drug targeting [12], the ASGPR is a key factor in the design and administration of glycoprotein pharmaceuticals more generally: the activity of the receptor can impact upon drug half life and thereby the window of therapeutic efficacy.

Human ASGPR is composed of two types of subunit: a major subunit (asialoglycoprotein receptor 1, ASGR1) and minor subunit (asialoglycoprotein receptor 2, ASGR2) [13]. Both subunits are type II, single pass proteins that broadly comprise a cytoplasmic domain, transmembrane domain, and extracellular carbohydrate recognition domain (CRD) [14, 15]. The subunits may exist as ASGR1-ASGR2 heterooligomers, ASGR1homotrimers and homotetramers, and ASGR2 homodimers and homotetramers. These different quaternary forms may allow for functional differences, such as, substrate specificity or rate of endocytosis [1, 16].

The genes encoding ASGR1 and ASGR2 (*ASGR1* and *ASGR2*, resp.) are located on the short arm of autosome 17, approximately 58.6 kilobases (kb) apart. The genes are evolutionarily related but differ significantly in their structural organization: *ASGR1* comprises 8 exons and is approximately 6 kb long, *ASGR2* contains 9 exons occupying 13.5 kb of DNA [14, 15].

Until recently, *ASGR1* was thought to yield one transcript encoding a single protein. However, in 2010, Liu et al. demonstrated the existence of two alternatively spliced transcripts. The longer transcript, T1, contains all 8 exons and is by far the more abundant; it encodes full-length ASGR1 (isoform a, 291 aminoacids). The shorter transcript, T2, has an in-frame deletion of exon 3 resulting in the loss of 39 residues (isoform b, 252 amino acids). Isoform b lacks the transmembrane domain and is secreted as a soluble protein [17].

ASGR2 gives rise to five transcripts (TH2′, T1, T2, T3, and T4) encoding four isoforms (a to d) that contain different in-frame deletions arising from alternative exon splicing events [15]. Isoforms a and c contain 5 aminoacids that serve as a proteolysis cleavage signal near the junction between the transmembrane domain and the CRD [18]. Cleavage at this site results in secretion of the CRD as a soluble protein [18, 19]. Isoforms b and d lack this signal therefore they are not proteolytically cleaved but rather remain membrane bound where they may oligomerize with ASGR1 isoform a to form native ASGPR at the cell surface.

The secreted forms of ASGR1 and ASGR2 are able to associate into soluble ASGPR [17, 19]. There is evidence to suggest that the soluble receptor may bind free substrates in the circulation and carry them to the liver for uptake and degradation [17]. Aside from membrane attachment/secretion, it is not known whether further functional variation, such as, substrate specificity, may be associated with the different isoforms of ASGR1 and ASGR2.

Based on the observation above that the ASGPR is expressed in rat peritoneal macrophages, it was considered possible that monocytes, the lineage precursor of tissue macrophages, may also express the receptor. To the best of our knowledge, this has not been reported in human monocytes. The present study investigated the expression of *ASGR1* and *ASGR2* in human peripheral blood monocytes. The data showed that both *ASGR1* and *ASGR2* are expressed in human monocytes in the circulation, that expression cannot be detected in lymphocytes and granulocytes, that the transcripts of both genes differ between individuals and that, in a given individual, the transcription profile for *ASGR2* is restricted to one of two patterns. The findings are of potential importance for health and disease in a variety of disciplines.

2. Design and Methods

2.1. Nomenclature. Genes, transcripts, and proteins are referred to by their formal scientific names as listed in the National Centre for Biotechnology Information (NCBI) [20], which complies with standardized international nomenclature. Table 1 provides a comparison with alternative names widely used in ASGPR literature. Gene sizes, exon numbering, and so forth were obtained from the reference sequences listed in Table 1.

2.2. Blood Samples. EDTA-anticoagulated whole blood in excess of that required for routine blood tests was anonymized and handled in accordance with Medical Research Council guidelines [21]. The study was reviewed by the local research ethics committee. All samples had normal haematological parameters.

2.3. Separation of White Blood Cells

2.3.1. Stage 1. Depletion of platelets. Citrate- or EDTA-anticoagulated whole blood (5 mL) was centrifuged at 200× g for 10 minutes, ambient temperature to prepare platelet-rich plasma (PRP). Two-thirds of the volume of the PRP was carefully removed from above the buffy coat, centrifuged at 1800× g for 10 minutes to pellet the platelets, and then the supernatant plasma returned to the original blood sample. This process was repeated three times.

2.3.2. Stage 2. Separation of cells. After the depletion of platelets as described above, the blood sample was loaded onto a discontinuous Histopaque (Sigma Aldrich, Dorset, UK) gradient comprising Histopaque 1119 (2 mL, lower layer) and Histopaque 1077 (3 mL, upper layer). Following centrifugation at 800× g for 1 h at ambient temperature, the cells at the interface between the plasma and Histopaque 1077 (monocytes and lymphocytes) were harvested; the cells

TABLE 1: ASGR1 and ASGR2 nomenclature and reference sequences.

Gene and transcript (NCBI)	Protein and isoform (NCBI)	Protein and isoform alternative name	Gene and transcript reference sequences (NCBI)	Protein and isoform reference sequences (NCBI)
ASGR1	ASGR1	H1	AC_000060.1	not applicable
T1	a	H1a	NM_001671.4	NP_001662.1
T2	b	H1b	NM_001197216.2	NP_001184145.1
ASGR2	ASGR2	H2		
T1	a	H2a	NG_029064.1	not applicable
T2	b	H2c (L-H2)	NM_001181.4	NP_001172.1
T3	c	None	NM_080913.3	NP_550435.1
T4	d	H2b	NM_080914.2	NP_550436.1
TH2′	a	H2a	NM_001201352.1NM_080912.3	NP_001188281.1NP_001172.1

NCBI abbreviates National Centre for Biotechnology Information [20].

at the interface between Histopaque 1077 and 1119 (principally granulocytes with a small proportion of lymphocytes) were also harvested.

Harvested cells were washed by resuspension in 10 mL phosphate buffered saline pH 7.2 (PBS) followed by centrifugation at 250× g for 10 minutes at ambient temperature. The cell pellet was resuspended in PBS (0.7 mL) and an aliquot (0.2 mL) was diluted 1 : 1 (v/v) with PBS and used for determination of cell counts on a Pentra 120 (Horiba ABX, Montpellier, France).

2.4. Cell Sorting Using Flow Cytometry.

Citrate- or EDTA-anticoagulated whole blood (5 mL) was depleted of platelets as described above (Separation of White Blood Cells). The blood sample (3 mL) was then loaded onto Histopaque 1077 (3 mL) and centrifuged to separate the white blood cells (WBC) (400× g, 30 min, ambient temperature). The WBC interface was harvested, washed, and an aliquot used for determination of cell counts as described above (Separation of White Blood Cells).

The remaining cells, suspended in PBS, were separated according to forward scatter and side scatter light characteristics using a MoFlo high speed cell sorter (Beckman Coulter, High Wycombe, Bucks, UK). Gating was restrictive rather than permissive: each gate was set only for the core population of cells that had the required scatter characteristics. In particular, the lymphocyte gate was set to minimise the possibility of small monocytes being harvested. Purity checks were subsequently performed by reanalysis of aliquots of the sorted cells in the flow cytometer; purity was typically around 98%.

2.5. Cell Culture.

All reagents and consumables for cell culture were supplied by Invitrogen Life Technologies Ltd, Paisley, Ireland. HepG2 (human hepatocyte carcinoma) [22] and THP1 (human acute monocytic leukemia) [23] cells were cultured in RPMI1640 medium + L-Glutamine, supplemented with 10% (v/v) fetal calf serum and containing penicillin (50 units/mL) and streptomycin (50 μg/mL). Cells were grown at 37°C in water-saturated air (95%, v/v), CO_2 (5%, v/v) in 25 cm^2 flasks. HepG2 cells (adherent cell line)

were liberated from the flask wall mechanically using a plastic scraper and were separated by exposure to shear force. THP1 cells (nonadherent) were harvested directly from the culture medium. Cells were counted using FastRead 10 disposable hemocytometers (Immune Systems, UK). HepG2 has been shown by others to express high levels of the Ashwell receptor [24] and was used as a positive control in all relevant experiments.

2.6. RNA Extraction and Reverse Transcription.

RNA was extracted using RNeasy kits (Qiagen, West Sussex, UK) according to the manufacturer's instructions. Yield was determined spectrophotometrically and then 1.0 μg was reverse transcribed into cDNA using random hexamers and the High Capacity cDNA Reverse Transcription kit (Life Technologies Ltd/Applied Biosystems, Paisley, UK) according to the manufacturer's instructions.

2.7. Polymerase Chain Reaction.

Standard polymerase chain reactions (PCRs) typically contained approximately 30 ng cDNA in a mixture of dNTPs (100 μmol/L each), Tris-HCl pH8.0 (10 mmol/L), $MgCl_2$ (1.5 mmol/L), *Taq* DNA polymerase (1U) (Applied Biosystems, Warwickshire, UK), and primers (Figure 1) in a final volume of 25 μL. All primers were used at 0.5 μmol/L. PCR was done using a 2720 DNA Thermal Cycler (Applied Biosystems, Warwickshire, UK).

The primer sequences (5′ to 3′) were as follows:

ASG1RTF gaaagatgaagtcgctagagt

ASG1RTR aggctccgcaggtcagacac

A1CDSF1 <u>gtagcgcgacggcc</u>agtactgaagaacctgggaatcagac

A1CDSR1 <u>cagggcgcagcgatg</u>acagctcctcaccttcggaacatca

A1TEST2 gaccaaggagtatcaagacctt

ASG2RTF cacacctggtggtcatcaac

ASG2RTR aattatctggctgagtgacag

ASGR2RTF3 agctgagctcggaggaaaatg

ASGR2RTR2 gcagctcggcttgcagctgtg

A2CDSF1 <u>gtagcgcgacggcc</u>agtcccagccctcagagcaacctca

A2CDSR1 <u>cagggcgcagcgatgactc</u>aacagagaagccagagctg-
gg

B2MRTF tccgtggccttagctgtgct

B2MRTR ccagtccttgctgaaagaca

N13F gtagcgcgacggccagt

N13R cagggcgcagcgatgac

Primers were designed for RT-PCR according to the following criteria: (1) they flanked at least one intron; (2) they were specific for *ASGR1* or *ASGR2* transcripts, cross hybridization was not possible despite the sequence homology between the coding sequences of the two genes. N13F and N13R are, respectively, modified M13 universal and reverse sequencing primers as previously described [25]. Underlined nucleotides are not part of *ASGR1* or *ASGR2* sequence but are tails corresponding to N13F or N13R to facilitate sequence analysis.

For nested PCR, first-round synthesis was as above. The first-round product was then diluted 10-fold with water and 1 μL of diluent was used as the substrate in the second round of PCR.

PCR conditions were as follows:

Standard PCR: 94°C for 30 s; 60°C for 30 s; 72°C for 1 min.

Nested PCR: 1st round 94°C for 30 s; 60°C for 30 s; 72°C for 1 min 30 s.

Nested PCR: 2nd round 94°C for 30 s; 60°C for 30 s; 72°C for 1 min.

Routinely, 30 cycles of amplification were used; however, for real time PCR, 55 cycles were employed.

2.8. Real-Time Quantitative PCR.

ASGR1 and *ASGR2* transcript levels were measured using LightCycler FastStart DNA Master SYBR Green I (Roche Products Ltd, Hertfordshire, UK) according to the manufacturer's instructions, with 3 mmol/L MgCl$_2$ final concentration in the reaction. Reactions contained 2 ng cDNA per 20 μL and relevant primers (Figure 1). Primers were used at 0.5 μmol/L. Analyses were done in duplicate in a LightCycler II using LightCycler v4.0 software (Perkin Elmer-Applied Biosystems, Warwickshire, UK) using the following conditions: 94°C for 10 min followed by 55 cycles comprising 94°C for 10 s; 64°C for 5 s; 72°C for 10 s.

2.9. Nucleotide Sequence Analysis.

PCR products were purified using the HighPure kit (Boehringer Mannheim, Nottingham, UK) and then sequenced using BigDye3.1 according to the manufacturer's instructions (Perkin Elmer-Applied Biosystems, Warrington, UK). Sequencing primers were either the 5′ PCR primer, 3′ PCR primer, N13F or N13R according to the amplification product. Sequencing reaction products were purified using QIAquick PCR purification columns (Qiagen, West Sussex, UK) and then electrophoresed on an Applied Biosystems 3130XL Genetic analyser.

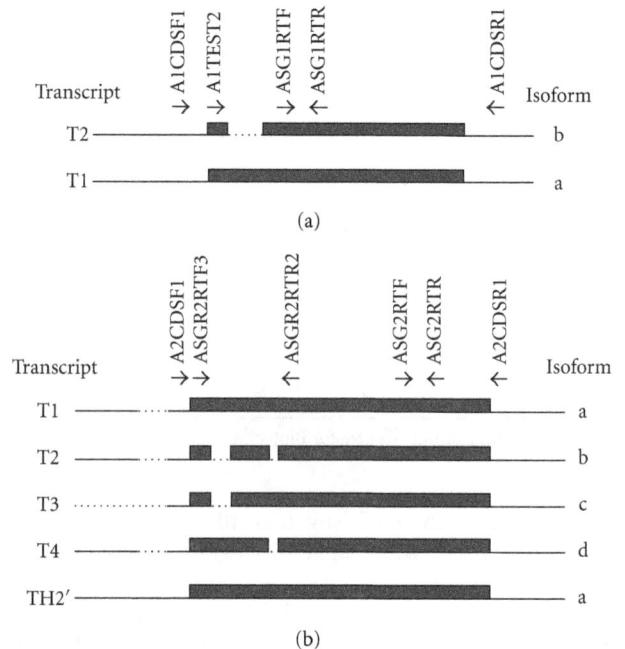

Figure 1: *ASGR1* and *ASGR2* transcripts and encoded protein isoforms. (a) *ASGR1*. (b) *ASGR2*. In both panels solid lines indicate nucleotide sequence that is present in a transcript, grey boxes represent coding sequence, dotted lines indicate sequence that is not present in a transcript. Arrows indicate location and direction of primers used in RT-PCR. Data for transcripts and isoforms are taken from NCBI [14, 15].

2.10. Gel Electrophoresis of PCR Products.

Agarose gel electrophoresis: the analyses used 2% (w/v) gels and 1x TBE buffer (tris (90 mmol/L), boric acid (90 mmol/L), EDTA (1.25 mmol/L), pH 8.0); visualization was done using ethidium bromide staining and UV light.

Polyacrylamide gel electrophoresis: gels comprised polyacrylamide (total acrylamide = 5%, w/v; cross link = 3.3%, w/v) and 1x TBE buffer, visualization was done using silver staining [26].

The DNA size standard pBR322/*Msp*I (New England Biolabs, Hertfordshire, UK) was used for all electrophoretic analyses.

3. Results

3.1. ASGR1 and ASGR2 Are Expressed in Peripheral Blood Monocytes.

Initially, peripheral blood mononuclear cells (PBMCs) were harvested and screened for *ASGR1* and *ASGR2* expression without fractionation into monocytes/ lymphocytes/granulocytes. RT-PCR of RNA extracted from PBMCs demonstrated the presence of both *ASGR1* and *ASGR2* transcripts (Figure 2(a)). These were also detected in the monocyte cell line THP1 (Figure 2(a)).

Following these initial findings, PBMCs were fractionated in order to localize the cells in which *ASGR1* and *ASGR2* expression was present. Histopaque gradients allowed separation of cells into monocytes + lymphocytes (M + L) and

ASGR1 and ASGR2, the Genes that Encode the Asialoglycoprotein Receptor (Ashwell Receptor), Are Expressed in Peripheral Blood Monocytes and Show Interindividual Differences in Transcript Profile

65

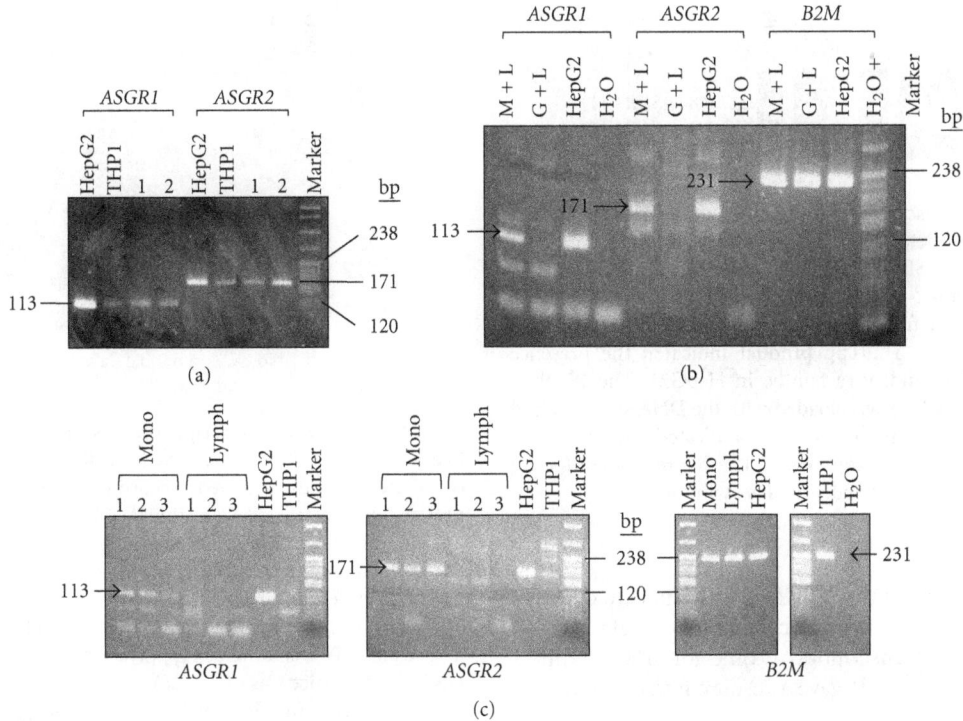

FIGURE 2: Detection of *ASGR1* and *ASGR2* transcripts in peripheral blood mononuclear cells and localization to monocytes. Transcripts were detected using RT-PCR, the presence of *ASGR1* transcripts was indicated by a 113 bp amplification product (primers ASG1RTF + ASG1RTR) and *ASGR2* transcripts by a 171 bp product (primers ASG2RTF + ASG2RTR). HepG2 was used as a positive control and water as a negative control. Where necessary, β-2-microglobulin (*B2M*) was used as a reference to demonstrate equivalent amplification for each RNA preparation, (primers B2MRTF + B2MRTR, product 231 bp). (a) Analysis of RNA from monocyte cell line THP1 and PBMCs from two unrelated individuals (1 and 2). *ASGR1* and *ASGR2* transcripts were detected in all cases. (b) Analysis of RNA from peripheral blood cell fractions: monocytes + lymphocytes (M + L) and granulocytes + lymphocytes (G + L). *ASGR1* and *ASGR2* transcripts were detected in M + L but not in G + L. (c) Analysis of RNA from cell-sorted monocytes (mono, lanes 1–3 triplicate analyses) and lymphocytes (lymph, lanes 1–3 triplicate analyses) (data for water control not shown for *ASGR1* and *ASGR2*). The data were obtained from a single experiment involving several gels: these are indicated by separate gel windows. For each panel, analysis was done using agarose gel electrophoresis with pBR322/*Msp*I size standard ("Marker", 238 bp and 120 bp indicated).

granulocytes + lymphocytes (G + L). These cell preparations typically contained 10% monocytes/90% lymphocytes (M + L) and 94% granulocytes/10% lymphocytes (G + L) (data not shown).

Using RT-PCR, *ASGR1* and *ASGR2* transcripts were detected in RNA from the M + L fraction but not in RNA from the G + L fraction (Figure 2(b)). These findings were reproducible for blood samples taken from 3 different individuals (data not shown). The G + L fraction did not yield a product for the transcripts even after 50 cycles of RT-PCR amplification (Figure 2(b)). Some nonspecific products were obtained as a result of this high number of cycles; however, these did not interfere with the main result of the experiment (Figure 2(b)).

Because lymphocytes were common to both cell fractions, and because the G + L fraction was reproducibly negative for *ASGR1* and *ASGR2* transcripts, the data suggested that expression of the two genes was localized to monocytes. To confirm this, monocytes and lymphocytes were formally cell sorted using flow cytometry, RNA extracted and screened using RT-PCR. Monocytes gave a positive result for both *ASGR1* and *ASGR2* transcripts, whilst lymphocytes did not

give a PCR product, even after 50 cycles of amplification (Figure 2(c)). This high cycle number resulted in the burst-through of some nonspecific products; however, these did not interfere with the interpretation (Figure 2(c)). These data provide strong evidence for expression of *ASGR1* and *ASGR2* in monocytes but not in lymphocytes or granulocytes.

3.2. Monocyte ASGR1 Transcripts Differ between Individuals. Two transcripts have been described for *ASGR1* [17]. The longer transcript (T1) encodes isoform a (full-length ASGR1), the shorter transcript (T2) has an in-frame deletion of 117 nucleotides in the coding sequence resulting in a shorter isoform (isoform b) (Figure 1(a)). The latter lacks the transmembrane domain and is secreted as a soluble protein. In the liver, T1 has been shown to be the predominant isoform, with very little T2 detectable [17]. Using a nested PCR approach to screen RNA from the M+L cell fraction of peripheral blood, different individuals showed the presence of either T1 or of both T1 and T2 (Figure 3). Among the five individuals screened, none showed the presence of transcript T2 on its own. In combination with the data in

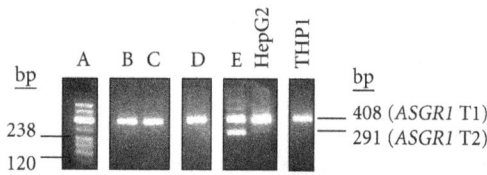

FIGURE 3: *ASGR1* transcript profiles in different individuals. Nested PCR was used to detect *ASGR1* transcripts T1 and T2 (first round primers A1CDSF1 + A1CDSR1; second round A1TEST2 + ASG1RTR). A product of 408 bp represented T1, present in RNA from the M + L cell fraction of five individuals (A–E) and also in HepG2 and THP1; a 291 bp product indicated the presence of T2 (individual E, and, very faintly, in HepG2). The RT-PCR product for individual A was coloaded with the DNA size standard (pBR322/*Msp*I, 238 bp and 120 bp sizes indicated). Analysis was done using agarose gel electrophoresis. The figure is a composite of various data (indicated by individual gel windows).

Figure 2 (which showed that *ASGR1* transcripts are detected in monocytes and not lymphocytes), these data suggest that the monocyte transcription profile for *ASGR1* differs between individuals. THP1 gave a result consistent with the presence of T1 (Figure 3).

3.3. Monocyte ASGR2 Transcripts Differ between Individuals. Five transcripts have been described for *ASGR2*, giving rise to four protein isoforms (Figure 1(b)) [15]. Using a RT-PCR designed to detect transcripts encoding different isoforms, the product profiles obtained for RNA extracted from the Histopaque M + L cell fraction differed among individuals. Two distinct profiles were obtained, the first corresponded to transcripts encoding isoforms a, b, c, and d, the second to transcripts encoding isoforms b and d only (Figure 4). ASGR2 isoforms a and c can give rise to soluble protein via proteolysis between the transmembrane domain and the CRD [18]. Isoforms b and d lack the proteolysis site and cannot produce the soluble form. The data therefore indicate that in some individuals, monocytes may produce both soluble and membrane-bound ASGR2, whilst in others, the soluble form is not produced by these cells.

3.4. Real-Time Quantitative PCR. Real-time PCR was optimized for primer pairs ASG1RTF + ASG1RTR, ASG2RTF + ASG2RTR (which, respectively, amplify all *ASGR1* and all *ASGR2* transcripts) and B2MRTF + B2MRTR (which amplify the reference target gene, *B2M*). For each primer pair, a specific product was obtained that had a characteristic melting curve (Figure 5).

Following PCR optimization, transcripts were measured using relative quantification, in which the target was *ASGR1* or *ASGR2*, the reference was *B2M*, the calibrator was HepG2, and the unknown was flow sorted monocytes or THP1. *B2M* is a recommended reference target for quantitative PCR, having the advantages of a single transcript, no processed pseudogene and expression at similar levels in a wide range of tissues [27]. Flow-sorted monocytes and THP1 cells gave similar results: relative to expression in HepG2, *ASGR1*

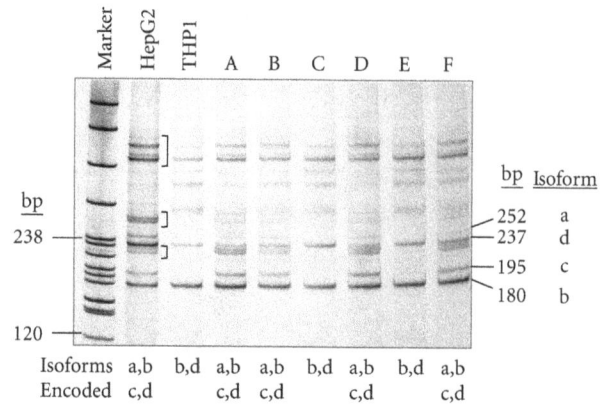

FIGURE 4: *ASGR2* transcript profiles in different individuals. Primer pair ASGR2RTF3 + ASGR2RTR2 gave RT-PCR products of different length (arrowed) according to the *ASGR2* transcripts present. RNA extracted from the M + L cell fraction of 6 different individuals (A–F) gave either a profile consistent with transcripts encoding all isoforms (a, b, c, and d) or just with isoforms b and d. The liver cell line HepG2 showed all transcripts, the monocyte cell line THP1 showed only transcripts encoding isoforms b and d. The analysis was done using polyacrylamide gel electrophoresis. DNA size reference was pBR322/*Msp*I ("Marker", 238 bp and 120 bp sizes indicated). In addition to the expected products, additional major bands were visualized on polyacrylamide gel electrophoresis of the RT-PCR screen for *ASGR2* transcripts (bracketed bands). It was considered possible that these represented heteroduplexes arising by cross-hybridization of complementary strands of the closely homologous true PCR products. To test this, each candidate heteroduplex band was excised from the gel, eluted into water, reamplified using the primers used in the initial PCR and the products electrophoresed on polyacrylamide. The results confirmed that the additional bands were hybrid duplexes containing complementary strands from nonidentical true PCR products: each hybrid yielded products consistent with amplification of two different template strands (data not shown).

transcripts were between 1.53E-07 to 7.53E-06-fold less, whilst *ASGR2* transcripts were between 7.16E-05 and 3.78E-04-fold less. Figure 5 illustrates relative quantification of *ASGR1* and *ASGR2* transcripts in cell-sorted monocytes. The ratio of expression of *ASGR1* : *ASGR2* in monocytes and the monocyte cell line THP1 was approximately 1 : 100.

3.5. Nucleotide Sequence Analysis. To ascertain whether THP1 would be a suitable cell line for future ASGPR studies in monocytes, the coding sequences of the *ASGR1* and *ASGR2* transcripts produced by these cells were determined. A nested PCR strategy was used as follows: the first round of synthesis amplified the entire coding sequence plus some 5′ and 3′ flanking sequence and then second round PCRs amplified nested portions within the first round product. The nested portions were sequenced in both directions. For *ASGR2*, one of the nested portions contained the region that differed between transcripts and gave products of different size according to the transcript. These products were electrophoresed, excised from the gel, reamplified

ASGR1 and ASGR2, the Genes that Encode the Asialoglycoprotein Receptor (Ashwell Receptor), Are Expressed in
Peripheral Blood Monocytes and Show Interindividual Differences in Transcript Profile

67

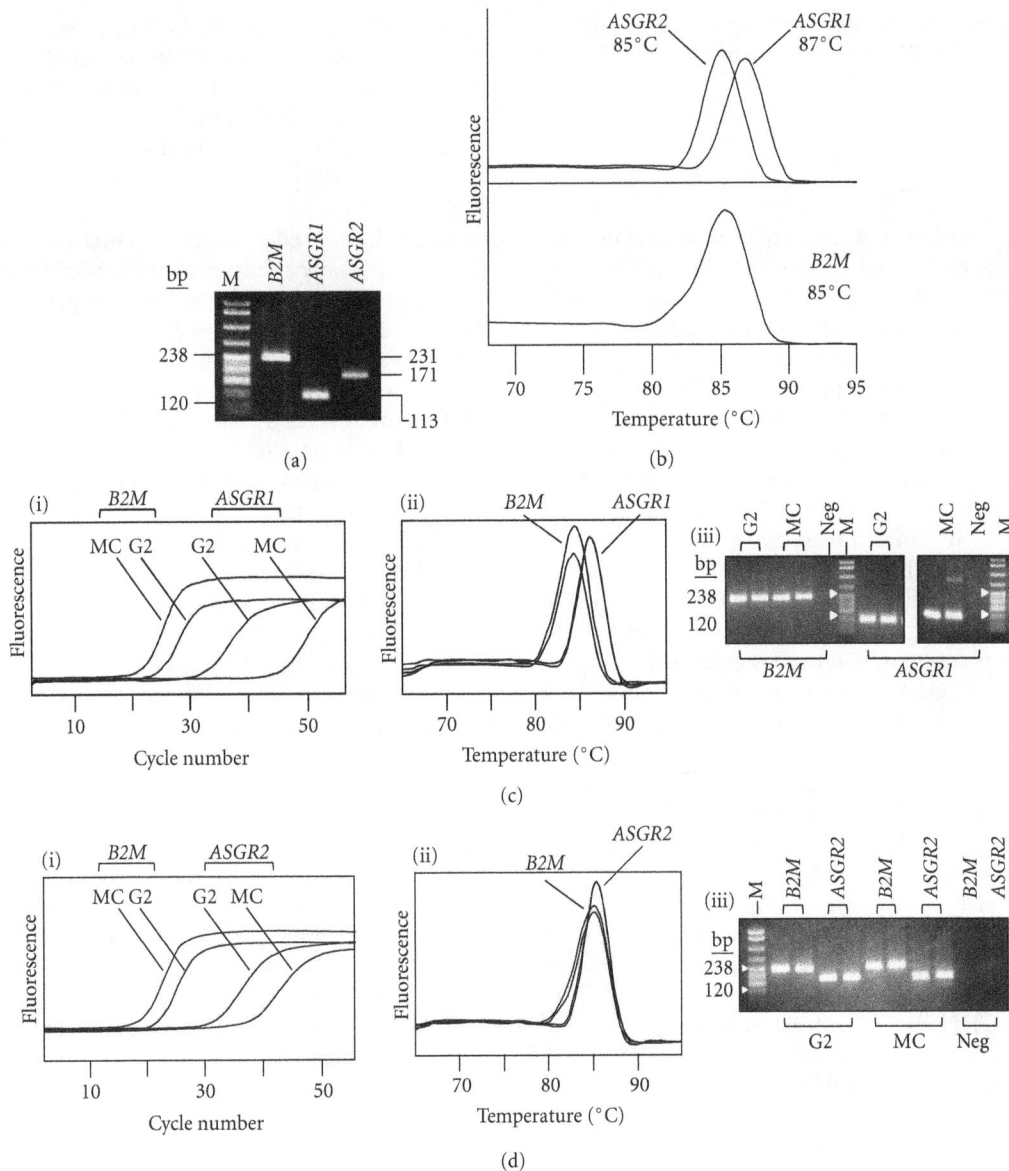

FIGURE 5: Real-time PCR analyses. (a) Agarose gel electrophoresis of products from optimized real-time PCRs for *ASGR1* (113 bp, primers
ASG1RTF + ASG1RTR), *ASGR2* (171 bp, primers ASG2RTF + ASG2RTR), and *B2M* (231 bp, primers B2MRTF + B2MRTR). Substrate
was HepG2 cDNA. (b) Melting curve analysis of real-time PCR products from panel (a). The Tm for each product is indicated. (c and d)
Relative quantification of *ASGR1* (c) and *ASGR2* (d) transcripts in cell-sorted monocytes (MC) compared with HepG2 (G2). *B2M* was
used as the reference. (i) Fluorescence profiles obtained during real-time PCR. Samples were analyzed in duplicate, for clarity only one of
each duplicate is shown. (ii) Melting curve analysis for the real-time PCR products in (i). (iii) Agarose gel electrophoresis of real-time PCR
products (duplicate analyses shown). Neg indicates real-time PCR control containing water instead of nucleic acid template. In panels a,
c(iii), and d(iii), M denotes pBR322/*Msp*I size standard (238 bp and 120 bp sizes indicated).

individually, and then sequenced, to give the sequence of
each of the coding regions that differed between transcripts.

THP1 *ASGR1* and *ASGR2* transcripts corresponded with
previously described splice variants found in the liver (data
not shown). The findings demonstrate that THP1 cells have
the capacity to transcribe and process correctly, transcripts
encoding functional isoforms of the ASGPR. Taken together
with the results in Figures 3 and 4, the data provide evidence
that monocytes express correctly processed transcripts from

both *ASGR1* and *ASGR2* and thereby have the potential to
produce functional ASGPR.

The coding sequence for *ASGR1* in THP1 differed from
the NCBI reference sequence (Table 1) by a single nucleotide:
c.267G>A, for which the monocyte cell line was homozygous
(data not shown). This change is silent at the protein level
(codon 89, AAG, changes to AAA, both encoding lysine). The
change is a naturally occurring variant (rs55714927) listed
in the international SNP database (dbSNP) [28]. The THP1

ASGR2 coding sequence for any transcript did not differ to the corresponding NCBI reference sequence (Table 1) (data not shown).

4. Discussion

In this study, expression of *ASGR1* and *ASGR2* was demonstrated in peripheral blood monocytes. For both genes, transcript profiles were obtained that corresponded with known splice variants found in the liver. To the best of our knowledge, this is the first report of the expression of correctly processed transcripts of *ASGR1* and *ASGR2* by human monocytes. In rat, a transcript encoding an asialoglycoprotein-binding protein was isolated from a peritoneal macrophage cDNA library and was highly homologous to that of the rodent hepatic ASGPR [5]. The data were interpreted to indicate that the rodent macrophage asialoglycoprotein-binding protein and the liver ASGPR were encoded by related genes [5]; however, it now seems possible that alternative splicing may underlie the differences. The findings of the present study do not suggest expression of alternative ASGPR-related genes in monocytes, rather they are entirely consistent with transcription of *ASGR1* and *ASGR2*, as occurs in the liver.

The various transcripts of *ASGR1* and *ASGR2* encode different isoforms of the proteins (Figure 1). It is not known whether the isoforms differ in their specificity or functionality, however it is predicted that certain isoforms have the potential to give rise to soluble forms of each protein. *ASGR1* transcript T2 has an in-frame deletion of 117 bp that encode the transmembrane domain; the resulting isoform "b" therefore is soluble and secreted [17]. *ASGR2* transcripts T1, TH2′ and T3 encode isoforms "a" (T1 and TH2′) and "c" (T3) which contain a proteolysis signal that, upon cleavage, gives rise to soluble protein [18]. Transcripts T2 and T4, respectively, encode isoforms b and d which lack the proteolysis signal and are membrane bound. The data presented here indicate that monocytes are able to express transcripts encoding soluble and membrane-bound isoforms of the ASGPR proteins. *ASGR2* was found to give rise to two transcript profiles, only one of which was present in the monocytes from any given individual. One profile corresponded with transcripts encoding both soluble and membrane-bound ASGR2 isoforms (a, b, c, and d), the other with membrane-bound ASGR2 isoforms (b and d) only. This novel finding indicates a fundamental difference between individuals. The underlying basis for this and whether or not it has physiologic significance in health or disease merit further exploration.

The level of *ASGR1* and *ASGR2* expression was considerably less in monocytes and the monocyte cell line THP1 compared with the liver cell line HepG2. It should be borne in mind that HepG2 expresses high levels of the receptor [24] and the monocyte expression was relative to this. That monocyte transcripts were not rare within the cells was indicated by the fact that they were readily detected using RT-PCR at routine cycle numbers. Low level of expression of the ASGPR has been reported previously in certain nonhepatic

tissues (rat macrophages [5], human intestinal epithelial cells [6], mouse testis [7], and rat thyroid gland [8]). It would be relevant to explore whether *ASGR1* and *ASGR2* transcription alters upon monocyte activation.

The relative expression of *ASGR1* and *ASGR2* in monocytes and THP1 (1 : 100) differed notably from ratios reported in liver (1 : 2 [1] and 1 : 6 [16]). The difference may be real, or it may reflect differences in methodology between studies or possibly a difference in the amplification efficiency of the primer pair used for *ASGR1* compared with that used for *ASGR2*. The true relative expression of the two genes by monocytes therefore remains to be established.

An important question arising from these results is whether monocytes translate *ASGR1* and *ASGR2* transcripts and produce functional ASGPR. The restricted expression of the genes in monocytes, but not in lymphocytes or granulocytes, suggests the transcripts are not produced randomly but specifically. Based on the fundamental principles of biology, this is likely to be for translation, it is difficult to think why else different transcripts encoding different isoforms may be produced by one specific cell type in the blood. The restricted tissue specificity for ASGPR expression in the body, and the lower expression level observed at the nonhepatic locations, could signal a specific role or function at those sites. In the case of monocytes, the receptor could, at the very least, serve a scavenger function where there is infection or tissue injury.

Expression of correctly processed transcripts of *ASGR1* and *ASGR2* by circulating monocytes has potentially significant implications in several important areas. These can be grouped into two main themes: normal physiology and drug design. If monocytes express functional ASGPR, they may contribute towards the normal physiological processes undertaken by the hepatocyte receptor. Whilst hepatocyte ASGPR is localized to the liver and relies upon the circulation to deliver ligands to it, the monocyte receptor would represent a mobile pool that can reach ligands in most parts of the body. Monocyte ASGPR may additionally interact with ligands in the blood, as does liver ASGPR. Thus, the monocyte receptor, if it is produced, may overlap functionally with that of the liver but may have additional physiological roles elsewhere in the body.

Hepatic ASGPR has been shown to be directly involved in the normal turnover of an important protein of primary hemostasis, VWF. Studies in knock-out mice indicated that the ASGPR is engaged in the constitutive control of VWF level and can, additionally, remove hemostatic components (VWF and platelets) desialylated by bacterial neuraminidase, a possible survival mechanism in disseminated intravascular coagulation [3]. The results in these studies were ascribed to the activity of the liver receptor; however, our data raise the possibility that monocytes may contribute to the relevant physiological processes via the ASGPR.

The findings have implications for the design of drugs targeting the liver via the ASGPR and drugs that are cleared by this receptor [29, 30]. Various strategies have been used, for example, conjugation of drug with galactosyl terminating molecules, as has been done with antiviral nucleoside analogues in the treatment of chronic viral hepatitis [31]. Besides drug targeting, strategies to deliver genes or other

ASGR1 and ASGR2, the Genes that Encode the Asialoglycoprotein Receptor (Ashwell Receptor), Are Expressed in
Peripheral Blood Monocytes and Show Interindividual Differences in Transcript Profile

69

nucleic acids to the liver have also made use of the ASGPR [10–12]. Relatively recently, successful *in vivo* delivery of siRNAs to the liver in mice has been achieved via simple intravenous injection employing ASGPR-specific conjugates containing GalNAc [11]. Pharmacological targeting of the ASGPR could not be considered to be predominantly liver-specific if monocytes express the receptor.

5. Conclusions

ASGR1 and *ASGR2* are expressed in peripheral blood monocytes in humans. For both genes, monocytes have the ability to produce correctly spliced transcripts encoding each of the known protein isoforms. However, there are interindividual differences in the transcript profiles; of particular note, despite several possible *ASGR2* transcripts, just two combinations were found to occur. This observation indicates that, in any given individual, *ASGR2* mRNA splicing is restricted to one of two profiles. Quantitative studies indicate that expression of both *ASGR1* and *ASGR2* is lower in monocytes than in the liver, but none-the-less is readily detected in monocytes (in comparison to other blood cells in which expression could not be demonstrated). The liver may represent a static site to which blood carries glycoconjugates for ASGPR-mediated removal, whilst monocytes may represent a mobile pool that can reach sites where the activity of the receptor is needed.

The data presented here open various important avenues for future research, a notable one of which is the measurement and characterization of monocyte ASGPR protein. The results for the cell line THP1 indicated it could be used as a model system for the investigation of monocyte ASPGR function and activity. Monocytes play a pivotal role in inflammation and immunity, the finding of ASGPR transcripts within these cells offers new insight into their biochemistry and functional potential.

Author's Contribution

D. J. Bowen and C. W. ven den d. Berg contributed to the design and organization of research, D. J. Bowen, C. W. ven den Berg and R. L. Harris contributed to the bench work, interpretation of data, and writing the paper.

Conflict of Interests

The authors declare that have no conflict of interests.

Acknowledgment

This work was supported by National Health Service Research and Development through the University Hospital of Wales.

References

[1] P. H. Weigel and J. H. N. Yik, "Glycans as endocytosis signals: the cases of the asialoglycoprotein and hyaluronan/
chondroitin sulfate receptors," *Biochimica et Biophysica Acta*, vol. 1572, no. 2-3, pp. 341–363, 2002.

[2] L. G. Ellies, D. Ditto, G. G. Levy et al., "Sialyltransferase ST3Gal-IV operates as a dominant modifier of hemostasis by concealing asialoglycoprotein receptor ligands," *Proceedings of the National Academy of Sciences of the United States of America*, vol. 99, no. 15, pp. 10042–10047, 2002.

[3] P. K. Grewal, S. Uchiyama, D. Ditto et al., "The Ashwell receptor mitigates the lethal coagulopathy of sepsis," *Nature Medicine*, vol. 14, no. 6, pp. 648–655, 2008.

[4] J. E. Zijderhand-Bleekemolen, A. L. Schwartz, J. W. Slot, G. J. Strous, and H. J. Geuze, "Ligand- and weak base-induced redistribution of asialoglycoprotein receptors in hepatoma cells," *Journal of Cell Biology*, vol. 104, no. 6, pp. 1647–1654, 1987.

[5] M. Ii, H. Kurata, N. Itoh, I. Yamashina, and T. Kawasaki, "Molecular cloning and sequence analysis of cDNA encoding the macrophage lectin specific for galactose and N-acetylgalactosamine," *The Journal of Biological Chemistry*, vol. 265, no. 19, pp. 11295–11298, 1990.

[6] J. Z. Mu, R. J. Fallon, P. E. Swanson, S. B. Carroll, M. Danaher, and D. H. Alpers, "Expression of an endogenous asialoglycoprotein receptor in a human intestinal epithelial cell line, Caco-2," *Biochimica et Biophysica Acta*, vol. 1222, no. 3, pp. 483–491, 1994.

[7] R. S. Monroe and B. E. Huber, "The major form of the murine asialoglycoprotein receptor: cDNA sequence and expression in liver, testis and epididymis," *Gene*, vol. 148, no. 2, pp. 237–244, Erratum in *Gene*, vol. 161, no. 2, pp. 307, 1995.

[8] F. Pacifico, D. Liguoro, R. Acquaviva, S. Formisano, and E. Consiglio, "Thyroglobulin binding and TSH regulation of the RHL-1 subunit of the asialoglycoprotein receptor in rat thyroid," *Biochimie*, vol. 81, no. 5, pp. 493–496, 1999.

[9] P. H. Weigel, "Galactosyl and N-acetylgalactosaminyl homeostasis: a function for mammalian asialoglycoprotein receptors," *BioEssays*, vol. 16, no. 7, pp. 519–524, 1994.

[10] C. Plank, K. Zatloukal, M. Cotten, K. Mechtler, and E. Wagner, "Gene transfer into hepatocytes using asialoglycoprotein receptor mediated endocytosis of DNA complexed with an artificial tetra-antennary galactose ligand," *Bioconjugate Chemistry*, vol. 3, no. 6, pp. 533–539, 1992.

[11] D. B. Rozema, D. L. Lewis, D. H. Wakefield et al., "Dynamic PolyConjugates for targeted in vivo delivery of siRNA to hepatocytes," *Proceedings of the National Academy of Sciences of the United States of America*, vol. 104, no. 32, pp. 12982–12987, 2007.

[12] J. Wu, M. H. Nantz, and M. A. Zern, "Targeting hepatocytes for drug and gene delivery: emerging novel approaches and applications," *Frontiers in Bioscience*, vol. 7, pp. d717–725, 2002.

[13] J. Bischoff and H. F. Lodish, "Two asialoglycoprotein receptor polypeptides in human hepatoma cells," *The Journal of Biological Chemistry*, vol. 262, no. 24, pp. 11825–11832, 1987.

[14] National Centre for Biotechnology Information Gene Database, "Asialoglycoprotein receptor 1 [homo sapiens]," http://www.ncbi.nlm.nih.gov/gene/432.

[15] National Centre for Biotechnology Information Gene Database, "Asialoglycoprotein receptor 2 [homo sapiens]," http://www.ncbi.nlm.nih.gov/gene/433.

[16] Y. I. Henis, Z. Katzir, M. A. Shia, and H. F. Lodish, "Oligomeric structure of the human asialoglycoprotein receptor: nature and stoichiometry of mutual complexes containing H1 and H2 polypeptides assessed by fluorescence photobleaching

recovery," *Journal of Cell Biology*, vol. 111, no. 4, pp. 1409–1418, 1990.

[17] J. Liu, B. Hu, Y. Yang et al., "A new splice variant of the major subunit of human asialoglycoprotein receptor encodes a secreted form in hepatocytes," *PLoS ONE*, vol. 5, no. 9, article e12934, 2010.

[18] S. Tolchinsky, M. H. Yuk, M. Ayalon, H. F. Lodish, and G. Z. Lederkremer, "Membrane-bound versus secreted forms of human asialoglycoprotein receptor subunits: role of a juxtamembrane pentapeptide," *The Journal of Biological Chemistry*, vol. 271, no. 24, pp. 14496–14503, 1996.

[19] H. Yago, Y. Kohgo, J. Kato, N. Watanabe, S. Sakamaki, and Y. Niitsu, "Detection and quantification of soluble asialoglycoprotein receptor in human serum," *Hepatology*, vol. 21, no. 2, pp. 383–388, 1995.

[20] National Centre for Biotechnology Information, http://www.ncbi.nlm.nih.gov/.

[21] Medical Research Council, http://www.mrc.ac.uk/Ourresearch/Ethicsresearchguidance/Useofhumantissue/index.htm.

[22] B. B. Knowles, C. C. Howe, and D. P. Aden, "Human hepatocellular carcinoma cell lines secrete the major plasma proteins and hepatitis B surface antigen," *Science*, vol. 209, no. 4455, pp. 497–499, 1980.

[23] S. Tsuchiya, M. Yamabe, and Y. Yamaguchi, "Establishment and characterization of a human acute monocytic leukemia cell line (THP-1)," *International Journal of Cancer*, vol. 26, no. 2, pp. 171–176, 1980.

[24] A. L. Schwartz, S. E. Fridovich, B. B. Knowles, and H. F. Lodish, "Characterization of the asialoglycoprotein receptor in a continuous hepatoma line," *The Journal of Biological Chemistry*, vol. 256, no. 17, pp. 8878–8881, 1981.

[25] S. Keeney, "Use of robotics in high-throughput DNA sequencing," *Methods in Molecular Biology*, vol. 688, pp. 227–237, 2011.

[26] B. Budowle and R. C. Allen, "Discontinuous polyacrylamide gel electrophoresis of DNA fragments," in *Protocols in Human Molecular Genetics, Methods in Molecular Biology*, C. G. Mathew, Ed., vol. 9, Humana Press, Clifton, NJ, USA, 1991.

[27] T. Lion, "Current recommendations for positive controls in RT-PCR assays," *Leukemia*, vol. 15, no. 7, pp. 1033–1037, 2001.

[28] National Centre for Biotechnology Information dbSNP Short Genetic Variations, http://www.ncbi.nlm.nih.gov/SNP/.

[29] E. V. Groman, P. M. Enriquez, C. Jung, and L. Josephson, "Arabinogalactan for hepatic drug delivery," *Bioconjugate Chemistry*, vol. 5, no. 6, pp. 547–556, 1994.

[30] Y. C. Lee, R. R. Townsend, M. R. Hardy et al., "Binding of synthetic oligosaccharides to the hepatic Gal/GalNAc lectin. Dependence on fine structural features," *The Journal of Biological Chemistry*, vol. 258, no. 1, pp. 199–202, 1983.

[31] L. Fiume, G. Di Stefano, C. Busi et al., "Liver targeting of antiviral nucleoside analogues through the asialoglycoprotein receptor," *Journal of Viral Hepatitis*, vol. 4, no. 6, pp. 363–370, 1997.

9

Hippo and *rassf1a* Pathways: A Growing Affair

Francesca Fausti,[1] **Silvia Di Agostino,**[2] **Andrea Sacconi,**[2]
Sabrina Strano,[1] **and Giovanni Blandino**[2]

[1] *Molecular Chemoprevention Group, Molecular Medicine Area, Regina Elena Cancer Institute, Via Elio Chianesi 53,*
   00143 Rome, Italy
[2] *Translational Oncogenomic Unit, Molecular Medicine Area, Regina Elena Cancer Institute, Via Elio Chianesi 53,*
   00143 Rome, Italy

Correspondence should be addressed to Giovanni Blandino, gblandino@activep53.eu

Academic Editor: Shairaz Baksh

First discovered in *Drosophila*, the Hippo pathway regulates the size and shape of organ development. Its discovery and study have helped to address longstanding questions in developmental biology. Central to this pathway is a kinase cascade leading from the tumor suppressor Hippo (Mst1 and Mst2 in mammals) to the Yki protein (YAP and TAZ in mammals), a transcriptional coactivator of target genes involved in cell proliferation, survival, and apoptosis. A dysfunction of the Hippo pathway activity is frequently detected in human cancers. Recent studies have highlighted that the Hippo pathway may play an important role in tissue homoeostasis through the regulation of stem cells, cell differentiation, and tissue regeneration. Recently, the impact of RASSF proteins on Hippo signaling potentiating its proapoptotic activity has been addressed, thus, providing further evidence for Hippo's key role in mammalian tumorigenesis as well as other important diseases.

1. Introduction

The Hippo pathway is a signaling pathway that regulates cell growth and cell death. It was discovered in *Drosophila melanogaster* as a pathway controlling organ size and of which mutations lead to tumorigenesis. This pathway is highly conserved, and its activation or repression could lead to the following most extreme outcomes: proliferation/transformation and death/tumor suppression. The Hippo pathway cross-talks with other signaling players such as Notch, Wnt, and Sonic hedgehog (Shh). It influences several biological events, and its dysfunction may possibly lie behind many human cancers. In this review, we discuss the complex data reported about *Drosophila* to date (schematic representation in Figure 1) and the human Hippo (schematic representation in Figure 2) pathways focusing on the relationship between the tumor suppression *rassf* protein family and the Hippo-like pathway in humans [1, 2].

2. The Hippo Signaling Network in *Drosophila*

Drosophila imaginal discs have facilitated molecular dissecting of signaling pathways controlling organ size during development. These imaginal discs allow to screen how organs grow several folds larger before differentiating into adult organs after proliferation in larval stages. By using the genetic analysis in *Drosophila*, Robin W. Justice and colleagues were the first to describe that loss of Wts (Warts), which encodes a kinase of Nuclear Dbf-2-related (NDR) family, results in a *Drosophila* phenotype characterized by tissue overgrowth [3]. Several years later many components of this pathway were characterized. Four tumor suppressors called Hippo (Hpo), Warts (Wts), Salvador (Sav), and Mats were established. These suppressors constitute the core linear kinase cassette of Hippo/Warts pathway whose products can affect proliferation without increasing apoptosis susceptibility [3–6] (Figure 1). Subsequent genetic screens identified at

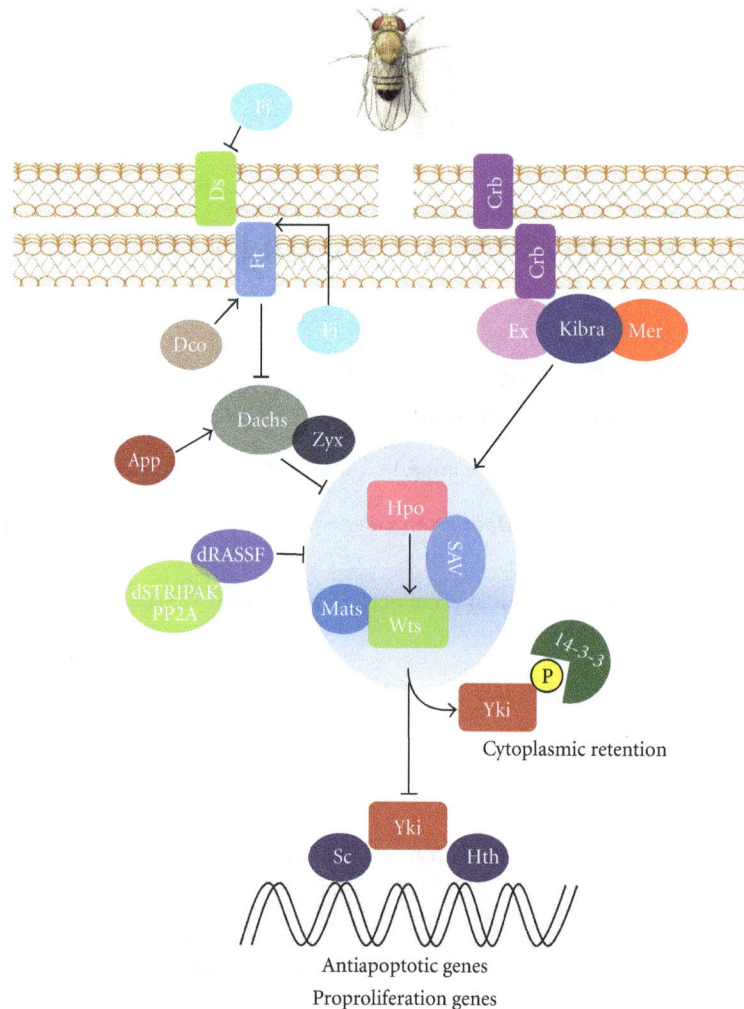

FIGURE 1: "Hpo signaling pathway in *Drosophila*." Schematic representation of Hippo kinases cascade and of its modulation by apical transmenbrame protein complexes.

least seven additional tumor suppressors whose biological functions converge on Hpo and/or Wts: the FERM domain proteins Merlin (Mer) and Expanded (Ex) [7–10], the protocadherins Fat (Ft) [11–14] and Dachsous (Ds) [15, 16], the CK1 family kinase Disc overgrown (Dco) [17, 18], the WW and C2 domain-containing protein Kibra [19–21], and the apical transmembrane protein Crumbs (Crb) [22–24]. All of these suppressors converge and act through a common downstream component, the transcriptional co-activator protein Yorkie (Yki) [25] (Figure 1). The mechanisms by which these upstream regulators signal towards the final player Yorkie are complex and are still focus of investigation. A great deal of evidence suggests that they work in a combinatorial or synergistic manner to regulate Hippo kinase activity.

2.1. The Apical Protein Complex: Kibra, Expanded, and Merlin.
The molecular link between upstream regulators and the core complex has not yet been clarified in mammals nor in *Drosophila*. In 2006, Hamaratoglu and collaborators proposed Mer (Merlin) and Ex (Expanded) as potential upstream regulators of the Hippo pathway [9], proteins which contain a FERM (4.1/ezrin/radixin/moesin) domain. Both proteins are considered tumor suppressors which cooperate to control organ growth. Their function seems to be partially redundant. In fact, while single mutation of each gene results in increased tissue growth, mutations in both genes give rise to a more strongly affected phenotype [9, 10]. Kibra, a third component of this apical complex, has recently been found. This protein possesses a WW domain which facilitates the interaction with other members of the Hippo pathway, such as Wts. It further interacts with a C2 domain that consists of a phospholipid-binding motif through which Kibra is believed to potentiate its membrane association [19–21]. WW domains are 35–40 amino acid protein–protein interaction domains that are characterized by a pair of conserved Trp residues, which generally interact with Pro-rich sequence motifs [26]. WW domain-Pro motif interactions appear to be particularly common in the Hpo

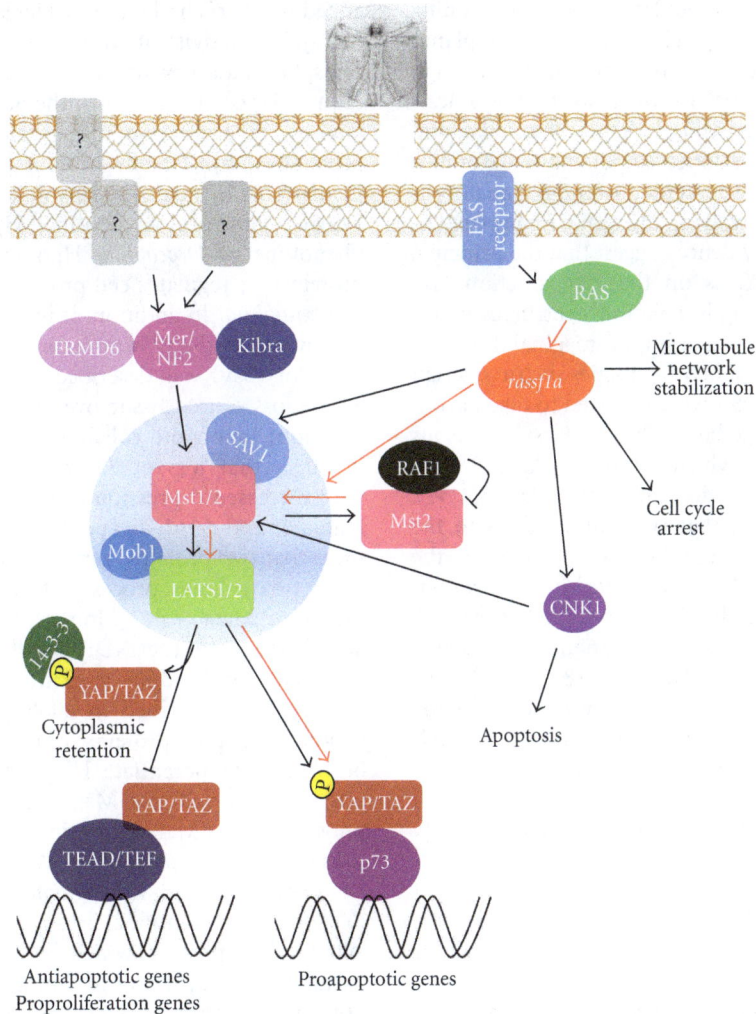

FIGURE 2: "Hpo signaling pathway in Mammals and the cross-talk with *rassf1a* signaling." Schematic representation of mammalian Hippo kinases cascade and interconnections between Hippo pathway and *rassf1a* protein signal. Red lines indicate the impact of *rassf1a* signaling in modulating activity of Hpo pathway components.

pathway. Three core components of Hpo signaling (Yki, Kibra, and Sav) contain WW domains, whereas three other components (Wts, Ex, and Hpo) hold PPxY motifs (reviewed in [27, 28]). While the formation of a ternary complex between Kibra, Ex, and Mer was observed, each protein was seen to localize to cellular membranes independently. Furthermore, it has been published that the Kibra-Mer-Ex complex is physically involved with the Hpo-Sav, constituting an apical protein complex required for associating the Hpo pathway to the cellular membranes [20, 21]. Studies on the Ex localization and function have led to the discovery of another important upstream regulator protein of Hpo, Crb (Crumbs) [22–24]. Crb is a transmembrane protein which normally localizes to the subapical membrane of epithelial cells that is responsible together with other apical complexes in *Drosophila* for organizing apical-basal polarity [29]. Crb binds to Ex through a short intracellular domain including a juxtamembrane FERM-binding motif (FBM). The FBM domain of Crb interacts with the FERM domain of Ex. This type of binding is necessary for Ex apical localization and stability. Furthermore, it has been published that Crb also works with Mer and Kibra [23]. The loss of Crb expression was shown to further determine a phenotype characterized by overgrowth, possibly to a lesser degree compared to the other members of Hpo signaling described until now [22–24]. Not long ago, this protein was proposed to have had an important function as a transmembrane receptor recognizing cell-cell contacts through Crb-Crb binding domains [22].

2.2. The Upstream Regulator: Transmembrane Protein Fat. The atypical cadherin FAT (Ft) was the first transmembrane protein shown to affect Hippo signaling. Fat is the first tumor suppressor gene isolated in *Drosophila*. In fact, the complete knock-out of the FAT protein induces death in *Drosophila* larvae with overgrown imaginal discs [11]. As previously mentioned, FAT is a large transmembrane protein, constitutively cleaved by unknown proteases. It contains

34 cadherin repeats in its extracellular domain, functioning as a receptor for Hippo signaling [12–14] as well as for planar cell polarity (PCP) [30, 31]. PCP is a mechanism through which cells orient themselves orthogonally to the apical-basal axis, as observed in the wing hairs of *Drosophila*, and the sensory hair cells in the inner ear of mouse. Notably, the mechanism by which FAT regulates Hippo signaling is different from the branch involving the ternary complex Ex-Mer-Kibra. Many lines of evidence suggest that the principal mechanism exerted by FAT is on the Wts function [18, 32]. Thus, FAT-Hpo signaling is genetically distinguishable, involved in Hippo pathway regulation of imaginal discs and neuroepithelial tissue, but not in other tissues such as ovarian tissue [14, 33, 34]. Many genes were reported to take part in this parallel mechanism together with FAT. First, Dachsous (Ds), an atypical cadherin which binds to FAT [15, 16]. FAT is regulated by an expression gradient of Ds [35, 36]. Four-jointed (Fj) is a kinase that typically localizes to the Golgi subcellular compartment and that phosphorylates the cadherin domains of FAT and Ds to mediate binding between these two proteins [37]. Another kinase responsible for FAT phosphorylation in its cytoplasmatic segment is a Casein I kinase, termed Discs overgrown (Dco) [17, 18]. The effective key mediator of FAT in the Hippo pathway seems to be Dachs, an unconventional myosin which antagonizes FAT, and whose activity is influenced by Approximated (App) [17]. App, in fact, antagonizes FAT signaling by modulating Dasch expression [38]. Another protein identified recently linked to the FAT branch in Hippo signaling is the LIM-domain protein Zyx102. It has been found to directly affect the core kinases of the Hippo pathway [39]. All of these components described above seem to be responsible for linking Hippo to extracellular stimuli [40].

Another so called "scaffold" protein that has been identified as a regulator of Hpo is called *Drosophila rassf* (*drassf*). This protein like its mammalian counterpart *rassf* can bind to Hpo through a conserved SARAH domain. But unlike in mammals, it hampers Hpo activity by competing with SAV to bind to Hpo [41] and by recruiting a Hpo-inactivating PP2A complex (dSTRIPAK) [42], thus showing a positive regulation of growth. Interestingly, Grzeschik and collaborators showed that the depletion of the *Drosophila* neoplastic tumor suppressor Lethal giant larvae (Lgl), which controls apical-basal cell polarity and proliferation, leads to upregulation of the Hippo pathway target Yki through a decreased phosphorylation and consecutively overprolif-eration of developing eyes, without affecting apical-basal polarity [43]. This mechanism is brought about by cellular mislocalization of Hpo and *rassf*. These both colocalize basolaterally leading to the deregulation of the Hippo kinase cascade, thereby preventing phosphorylation and inactivation of Yki. This concurs with data previously discussed wherein *rassf* is able to bind to Hpo precluding its interaction with SAV [41].

2.3. The Key Effectors of Growth Control: Hippo, Warts, Sal-vador, and Yorkie. Warts is crucial in the phosphorylation-dependent regulation of Yki [25, 44, 45]. Warts (Wts) encodes a Ser/Thr kinase of Nuclear Dbf-2-related (NDR) family. The activity of Warts is controlled through a series of phosphorylation events. Warts is directly phosphorylated by Hippo (Hpo), a member of the Sterile-20 family of Ser/Thr kinases, in a reaction that is facilitated by the Salvador protein [4, 5]. The fly protein Hippo (Hpo) is the first mediator of this pathway characterized by a kinase cascade. Wu and collaborators identified Hpo through analysing the phenotype of *Drosophila* Hpo mutants. Hpo is a kinase protein that regulates cell proliferation as well as apoptosis in *Drosophila*. In addition, it interacts, phosphorylates, and is activated by the WW domain-containing protein Salvador. Salvador (Sav) was described as a tumor suppressor gene, whose loss caused tissue overgrowth, similar to Wts loss of function. Tapon and collaborators were the first to observe, in 2002, that loss of Sav or Wts was strictly associated with increased expression of *cyc e*, a cell cycle progression regulator and *diap1*, an apoptosis inhibitor, thus, confirming these that two proteins' very important role in coordinating these two cellular processes [4]. Similar to Sav function on Hpo, Mats' role (Mob as tumor suppressor) which also belongs to the NDR family, as well as its kinase-like behavior binding to and potentiating Wts intrinsic activity, was described in 2005 [6]. Thus, Sav and Mats action as adaptor proteins, often termed scaffold proteins, both serve to potentiate Hippo signaling. Interestingly, it was also reported that Mats is a Hpo substrate. The latter phosphorylates Mats increasing its affinity for Wts binding, thus inducing potentiation of Wts kinase activity [46].

The downstream key regulator of Hpo signaling is Yorkie (Yki). It was identified in a yeast two-hybrid screen for Wts-binding protein, which is the final step in the Hippo pathway, driving its transcriptional regulation [25]. Yki is not a direct transcriptional factor because it does not possess its own consensus DNA-binding motif but is known as a potent transcriptional co-activator by cooperating with different DNA-binding proteins. Wts directly phosphorylates Yki at Ser 168, thus creating a binding site for 14-3-3 proteins which sequester Yki in the cytoplasm and prevent its nuclear import [44, 45]. In actual fact, the loss of Hippo signaling as well as mutations in 14-3-3 binding site for Yki was shown to produce strong nuclear accumulation, a common feature, coupled with aberrant activity of Yki [47]. Another two residues of Yki are believed to be targets of Wts phosphorylation (Ser111 and Ser250); however, little is known about the underlying mechanisms. As mentioned before, Yki cooperates with many DNA-binding proteins which act as transcription factors, potentiating their function. It is worth noting that some binding partners of Yki are the same kinases that function upstream to it in the Hippo pathway. Thus, through the PY (PPxY)-WW domain interactions, Yki is able to bind to Ex, Wts, and Hpo that sequester Yki at a cytoplasmatic level, independently from its phopshorylated state [48, 49]. Loss of Hippo signaling and consecutive aberrant Yki activation leads to deregulation of some gene class transcriptions. One class includes genes involved in cell survival and proliferation. One of the Yki partners, Scalloped (Sc), a member of TEAD/TEFs family, is responsible for Yki overexpression induced tissue overgrowth [50, 51]. Another partner of Yki in *Drosophila* is Homothorax (Hth) that promotes cell survival and cell proliferation in eye development from eye imaginal discs [52]. Both Sc and Hth are able to bind a Hippo consensus DNA motif, termed Hippo response element (HRE), which is present in many Hippo target genes. Particularly, Sc together with Yki bind to the HRE present in a very well-known target gene, *diap1* [50], an apoptosis inhibitor, as mentioned above. Hth has only little influence on *diap1* transcription. It is very important in regulating the transcription of another Yki target, the growth promoting microRNA gene *bantam*. Other Yki targets in this class are the cell-cycle regulators *cyc e*, *e2f1* [4, 53], and *Drosophila* Myc (dMyc) whose expression seems to be positively regulated by Yki [54, 55]. Another important class is made up of components from other signaling pathways, such as ligands for Notch, Wnt, EGFR, and Jak-Stat pathways. In fact, other known Yki partners are believed to be Smad proteins [56]. This interaction appears to potentiate the transcriptional response to BMP/TGF-β signaling, addressing a possible crosstalk between Hippo and BMP/TGF-β pathway. Finally, a third class of Yki targets consisting of several proteins from its own Hippo cascade, such as Ex, Mer, Kibra, Crb, and Fj. These are downstream transcriptional targets of Yki [9, 17, 20, 57] and define a sort of positive feedback loop which characterizes most signal pathways.

3. The Hippo Kinase Signaling in Mammals

3.1. YAP and TAZ: Mammalian Effectors of Hippo Pathway. The Hippo pathway is highly conserved in mammalian systems. It was demonstrated that loss of function of mutant flies can be rescued by expressing their respective human counterparts [5, 6]. These data strongly correlate with the importance of Hippo signaling in controlling organ size, tumorigenesis as well as the insurgence of other important diseases in mammals. The ortholog human counterparts of core kinases Hpo and Warts are represented by the pro-apoptotic MST1/2 and LATS1/2 kinases [58, 59] (Figure 2). One ortholog exists for the adaptor protein Sav, termed WW45 or SAV1, and the other two orthologs for Mats are termed MOBKL1A and MOBKL1B (referred to as Mob1). These proteins form a conserved kinase cassette that phosphorylates and inactivates the mammalian Yki homologs YAP and TAZ [25, 47, 60] in response to cell density. This cell density-dependent activation of the Hippo pathway is required in contacting inhibition of cultured mammalian cells [47]. Similar to *Drosophila* Hippo signaling, all the mammalian components of the Hippo pathway clearly show tumor suppression activity. In fact, transgenic overexpression of YAP [61, 62] and liver-specific knockout of *Mst1/2* or *Sav1* [63–66] induce abnormal liver expansion in terms of size, and eventually hepatocellular carcinoma formation (HCC). YAP was initially identified as a 65 kDa binding partner of *c*-Yes from Sudol and collaborators [67]. YAP is a transcriptional co-activator of many transcription factors via its own WW-domain (reviewed in [68]). The TEAD/TEF family of transcription factors, whose homolog is represented by Sc in *Drosophila*, is considered the major partner of both YAP and TAZ in executing their activities within the Hippo pathway. The 4 mammalian TEF/TAED transcription factors are widely expressed and regulate transcription in specific tissues during certain development stages [69]. It was shown that TAED1/TEF2 and YAP share a large number of target genes [51, 70, 71]. In support of this evidence, TEAD1 and TEAD2 double-knockout mice display similar phenotypes to YAP knockouts [69]. Furthermore, ablation of TAED/TEF expression decreases the ability of YAP/TAZ in promoting anchorage independent growth and EMT (epithelial to mesenchymal transition) [51, 71, 72]. Recently Dupont and collaborators have identified YAP and TAZ as the nuclear principal complex of mechanical signals exerted by extracellular matrix (ECM) rigidity and cell shape. This regulation requires Rho GTPase activity and tension of the actomyosin cytoskeleton but is independent from the Hippo/LATS cascade. YAP/TAZ is required for differentiation of mesenchymal stem cells induced by ECM stiffness and for survival of endothelial cells regulated by cell geometry [73].

The exact role of YAP has yet to be defined since it appears to be able to act as an oncogene or as a tumor suppressor depending on the cellular context. YAP1 was shown to bind long forms of p73 and p63, while not to wt p53, thereby potentiating p73- and p63-induced apoptosis [74, 75]. In particular, p73 recapitulates the most well-characterized p53 antitumoral effects, from growth arrest and apoptosis to senescence. YAP imparts transcriptional target specificity to p73 in promoting either growth arrest or apoptosis in response to different stimuli [76–78].

3.2. The Complexity of Upstream Regulators: FRMD6, Mer, and Kibra. As mentioned above, the complexity of molecular links between the upstream regulators and the core kinases in mammals has not been clarified either for *Drosophila*. The mammalian genome contains homologs for all the reported upstream regulators of the Hippo pathway. Notably, it encodes more than one paralogue for each *Drosophila* component, thus increasing complexity and the need for further investigation. Two homologs for Kibra, KIBRA/WWC1 and WWC2 and for Expanded, FRMD6 and FRMD1, while only one for Merlin, NF2, were identified. Interestingly, they often differ in protein structure compared to *Drosophila* counterparts. One Ex homolog for FRMD6 does not possess the extended C-terminal portion that is required for growth inhibition activity of Ex and binding to Kibra [20, 79]. No interaction between FRM6 and MST1/2 has been confirmed, in contrast to the described interaction between Ex and Hippo [21]. Also Mer/NF2 is a FERM domain-containing protein and the most investigated. It is a tumor suppressor, whose mutations trigger neurofibromatosis 2, mainly characterized by tumor insurgence in the nervous system [80, 81]. It has a prominent role in growth inhibition triggered by C-adherin-based cell contact. Growth inhibitory action of Mer/NF2 appears to stem from controlling the distribution and signaling of membrane receptors. In fact, in Merlin K/D cells the activation and internalization of the EGF receptor are also maintained in high-cell-density conditions [82].

Furthermore, contrasting data for Mer/NF2 involvement in developing hepatocellular carcinoma (HCC) and tumors of the bile duct were reported. It is worthy to note that in specific *Merlin* $^{-/-}$ liver an increased proliferation of hepatocytes and of bile ducts was reported, coupled with minor LATS and YAP phopshorylation and increased YAP nuclear export [83]. Conversely, in this context, other authors did not observe any alterations in YAP phosphorylation and localization [84].

3.3. The Core Kinases: MST, LATS, and MOB. The ortholog human counterparts of core kinases Hpo and Warts are represented by the proapoptotic MST1/2 and LATS1/2 kinases [58, 59]. MST1/2 are serine-threonine kinases, better known for their ability to initiate apoptosis when overexpressed through a combination of p53- as well as JNK-mediated pathways [85, 86]. Generally, apoptosis induced by different stimuli is coupled with the activation of kinases MST1/2, which result themselves as substrates for caspases 3, 6, and 7 cleavage. This produces highly active catalytic fragments, which are mainly localized in the nucleus, where they exert their proapoptotic function [85–87]. As mentioned above, loss of function of the MST1/2 ortholog Hpo shows a phenotype characterized by a marked overgrowth due to accelerated cell-cycle progression and deregulated apoptosis. Exogenous MST2 expression can successfully rescue this phenotype. MSTs become activated by autophosphorylation in the threonine residues within their activation loop domain. Inhibition of dimerization and autophosphorylation of MST2 exerted by RAF1 was reported [88]. In this latter context, expression of *rassf1a* is able to release MST2 from RAF1 inhibition, thus inducing apoptosis [77]. Moreover, PP2A phosphatase dephosphorylates MST1/2 kinases as shown by two different groups [42, 89]. How autophosphorylation and activation of MST kinases are triggered by unknown extracellular stimuli remain to be elucidated, and okadaic acid treatment or siRNA-mediated knockdown of PP2A promote MST1/2 phopshorylation and activation. Interestingly, Guo and collaborators very recently showed that *rassf1a* activates MST1 and MST2 by preventing their dephosphorylation. Specifically, they observed that *rassf1a* knockdown, which is a frequent phenomenon in human tumors, leads to a dramatic decreased in MST1/2 levels exerted by phosphates. They also observed that restoring *rassf1a* expression and function promotes the formation of active MST1/2 by counteracting the role of phosphates. This is one of the first examples of a tumor suppressor acting as an inhibitor of a specific dephosphorylation pathway.

In the Hippo pathway context, MST substrates include LATS and MOB1. LATS1/2 kinases control cellular homeostasis, negatively regulating cell division cycle 2 (CDC2) and favoring G2/M arrest [90–92]. LATS2 was also reported to induce G1/S arrest [93]. In fact, both overexpressions of LATS1 and 2 dramatically inhibit both cell proliferation and anchorage-independent growth [47, 94] in various cell lines. It is also true that loss of LATS1/2 leads to a broad variety of tumors, such as soft tissue sarcoma and leukemia [95]. In light of these data, these proteins are believed to be strong tumor suppressors. Recent data addressed LATS involvement

in tumor suppressive as well as oncogenic pathways, such as p53, RAS, and Akt signaling pathways. Interestingly, LATS2 can bind to MDM2 protein, thus inhibiting its E3 ubiquitin ligase activity to stabilize p53, which in turn favors the transcription of LATS2 [96]. Up until now, YAP and TAZ are the main LATS substrates identified in its kinase activity, but yet they only mediate some of the effects of LATS, thus indicating the existence of other substrates, such as Snail [97], DYRK1A [98], and LATS1 and LATS2 [99].

In the Hippo pathway context, LATS activity is supported by MOB1. This protein, which corresponds to the human ortholog of the Mats adaptor protein, binds to and phosphorylates LATS kinases, favoring YAP and TAZ proto-oncogenes phosphorylation and inhibiting their nuclear activity. MOB1 binding to LATS kinases is strongly enhanced upon phosphorylation of MOB1 by MST1/2 kinases [46]. Loss of MOB1 function results in increased cell proliferation and decreased cell death, suggesting that MOB1 functions, as well as the other Hippo pathway components, as a tumor suppressor protein.

4. *rassf1a* Signaling into Hippo Pathway

Due to the absence of enzyme activity, Ras-Association Domain Family (*rassf*) are noncatalytic-proteins. They are often referred to as "scaffold proteins," which are ubiquitously expressed in normal tissue and described in literature as a strong tumor suppressor family of proteins (reviewed in [100]). The *rassf*'s family comprise ten members from *rassf1* to *rassf10*. Among them only *rassf1a* shares the closest homology to *Drosophila rassf* (*drassf*) (reviewed in [101]). *rassf1a* exhibits strong tumor suppressor function [102]. Loss of *rassf1a* allele is a frequent occurrence in primary human cancers [103, 104]. Furthermore, hypermethylation of *rassf1a* promoter is very often correlated with oncogenic phenotypes. Concomitantly, the identification of specific point mutations of *rassf1a* impinges on the ability of this protein to inhibit tumor cell growth [105, 106]. About 15% of primary tumors show point mutations of *rassf1a* [107]. Two independent research groups generated *rassf1a* knockout mice [108, 109]. Both these mice showed a phenotype with greatly increased susceptibility to tumor formation. Pursuing the hypothesis that the protein-protein interaction of YAP pattern changes as a consequence of different stimuli, Matallanas and colleagues followed the behavior of *rassf1a* after triggering apoptosis [77]. They showed that *rassf1a* disrupts the inhibitory complex between RAF1 and MST2 and favors the physical association between MST2 and LATS1 concomitantly, therefore, leading to YAP1 phosphorylation and nuclear relocalization where it binds to p73 and potentiates its apoptotic activity (Figure 2). It was also shown that the FAS active receptor induces *rassf1a* to compete with RAF1 in binding to MST2, thus promoting the formation of a LATS1 complex. This results in the translocation of YAP from the cytoplasm to the nucleus. These findings may suggest that the activation of the *rassf1a* complex indirectly diverts LATS1 from phosphorylating YAP, thus making it available for different phosphorylation events.

In addition, it is also able to enter into the nucleus where it can activate the transcription of p73 target genes involved in apoptosis.

It is worthy to note that in 2009, Hamilton and collaborators identified a novel DNA damage pathway that is activated by ATM kinase, involving *rassf1a* and Hippo pathway members [110]. They showed that, upon DNA damage, *rassf1a* becomes phopshorylated by ATM on Ser131. This event seems to be necessary in promoting MST2 binding to *rassf1a*, potentiating MST2 and LATS1 proapoptotic activity leading to p73 stabilization. Thus, this confirms findings observed in previous *in vitro* experiments showing that the *rassf1a* peptide containing an ATM putative domain is a substrate for ATM phosphorylation [111, 112].

More recently, the interaction, between *rassf1a* and SAV Hippo pathway member [113], was shown to potentiate p73-dependent apoptosis [114]. While this effect does not seem to require direct interaction between *rassf1a* and MST kinases, it was shown to trigger apoptosis via the MST/LATS pathway [77]. It is also true that SAV acts as a scaffold protein connecting MST kinases with LATS kinases [115] and that the expression of exogenous SAV can greatly enhance this proapoptotic signal [113]. Consequently, it is reasonable for authors to speculate the existence of a functional axis involving *rassf1a*-MST-SAV-LATS-YAP in promoting p73-induced apoptosis. Altogether, these findings show a close functional interconnection between *rassf1a*, Hippo, and p53 family tumor suppressor effects.

RASFF1A functions as a negative regulator of cardiomyocyte hypertrophy [116]. The latter displays an enlargement in size of cardiomyocytes, which is very often associated with heart failure [117]. It was proposed that a large number of protooncogenes, which are expressed in the heart, could possibly mediate this aberrant process [118]. *rassf1a* exon1α knockout mice exhibit normal cardiac morphology at 12 weeks of age. Notably, the application of a pressure overloaded the transverse aortic constriction causing massive cardiac hypertrophy, among the severest reactions ever to be reported [116]. This may suggest that *rassf1a* plays a role in contrasting overproliferation of cardiomyocytes. Interestingly, the authors observed that *rassf1a* in this cellular system greatly opposes the RAS-RAF1-ERK1/2 signal pathway. Not long ago, it was proposed that the activation of RAF by RAS requires a complex regulation of many adaptor molecules including the involvement of CNK1 (connector enhancer of kinase suppressor of RAS). This protein is able to form a complex with *rassf1a*, increasing *rassf1a*-induced cell death [119]. In light of these data authors speculated about a possible imbalance in the ratio of the components of the scaffold complex required for RAS signal transmission. CNK1 was also found to interact with MST1 and MST2, requiring MST kinases to induce apoptosis. Deleting the MST1 segment that mediates binding to *rassf1a* also eliminates the physical association between MST1 and CNK1. To sum up, CNK1 binds to *rassf1a* and promotes apoptosis through a pathway that requires *rassf1a* and MST kinases [119]. This mechanism may be the underlying factor behind *rassf1a*'s action in preventing cardiomyocytes hypertrophy. Supporting this, Del Re and collaborators showed that *rassf1a*

is an endogenous activator of MST1 in the heart. They also found that in cardiac fibroblasts the *rassf1a*/MST1 pathway negatively regulates TNF-α that is believed to be a key mediator of hypertrophy and consecutive cardiac dysfunction [120]. Altogether, these findings highlight the importance of a crosstalk between *rassf1a* and components of the human Hippo pathway in preventing cardiac dysfunction due to aberrant overproliferation of cardiomyocytes. Of note, other Hippo pathway members were shown to be involved in heart development and size, such as YAP [121], Dch1-FAT [122], LATS2 [123], and SAV [124].

5. *rassf5* and *rassf6*

Other *rassf* family members were involved in modulating the activity of Hippo pathway components. The first RAS interactor discovered within this family was *rassf5* [125], often called Novel Ras Effector 1 (NORE1). This isoform that shares up to 60% homology with *rassf1*, is the most common isoform. As for many *rassf*s, it was demonstrated to be a centrosomal protein that can bind to the microtubule scaffold structure. This event appears to be required for growth inhibition and consequently tumor suppression activity, which is achieved through the inhibition of ERK signaling [126]. Furthermore, it has been reported that active RAS binds to *rassf5*-MST1 complex thereby conferring the role of the RAS effector complex in mediating the proapoptotic function of KiRASG12V [127]. RASFF5 and the MST1 pro-apoptotic kinase are involved in a physical interaction, thus forming an active complex where RAS interacts upon serum stimulation consequently leading to its proapoptotic function. Furthermore, the interaction of *rassf1a* and NORE1 with MST1 appears to be controversial. In fact, an inhibition of MST kinases activity by coexpression with the complex NORE1-*rassf1a* in excess was reported [128]. At the same time, by *in vivo* experiments, overexpression of *rassf1a* together with MST2 was shown to increase kinase activity of MST2 consequently potentiating its pro-apoptotic effect [77, 113, 129].

In 2009, Ikeda and collaborators showed that another *rassf* member, *rassf6*, can bind to MST2 kinase. This protein is known to induce apoptosis [130, 131]. When *rassf6* is bound to MST2, *rassf6* inhibits MST2 activity, thus, inhibiting its role in the Hippo pathway. Conversely, the release of MST2 from *rassf6* causes apoptosis in a WW45-dependent manner (*Drosophila* SAV). Therefore, *rassf6* impinges the Hippo proapoptotic pathway by inhibiting MST2, but it is *per se* able to induce apoptosis through a parallel Hippo mechanism. In fact, MST2 is responsible for apoptosis induced through Hippo signaling and through a *rassf6*-WW45-mediated pathway [131].

6. Concluding Remarks and Future Perspectives

In conclusion, the Hippo pathway is a signaling pathway that regulates cell proliferation and cell death. It is a kinase cascade that phosphorylates and negatively regulates transcription by transcriptional coactivators. As summarized above, the loss of function of the Hippo pathway triggers tumorigenesis. Accordingly, the downregulation of the Hippo pathway is frequently observed in human cancers. Aberrant activation of Hippo downstream executors, YAP1 and TAZ, induce epithelial-mesenchymal transition and the expression of stem-cell markers in cancer cells. Quite recently, the Hippo and the *rassf* pathways have emerged to be closely linked. The tumor suppressor *rassf* proteins were shown to induce cell-cycle arrest and apoptosis. Stimuli activating the Hippo pathway simultaneously induce *rassf*-dependent biological events. Thereby, the Hippo and *rassf* pathways cooperate in preventing tumorigenesis. Reintegration of the Hippo pathway and *rassf* functions should be implemented in cancer therapy. However, it is also true that if this cross-talk results disproportionate, the consequence will be excessive apoptosis and consecutive organ dysfunction. In such cases, the involvement of the Hippo/*rassf* inhibitors will be useful. The relationship between the Hippo and *rassf* pathways is probably not restricted to cancer biology since many of the Hippo components also regulate adipogenesis, osteogenesis, and myogenesis. As discussed above, a growing body of evidence shows that this relationship between *rassf* and the Hippo pathways also occurs in cardiac tissue inhibiting cardiac hypertrophy and playing a critical role in preventing heart failure. Based on what has been described and in light of the synergistic effects observed on the interaction within *rassf* and components of Hippo signaling in preventing defects of proper biological development such as insurgence of many human diseases, much more work is needed to further investigate the importance of this physiological relationship.

Acknowledgment

The authors thank MS Tania Merlino (Regina Elena Cancer Institute) for revising and proofreading the review.

References

[1] D. Pan, "The hippo signaling pathway in development and cancer," *Developmental Cell*, vol. 19, no. 4, pp. 491–505, 2010.

[2] B. Zhao, K. Tumaneng, and K. L. Guan, "The hippo pathway in organ size control, tissue regeneration and stem cell self-renewal," *Nature Cell Biology*, vol. 13, no. 8, pp. 877–883, 2011.

[3] R. W. Justice, O. Zilian, D. F. Woods, M. Noll, and P. J. Bryant, "The *Drosophila* tumor suppressor gene warts encodes a homolog of human myotonic dystrophy kinase and is required for the control of cell shape and proliferation," *Genes and Development*, vol. 9, no. 5, pp. 534–546, 1995.

[4] N. Tapon, K. F. Harvey, D. W. Bell et al., "salvador promotes both cell cycle exit and apoptosis in *Drosophila* and is mutated in human cancer cell lines," *Cell*, vol. 110, no. 4, pp. 467–478, 2002.

[5] S. Wu, J. Huang, J. Dong, and D. Pan, "hippo encodes a Ste-20 family protein kinase that restricts cell proliferation and promotes apoptosis in conjunction with salvador and warts," *Cell*, vol. 114, no. 4, pp. 445–456, 2003.

[6] Z. C. Lai, X. Wei, T. Shimizu et al., "Control of cell proliferation and apoptosis by mob as tumor suppressor, mats," *Cell*, vol. 120, no. 5, pp. 675–685, 2005.

[7] M. Boedigheimer, P. Bryant, and A. Laughon, "Expanded, a negative regulator of cell proliferation in *Drosophila*, shows homology to the NF2 tumor suppressor," *Mechanisms of Development*, vol. 44, no. 2-3, pp. 83–84, 1993.

[8] D. R. LaJeunesse, B. M. McCartney, and R. G. Fehon, "Structural analysis of *Drosophila* Merlin reveals functional domains important for growth control and subcellular localization," *Journal of Cell Biology*, vol. 141, no. 7, pp. 1589–1599, 1998.

[9] F. Hamaratoglu, M. Willecke, M. Kango-Singh et al., "The tumour-suppressor genes NF2/Merlin and expanded act through hippo signalling to regulate cell proliferation and apoptosis," *Nature Cell Biology*, vol. 8, no. 1, pp. 27–36, 2006.

[10] S. Maitra, R. M. Kulikauskas, H. Gavilan, and R. G. Fehon, "The tumor suppressors Merlin and expanded function cooperatively to modulate receptor endocytosis and signaling," *Current Biology*, vol. 16, no. 7, pp. 702–709, 2006.

[11] P. A. Mahoney, U. Weber, P. Onofrechuk, H. Biessmann, P. J. Bryant, and C. S. Goodman, "The fat tumor suppressor gene in *Drosophila* encodes a novel member of the cadherin gene superfamily," *Cell*, vol. 67, no. 5, pp. 853–868, 1991.

[12] F. C. Bennett and K. F. Harvey, "Fat cadherin modulates organ size in *Drosophila* via the Salvador/Warts/hippo signaling pathway," *Current Biology*, vol. 16, no. 21, pp. 2101–2110, 2006.

[13] E. Silva, Y. Tsatskis, L. Gardano, N. Tapon, and H. McNeill, "The tumor-suppressor gene fat controls tissue growth upstream of expanded in the hippo signaling pathway," *Current Biology*, vol. 16, no. 21, pp. 2081–2089, 2006.

[14] M. Willecke, F. Hamaratoglu, M. Kango-Singh et al., "The fat cadherin acts through the hippo tumor-suppressor pathway to regulate tissue size," *Current Biology*, vol. 16, no. 21, pp. 2090–2100, 2006.

[15] H. Matakatsu and S. S. Blair, "Interactions between Fat and Dachsous and the regulation of planar cell polarity in the *Drosophila* wing," *Development*, vol. 131, no. 15, pp. 3785–3794, 2004.

[16] H. Matakatsu and S. S. Blair, "Separating the adhesive and signaling functions of the Fat and Dachsous protocadherins," *Development*, vol. 133, no. 12, pp. 2315–2324, 2006.

[17] E. Cho, Y. Feng, C. Rauskolb, S. Maitra, R. Fehon, and K. D. Irvine, "Delineation of a Fat tumor suppressor pathway," *Nature Genetics*, vol. 38, no. 10, pp. 1142–1150, 2006.

[18] Y. Feng and K. D. Irvine, "Processing and phosphorylation of the Fat receptor," *Proceedings of the National Academy of Sciences of the United States of America*, vol. 106, no. 29, pp. 11989–11994, 2009.

[19] R. Baumgartner, I. Poernbacher, N. Buser, E. Hafen, and H. Stocker, "The WW domain protein kibra acts upstream of hippo in *Drosophila*," *Developmental Cell*, vol. 18, no. 2, pp. 309–316, 2010.

[20] A. Genevet, M. C. Wehr, R. Brain, B. J. Thompson, and N. Tapon, "Kibra is a regulator of the Salvador/Warts/hippo signaling network," *Developmental Cell*, vol. 18, no. 2, pp. 300–308, 2010.

[21] J. Yu, Y. Zheng, J. Dong, S. Klusza, W. M. Deng, and D. Pan, "Kibra functions as a tumor suppressor protein that regulates hippo signaling in conjunction with Merlin and Expanded," *Developmental Cell*, vol. 18, no. 2, pp. 288–299, 2010.

[22] C. L. Chen, K. M. Gajewski, F. Hamaratoglu et al., "The apical-basal cell polarity determinant Crumbs regulates

hippo signaling in *Drosophila*," *Proceedings of the National Academy of Sciences of the United States of America*, vol. 107, no. 36, pp. 15810–15815, 2010.

[23] C. Ling, Y. Zheng, F. Yin et al., "The apical transmembrane protein Crumbs functions as a tumor suppressor that regulates hippo signaling by binding to expanded," *Proceedings of the National Academy of Sciences of the United States of America*, vol. 107, no. 23, pp. 10532–10537, 2010.

[24] B. S. Robinson, J. Huang, Y. Hong, and K. H. Moberg, "Crumbs regulates Salvador/Warts/hippo signaling in *Drosophila* via the FERM-domain protein expanded," *Current Biology*, vol. 20, no. 7, pp. 582–590, 2010.

[25] J. Huang, S. Wu, J. Barrera, K. Matthews, and D. Pan, "The hippo signaling pathway coordinately regulates cell proliferation and apoptosis by inactivating Yorkie, the *Drosophila* homolog of YAP," *Cell*, vol. 122, no. 3, pp. 421–434, 2005.

[26] M. Sudol, H. I. Chen, C. Bougeret, A. Einbond, and P. Bork, "Characterization of a novel protein-binding module—the WW domain," *FEBS Letters*, vol. 369, no. 1, pp. 67–71, 1995.

[27] M. Sudol, "Newcomers to the WW domain-mediated network of the hippo tumor suppressor pathway," *Genes and Cancer*, vol. 1, no. 11, pp. 1115–1118, 2010.

[28] M. Sudol and K. F. Harvey, "Modularity in the hippo signaling pathway," *Trends in Biochemical Sciences*, vol. 35, no. 11, pp. 627–633, 2010.

[29] U. Tepass, C. Theres, and E. Knust, "*Crumbs* encodes an EGF-like protein expressed on apical membranes of *Drosophila* epithelial cells and required for organization of epithelia," *Cell*, vol. 61, no. 5, pp. 787–799, 1990.

[30] C. H. Yang, J. D. Axelrod, and M. A. Simon, "Regulation of Frizzled by Fat-like cadherins during planar polarity signaling in the *Drosophila* compound eye," *Cell*, vol. 108, no. 5, pp. 675–688, 2002.

[31] J. Casal, P. A. Lawrence, and G. Struhl, "Two separate molecular systems, Dachsous/Fat and Starry night/Frizzled, act independently to confer planar star polarity," *Development*, vol. 133, no. 22, pp. 4561–4572, 2006.

[32] Y. Feng and K. D. Irvine, "Fat and expanded act in parallel to regulate growth through Warts," *Proceedings of the National Academy of Sciences of the United States of America*, vol. 104, no. 51, pp. 20362–20367, 2007.

[33] C. Polesello and N. Tapon, "Salvador-Warts-hippo signaling promotes *Drosophila* posterior follicle cell maturation downstream of notch," *Current Biology*, vol. 17, no. 21, pp. 1864–1870, 2007.

[34] B. V. V. G. Reddy, C. Rauskolb, and K. D. Irvine, "Influence of Fat-hippo and Notch signaling on the proliferation and differentiation of *Drosophila* optic neuroepithelia," *Development*, vol. 137, no. 14, pp. 2397–2408, 2010.

[35] D. Rogulja, C. Rauskolb, and K. D. Irvine, "Morphogen control of wing growth through the Fat signaling pathway," *Developmental Cell*, vol. 15, no. 2, pp. 309–321, 2008.

[36] M. Willecke, F. Hamaratoglu, L. Sansores-Garcia, C. Tao, and G. Halder, "Boundaries of Dachsous Cadherin activity modulate the hippo signaling pathway to induce cell proliferation," *Proceedings of the National Academy of Sciences of the United States of America*, vol. 105, no. 39, pp. 14897–14902, 2008.

[37] H. O. Ishikawa, H. Takeuchi, R. S. Haltiwanger, and K. D. Irvine, "Four-jointed is a Golgi kinase that phosphorylates a subset of cadherin domains," *Science*, vol. 321, no. 5887, pp. 401–404, 2008.

[38] H. Matakatsu and S. S. Blair, "The DHHC palmitoyltransferase approximated regulates Fat signaling and Dachs

localization and activity," *Current Biology*, vol. 18, no. 18, pp. 1390–1395, 2008.

[39] C. Rauskolb, G. Pan, B. V. V. G. Reddy, H. Oh, and K. D. Irvine, "Zyxin links fat signaling to the hippo pathway," *PLoS Biology*, vol. 9, no. 6, Article ID e1000624, 2011.

[40] K. Harvey and N. Tapon, "The Salvador-Warts-hippo pathway—an emerging tumour-suppressor network," *Nature Reviews Cancer*, vol. 7, no. 3, pp. 182–191, 2007.

[41] C. Polesello, S. Huelsmann, N. Brown, and N. Tapon, "The *Drosophila rassf* homolog antagonizes the hippo pathway," *Current Biology*, vol. 16, no. 24, pp. 2459–2465, 2006.

[42] P. S. Ribeiro, F. Josué, A. Wepf et al., "Combined functional genomic and proteomic approaches identify a PP2A complex as a negative regulator of hippo signaling," *Molecular Cell*, vol. 39, no. 4, pp. 521–534, 2010.

[43] N. A. Grzeschik, L. M. Parsons, M. L. Allott, K. F. Harvey, and H. E. Richardson, "Lgl, aPKC, and Crumbs regulate the Salvador/Warts/hippo pathway through two distinct mechanisms," *Current Biology*, vol. 20, no. 7, pp. 573–581, 2010.

[44] S. Dong, S. Kang, T. L. Gu et al., "14-3-3 Integrates prosurvival signals mediated by the AKT and MAPK pathways in ZNF198-FGFR1-transformed hematopoietic cells," *Blood*, vol. 110, no. 1, pp. 360–369, 2007.

[45] H. Oh and K. D. Irvine, "In vivo regulation of Yorkie phosphorylation and localization," *Development*, vol. 135, no. 6, pp. 1081–1088, 2008.

[46] M. Praskova, F. Xia, and J. Avruch, "MOBKL1A/MOBKL1B Phosphorylation by MST1 and MST2 Inhibits Cell Proliferation," *Current Biology*, vol. 18, no. 5, pp. 311–321, 2008.

[47] B. Zhao, X. Wei, W. Li et al., "Inactivation of YAP oncoprotein by the hippo pathway is involved in cell contact inhibition and tissue growth control," *Genes and Development*, vol. 21, no. 21, pp. 2747–2761, 2007.

[48] C. Badouel, L. Gardano, N. Amin et al., "The FERM-domain protein Expanded regulates hippo pathway activity via direct interactions with the transcriptional activator Yorkie," *Developmental Cell*, vol. 16, no. 3, pp. 411–420, 2009.

[49] H. Oh, B. V. V. G. Reddy, and K. D. Irvine, "Phosphorylation-independent repression of Yorkie in Fat-hippo signaling," *Developmental Biology*, vol. 335, no. 1, pp. 188–197, 2009.

[50] S. Wu, Y. Liu, Y. Zheng, J. Dong, and D. Pan, "The TEAD/TEF family protein Scalloped mediates transcriptional output of the hippo growth-regulatory pathway," *Developmental cell*, vol. 14, no. 3, pp. 388–398, 2008.

[51] L. Zhang, F. Ren, Q. Zhang, Y. Chen, B. Wang, and J. Jiang, "The TEAD/TEF family of transcription factor Scalloped mediates hippo signaling in organ size control," *Developmental cell*, vol. 14, no. 3, pp. 377–387, 2008.

[52] H. W. Peng, M. Slattery, and R. S. Mann, "Transcription factor choice in the hippo signaling pathway: homothorax and yorkie regulation of the microRNA bantam in the progenitor domain of the *Drosophila* eye imaginal disc," *Genes and Development*, vol. 23, no. 19, pp. 2307–2319, 2009.

[53] Y. Goulev, J. D. Fauny, B. Gonzalez-Marti, D. Flagiello, J. Silber, and A. Zider, "SCALLOPED interacts with YORKIE, the nuclear effector of the hippo tumor-suppressor pathway in *Drosophila*," *Current Biology*, vol. 18, no. 6, pp. 435–441, 2008.

[54] R. M. Neto-Silva, S. de Beco, and L. A. Johnston, "Evidence for a growth-stabilizing regulatory feedback mechanism between Myc and Yorkie, the *Drosophila* homolog of Yap," *Developmental Cell*, vol. 19, no. 4, pp. 507–520, 2010.

[55] M. Ziosi, L. A. Baena-López, D. Grifoni et al., "dMyc functions downstream of yorkie to promote the supercompetitive behavior of hippo pathway mutant cells," *PLoS Genetics*, vol. 6, no. 9, Article ID e1001140, 2010.

[56] C. Alarcón, A. I. Zaromytidou, Q. Xi et al., "Nuclear CDKs drive smad transcriptional activation and turnover in BMP and TGF-β pathways," *Cell*, vol. 139, no. 4, pp. 757–769, 2009.

[57] A. Genevet, C. Polesello, K. Blight et al., "The hippo pathway regulates apical-domain size independently of its growth-control function," *Journal of Cell Science*, vol. 122, pp. 2360–2370, 2009.

[58] A. Hergovich and B. A. Hemmings, "Mammalian NDR/LATS protein kinases in hippo tumor suppressor signaling," *BioFactors*, vol. 35, no. 4, pp. 338–345, 2009.

[59] M. Radu and J. Chernoff, "The DeMSTification of mammalian Ste20 kinases," *Current Biology*, vol. 19, no. 10, pp. R421–R425, 2009.

[60] Q. Y. Lei, H. Zhang, B. Zhao et al., "TAZ promotes cell proliferation and epithelial-mesenchymal transition and is inhibited by the hippo pathway," *Molecular and Cellular Biology*, vol. 28, no. 7, pp. 2426–2436, 2008.

[61] F. D. Camargo, S. Gokhale, J. B. Johnnidis et al., "YAP1 Increases Organ Size and Expands Undifferentiated Progenitor Cells," *Current Biology*, vol. 17, no. 23, pp. 2054–2060, 2007.

[62] J. Dong, G. Feldmann, J. Huang et al., "Elucidation of a Universal Size-Control Mechanism in *Drosophila* and Mammals," *Cell*, vol. 130, no. 6, pp. 1120–1133, 2007.

[63] D. Zhou, C. Conrad, F. Xia et al., "Mst1 and Mst2 Maintain Hepatocyte Quiescence and Suppress Hepatocellular Carcinoma Development through Inactivation of the Yap1 Oncogene," *Cancer Cell*, vol. 16, no. 5, pp. 425–438, 2009.

[64] K. P. Lee, J. H. Lee, T. S. Kim et al., "The hippo-Salvador pathway restrains hepatic oval cell proliferation, liver size, and liver tumorigenesis," *Proceedings of the National Academy of Sciences of the United States of America*, vol. 107, no. 18, pp. 8248–8253, 2010.

[65] L. Lu, Y. Li, S. M. Kim et al., "hippo signaling is a potent in vivo growth and tumor suppressor pathway in the mammalian liver," *Proceedings of the National Academy of Sciences of the United States of America*, vol. 107, no. 4, pp. 1437–1442, 2010.

[66] H. Song, K. K. Mak, L. Topol et al., "Mammalian Mst1 and Mst2 kinases play essential roles in organ size control and tumor suppression," *Proceedings of the National Academy of Sciences of the United States of America*, vol. 107, no. 4, pp. 1431–1436, 2010.

[67] M. Sudol, P. Bork, A. Einbond et al., "Characterization of the mammalian YAP (Yes-associated protein) gene and its role in defining a novel protein module, the WW domain," *Journal of Biological Chemistry*, vol. 270, no. 24, pp. 14733–14741, 1995.

[68] E. Bertini, T. Oka, M. Sudol, S. Strano, and G. Blandino, "At the crossroad between transformation and tumor suppression," *Cell Cycle*, vol. 8, no. 1, pp. 49–57, 2009.

[69] A. Sawada, H. Kiyonari, K. Ukita, N. Nishioka, Y. Imuta, and H. Sasaki, "Redundant roles of Tead1 and Tead2 in notochord development and the regulation of cell proliferation and survival," *Molecular and Cellular Biology*, vol. 28, no. 10, pp. 3177–3189, 2008.

[70] M. Ota and H. Sasaki, "Mammalian Tead proteins regulate cell proliferation and contact inhibition as transcriptional mediators of hippo signaling," *Development*, vol. 135, no. 24, pp. 4059–4069, 2008.

[71] B. Zhao, X. Ye, J. Yu et al., "TEAD mediates YAP-dependent gene induction and growth control," *Genes and Development*, vol. 22, no. 14, pp. 1962–1971, 2008.

[72] S. W. Chan, C. J. Lim, L. S. Loo, Y. F. Chong, C. Huang, and W. Hong, "TEADs mediate nuclear retention of TAZ to promote oncogenic transformation," *Journal of Biological Chemistry*, vol. 284, no. 21, pp. 14347–14358, 2009.

[73] S. Dupont, L. Morsut, M. Aragona et al., "Role of YAP/TAZ in mechanotransduction," *Nature*, vol. 474, no. 7350, pp. 179–183, 2011.

[74] M. Sudol and T. Hunter, "New wrinkles for an old domain," *Cell*, vol. 103, no. 7, pp. 1001–1004, 2000.

[75] S. Strano, E. Munarriz, M. Rossi et al., "Physical Interaction with Yes-associated Protein Enhances p73 Transcriptional Activity," *Journal of Biological Chemistry*, vol. 276, no. 18, pp. 15164–15173, 2001.

[76] S. Strano, O. Monti, N. Pediconi et al., "The transcriptional coactivator yes-associated protein drives p73 gene-target specificity in response to DNA damage," *Molecular Cell*, vol. 18, no. 4, pp. 447–459, 2005.

[77] D. Matallanas, D. Romano, K. Yee et al., "*rassf1a* Elicits Apoptosis through an MST2 Pathway Directing Proapoptotic Transcription by the p73 Tumor Suppressor Protein," *Molecular Cell*, vol. 27, no. 6, pp. 962–975, 2007.

[78] E. Lapi, S. Di Agostino, S. Donzelli et al., "PML, YAP, and p73 Are Components of a Proapoptotic Autoregulatory Feedback Loop," *Molecular Cell*, vol. 32, no. 6, pp. 803–814, 2008.

[79] M. J. Boedigheimer, K. P. Nguyen, and P. J. Bryant, "expanded functions in the apical cell domain to regulate the growth rate of imaginal discs," *Developmental Genetics*, vol. 20, no. 2, pp. 103–110, 1997.

[80] G. A. Rouleau, P. Merel, M. Lutchman et al., "Alteration in a new gene encoding a putative membrane-organizing protein causes neuro-fibromatosis type 2," *Nature*, vol. 363, no. 6429, pp. 515–521, 1993.

[81] Trofatter, "Erratum: A novel moesin-, ezrin-, and radixin-like gene is a candidate for the neurofibromatosis 2 tumor suppressor," *Cell*, vol. 75, no. 4, pp. 791–800, 1993.

[82] M. Curto, B. K. Cole, D. Lallemand, C. H. Liu, and A. I. McClatchey, "Contact-dependent inhibition of EGFR signaling by Nf2/Merlin," *Journal of Cell Biology*, vol. 177, no. 5, pp. 893–903, 2007.

[83] N. Zhang, H. Bai, K. K. David et al., "The Merlin/NF2 Tumor Suppressor Functions through the YAP Oncoprotein to Regulate Tissue Homeostasis in Mammals," *Developmental Cell*, vol. 19, no. 1, pp. 27–38, 2010.

[84] S. Benhamouche, M. Curto, I. Saotome et al., "Nf2/Merlin controls progenitor homeostasis and tumorigenesis in the liver," *Genes and Development*, vol. 24, no. 16, pp. 1718–1730, 2010.

[85] Y. Lin, A. Khokhlatchev, D. Figeys, and J. Avruch, "Death-associated protein 4 binds MST1 and augments MST1-induced apoptosis," *Journal of Biological Chemistry*, vol. 277, no. 50, pp. 47991–48001, 2002.

[86] S. Ura, H. Nishina, Y. Gotoh, and T. Katada, "Activation of the c-Jun N-terminal kinase pathway by MST1 is essential and sufficient for the induction of chromatin condensation during apoptosis," *Molecular and Cellular Biology*, vol. 27, no. 15, pp. 5514–5522, 2007.

[87] R. Anand, A. Y. Kim, M. Brent, and R. Marmorstein, "Biochemical analysis of MST1 kinase: Elucidation of a C-terminal regulatory region," *Biochemistry*, vol. 47, no. 25, pp. 6719–6726, 2008.

[88] E. O'Neill, L. Rushworth, M. Baccarini, and W. Kolch, "Role of the kinase MST2 in suppression of apoptosis by the proto-oncogene product Raf-1," *Science*, vol. 306, no. 5705, pp. 2267–2270, 2004.

[89] C. Guo, X. Zhang, and G. P. Pfeifer, "The tumor suppressor *rassf1a* prevents dephosphorylation of the mammalian STE20-like kinases MST1 and MST2," *Journal of Biological Chemistry*, vol. 286, no. 8, pp. 6253–6261, 2011.

[90] X. Yang, D. M. Li, W. Chen, and T. Xu, "Human homologue of *Drosophila* lats, LATS1, negatively regulate growth by inducing G2/M arrest or apoptosis," *Oncogene*, vol. 20, no. 45, pp. 6516–6523, 2001.

[91] H. Xia, H. Qi, Y. Li et al., "LATS1 tumor suppressor regulates G2/M transition and apoptosis," *Oncogene*, vol. 21, no. 8, pp. 1233–1241, 2002.

[92] N. Yabuta, N. Okada, A. Ito et al., "LATS2 is an essential mitotic regulator required for the coordination of cell division," *Journal of Biological Chemistry*, vol. 282, no. 26, pp. 19259–19271, 2007.

[93] Y. Li, J. Pei, H. Xia, H. Ke, H. Wang, and W. Tao, LATS2, a putative tumor suppressor, inhibits G1/S transition," *Oncogene*, vol. 22, no. 28, pp. 4398–4405, 2003.

[94] Y. Aylon, N. Yabuta, H. Besserglick et al., "Silencing of the LATS2 tumor suppressor overrides a p53-dependent oncogenic stress checkpoint and enables mutant H-Ras-driven cell transformation," *Oncogene*, vol. 28, no. 50, pp. 4469–4479, 2009.

[95] M. A. R. St John, W. Tao, X. Fei et al., "Mice deficient of Lats1 develop soft-tissue sarcomas, ovarian tumours and pituitary dysfunction," *Nature Genetics*, vol. 21, no. 2, pp. 182–186, 1999.

[96] Y. Aylon, D. Michael, A. Shmueli, N. Yabuta, H. Nojima, and M. Oren, "A positive feedback loop between the p53 and LATS2 tumor suppressors prevents tetraploidization," *Genes and Development*, vol. 20, no. 19, pp. 2687–2700, 2006.

[97] K. Zhang, E. Rodriguez-Aznar, N. Yabuta et al., LATS2 kinase potentiates Snail1 activity by promoting nuclear retention upon phosphorylation," *EMBO Journal*, vol. 31, no. 1, pp. 29–43, 2012.

[98] K. Tschöp, A. R. Conery, L. Litovchick et al., "A kinase shRNA screen links LATS2 and the pRB tumor suppressor," *Genes and Development*, vol. 25, no. 8, pp. 814–830, 2011.

[99] T. Hori, A. Takaori-Kondo, Y. Kamikubo, and T. Uchiyama, "Molecular cloning of a novel human protein kinase, kpm, that is homologous to warts/lats, a *Drosophila* tumor suppressor," *Oncogene*, vol. 19, no. 27, pp. 3101–3109, 2000.

[100] L. van der Weyden and D. J. Adams, "The Ras-association domain family (RASSF) members and their role in human tumourigenesis," *Biochimica et Biophysica Acta - Reviews on Cancer*, vol. 1776, no. 1, pp. 58–85, 2007.

[101] M. Gordon and S. Baksh, "*rassf1a*: Not a prototypical Ras effector," *Small GTPases*, vol. 2, no. 3, pp. 148–157, 2011.

[102] S. Baksh, S. Tommasi, S. Fenton et al., "The tumor suppressor *rassf1a* and MAP-1 link death receptor signaling to bax conformational change and cell death," *Molecular Cell*, vol. 18, no. 6, pp. 637–650, 2005.

[103] R. Dammann, C. Li, J. H. Yoon, P. L. Chin, S. Bates, and G. P. Pfeifer, "Epigenetic inactivation of a RAS association domain family protein from the lung tumour suppressor locus 3p21.3," *Nature Genetics*, vol. 25, no. 3, pp. 315–319, 2000.

[104] D. G. Burbee, E. Forgacs, S. Zöchbauer-Müller et al., "Epigenetic inactivation of *rassf1a* in lung and breast cancers and malignant phenotype suppression," *Journal of the National Cancer Institute*, vol. 93, no. 9, pp. 691–699, 2001.

[105] I. Kuzmin, J. W. Gillespie, A. Protopopov et al., "The *rassf1a* tumor suppressor gene is inactivated in prostate tumors and suppresses growth of prostate carcinoma cells," *Cancer Research*, vol. 62, no. 12, pp. 3498–3502, 2002.

[106] L. Shivakumar, J. Minna, T. Sakamaki, R. Pestell, and M. A. White, "The *rassf1a* tumor suppressor blocks cell cycle progression and inhibits cyclin D1 accumulation," *Molecular and Cellular Biology*, vol. 22, no. 12, pp. 4309–4318, 2002.

[107] Z. G. Pan, V. I. Kashuba, X. Q. Liu et al., "High frequency somatic mutations in *rassf1a* in nasopharyngeal carcinoma," *Cancer Biology and Therapy*, vol. 4, no. 10, pp. 1116–1122, 2005.

[108] S. Tommasi, R. Dammann, Z. Zhang et al., "Tumor susceptibility of *rassf1a* knockout mice," *Cancer Research*, vol. 65, no. 1, pp. 92–98, 2005.

[109] L. van der Weyden, K. K. Tachibana, M. A. Gonzalez et al., "The *rassf1a* isoform of RASSF1 promotes microtubule stability and suppresses tumorigenesis," *Molecular and Cellular Biology*, vol. 25, no. 18, pp. 8356–8367, 2005.

[110] G. Hamilton, K. S. Yee, S. Scrace, and E. O'Neill, "ATM Regulates a *rassf1a*-Dependent DNA Damage Response," *Current Biology*, vol. 19, no. 23, pp. 2020–2025, 2009.

[111] S. T. Kim, D. S. Lim, C. E. Canman, and M. B. Kastan, "Substrate specificities and identification of putative substrates of ATM kinase family members," *Journal of Biological Chemistry*, vol. 274, no. 53, pp. 37538–37543, 1999.

[112] T. O'Neill, A. J. Dwyer, Y. Ziv et al., "Utilization of oriented peptide libraries to identify substrate motifs selected by ATM," *Journal of Biological Chemistry*, vol. 275, no. 30, pp. 22719–22727, 2000.

[113] C. Guo, S. Tommasi, L. Liu, J. K. Yee, R. Dammann, and G. Pfeifer, "*rassf1a* Is Part of a Complex Similar to the *Drosophila* hippo/Salvador/Lats Tumor-Suppressor Network," *Current Biology*, vol. 17, no. 8, pp. 700–705, 2007.

[114] H. Donninger, N. Allen, A. Henson et al., "Salvador protein is a tumor suppressor effector of *rassf1a* with hippo pathway-independent functions," *Journal of Biological Chemistry*, vol. 286, no. 21, pp. 18483–18491, 2011.

[115] L. J. Saucedo and B. A. Edgar, "Filling out the hippo pathway," *Nature Reviews Molecular Cell Biology*, vol. 8, no. 8, pp. 613–621, 2007.

[116] D. Oceandy, A. Pickard, S. Prehar et al., "Tumor suppressor ras-association domain family 1 Isoform A Is a novel regulator of cardiac hypertrophy," *Circulation*, vol. 120, no. 7, pp. 607–616, 2009.

[117] G. W. Dorn, J. Robbins, and P. H. Sugden, "Phenotyping hypertrophy: Eschew obfuscation," *Circulation Research*, vol. 92, no. 11, pp. 1171–1175, 2003.

[118] E. Marban and Y. Koretsune, "Cell calcium, oncogenes, and hypertrophy," *Hypertension*, vol. 15, no. 6, pp. 652–658, 1990.

[119] S. Rabizadeh, R. J. Xavier, K. Ishiguro et al., "The scaffold protein CNK1 interacts with the tumor suppressor *rassf1a* and augments *rassf1a*-induced cell death," *Journal of Biological Chemistry*, vol. 279, no. 28, pp. 29247–29254, 2004.

[120] D. P. Del Re, T. Matsuda, P. Zhai et al., "Proapoptotic *rassf1a*/Mst1 signaling in cardiac fibroblasts is protective against pressure overload in mice," *Journal of Clinical Investigation*, vol. 120, no. 10, pp. 3555–3567, 2010.

[121] A. von Gise, Z. Lin, K. Schlegelmilch et al., "YAP1, the nuclear target of hippo signaling, stimulates heart growth through cardiomyocyte proliferation but not hypertrophy,"

Proceedings of the National Academy of Sciences of the United States of America, vol. 109, no. 7, pp. 2394–2399, 2012.

[122] Y. Mao, J. Mulvaney, S. Zakaria et al., "Characterization of a Dchs1 mutant mouse reveals requirements for Dchs1-Fat4 signaling during mammalian development," *Development*, vol. 138, no. 5, pp. 947–957, 2011.

[123] Y. Matsui, N. Nakano, D. Shao et al., "LATS2 is a negative regulator of myocyte size in the heart," *Circulation Research*, vol. 103, no. 11, pp. 1309–1318, 2008.

[124] T. Heallen, M. Zhang, J. Wang et al., "hippo pathway inhibits wnt signaling to restrain cardiomyocyte proliferation and heart size," *Science*, vol. 332, no. 6028, pp. 458–461, 2011.

[125] D. Vavvas, X. Li, J. Avruch, and X. F. Zhang, "Identification of Nore1 as a potential Ras effector," *Journal of Biological Chemistry*, vol. 273, no. 10, pp. 5439–5442, 1998.

[126] A. Moshnikova, J. Frye, J. W. Shay, J. D. Minna, and A. V. Khokhlatchev, "The growth and tumor suppressor NORE1A is a cytoskeletal protein that suppresses growth by inhibition of the ERK pathway," *Journal of Biological Chemistry*, vol. 281, no. 12, pp. 8143–8152, 2006.

[127] A. Khokhlatchev, S. Rabizadeh, R. Xavier et al., "Identification of a novel Ras-regulated proapoptotic pathway," *Current Biology*, vol. 12, no. 4, pp. 253–265, 2002.

[128] M. Praskova, A. Khoklatchev, S. Ortiz-Vega, and J. Avruch, "Regulation of the MST1 kinase by autophosphorylation, by the growth inhibitory proteins, RASSF1 and NORE1, and by Ras," *Biochemical Journal*, vol. 381, pp. 453–462, 2004.

[129] H. J. Oh, K. K. Lee, S. J. Song et al., "Role of the tumor suppressor *rassf1a* in Mst1-mediated apoptosis," *Cancer Research*, vol. 66, no. 5, pp. 2562–2569, 2006.

[130] N. P. C. Allen, H. Donninger, M. D. Vos et al., "RASSF6 is a novel member of the RASSF family of tumor suppressors," *Oncogene*, vol. 26, no. 42, pp. 6203–6211, 2007.

[131] M. Ikeda, A. Kawata, M. Nishikawa et al., "hippo pathway-dependent and-independent roles of *rassf6*," *Science Signaling*, vol. 2, no. 90, Article ID ra59, 2009.

10

RASSF1 Polymorphisms in Cancer

Marilyn Gordon,[1, 2] Mohamed El-Kalla,[1, 2] and Shairaz Baksh[1, 2]

[1] Department of Pediatrics, Faculty of Medicine and Dentistry, University of Alberta, 3-055 Katz Group Centre for Pharmacy and Health Research, 113 Street 87 Avenue, Edmonton, AB, Canada T6G 2E1
[2] Women and Children's Health Research Institute, University of Alberta, 4-081 Edmonton Clinic Health Academy, 11405-87 Avenue, Edmonton, AB, Canada T6G 1C9

Correspondence should be addressed to Shairaz Baksh, sbaksh@ualberta.ca

Academic Editor: Geoffrey J. Clark

Ras association domain family 1A (RASSF1A) is one of the most epigenetically silenced elements in human cancers. Localized on chromosome 3, it has been demonstrated to be a bone fide tumor suppressor influencing cell cycle events, microtubule stability, apoptosis, and autophagy. Although it is epigenetically silenced by promoter-specific methylation in cancers, several somatic nucleotide changes (polymorphisms) have been identified in RASSF1A in tissues from cancer patients. We speculate that both nucleotide changes and epigenetic silencing result in loss of the RASSF1A tumor suppressor function and the appearance of enhanced growth. This paper will summarize what is known about the origin of these polymorphisms and how they have helped us understand the biological role of RASSF1A.

1. Introduction

Cancer is a disease affecting 1 in 3 adults worldwide and is considered to be the second leading cause of death in both Canada and the United States behind heart disease [1, 2]. It is thought that cancer arises due to the occurrence of 2–5 genetic events to potentiate tumor formation and sustain abnormal growth [3]. These genetic changes occur in passenger genes (to support the cancer phenotype) and driver genes (to promote the cancer phenotype) [4]. About 10% of driver genes code for oncogenes that promote accelerated growth. However, about 90% of the driver genes code for tumor suppressor genes that inhibit accelerated growth [3], suggesting that tumor suppressor genes play an integral part in the origin of cancer. Evidence also suggests that the mutation rate of tumor suppressor genes are much higher than oncogenes supporting their importance in cancer formation [3, 5].

In 2000, Hanahan and Weinberg systematically described several key features or "hallmarks" of cancer that defined the behavior of a cancer cell [6]. These defining features of a cancer cell included the unique properties of limitless replicative potential, evasion of apoptosis, ability to stimulate neo-vascularization, invasion and metastasis, inhibition of suppressor pathways, and sustained proliferation. As described in their seminal paper, the aforementioned hallmarks are acquired through a "multistep process" that allows the cancer cells to acquire key survival traits while avoiding the watchful eye of established molecular "checkpoints" to inhibit abnormal growth [7]. It was around this time that the RASSF1 was identified as a potential tumor suppressor gene on chromosome 3, at 3p21.23 [8, 9]. Now more than a decade later, RASSF1A has been demonstrated using numerous approaches to be a tumor suppressor gene and an important driver gene in cancer influencing/intersecting with many of the hallmarks of cancer [8, 10]. It is epigenetically silenced in the majority of cancers by promoter specific methylation, resulting in loss of expression of the RASSF1A protein [11]. Although expression loss of RASSF1A by methylation occurs frequently in cancer, nucleotide changes by somatic mechanisms have also been detected in patients from several cancer subtypes. Several studies have tried to elucidate the importance of these polymorphic changes and how it may affect the tumor suppressor function of RASSF1A. They have also revealed interesting and surprising influences on numerous aspects of biology.

2. The Origin of RASSF1A Polymorphisms

The RASSF1 gene consists of eight exons alternatively spliced to produce 8 isoforms, RASSF1A-H, that have distinct functional domains including the Ras association (RA) domain [9, 14]. Of these, RASSF1A and RASSF1C are the predominant ubiquitously expressed forms in normal tissues [9, 11]. RASSF1C has been demonstrated to be perinuclear in appearance in NCI H1299 lung cancer cells [15], nuclear in HeLa cells with translocation to the cytosol upon DNA damage [16], and localized to microtubules in a similar fashion to RASSF1A in 293T cells [17, 18]. Thus, the localization of RASSF1C is varied and controversial. This is not the case for RASSF1A as it has been demonstrated by our group and several others to be a microtubule binding protein having a microtubule like localization and functioning to stabilize tubulin in a taxol like manner [13, 18, 19]. To date, a crystal structure for RASSF1A or RASSF1C has not been identified, but Foley et al. [20] provided a molecular model of the N-terminal C1 domain containing four zinc finger motifs which is very similar to the one found on RASSF5A/Nore1A [21]. The zinc finger motifs have now been demonstrated to be involved in death receptor associations and possible associations with other receptors or signaling components [20]. In addition to the C1 domain, RASSF1A has been noted to have a sequence specificity motifs to associate with SH3 domain (PxxP); motifs for 14-3-3 associations; a Ras association (RA) domain (although association is weak or indirect for K-Ras) [10]; associations with the anaphase promoting complex protein cdc20 and the autophagy protein C19ORF5/MAP1S; and heterotrophic associations with the Hippo proapoptotic kinase (MST1/2) and the BH3-like protein modulator of apoptosis 1 (MOAP-1) through the Salvador/RASSF/Hippo (SARAH) domain (both reviewed elsewhere in this issue) (please see Figure 1 for schematic summary of RASSF1A protein associations).

RASSF1A polymorphisms have been identified in several cancers as listed in Table 1 and can be mapped to specific protein interaction domains (Figure 1). These polymorphisms have been found in tumors from numerous cancer patients and cell lines [22]. The population distribution and significance of these alterations in tumorigenesis remain to be determined but do vary from 9% to 33% of the specific cancer population. The majority of RASSF1A polymorphisms have been confirmed using several approaches as outlined by the 1000Genome project (http://www.1000genomes.org/), HapMap project (http://hapmap.ncbi.nlm.nih.gov/) and submitted by multiple sources (Table 1 and NCBI SNP database [http://www.ncbi.nlm.nih.gov/projects/SNP/snp_ref.cgi?showRare=on&chooseRs=coding&go=Go&locusId=11186] and University of Maryland SNP database [http://bioinf.umbc.edu/dmdm/gene_prot_page.php?search_type=gene&id=11186, NP_009113]). Recently, a comprehensive study of 400 lung, renal, breast, cervical, and ovarian cancers by Kashuba et al. [22] revealed frequent loss of genetic material on

chromosome 3p in 90% of the tumors investigated. Furthermore, they determined that the mutation rate in cancer for RASSF1A was 0.42 mutation frequency/100 base pair whereas in the "normal" population was about 0.10 mutations/100 base pairs. They speculate that RASSF1A has a 73% GC content within exons 1–2 which may explain the high mutation rate of RASSF1A within cancer cells. Within cell lines, RASSF1A was found to carry 0.7 mutations/100 bp in the Burkitt's lymphoma-derived cell lines, BL2 and RAMOS, whereas it was 0.14 in the renal carcinoma cell line KRC/Y and, with each division of the BL2 lymphoma line, transitional mutations were observed. Interestingly, codon changes in RASSF1A were also observed in 15 normal human hearts that included two nucleotide changes (CTA to CTG and GTA to GTG) but no amino acid changes [22]. They speculate that RASSF is simply located in an area that is "extensively damaged" and susceptible to mutational pressures in 90% of epithelial cancers [22].

The most common polymorphism is the alanine (A) to serine (S) at amino acid 133 (A133S) located within the ATM DNA damage checkpoint kinase site (please see below sections). This has been identified as a single nucleotide germ line polymorphism (SNP) on both alleles in some breast cancer patients and is significantly associated with BRAC1/2 mutations. Patients with wild-type BRAC1/2 and RASSF1A A133S have a +15-year better survival period than those harboring both BRAC1/2 mutations and RASSF1A A133S [23, 29]. The RASSF1A A133S SNP has been found in 20.6% of patients with breast carcinomas [23, 29], 19.8% in lung cancer [29, 32], 11.1% in head and neck cancer [32], 6.9% in colorectal cancer [32], 14.3% in esophageal cancer [32], 24.3% in patients with fibroadenoma and in 2.9%–10% of healthy controls [23, 32]. Interestingly, Gao et al. [29] also revealed the presence of the A133S polymorphism in brain and kidney cancer patients and Bergqvist et al. (2010) detected the presence of the A133S SNP in 18.4% of the white British female population [28]. The high percent obtained for the latter is surprising and requires further validation. The prevalence of the rest of the RASSF1A polymorphisms has not been determined yet, and functional studies to systematically determine influence of these polymorphisms on RASSF1A biological function are yet to be done. However, in this paper we will only summarize what has been carried out already to ascertain the consequences of polymorphisms to RASSF1.

3. RASSF1A: A Key Element in Cellular Stability

One of the most striking features of RASSF1A is its microtubule appearance. Numerous tagged versions of RASSF1A have all revealed similar microtubule-like appearance as seen in MCF-7 breast cancer cells in Figure 2. This appearance has been observed in many other cell lines with similar appearances. It has also been determined that both N- and C-terminal residues of RASSF1A are required for the microtubule appearance of RASSF1A [13, 18]. Several groups have characterized the appearance and function of the microtubule localization of RASSF1A. It has been demonstrated that the microtubule localization of RASSF1A mainly

TABLE 1: RASSF1A single nucleotide polymorphisms. Several RASSF1A polymorphic changes have been identified as outlined in Table 1. SNP sites consulted to draft this table include NCBI (at http://www.ncbi.nlm.nih.gov/projects/SNP/snp_ref.cgi?showRare=on&chooseRs=coding&go=Go&locusId=11186) and DMDM (at http://bioinf.umbc.edu/dmdm/gene_prot_page.php?search_type=gene&id=11186).

Polymorphism	Tissue or cell line origin	SNP ID% and other information	References
K21Q (AAG → CAG)	Breast (tumor) Kidney (renal carcinoma cell TK10 and KRC/Y) Lung (Non small cell Lung cancer cell line)	rs4688725*,**,	Schagdarsurengin et al. [23]; Dammann et al. [24]; Agathanggelou et al. [25]; Burbee et al. [26]
R28H (CGT → CAT)	Breast (Tumor) Lung (nonsmall cell lung cancer cell line)	Presence in lung carcinomas are rare	Schagdarsurengin et al. [23]; Dammann et al. [24]; Burbee et al. [26]
V47F (GTC → TTC)	Not listed	rs61758759*,**, ,	NCB1%
R53C (CGC → TGC)	Breast (tumor) Lung (nonsmall cell lung cancer cell line)	Q9NS23	Schagdarsurengin et al. [23]; Dammann et al. [24]; Burbee et al. [26]
A60T (GCA → ACA)	Breast	No SNP ID found	Agathanggelou et al. [25]
C65R (TGC → CGT)	Breast (tumor)	No SNP ID found	Dallol et al. [27]
S131F (TCT → TTT)	Breast (tumor) Kidney (Wilm's tumor)	No SNP ID found	Schagdarsurengin et al. [23]; Dammann et al. [24]
A133S (GCT → TCT)	Breast (tumor, fibroademonas), 33% Kidney (Wilm's tumor), 21% Brain (medulloblastoma), 9% Muscle (rhadomyosarcoma), 19% Lung (nonsmall cell lung cancer cell line)	rs52807901 and rs2073498 Association with BRAC1/2 mutations Homozygous in breast cancer 21% of kids with germ line mutation, maternal in origin	Schagdarsurengin et al. [23]; Dammann et al. [24]; Bergqvist et al. [28]; Gao et al. [29]; Burbee et al. [26]; Lusher et al. [30]
I135T (ATT → ACT)	Lung (nonsmall cell lung cancer cell line) Breast (tumor cell line)	No SNP ID found	Dammann et al. [24]; Agathanggelou et al. [25]
V211A (GTC → GCC)	Breast	No SNP ID found	Agathanggelou et al. [25]
R201H (CGC → CAC)	ENT (nasopharyngeal carcinoma)	In 23 tumor samples, 34 other polymorphisms were detected (not listed in this table) with 30 transitions, 2 transversions, and 2 deletions (6 in SH3/C1 domain and 6 in RA domain)	Zhi-Gang Pan et al. [31]
E246K (GAA → AAG)	Breast (tumor)	No SNP ID found	Agathanggelou et al. [25]
R257Q (CGG → CAG)	Breast (Tumor) Lung (nonsmall cell lung cancer cell line	No SNP ID found	Schagdarsurengin et al [23]; Dammann et al. [24]; Agathanggelou et al. [25]; Dallol et al. [27]
H315R (CAC → CGC)	NCBI SNP database, source unknown	rs52792349 and rs12488879	Geoffery Clark (personnel communication)
Y325C (TAT → TGT)	Breast (tumor) Lung (nonsmall cell lung cancer cell line	No SNP ID found	Schagdarsurengin et al. [23]; Dammann et al. [24]; Burbee et al. [26]
L270V (CTG → GTG)	Cervical (tumor)	No SNP ID found	Schagdarsurengin et al. [23]; Dammann et al. [24]
A336T (GCC → ACC)	Lung (nonsmall cell lung cancer cell line	No SNP ID found	Dammann et al. [24]

% http://bioinf.umbc.edu/dmdm/gene_prot_page.php?search_type=gene&id=11186.
$UniProtKB/Swiss-Prot.
*Validated by multiple, independent submissions to the refSNP cluster.
**Validated by frequency or genotype data: minor alleles observed in at least two chromosomes.
Validated by the 1000 Genomes Project, http://www.1000genomes.org/.

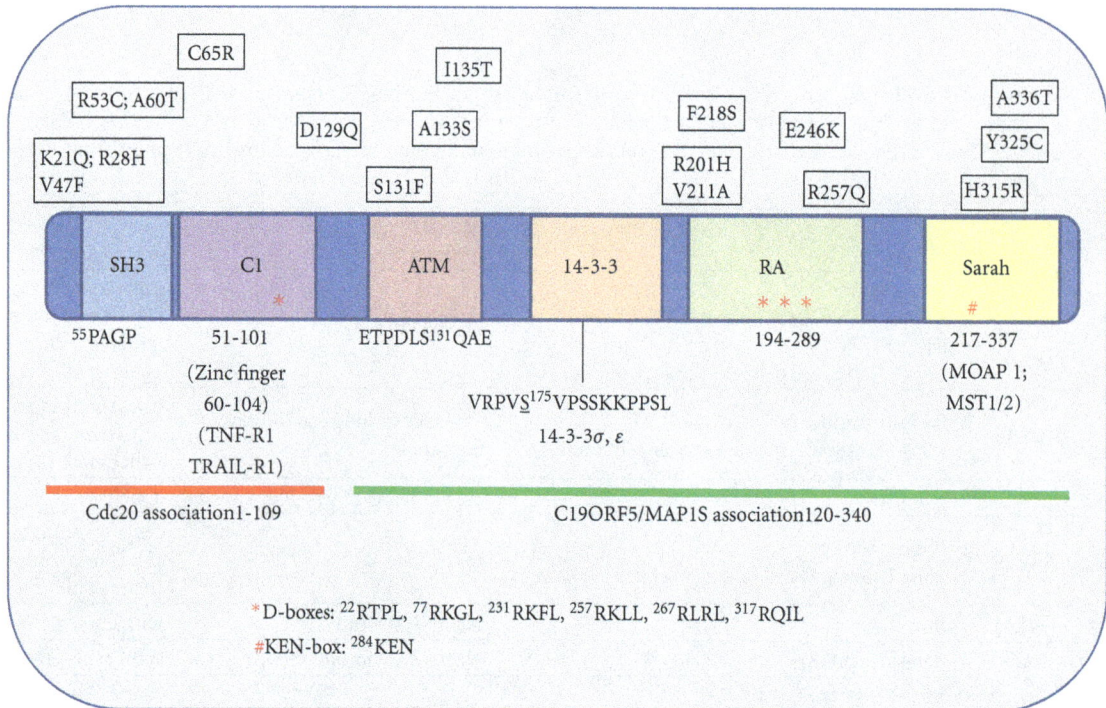

FIGURE 1: Schematic of RASSF1A with location of identified polymorphisms. Location of identified RASSF1A polymorphisms is indicated with respect to amino acid location, changed amino acid, and exon location. A potential binding sequence to an SH3 domain has been identified with a PxxP motif. The ATM phosphorylation site is underlined with surrounding residues shown. The docking sites for several RASSF1A effector proteins are shown including the location of potential D- and KEN-boxes for protein association (D1 to D6). The latter boxes are thought to be important for associations with APC/cdc-20 [12]. The Ras association domain (RA) is present in RASSF1A but has not been convincingly demonstrated to associate with the Ras family of oncogenes [10]. The SARAH domain modulates heterotypic associations with the sterile-20-like kinases, MST1 and MST2 (adapted from El-Kalla et al. (2010)) [13] and Gordon and Baksh (2011) [10].

functions to stabilize tubulin both in interphase and in mitosis even in the presence of the microtubule destabilizer, nocodazole [13, 27, 33, 34]. To date, RASSF1A has not been demonstrated to colocalize to actin or intermediate filaments. RASSF1A associations function to stabilize tubulin in a paclitaxel (taxol)-like manner [17, 27] especially during mitosis allowing sister chromatid segregation. This function is governed by associations with γ-tubulin at spindle poles and centromeric areas during metaphase and anaphase and near the microtubule organizing center (MTOC) where microtubules emerge and nucleate [35–37]. If the microtubule spindle complex is not properly formed, cell death proceeds to prevent inheritable aneuploidy. In the absence of cell death pathways chromosomal missegregation and inheritable aneuploidy arise which can lead to malignancy. Several of the effects on microtubule biology have been observed in mouse embryonic fibroblasts (MEFs) obtained from $Rassf1a^{-/-}$ mice developed by two separate groups [19, 38]. $Rassf1a^{-/-}$ mice are viable, fertile and retain expression of isoform 1C. However, by 12–16 months of age they have increased tumor incidence, especially in the breast, lung, and immune system (gastrointestinal carcinomas and B-cell-related lymphomas) [19, 38]. These data suggest a tumor suppressor function specific for the RASSF1A isoform. MEFs obtained from $Rassf1a^{-/-}$ mice are more susceptible to

nocodazole-induced microtubule depolymerization suggesting a protective effect of RASSF1A on microtubule stability similar to what has been observed using tissue culture approaches [19].

It has now been demonstrated that RASSF1A disease associated polymorphisms may affect the function of RASSF1A as a microtubule stabilizer. It was demonstrated that the S131F mutant of RASSF1A continued to maintain the ability to promote tubulin stability as determined by immunofluorescence microscopy and acetylation status of tubulin [17]. Furthermore, it was demonstrated by Vos et al. [17] that RASSF1C could not function in a similar manner to RASSF1A to stabilize tubulin. This provided one of the first evidences for differential function for these two prominent isoforms of the RASSF1 loci. A comprehensive analysis of several other RASSF1A polymorphisms was carried out by Liu et al. [39]. They demonstrated that polymorphisms around the ATM phosphorylation site (A133S, S131F, and I135T) maintained the microtubule appearance of RASSF1A. A second comprehensive study revealed that the C65R and R257Q polymorphisms of RASSF1A resulted in "atypical localizations" of RASSF1A away from a microtubular appearance [27]. Furthermore, both C65R (a residue within the C1 domain) and R257Q (a residue within the RA domain) promoted enhanced BrdU incorporation into NCI-H1299

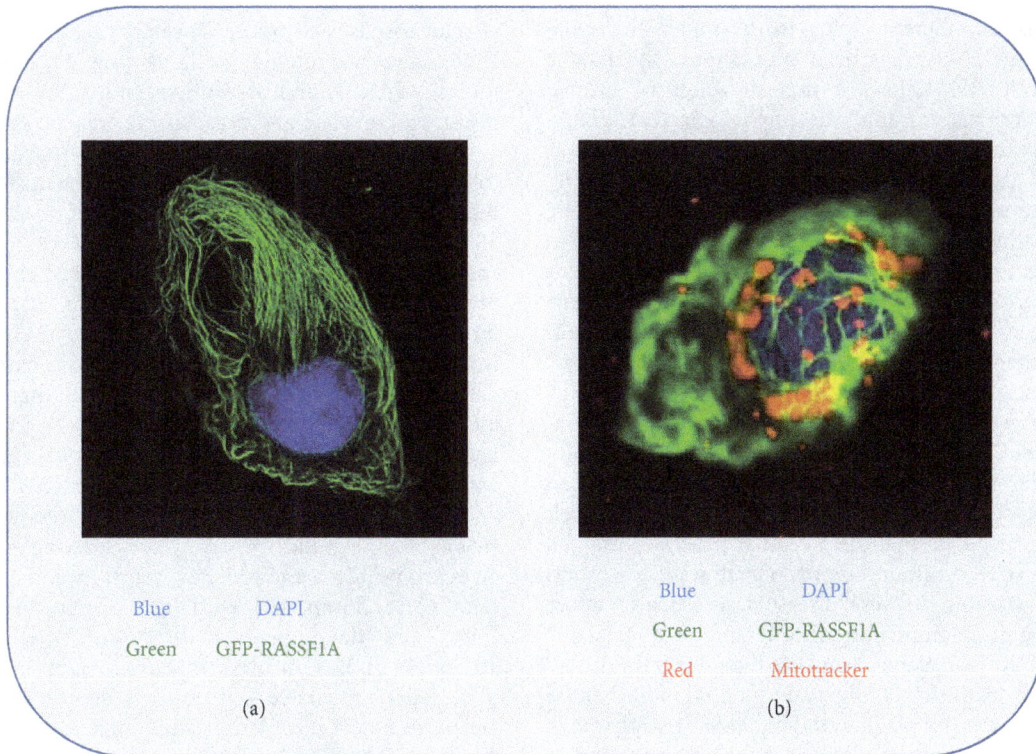

FIGURE 2: Microtubule localization of RASSF1A. GFP-RASSF1A was expressed in U2OS osteosarcoma cells (a and b) and costained with DAPI to reveal the nucleus (a and b) and with mitotracker red to reveal mitochondrial localization (b). Areas of yellow reveal colocalization and all images were acquired using confocal microscopy using a Zeiss system and a 63x oil immersion lens.

nonsmall cell lung cancer cells suggesting loss of tumor suppressor function. Recently, we have also observed a complete loss of the microtubule localization of RASSF1A in the presence of a C65R change and "oncogenic" properties of this polymorphism in a classical xenograft assay in athymic mice [13]. The C65R polymorphism acquired a nuclear localization for unexplained reasons and also failed to stabilize tubulin in the presence of the microtubule depolymerizing agent nocodazole [13]. It clearly lost the tumor suppressor function of RASSF1A in a xenograft assay and can robustly drive enhanced growth [13]. Similarly, both the A133S and E246K mutants maintained microtubule localization and lost tumor suppressor function but not to the level of the C65R polymorphism (xenograft assays were carried out in HCT116 colon cancer cells) [13]. We are currently characterizing many of the other polymorphisms in Table 1 for their ability to behave as tumor suppressor, inhibit abnormal growth, and affect microtubule stability and protein interaction with established RASSF1A effectors.

Interestingly, it has been reported that Epstein-Barr virally encoded protein, latent membrane protein 1 (LMP1)

can function to transcriptionally decrease RASSF1A levels and promote tubulin depolymerization and mitotic instability in human epithelial cells (HeLa and HaCaT) [40]. Punctuate structures of tubulin were observed in the cytoplasm indicative of tubulin depolymerization [40]. Decreased RASSF1A levels resulted in increased phosphorylation of $I\kappa B\alpha$ and elevated $NF\kappa B$ activity. Cause and effect of changes in $NF\kappa B$ activity were not fully elucidated. However, we have evidence that the loss of RASSF1A can lead to enhanced $NF\kappa B$ activity (El-Kalla et al., unpublished observations) suggesting that the decreased expression of RASSF1A induced by LMP1 production may have resulted from the loss of the ability of RASSF1A to restrict $NF\kappa B$ function. EPV infection is closely related to the appearance of nasopharyngeal cancers and we speculate that a precondition characterized by enhanced $NF\kappa B$ activity (and hence inflammation) may promote tumorigenesis and the appearance of nasopharyngeal cancers upon EBV infection. We are currently exploring the role of RASSF1A as a molecular link between inflammation and tumorigenesis.

4. RASSF1A: Linking Extrinsic Death Receptor Stimulation to Bax Activation

Every cell has an inherent ability to die under abnormal conditions. This ability has been programmed by nature into every cell and follows a defined series of events. Apoptosis is critical for multiple physiological processes, including organ formation, immune cell selection, and inhibition of tumor formation [41]. Two types of signaling pathways promote apoptosis using the mitochondria. The "intrinsic" pathway is activated by noxious factors such as DNA damage, unbalanced proliferative stimuli, and nutrient or energy depletion. Components of intrinsic-dependent apoptosis are still unclear, although Bcl-2-homlogy domain 3 (BH3) proteins are required. In contrast, the "extrinsic" pathway is stimulated by specific death receptors (e.g., TNFα receptor R1 (TNF-R1), TNFα-related apoptosis-inducing ligand receptor (TRAIL-R1) or Fas (CD95)) [42–44]. Molecular mechanisms modulating programmed cell death (apoptosis) impinge on growth and immune cell function. We speculate that these cellular processes may be regulated in part by tumor suppressor pathways, pathways frequently inactivated in several disease states (such as cancer and autoimmune/inflammatory disorders).

RASSF1A is one element involved in death receptor-dependent cell death that is epigenetic-silenced in numerous cancers. In the majority of these studies, RASSF1A epigenetic silencing strongly correlates with the epigenetic silencing of three other genes—p16^{INK4a}, death associated protein kinase (DAPK), and caspase 8 [45–48]. Two of these genes are involved in proapoptotic pathways, DAPK and caspase-8 [43, 49, 50]. DAPK is a unique calcium/calmodulin activated serine/threonine kinase involved in several cell death-related signaling pathways including tumor necrosis factor α receptor 1 (TNF-R1) cell death and autophagy [50, 51]. It is a tumor suppressor protein [50] that has also been demonstrated to be involved in associations with and the regulation of pyruvate kinase, a key glycolytic enzyme that may be influential in the link between metabolism and cancer [52]. We have evidence to demonstrate association of RASSF1A and DAPK (Baksh et al., unpublished observations) and RASSF1A has two potential phosphorylation sites for DAPK within the RA domain at ^{193}GRGTSVRRRTSFYLPK [53]. Curiously, these sites have also been demonstrated to be sites for protein kinase C [54] and aurora kinases [55]. In the presence of S197A or S203A mutant of RASSF1A, PKC failed to phosphorylate RASSF1A resulting in the loss of microtubule organization in COS-7 cells. Similarly, Aurora B kinase failed to phosphorylate RASSF1A in the presence of S203A resulting in a failed cytokinesis [55]. It remains to be determined the physiological importance of these potential DAPK phosphorylation sites.

Caspase 8 is cysteine-dependent aspartate-directed protease and an initiator caspase, and targeted activation of caspase 8 is driven by the disc inducing signaling complex (DISC) [43, 49]. DISC-dependent activation of caspase 8 triggers a series of events resulting in the cleavage of Bid and insertion of Bid on the outer mitochondrial membrane, the release of small molecules (such as cytochrome c) from the mitochondria into the cytosol, and the activation of downstream effector caspases (such as caspase-3) [56]. Intrinsic pathway stimulation also leads to cytochrome c release and effector caspase activation. Once activated, effector caspases cleave several proteins (such as poly(ADP-ribose) polymerase (PARP)) and activate specific DNA endonucleases resulting in nuclear and cytoplasmic breakdown [57].

Our research group was the first to define and continues to define some of the molecular mechanisms of death receptor-dependent apoptotic regulation by RASSF1A [20, 58, 59]. Ectopic expression of RASSF1A (but not RASSF1C) specifically enhanced death receptor-evoked apoptosis stimulated by TNFα that does not require caspase 8 activity or Bid cleavage [20, 58]. We have also shown that RASSF1A does not influence the intrinsic pathway of cell death [58]. Furthermore, we demonstrated that microtubule localization was required for association with death receptors and for the role of RASSF1A in apoptosis [13, 20]. In contrast, RASSF1A knockdown cells (by RNA interference) and Rassf1a$^{-/-}$ knockout mouse embryonic fibroblasts (MEFs) have significantly reduced caspase activity, defective cytochrome c release and Bax translocation (but not Bid cleavage), and impaired death receptor-dependent apoptosis [58]. These data suggest a direct link of death receptor activation of Bax through RASSF1A. Our current model of RASSF1A-mediated cell death is described in Figure 3. Death receptor stimulation functions to bring RASSF1A (and not RASSF1C or RASSF5A/Nore1A) and modulator of apoptosis 1 (MOAP-1) to TNF-R1 in order to promote a more "open" MOAP-1 to subsequently associate and promote Bax conformational change and translocation to the mitochondria to activate cell death (Figure 3) [20, 58–60]. We have evidence that the 14-3-3 may keep RASSF1A in check and inhibit it from promoting cell death or associating with other unexplored signaling components [59]. We are currently characterizing the primary and secondary signals required for MOAP-1 induced Bax conformational change and the apoptotic regulation of MOAP-1 by ubiquitination (Law et al., unpublished observations).

To date, very little is known about the cell death properties of numerous RASSF1A polymorphisms. Dallol et al. demonstrated that both C65R and R257Q promoted enhanced BrdU incorporation into NCI-H1299 non-small cell lung cancer cells suggesting loss of tumor suppressor function and possible loss of cell death properties [27]. We have observed partial activation of apoptosis in the presence of several RASSF1A polymorphisms (such as C65R, A133S, I135T, and A336T) suggesting importance to death receptor-dependent apoptosis via ATM site and SARAH domain associations (El-Kalla et al., unpublished observations). Further analysis is warranted to explore how RASSF1A polymorphisms may affect death receptor-dependent apoptosis.

Although not discussed in great detail here, RASSF1A can also promote cell death utilizing the autophagic protein, C19ORF5/MAP1S, [27, 33, 61] the Hippo pathway components MST1/2 and possibly Salvador [14, 62], and, in melanoma cells, influence Bcl-2 levels and activate apoptosis signal regulating kinase 1 (ASK-1) [63]. Min et al. [33] demonstrated that the ability of RASSF1A to efficiently

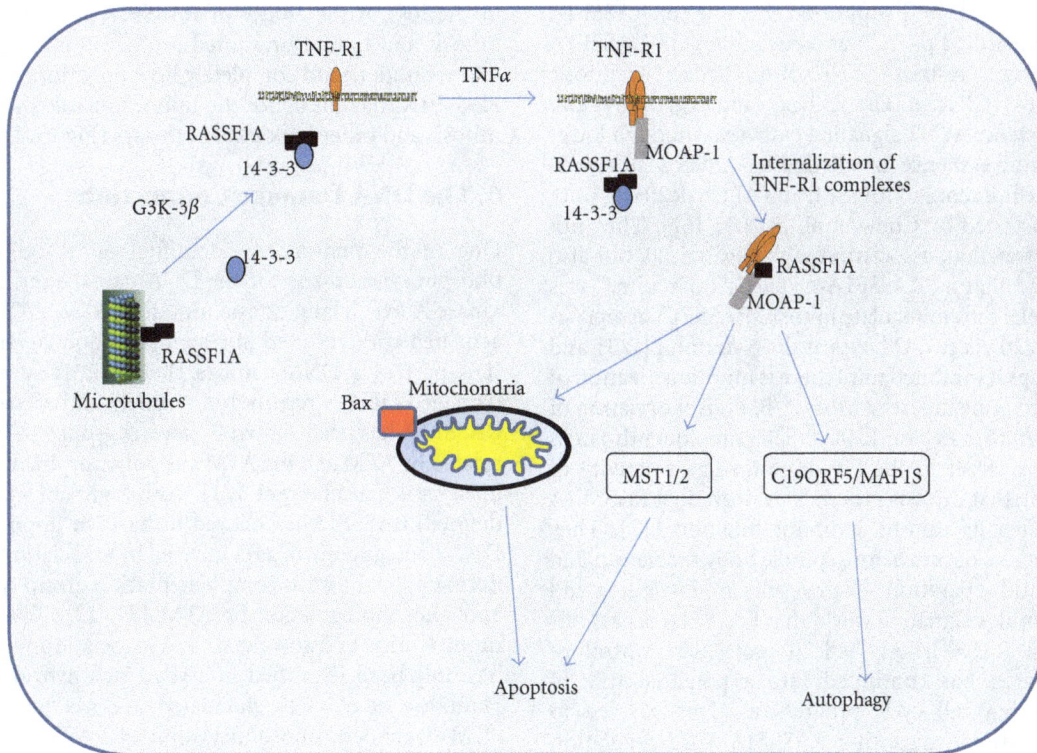

FIGURE 3: Model for the RASSF1A/MOAP-1 proapoptotic pathway. Death receptor-induced cell death (TNFα is used as an example) can result in the recruitment of protein complexes to activate Bax and promote apoptosis. Basally, RASSF1A is kept complexed with 14-3-3 by GSK-3β phosphorylation in order to prevent unwanted recruitment of RASSF1A to death receptor and uncontrolled stimulation of Bax and apoptosis. Once a death receptor stimuli have been received (TNFα as shown above), the TNF-R1/MOAP-1/RASSF1A complex promotes the "open" form of MOAP-1 to associate with Bax. This in turn results in Bax conformational change and recruitment to the mitochondria to initiate cell death. Following release from TNF-R1/MOAP-1 complex, RASSF1A may reassociate with 14-3-3 to prevent continued stimulation of this cell death pathway (unpublished observations). Please see text for further details.

inhibit APC/cdc20 activity during mitosis (please see next section) is dependent on the recruitment of RASSF1A to spindle poles via C19ORF5/MAP1S. C19ORF5/MAP1S was also shown to regulate mitotic progression by stabilizing mitotic cyclins in a RASSF1A-dependent manner. Recently, C19ORF5/MAP1S was demonstrated by Lui et al. [61] to associate with a component of the autophagosome, LC3, and the mitochondria-associatedv leucine-rich PPR-motif containing protein (LRPPRC) protein. These associations suggest that C19ORF5/MAP1S may serve as a potential link between autophagic cell death, mitochondria, and micro-tubules and appears to require RASSF1A. It will be essential to determine associations of RASSF1A polymorphisms with key cell death mediators, such as MOAP-1, TNF-R1, DAPK, C19ORF5/MAP1S, and MST1/2 in order to ascertain their importance in influencing the tumor suppressor function of RASSF1A. A detailed discussion about the Hippo and RASSF1A/MOAP-1 pathways of cell death is presented in this special review.

5. Cell Cycle Control Pathways Influenced by RASSF1A

As mentioned previously, RASSF1A is a microtubule binding protein that colocalizes with α- and β-tubulin, and with γ-tubulin on centromeres [35–37]. RASSF1A is thought be an important component of mitotic spindles and can influence the separation of sister chromatids at the metaphase plate. This observation has held true five years later and reinforced the findings of Song et al. [64] of the possible involvement of RASSF1A in cell cycle control. Although, very limited knowledge of the cell cycle effects of polymorphic forms of RASSF1A are known, several lines of evidence do suggest a role in cell cycle control. In 2004, RASSF1A was identified as an interacting protein with the anaphase promoting complex (APC)/cdc20 and prevented the ability of APC/cdc20 to degrade cyclins A and B in order to exit mitosis [64]. In the absence of RASSF1A, cyclins A and B were rapidly degraded due to increased ubiquitination of the cyclins to allow exit from mitosis.

Whitehurst et al. [65] supported this role for RASSF1A and further identified β-TrCP as associating with RASSF1A and functioning to restrict the role of APC-cdc20 in mitotic progression. β-TrCP is an IκBα E3 ligase and negative regulator of the β-catenin/WNT signaling pathway. Although Liu et al. could not find evidence for a RASSF1A-APC/cdc20 association [66], the influence of RASSF1A on APC/cdc20 was once again demonstrated by Chow et al. in 2011 [67]. They not only demonstrated an association with APC/cdc20, but also clearly showed that a "RASSF1A-APC/cdc20 circuitry" was in place in HeLa cells to regulate mitosis. RASSF1A associates with APC/cdc20 via two D boxes at the N-termini (DB1 and DB2) and keeps it inhibited until there is mitotic activation of the serine/threonine kinases Aurora A/B. Phosphorylation of RASSF1A by Aurora A/B on T202 or S203 subsequently labels RASSF1A as a target to the E3-ubiquitin ligase activity of APC, ensuring that mitosis proceeds by degrading RASSF1A and suppressing its mitotic inhibitor function [12]. They speculate that this occurs before spindle body formation and sister chromatid separation. Their results are intriguing and reveal the complex signaling world that RASSF1A is part of.

Beyond a RASSF1A-APC/cdc20 molecular control of mitosis, research has continued into a potential role of RASSF1A during cell cycle progression. This has led to several observations suggesting RASSF1A G1/S regulation of cyclin D1 [63, 65, 68] in melanoma and HeLa cells (resp.), interaction with the transcriptional regulator p120^{E4F} at the G1/S phase transition resulting in inhibition of passage from G1 [69], DNA damage control regulation by ATM and by the DNA damage binding protein 1 (DDB1) that can associate with RASSF1A linking to the E3-ligase cullin 4A during mitosis [70]. The p120^{E4F} transcription factor was determined to be involved in inhibiting the transcription of cyclin A, resulting in the failure of cyclin A to associate with CDK2 to allow for progression through S phase. RASSF1A cooperates with p120^{E4F} to repress cyclin A expression by enhancing its binding at the promoter region [69]. ATM and DDB1 are important DNA damage control elements during ultraviolet and gamma irradiation which have evolved to repair damage DNA and will be discussed elsewhere in this special RASSF issue.

Shivakumar et al. revealed that in both H1299 non-small cell lung cancer and in the human mammary epithelial telomerase immortalized (HME50-hTERT) cell line, over-expression of RASSF1A wild-type expression construct can reduce BrDU accumulation and cyclin D1 expression [68]. The ability of RASSF1A to inhibit growth, and cyclin D1 expression was lost in the presence of the A133S and S131F ATM site mutants of RASSF1A suggesting an important role in tumor suppression [68]. Other polymorphic forms of RASSF1A have not been explored with respect to their abilities to regulate mitosis. What these studies reveal is how highly regulated RASSF1A is, not only in interphase cells, but especially in cells undergoing active cell division. It can then be appreciated how devastating the functional consequence of the loss of RASSF1A would be resulting in an unregulated and unwanted increase in mitotic cyclins, accelerated mitosis, enhanced growth and tumor formation. It would be interesting to speculate that they may result in the loss of the ability of RASSF1A to properly regulate mitosis and inhibit unwanted proliferation. It is imperative that we understand completely how polymorphic changes in RASSF1A may influence the important role of RASSF1A in mitosis and other biological pathways (Figure 4).

6. The DNA Damage Connection

One of the first motifs identified on RASSF1A was the phosphorylation site for the DNA damage serine/threonine kinase Ataxia telangiectasia mutated (ATM). ATM is usually activated and recruited in response to double strand breaks. It is part of a DNA damage checkpoint that ensures that damaged DNA is repaired in a timely and efficient manner. RASSF1A has been shown by several groups to be phosphorylated by ATM and the ATM site polymorphisms are present in several cancer types [71]. Although not currently well defined, RASSF1A is believed to have an important role in DNA damage control as evidenced by associations with xeroderma pigmentosum complementation group A (XPA) [72] and phosphoregulation by ATM [71, 73]. XPA is involved in nucleotide excision repair and association with RASSF1A has only been identified in a yeast two-hybrid screen [24]. Hamilton et al. [71] elucidated a novel pathway linking ATM-dependent phosphorylation of RASSF1A in response to gamma irradiation on serine-131 followed by MST/LATS activation resulting in Yes associated protein (YAP)/p73-dependent transcriptional program to promote cell death. The S131F mutant of RASSF1A lacked the ability to carry out the transactivation of YAP/p73. Curiously, RASSF1C has been demonstrated to be constitutively anchored to the death domain-associated protein (DAXX) in the nucleus and is released upon UV-induced DNA damage [16]. Localization with DAXX occurs on promyelocytic leukaemia-nuclear bodies (PML-NBs). DNA damage promotes the degradation and ubiquitination of DAXX, release of RASSF1C to allow the nucleocytoplasmic shuttling of RASSF1C to cytoplasmic microtubules, and the activation of the SAPK/JNK pathway in HeLa cells. RASSF1A was shown to only associate weakly with DAXX suggesting a specific role for RASSF1C [16]. Recently, it was demonstrated that the E3 ligase, Mule, can ubiquitinate RASSF1C under normal conditions, and both Mule and β-TrCP can ubiquitinate RASSF1C under UV exposure [74]. These studies and others have continued to demonstrate the diverse role that the splice variants of RASSF1 may function in biology. A detailed discussion about the role of RASSF1A during DNA damage repair will be presented in this special review.

7. RASSF1C: The Other RASSF1 Isoform

Very little is known about the biological role for the other major splice variant of the RASSF1 gene family. Several lines of evidence suggest that RASSF1C may be a tumor suppressor gene in prostate and renal carcinoma cells but not in lung cancer cells [75]. In fact, it has been demonstrated by Amaar et al. that the loss of RASSF1C actually results in the loss of proliferation of lung and breast cancer cells suggesting a prosurvival (not tumor suppressor) role for

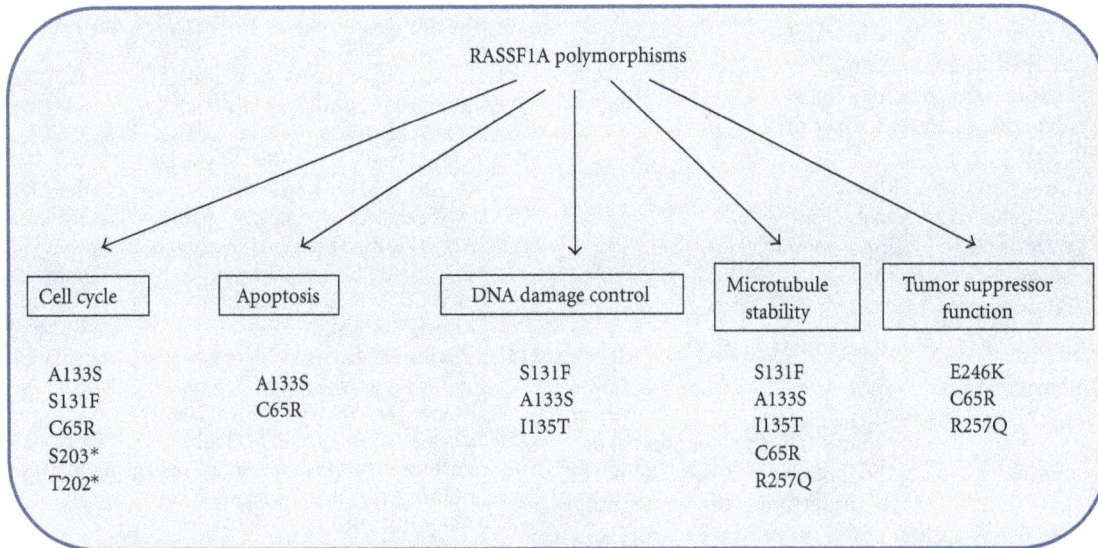

FIGURE 4: Identified biological roles for RASSF1A polymorphisms. Several polymorphisms have been identified for RASSF1A over the past decade since it was first cloned. Biological analyses of the *in vivo* role have identified the importance of RASSF1A over numerous pathways. This figure summarizes what is known about RASSF1A polymorphisms. *denotes a nonpolymorphic but mutational change. This change does not naturally exist in the cancer patient population to our knowledge.

RASSF1C [76, 77]. Furthermore, RASSF1C can associate with the E3 ligase β-TrCP via the $SS_{18}GYXS_{19}$ motif (where X is any amino acid and numbers correspond to amino acid sequence in RASSF1C) at the N-terminus (i.e., not present in RASSF1A) [78] and promote the accumulation and transcriptional activation of β-catenin [78]. Activation of β-catenin would result in enhanced proliferation by transcriptional upregulation of genes such as cyclin D1, Myc, and TCF-1. Thus, either the lack of RASSF1A expression or the overexpression of RASSF1C perturbs β-TrCP E3 ligase/β-catenin homeostasis and WNT signaling pathways.

Unlike RASSF1A, RASSF1C has not been found to be significantly epigenetically silenced in cancer. Polymorphisms to RASSF1C have not been uncovered yet, but a C61F mutation in RASSF1C (equivalent to the S131F mutation in RASSF1A) resulted in the failure of RASSF1C to protect microtubules against nocodazole-induced depolymerization [17]. This would again suggest importance of serine residue within the ATM site found on both RASSF1A and 1C. Recently, it has been suggested that a possible pathogenic role for RASSF1C in cancer may exist as its expression was more than eleven-fold greater in pancreatic endocrine tumors than in normal tissue [79]. It remains to be determined the exact biological role for RASSF1C, but the ability of RASSF1C to function as a tumor suppressor is cell specific and remains to be further investigated and confirmed.

8. The Future of Understanding RASSF Polymorphisms

Knudson stated in 1971 that cancer is the result of accumulated mutations to the DNA of cells and that multiple "hits" to DNA were necessary to cause cancer [80]. It is generally known that the loss of function in a tumor suppressor protein typically requires the inactivation of both alleles of its gene in contrast to proto-oncogenes which promote tumorigenesis due to dominant acting mutations affecting one gene copy. Similar to what Knudson discovered for retinoblastoma, the RASSF1A tumor suppressor may become inactivated by the epigenetic loss by promoter specific methylation of both allele or by a combination of epigenetic silencing and loss of function polymorphic changes. Most cancers investigated to date have >50% of the disease population containing epigenetic silencing of RASSF1A [11, 81]. However, numerous cancers such as cervical, head and neck, myeloma, and leukemia have <25% of the disease population containing epigenetic silencing of RASSF1A. It may be speculated that polymorphic changes to RASSF1A may exist in the latter patients that, in agreement with the Knudson two hit hypothesis, resulting in the loss of function of the RASSF1A tumor suppressor and causing cancer. A systematic and functional analysis of RASSF1A polymorphism is therefore necessary to allow physicians to carry out personalized medicine on patients harboring polymorphic changes to RASSF1A.

Abbreviations

APC:	Anaphase promoting complex
ATM:	Ataxia telangiectasia mutated
BH3:	Bcl-2-homlogy domain 3 (BH3)
DAPK:	Death-associated protein kinase
C19ORF5/MAP1S:	Chromosome 19 open reading frame 5/microtubule-associated protein 1S
Nore:	Novel ras effector
PKC:	Protein kinase c
RAS:	Rat Sarcoma

RASSF: Ras association domain family
RA: Ras association domain
NFκB: Nuclear factor of kappa light
            polypeptide gene enhancer in
            B-cells
MAP: Mitogen activated kinase
MOAP-1:   Modulator of apoptosis
SARAH: Salvador/RASSF/Hippo domain
TNFα: Tumor necrosis factor α
TNF-R1:   TNFα receptor 1.

Acknowledgments

The authors would like to thank all the members of the Baksh laboratories, past and present, for their helpful discussions and dedication to understand the biological role of RASSF1A. They are grateful for the support of the Department of Pediatrics, especially the Divisions of Hematology/Oncology/Palliative Care/Epidemiology under the guidance of Dr. Paul Grundy. Investigation of the biological role of RASSF1A has been supported by grants from the Canadian Institutes of Health Research, Women and Children's Health Research Institute, Alberta Heritage Foundation for Medical Research/Alberta Innovates-Health Solutions, Canadian Foundation for Innovation/Alberta Small Equipment Grants Program, Canadian Breast Cancer Foundation (Prairies/NWT Region) and The Stollery Children's Foundation/Hair Massacure Donation Grant generously donated by the MacDonald family.

References

[1] S. Canada, *Leading Causes of Death in Canada*, Government of Canada Publications, 2011.

[2] A. M. Minino, "Death in the United States," Tech. Rep. 64, NCHS Data Brief, Hyattsville, Md, USA, 2009.

[3] I. Bozic, T. Antal, H. Ohtsuki et al., "Accumulation of driver and passenger mutations during tumor progression," *Proceedings of the National Academy of Sciences of the United States of America*, vol. 107, no. 43, pp. 18545–18550, 2010.

[4] B. Vogelstein and K. W. Kinzler, "Cancer genes and the pathways they control," *Nature Medicine*, vol. 10, no. 8, pp. 789–799, 2004.

[5] F. Wang, W. Tan, D. Guo, X. Zhu, K. Qian, and S. He, "Altered expression of signaling genes in jurkat cells upon FTY720 induced apoptosis," *International Journal of Molecular Sciences*, vol. 11, no. 9, pp. 3087–3105, 2010.

[6] D. Hanahan and R. A. Weinberg, "The hallmarks of cancer," *Cell*, vol. 100, no. 1, pp. 57–70, 2000.

[7] D. Hanahan and R. A. Weinberg, "Hallmarks of cancer: the next generation," *Cell*, vol. 144, no. 5, pp. 646–674, 2011.

[8] A. M. Richter, G. P. Pfeifer, and R. H. Dammann, "The RASSF proteins in cancer; from epigenetic silencing to functional characterization," *Biochimica et Biophysica Acta*, vol. 1796, no. 2, pp. 114–128, 2009.

[9] L. van der Weyden and D. J. Adams, "The Ras-association domain family (RASSF) members and their role in human tumourigenesis," *Biochimica et Biophysica Acta*, vol. 1776, no. 1, pp. 58–85, 2007.

[10] M. Gordon and S. Baksh, "RASSF1A: not a prototypical Ras effector," *Small Gtpases*, vol. 2, no. 3, pp. 148–157, 2011.

[11] H. Donninger, M. D. Vos, and G. J. Clark, "The RASSF1A tumor suppressor," *Journal of Cell Science*, vol. 120, no. 18, pp. 3163–3172, 2007.

[12] T. Mochizuki, S. Furuta, J. Mitsushita et al., "Inhibition of NADPH oxidase 4 activates apoptosis via the AKT/apoptosis signal-regulating kinase 1 pathway in pancreatic cancer PANC-1 cells," *Oncogene*, vol. 25, no. 26, pp. 3699–3707, 2006.

[13] M. El-Kalla, C. Onyskiw, and S. Baksh, "Functional importance of RASSF1A microtubule localization and polymorphisms," *Oncogene*, vol. 29, no. 42, pp. 5729–5740, 2010.

[14] J. Avruch, R. Xavier, N. Bardeesy et al., "Rassf family of tumor suppressor polypeptides," *Journal of Biological Chemistry*, vol. 284, no. 17, pp. 11001–11005, 2009.

[15] G. Pelosi, C. Fumagalli, M. Trubia et al., "Dual role of RASSF1 as a tumor suppressor and an oncogene in neuroendocrine tumors of the lung," *Anticancer Research*, vol. 30, no. 10, pp. 4269–4281, 2010.

[16] D. Kitagawa, H. Kajiho, T. Negishi et al., "Release of *RASSF1C* from the nucleus by Daxx degradation links DNA damage and SAPK/JNK activation," *The EMBO Journal*, vol. 25, no. 14, pp. 3286–3297, 2006.

[17] M. D. Vos, A. Martinez, C. Elam et al., "A role for the RASSF1A tumor suppressor in the regulation of tubulin polymerization and genomic stability," *Cancer Research*, vol. 64, no. 12, pp. 4244–4250, 2004.

[18] R. Rong, W. Jin, J. Zhang, M. S. Sheikh, and Y. Huang, "Tumor suppressor RASSF1A is a microtubule-binding protein that stabilizes microtubules and induces G_2/M arrest," *Oncogene*, vol. 23, no. 50, pp. 8216–8230, 2004.

[19] L. Van Der Weyden, K. K. Tachibana, M. A. Gonzalez et al., "The RASSF1A isoform of RASSF1 promotes microtubule stability and suppresses tumorigenesis," *Molecular and Cellular Biology*, vol. 25, no. 18, pp. 8356–8367, 2005.

[20] C. J. Foley, H. Freedman, S. L. Choo et al., "Dynamics of RASSF1A/MOAP-1 association with death receptors," *Molecular and Cellular Biology*, vol. 28, no. 14, pp. 4520–4535, 2008.

[21] E. Harjes, S. Harjes, S. Wohlgemuth et al., "GTP-Ras disrupts the intramolecular complex of C1 and RA domains of Nore1," *Structure*, vol. 14, no. 5, pp. 881–888, 2006.

[22] V. I. Kashuba, T. V. Pavlova, E. V. Grigorieva et al., "High mutability of the tumor suppressor genes RASSF1 and RBSP3 (CTDSPL) in cancer," *PLoS One*, vol. 4, no. 5, Article ID e5231, 2009.

[23] U. Schagdarsurengin, C. Seidel, E. J. Ulbrich, H. Kölbl, J. Dittmer, and R. Dammann, "A polymorphism at codon 133 of the tumor suppressor RASSF1A is associated with tumorous alteration of the breast," *International Journal of Oncology*, vol. 27, no. 1, pp. 185–191, 2005.

[24] R. Dammann, C. Li, J. H. Yoon, P. L. Chin, S. Bates, and G. P. Pfeifer, "Epigenetic inactivation of a RAS association domain family protein from the lung tumour suppressor locus 3p21.3," *Nature Genetics*, vol. 25, no. 3, pp. 315–319, 2000.

[25] A. Agathanggelou, S. Honorio, D. P. Macartney et al., "Methylation associated inactivation of RASSF1A from region 3p21.3 in lung, breast and ovarian tumours," *Oncogene*, vol. 20, no. 12, pp. 1509–1518, 2001.

[26] D. G. Burbee, E. Forgacs, S. Zochbauer-Muller et al., "Epigenetic inactivation of RASSF1A in lung and breast cancers and malignant phenotype suppression," *Journal of the National Cancer Institute*, vol. 93, no. 9, pp. 691–699, 2001.

[27] A. Dallol, A. Agathanggelou, S. L. Fenton et al., "RASSF1A interacts with microtubule-associated proteins and modulates

microtubule dynamics," *Cancer Research*, vol. 64, no. 12, pp. 4112–4116, 2004.

[28] J. Bergqvist, A. Latif, S. A. Roberts et al., "RASSF1A polymorphism in familial breast cancer," *Familial Cancer*, vol. 9, no. 3, pp. 263–265, 2010.

[29] B. Gao, X. J. Xie, C. Huang et al., "RASSF1A polymorphism A133S is associated with early onset breast cancer in BRCA1/2 mutation carriers," *Cancer Research*, vol. 68, no. 1, pp. 22–25, 2008.

[30] M. E. Lusher, J. C. Lindsey, F. Latif et al., "Biallelic epigenetic inactivation of the RASSF1A tumor suppressor gene in medulloblastoma development," *Cancer Research*, vol. 62, no. 20, pp. 5906–5911, 2002.

[31] Z. G. Pan, V. I. Kashuba, X. Q. Liu et al., "High frequency somatic mutations in RASSF1A in nasopharyngeal carcinoma," *Cancer biology & therapy*, vol. 4, no. 10, pp. 1116–1122, 2005.

[32] H. Kanzaki, H. Hanafusa, H. Yamamoto et al., "Single nucleotide polymorphism at codon 133 of the RASSF1 gene is preferentially associated with human lung adenocarcinoma risk," *Cancer Letters*, vol. 238, no. 1, pp. 128–134, 2006.

[33] S. S. Min, S. C. Jin, J. S. Su, T. H. Yang, H. Lee, and D. S. Lim, "The centrosomal protein RAS association domain family protein 1A (RASSF1A)-binding protein 1 regulates mitotic progression by recruiting RASSF1A to spindle poles," *Journal of Biological Chemistry*, vol. 280, no. 5, pp. 3920–3927, 2005.

[34] L. Liu, S. Tommasi, D. H. Lee, R. Dammann, and G. P. Pfeifer, "Control of microtubule stability by the RASSF1A tumor suppressor," *Oncogene*, vol. 22, no. 50, pp. 8125–8136, 2003.

[35] A. Moshnikova, J. Frye, J. W. Shay, J. D. Minna, and A. V. Khokhlatchev, "The growth and tumor suppressor NORE1A is a cytoskeletal protein that suppresses growth by inhibition of the ERK pathway," *Journal of Biological Chemistry*, vol. 281, no. 12, pp. 8143–8152, 2006.

[36] L. Cuschieri, T. Nguyen, and J. Vogel, "Control at the cell center: the role of spindle poles in cytoskeletal organization and cell cycle regulation," *Cell Cycle*, vol. 6, no. 22, pp. 2788–2794, 2007.

[37] B. Raynaud-Messina and A. Merdes, "Gamma-tubulin complexes and microtubule organization," *Current Opinion in Cell Biology*, vol. 19, no. 1, pp. 24–30, 2007.

[38] S. Tommasi, R. Dammann, Z. Zhang et al., "Tumor susceptibility of RASSF1A knockout mice," *Cancer Research*, vol. 65, no. 1, pp. 92–98, 2005.

[39] L. Liu, A. Vo, and W. L. McKeehan, "Specificity of the methylation-suppressed A isoform of candidate tumor suppressor RASSF1 for microtubule hyperstabilization is determined by cell death inducer C19ORF5," *Cancer Research*, vol. 65, no. 5, pp. 1830–1838, 2005.

[40] C. Man, J. Rosa, L. T. O. Lee et al., "Latent membrane protein 1 suppresses RASSF1A expression, disrupts microtubule structures and induces chromosomal aberrations in human epithelial cells," *Oncogene*, vol. 26, no. 21, pp. 3069–3080, 2007.

[41] C. Mondello and A. I. Scovassi, "Apoptosis: a way to maintain healthy individuals," *Sub-Cellular Biochemistry*, vol. 50, pp. 307–323, 2010.

[42] F. Van Herreweghe, N. Festjens, W. Declercq, and P. Vandenabeele, "Tumor necrosis factor-mediated cell death: to break or to burst, that's the question," *Cellular and Molecular Life Sciences*, vol. 67, no. 10, pp. 1567–1579, 2010.

[43] B. Pennarun, A. Meijer, E. G. E. de Vries, J. H. Kleibeuker, F. Kruyt, and S. de Jong, "Playing the DISC: turning on TRAIL death receptor-mediated apoptosis in cancer," *Biochimica et Biophysica Acta*, vol. 1805, no. 2, pp. 123–140, 2010.

[44] A. Thorburn, "Death receptor-induced cell killing," *Cellular Signalling*, vol. 16, no. 2, pp. 139–144, 2004.

[45] J. Yu, M. Ni, J. Xu et al., "Methylation profiling of twenty promoter-CpG islands of genes which may contribute to hepatocellular carcinogenesis," *BMC Cancer*, vol. 2, no. 1, article 29, 2002.

[46] U. Schagdarsurengin, L. Wilkens, D. Steinemann et al., "Frequent epigenetic inactivation of the RASSF1A gene in hepatocellular carcinoma," *Oncogene*, vol. 22, no. 12, pp. 1866–1871, 2003.

[47] F. Jing, W. Yuping, C. Yong et al., "CpG island methylator phenotype of multigene in serum of sporadic breast carcinoma," *Tumor Biology*, vol. 31, no. 4, pp. 321–331, 2010.

[48] J. R. Fischer, U. Ohnmacht, N. Rieger et al., "Promoter methylation of RASSF1A, RAR beta and DAPK predict poor prognosis of patients with malignant mesothelioma," *Lung Cancer*, vol. 54, no. 1, pp. 109–116, 2006.

[49] T. B. Kang, T. Ben-Moshe, E. E. Varfolomeev et al., "Caspase-8 serves both apoptotic and nonapoptotic roles," *Journal of Immunology*, vol. 173, no. 5, pp. 2976–2984, 2004.

[50] S. Bialik and A. Kimchi, "The death-associated protein kinases: structure, function, and beyond," *Annual Review of Biochemistry*, vol. 75, pp. 189–210, 2006.

[51] D. Gozuacik and A. Kimchi, "DAPk protein family and cancer," *Autophagy*, vol. 2, no. 2, pp. 74–79, 2006.

[52] I. Mor, R. Carlessi, T. Ast, E. Feinstein, and A. Kimchi, "Death-associated protein kinase increases glycolytic rate through binding and activation of pyruvate kinase," *Oncogene*, vol. 31, pp. 683–693, 2011.

[53] S. Bialik, H. Berissi, and A. Kimchi, "A high throughput proteomics screen identifies novel substrates of death-associated protein kinase," *Molecular and Cellular Proteomics*, vol. 7, no. 6, pp. 1089–1098, 2008.

[54] S. K. Verma, T. S. Ganesan, and P. J. Parker, "The tumour suppressor RASSF1A is a novel substrate of PKC," *FEBS Letters*, vol. 582, no. 15, pp. 2270–2276, 2008.

[55] S. J. Song, S. J. Kim, M. S. Song, and D. S. Lim, "Aurora B-mediated phosphorylation of RASSF1A maintains proper cytokinesis by recruiting syntaxin16 to the midzone and midbody," *Cancer Research*, vol. 69, no. 22, pp. 8540–8544, 2009.

[56] D. Demon, P. Van Damme, T. V. Berghe et al., "Caspase substrates: easily caught in deep waters?" *Trends in Biotechnology*, vol. 27, no. 12, pp. 680–688, 2009.

[57] S. J. Riedl and Y. Shi, "Molecular mechanisms of caspase regulation during apoptosis," *Nature Reviews Molecular Cell Biology*, vol. 5, no. 11, pp. 897–907, 2004.

[58] S. Baksh, S. Tommasi, S. Fenton et al., "The tumor suppressor RASSF1A and MAP-1 link death receptor signaling to bax conformational change and cell death," *Molecular Cell*, vol. 18, no. 6, pp. 637–650, 2005.

[59] H. A. Ghazaleh, R. S. Chow, S. L. Choo et al., "14-3-3 mediated regulation of the tumor suppressor protein, RASSF1A," *Apoptosis*, vol. 15, no. 2, pp. 117–127, 2010.

[60] M. D. Vos, A. Dallol, K. Eckfeld et al., "The RASSF1A tumor suppressor activates bax via MOAP-1," *Journal of Biological Chemistry*, vol. 281, no. 8, pp. 4557–4563, 2006.

[61] L. Liu, R. Xie, C. Yang, and W. L. McKeehan, "Dual function microtubule- and mitochondria-associated proteins mediate mitotic cell death," *Cellular Oncology*, vol. 31, no. 5, pp. 393–405, 2009.

[62] H. Donninger, N. Allen, A. Henson et al., "Salvador protein is a tumor suppressor effector of RASSF1A with hippo pathway-independent functions," *Journal of Biological Chemistry*, vol. 286, no. 21, pp. 18483–18491, 2011.

[63] M. Yi, J. Yang, X. Chen et al., "RASSF1A suppresses melanoma development by modulating apoptosis and cell-cycle progression," *Journal of Cellular Physiology*, vol. 226, no. 9, pp. 2360–2369, 2011.

[64] M. S. Song, S. J. Song, N. G. Ayad et al., "The tumour suppressor RASSF1A regulates mitosis by inhibiting the APC-Cdc20 complex," *Nature Cell Biology*, vol. 6, no. 2, pp. 129–137, 2004.

[65] A. W. Whitehurst, R. Ram, L. Shivakumar, B. Gao, J. D. Minna, and M. A. White, "The RASSF1A tumor suppressor restrains anaphase-promoting complex/cyclosome activity during the G_1/S phase transition to promote cell cycle progression in human epithelial cells," *Molecular and Cellular Biology*, vol. 28, no. 10, pp. 3190–3197, 2008.

[66] L. Liu, K. Baier, R. Dammann, and G. P. Pfeifer, "The tumor suppressor RASSF1A does not interact with Cdc20, an activator of the anaphase–promoting complex," *Cell Cycle*, vol. 6, no. 13, pp. 1663–1665, 2007.

[67] C. Chow, N. Wong, M. Pagano et al., "Regulation of APC/C(Cdc20) activity by RASSF1A-APC/C(Cdc20) circuitry," *Oncogene*, vol. 31, no. 15, pp. 1975–1987, 2011.

[68] L. Shivakumar, J. Minna, T. Sakamaki, R. Pestell, and M. A. White, "The RASSF1A tumor suppressor blocks cell cycle progression and inhibits cyclin D1 accumulation," *Molecular and Cellular Biology*, vol. 22, no. 12, pp. 4309–4318, 2002.

[69] S. L. Fenton, A. Dallol, A. Agathanggelou et al., "dentification of the E1A-regulated transcription factor p120 E4F as an interacting partner of the RASSF1A candidate tumor suppressor gene," *Cancer Research*, vol. 64, no. 1, pp. 102–107, 2004.

[70] L. Jiang, R. Rong, M. S. Sheikh, and Y. Huang, "Cullin-4A.DNA damage-binding protein 1 E3 ligase complex targets tumor suppressor RASSF1A for degradation during mitosis," *Journal of Biological Chemistry*, vol. 286, no. 9, pp. 6971–6978, 2011.

[71] G. Hamilton, K. S. Yee, S. Scrace, and E. O'Neill, "ATM regulates a RASSF1A-dependent DNA damage response," *Current Biology*, vol. 19, no. 23, pp. 2020–2025, 2009.

[72] K. Dreijerink, E. Braga, I. Kuzmin et al., "The candidate tumor suppressor gene, RASSF1A, from human chromosome 3p21.3 is involved in kidney tumorigenesis," *Proceedings of the National Academy of Sciences of the United States of America*, vol. 98, no. 13, pp. 7504–7509, 2001.

[73] S. T. Kim, D. S. Lim, C. E. Canman, and M. B. Kastan, "Substrate specificities and identification of putative substrates of ATM kinase family members," *Journal of Biological Chemistry*, vol. 274, no. 53, pp. 37538–37543, 1999.

[74] X. Zhou, T. T. Li, X. Feng et al., "Targeted polyubiquitylation of RASSF1C by the Mule and SCF beta-TrCP ligases in response to DNA damage," *Biochemical Journal*, vol. 441, no. 1, pp. 227–236, 2012.

[75] J. Li, F. Wang, A. Protopopov et al., "Inactivation of RASSF1C during *in vivo* tumor growth identifies it as a tumor suppressor gene," *Oncogene*, vol. 23, no. 35, pp. 5941–5949, 2004.

[76] Y. G. Amaar, M. G. Minera, L. K. Hatran, D. D. Strong, S. Mohan, and M. E. Reeves, "Ras association domain family 1C protein stimulates human lung cancer cell proliferation," *American Journal of Physiology*, vol. 291, no. 6, pp. L1185–L1190, 2006.

[77] M. E. Reeves, S. W. Baldwin, M. L. Baldwin et al., "Ras-association domain family 1C protein promotes breast cancer cell migration and attenuates apoptosis," *BMC Cancer*, vol. 10, article 562, 2010.

[78] E. Estrabaud, I. Lassot, G. Blot et al., "RASSF1C, an isoform of the tumor suppressor RASSF1A, promotes the accumulation of beta-catenin by interacting with betaTrCP," *Cancer Research*, vol. 67, no. 3, pp. 1054–1061, 2007.

[79] G. Malpeli, E. Amato, M. Dandrea et al., "Methylation-associated down-regulation of ,RASSF1A and up-regulation of RASSF1C in pancreatic endocrine tumors," *BMC Cancer*, vol. 11, article 351, 2011.

[80] A. G. Knudson, "Mutation and cancer: statistical study of retinoblastoma," *Proceedings of the National Academy of Sciences of the United States of America*, vol. 68, no. 4, pp. 820–823, 1971.

[81] R. Dammann, U. Schagdarsurengin, C. Seidel et al., "The tumor suppressor RASSF1A in human carcinogenesis: an update," *Histology and Histopathology*, vol. 20, no. 2, pp. 645–663, 2005.

11

The Continuing Evolution of HIV-1 Therapy: Identification and Development of Novel Antiretroviral Agents Targeting Viral and Cellular Targets

Tracy L. Hartman and Robert W. Buckheit Jr.

Anti-Infective Research Department, ImQuest BioSciences, Inc., 7340 Executive Way, Suite R, Frederick, MD 21704, USA

Correspondence should be addressed to Robert W. Buckheit Jr., rbuckheit@imquestbio.com

Academic Editor: Gilda Tachedjian

During the past three decades, over thirty-five anti-HIV-1 therapies have been developed for use in humans and the progression from monotherapeutic treatment regimens to today's highly active combination antiretroviral therapies has had a dramatic impact on disease progression in HIV-1-infected individuals. In spite of the success of AIDS therapies and the existence of inhibitors of HIV-1 reverse transcriptase, protease, entry and fusion, and integrase, HIV-1 therapies still have a variety of problems which require continued development efforts to improve efficacy and reduce toxicity, while making drugs that can be used throughout both the developed and developing world, in pediatric populations, and in pregnant women. Highly active antiretroviral therapies (HAARTs) have significantly delayed the progression to AIDS, and in the developed world HIV-1-infected individuals might be expected to live normal life spans while on lifelong therapies. However, the difficult treatment regimens, the presence of class-specific drug toxicities, and the emergence of drug-resistant virus isolates highlight the fact that improvements in our therapeutic regimens and the identification of new and novel viral and cellular targets for therapy are still necessary. Antiretroviral therapeutic strategies and targets continue to be explored, and the development of increasingly potent molecules within existing classes of drugs and the development of novel strategies are ongoing.

1. Introduction

Since the approval of AZT for the treatment of HIV-1 infection, twenty-three additional therapeutic agents have been approved for use in humans [1]. The first drugs approved in the United States to treat HIV-1 infection inhibit the specific activity of the virally encoded reverse transcriptase, the viral enzyme essential for conversion of the viral RNA genome into a DNA provirus that integrates itself into the host genome. Two classes of reverse transcriptase inhibitors are currently marketed—nonnucleoside reverse transcriptase inhibitors (NNRTIs) and nucleoside/nucleotide reverse transcriptase inhibitors (N(t)RTIs) [2]. Another approved and marketed class of HIV-1 antiviral therapeutics inhibits the HIV-1 protease, a viral enzyme required to process newly synthesized viral polyproteins into the mature viral gene products, enabling the virus to assemble itself into

new infectious virus particles [3]. A third class of HIV-1 therapeutics inhibits viral infection by preventing virus attachment to the host cell CCR5 chemokine receptor or prevents the fusion of the viral and cellular membranes [4]. Most recently, compounds which prevent the integration of the HIV-1 proviral precursor into cellular DNA have been successfully developed and utilized. Clinical experience with all HIV-1 agents has clearly demonstrated the ability of HIV-1 to easily evade the antiviral effects of any monotherapeutic drug administration strategy through the rapid accumulation of amino acid changes in the targeted proteins—reverse transcriptase, protease, envelope, and integrase [5]. The high turnover rate of virus replication along with the highly error prone HIV-1 reverse transcriptase, with its lack of proof-reading capability, generates significant heterogeneity within the highly related but nonidentical populations (or quasis-pecies) of viruses circulating in a patient [6]. It is widely accepted that most drug-resistant viruses preexist within the population of viruses and are selected from within this heterogeneous environment upon application of selective drug pressure [7]. In addition to the high levels of resistance possible to single agents, each of the anti-HIV-1 agents employed to date has had significant dose limiting and long-term toxicities that render successful long term therapy for HIV-1 disease difficult to achieve [8].

In much of the developing world, antiretroviral therapy has successfully suppressed HIV-1 replication in patients, allowing significant delays to the progression of AIDS and in some cases completely normal life spans. However, HIV-1 therapies in general are plagued by patient compliance issues reflective of difficult treatment regimens, involving up to four antiretroviral drugs, significant class-specific toxicity [9], and the emergence and spread of virus isolates selected for resistance to single or multiple antiretroviral agents [10]. In the developing world many of these therapeutic strategies are uniformly unavailable due to the prohibitive cost of the drugs. The absence of an effective vaccine and the lack of effective therapy means that sub-Saharan Africa and Southeast Asia, among other developing regions of the world, remain epicenters for the continued spread of HIV-1, especially among heterosexual women [11]. In these areas of extremely high HIV-1 transmission rates, the opportunities to derail the AIDS pandemic rest on the processes of education and the development of effective topical microbicides, a specific HIV-1 prevention strategy employing HIV-1 drugs to prevent the sexual transmission of HIV-1 [12].

2. Identification and IND-Directed Development of New Antiretroviral Agents

The FDA has published guidance documents that relate to the development of systemic HIV-1 inhibitors [1]. These documents define the preclinical pharmacologic data that must be provided in an IND submission to begin human testing of a new antiretroviral agent. The submitted data package must specifically address the efficacy and toxicity of the test compound in a relevant cell-based assay system. In addition studies should be initiated that adequately address the range and mechanism of action of the test compound. With the wide variety of approved anti-HIV-1 drugs already on the market and the demonstrated efficacy of highly active antiretroviral therapies (HAARTs) [13], the ability of test compounds to be utilized as a component of combination drug therapies with the approved HIV-1 drugs should be evaluated in detail. Finally, drug resistance should be evaluated to define the ease of selection of resistant strains and to define diagnostic resistance-engendering mutations prior to clinical trials. Animal models to evaluate the effectiveness of HIV-1 therapies are available but their predictability for clinical efficacy is still highly debated, and thus most drug development programs bypass these animal models and move directly to Phase I human safety trials.

It is clear that highly active antiretroviral therapy (HAART) has significantly decreased morbidity and mortality among patients infected with HIV-1 and has prolonged the life of infected individuals. HAART has transformed HIV-1 infection from a lethal infection to a chronic disease much like diabetes. However, new anti-HIV-1 agents are still needed to confront the emergence of drug resistance and various adverse effects associated with long-term use of antiretroviral therapy. New antiviral agents that inhibit an increasing number of viral and cellular processes are critical for treating infected patients, as well as for prophylactic use, and all possible targets for countering HIV-1 infection, replication, and persistence need to be considered. Finally, efforts to eradicate HIV-1 from latent reservoirs within the body have gained increasing traction with the success of HAARTs and eradication efforts will require novel drugs and new ways of thinking about antiretroviral therapy of latent and silent HIV-1 infection.

The identification of new antiretroviral agents typically involves either cell-based or biochemical/enzymatic target-based screening programs. The end result of these screening programs is a lead compound which provides a pharmacophore for medicinal chemistry structure-activity relationships (SARs) efforts to enhance the potency (increased efficacy and/or reduced toxicity) of the lead molecule, subsequently yielding candidates which progress through the IND-directed clinical development pathway to human clinical trials. Historically, new drug entities have been highly specific virus-targeted agents which inhibited critical steps in HIV-1 replication. Current development efforts continue to exploit the known targets for antiretroviral intervention, but have been expanded to include agents which target cellular processes that are essential for HIV-1 replication. Genomic, proteomic, and metabolomics approaches have identified large numbers of cellular products and pathways that are positively and negatively impacted by HIV infection, providing a large number of potential novel anti-HIV targets [14]. Herein we provide an overview of current and continuing drug development in the HIV-1 field based on the target of the antiretroviral agent as well as an overview of the methodology utilized to identify and confirm the new molecules as potential drug candidates for clinical development. Methods available to identify and characterize the mechanism of action of new antiretroviral agents are summarized in Table 1.

3. HIV-1 Entry Inhibitors

Though a range of hematopoietic cells, including monocyte-macrophages, B lymphocytes, eosinophils, and dendritic cells, as well as columnar epithelial cells, have been found to be infected by HIV-1, the CD4-positive helper T lymphocyte has been identified as the primary target for HIV-1infection [15]. HIV-1 enters CD4-positive T cells through direct interaction of the viral envelope gp120 with the D1 region of the CD4 receptor on the cell surface of target cells. The interaction of gp120 with CD4 causes a conformational change in the viral envelope gp120, resulting in exposure of the gp41 transmembrane envelope protein which subsequently inserts into and fuses with the target cell membrane. HIV-1 envelope proteins interact with coreceptor molecules on the surface of

The Continuing Evolution of HIV-1 Therapy: Identification and Development of Novel Antiretroviral Agents Targeting Viral and Cellular Targets

97

TABLE 1: Anti-HIV-1 screening assays.

Replication event	Assay	Method
Virus attachment	HeLa-CD4-LT4-β-gal cells	HeLa-cell based assay measuring reduction in chemiluminescence of HIV-1-infected cells [16]
	TZM-bl cells	
	gp120/CD4 Ab binding inhibition	Cell based HIV-1 neutralization assay
	gp120 : CD4 ELISA	Biochemical assay with soluble CD4 and monoclonal gp120 antibodies [17]
	gp120/CD4/coreceptor	Cell based, temperature sensitive fusion assay
Fusion and chemokine coreceptor interaction	HL2/3 cells + HeLa-CD4-LTR-β-gal cells	Cell based assay measuring reduction in chemiluminesence [16]
	Coreceptor inhibiton	GHOST-cell based assays measuring reduction in virus replication
	Coreceptor typing	PBMC and Macrophage cell-based assays with tropism-specific clinical HIV-1 isolates [18]
	Compound displacement of chemokine ligands	
	Ca^{++} flux	
Reverse transcription	Homopolymer and heteropolymer RT inhibition	Biochemical assay measuring reduction in dGTP-[P32] incorporation [19]
	E/intermediate/late RT products	PCR amplification
	RNaseH inhibition	Biochemical assay [20]
	RT inhibition assays using enzymes with specific mutations	Biochemical dGTP-[P32] incorporation assay [19]
Nuclear localization	2 LTR product in cell nucleus	PCR detection
Integration	Provirus in genomic DNA	PCR detection [21]
	Integrase Complementation	Cell based IN-mutant and Vpr-IN transfection [22]
	Integrase inhibition	Biochemical SPA assay [23]
	Integrase negative virus	
Protein expression	Northern, Western and flow cytometry	Cell based assays with molecular biology endpoints [24, 25]
	Tat, Rev, and Nef inhibition	Biochemical assays [26, 27]
	Cell-based reporter assay for Rev and Tat function	
	Intracellular p24	CEM-SS cells infected with HIV-1 and lysed to quantitate p24 by ELISA
	LTR-mediated transcriptional activation	
Protease	Intracellular and virion protein processing	Cell based assay with Western analysis [28]
	Polyprotein cleavage	Biochemical FRET assay [29]

CD4-positive cells, typically either the α-chemokine receptor CXCR4 or the β-chemokine receptor CCR5 or both, to trigger the fusion of the viral and cellular membranes. The HIV-1 fusion inhibitor, enfuvirtide, developed by Hoffmann-La Roche and Trimeris, was the first therapeutic in its class to be approved for use in humans by the FDA [30]. Enfuvirtide binds to gp41 and prevents the pore formation required for the capsid of the virus to enter the target cell. Since enfuvirtide is a peptide, the drug is marketed in an injectable form, which has somewhat limited its therapeutic utility. In addition, primary resistance mutations in the HR1 region of gp41 have been identified in 10.5% of enfuvirtide-naïve patients which allow the virus to evade the antiviral effects of the drug [31]. The CCR5 coreceptor antagonist, maraviroc, was developed by Pfizer, and the FDA approved maraviroc for combination therapy in 2007 [32]. By blocking the HIV-1 gp120 protein from associating with the CCR5 coreceptor, maraviroc prevents HIV-1 from entering the target cell.

However, since HIV-1 can use other coreceptors for entry, an HIV-1 tropism test must be performed to determine if the drug will be effective in a particular patient. CCR5-tropic HIV-1 strains are more common than CXCR4-tropic strains and have been identified as the strain which is predominantly transmitted, suggesting that maraviroc will be useful for both prevention of virus transmission (topical microbicide use) and treatment. Additionally, the CXCR4 coreceptor is more critical for immune function and cannot be safely blocked, indicating that CXCR4-targeted inhibitors would be immune-toxic in the host. Among individuals found mostly in Northern Europe, there is a polymorphism in CCR5, involving a 32-base pair inactivating deletion known as delta32 (Δ32), which reduces or completely eliminates cell surface expression of CCR5 [33]. Individuals with one CCR5-Δ32 polymorphism exhibit reduced disease progression, while those homozygous for the deletion appear to have natural resistance to HIV-1 infection [33]. There appears to be no obvious immunologic detriment from the Δ32 deletion, making CCR5 a highly credible antiviral target. Other CCR5 antagonists completed or currently in Phase II clinical development include INCB9471 (Incyte), HGS004 (Human Genome Sciences), PRO140 (Progenics), PF232798 (ViiV Healthcare), cenicriviroc (Tobira Therapeutics), and VCH286 (ViroChem Pharma). CCR5 antagonists in Phase I clinical development include AK602 (Kumamoto University), SCH532706 (Schering), GSK706769 (ViiV Healthcare), and VIR576 (Viro Pharmaceuticals). Other entry inhibitors in development include SP01A (Samaritan Pharmaceuticals) in Phase III clinical trials and ibalizumab (Taimed Biologics), a nonimmunosuppressive monoclonal antibody that binds CD4 [34], in Phase II studies. HIV-1 can enter and bud from lipid rafts of plasma membranes of infected cells. Lipid rafts play a crucial role in colocalizing CD4 and chemokine receptors for entry of HIV-1 into T cells. Depletion of plasma membrane cholesterol relocalizes raft-resident markers to a nonraft environment and inhibits productive infection by HIV-1 [35]. SP01A affects cholesterol synthesis by reducing 3-hydroxy-3-methylglutary coenzyme A (HMG-CoA) reductase mRNA expression. Inhibition of cholesterol synthesis by SP01A modifies the cholesterol content of the host cell membrane lipid rafts and prevents HIV-1 fusion with CD4-positive cells. SP10A is a second generation oral entry inhibitor that is being developed by Samaritan Pharmaceuticals [35].

A number of preclinical assays have been developed to identify potential inhibitors of HIV-1 entry utilizing both replication competent wild type viruses and pseudotype viruses. Compounds may be evaluated for inhibitory activity in CD4-dependent and CD4-independent virus transmission assays. A variety of cell lines have been made which stably express CD4, CCR5, or both coreceptors on the cell surface to evaluate coreceptor inhibitors of HIV-1 infection. Evaluations can be performed to define the specific mechanism of antiviral action of compounds which are directly virucidal, or which inhibit virus attachment, or virus-cell fusion, or virus entry to the target cell using indicator cell lines with reporter gene endpoints (colorimetric, chemiluminescent, and fluorescent) to measure virus replication [16]. Compounds which directly interfere with binding of gp120 to CD4 can be evaluated via ELISA using purified proteins [17]. The effect of inhibitors on the virus gp120/CD4/coreceptor complex that would target gp41 can be evaluated using an indicator cell line, such as HeLa-LTR-CD4-β-galactosidase cells which employ a tat protein-induced transactivation of the reporter gene driven by the HIV-1 long terminal repeat promoter, and manipulating the fusion step with temperature changes. Varying the time of drug addition in high multiplicity of infection (MOI), single round of infection anti-HIV-1 assays is often useful in demonstrating that entry inhibitors must be present prior to 2 hours post-infection of target cells in order to provide antiviral activity.

4. HIV-1 Reverse Transcriptase Inhibitors

Inhibitors of HIV-1 reverse transcriptase can be divided into two classes: nucleoside/nucleotide reverse transcriptase inhibitors (NRTI/NtRTIs) and nonnucleoside reverse transcriptase inhibitors (NNRTIs). NRTIs are analogues of the naturally occurring deoxynucleotides required for synthesis of viral DNA and following phosphorylation to the active triphosphate form by cellular kinases; they compete with the natural deoxynucleotides for incorporation into the growing viral DNA chain. However, unlike the natural deoxynucleotide substrates, NRTIs and NtRTIs lack a 3′-hydroxyl group on the deoxyribose moiety or are pseudosugars unable to be extended. As a result, following incorporation of an NRTI or an NtRTI, the next incoming deoxynucleotide cannot form the 5′-3′ phosphodiester bond needed to extend the DNA chain and thus causing chain termination. Mitochondrial toxicity is recognized as a major adverse effect of nucleoside analogue treatment. Nucleoside analogues are effective in inhibiting HIV-1 replication due to their high affinity for the viral RT enzyme. However, NRTIs can also bind to other human DNA polymerases, like DNA polymerase beta, necessary for repair of nuclear DNA, and mitochondrial DNA polymerase gamma, which is exclusively responsible for the replication of mitochondrial DNA. NRTIs and NtRTIs comprise the first class of antiretroviral drugs developed and approved for use in humans to treat HIV-1 infection. There are a number of FDA-approved nucleoside/nucleotide reverse transcriptase inhibitors. Cytidine analogs include zalcitabine (ddC), which is no longer marketed, emtricitabine (FTC), and lamivudine (3TC). Thymidine analogs include zidovudine (AZT) and stavudine (d4T). Didanosine (ddI) and tenofovir disoproxil fumarate (TDF) are analogs of adenosine, and abacavir sulfate (ABC) is an analog of guanine. HIV-1 can become resistant to NRTIs by two mechanisms. The first resistance mechanism involves the reduced incorporation of the nucleotide analog into DNA over the normal nucleotide. This resistance mechanism results from mutations in the N-terminal polymerase domain of the reverse transcriptase that reduce the enzyme's affinity or ability to bind to the drug. A prime example of this mechanism is the M184V mutation that confers resistance to lamivudine (3TC) and emtricitabine (FTC). Another well-characterized set of mutations is the Q151M complex found in multidrug-resistant HIV-1 which decreases reverse transcriptase's efficiency at incorporating NRTIs but does not affect natural nucleotide incorporation. The complex includes the Q151M mutation along with amino acid changes A62V, V75I, F77L, and F116Y. A virus with Q151M alone is moderately resistant to zidovudine (AZT), didanosine (ddI), zalcitabine (ddC), stavudine (d4T), and slightly resistant to abacavir (ABC). A virus with Q151M in concert with one or more of the other four noted mutations becomes highly resistant to those drugs and is additionally resistant to lamivudine (3TC) and emtricitabine (FTC) [36]. A virus with the Q151M complex in addition to the K70Q mutation significantly enhanced resistance to several approved NRTIs and also resulted in 10-fold resistance to TDF [37]. The K65R mutation emerges in response to treatment with TDF, ABC, ddI, or d4T and has been shown to have an increased frequency in subtype C HIV-1 [38]. The second resistance mechanism involves the ATP-based excision of the incorporated drug by 3′ → 5′ exonuclease activity, which allows

The Continuing Evolution of HIV-1 Therapy: Identification and Development of Novel Antiretroviral Agents Targeting Viral and Cellular Targets

99

the DNA chain to be extended and polymerization to continue [39]. Excision enhancement mutations, typically M41L, D67N, K70R, L210W, T215Y/F, and K219E/Q, are selected by thymidine analogs AZT and D4T and are therefore referred to as thymidine analog mutations (TAMs). The excision-based mutations improve the ability of the RT to bind ATP. ATP-dependent pyrophosphorylation removes the drug and releases a dinucleotide tetraphosphate. The goal of next generation reverse transcriptase inhibitors is to treat patients with multidrug-resistant HIV-1, prolong the time to emergence of drug resistance to the new inhibitors, and to increase drug adherence by minimizing pill burden and side effects. Several NRTIs are in development to treat HIV-1-infected patients. Entecavir (ETV), a guanine analog for HIV-1 infection, is currently in development by Bristol-Myers Squibb and has been FDA approved for treatment of HBV infection since 2005. Apricitabine (ATC), a cytidine analog with antiviral activity against 3TC and AZT-resistant HIV-1 being developed by Avexa Pharmaceuticals, was given fast track approval by the FDA in March 2011. Dexelvucitabine (DFC) and racivir are cytidine analogs in development by Pharmasset. DFC is active against drug-resistant HIV-1 containing the M184V, K65R, L74V and TAMS mutations. However, Incyte discontinued co-development of DFC due to increased incidence of grade 4 hyperlipasemia, a marker of pancreatic inflammation, in a Phase IIb clinical trial. Racivir has completed a Phase II clinical trial in comparison with lamivudine in patients with the M184V lamivudine-resistant virus. Elvucitabine is a cytosine nucleoside analog of stavudine which was evaluated in a Phase II clinical trial by Achillion Pharmaceuticals. Unimpressive clinical results did not provide a rationale for further development of the drug. Another derivative of stavudine, festinavir, is being developed by Bristol-Myers Squibb and has antiviral activity against multidrug-resistant HIV-1 with less toxicity compared to stavudine. Chimerix, Inc. developed a lipid conjugate of tenofovir and unlike tenofovir, disoproxil fumarate and most prodrugs, the CMX157 prodrug is not efficiently cleaved in plasma thus increasing the levels of active tenofovir in target cells. CMX157 is greater than 300 times more potent than tenofovir with increased oral bioavailability [40]. Following a favorable Phase I clinical trial, Chimerix is seeking to outlicense the compound for further development. Another prodrug of tenofovir, GS-7340, is being developed by Gilead Sciences to better target lymphoid tissues and cells [41]. GS-7340 has increased plasma stability compared with tenofovir. A recent Phase I clinical trial resulted in no serious adverse effects. Investigators at the University of Georgia identified 1-(β-D-dioxolane) thymine (DOT) as a potent inhibitor of AZT- and 3TC-resistant HIV-1 strains, and this compound is currently in a Phase I clinical trial [42]. Medivir is developing MIV-210, a nucleoside analog with potent antiviral activity versus drug resistant HIV-1 as well as hepatitis B virus. Following favorable plasma levels of MIV210 and good oral bioavailability in Phase I studies, a Phase IIa clinical trial has been initiated with multi-drug-resistant HIV-1 infected patients.

In contrast, to the NRTIs and NtRTIs, NNRTIs have a completely different mode of action. NNRTIs allosterically block reverse transcriptase by binding at a different site on the enzyme as compared to the chain terminating analogs. NNRTIs are not incorporated into the viral DNA but instead inhibit the movement of protein domains of reverse transcriptase essential for DNA synthesis. Since the hydrophobic binding area found in HIV-1 reverse transcriptase does not appear in HIV-2, NNRTIs are specific to inhibition of HIV-1 replication. NNRTIs do not bind to the active site of the polymerase but bind to a less conserved area near the active site in the p66 subdomain. NNRTI binding results in a conformational change in the reverse transcriptase that distorts the positioning of the residues that bind DNA, inhibiting polymerization. NNRTI resistance is conferred by mutations that decrease the binding of the drug to this pocket. Treatment with a regimen including efavirenz (EFV) and nevirapine (NVP) typically results in the appearance of mutations L100I, Y181C/I, K103N, V106A/M, V108I, Y188C/H/L, and G190A/S. Current FDA-approved NNRTIs also include delavirdine (Pfizer) and three diarylpyrimidines developed by Tibotec Therapeutics, dapivirine, etravirine and rilpivirine. The second-generation NNRTIs by Tibotec have better potency, longer half-life, and reduced side effects compared with the older NNRTIs, such as efavirenz. Delavirdine is not recommended for use as part of initial therapy due to its lower efficacy compared to other NNRTIs, interactions with other medications due to its inhibition of CYP3A4, and higher pill burden. As patients live longer on HAART and the pool of NNRTI-resistant virus increases, so does the need for the development of new NNRTIs with antiviral activity against both wild-type and the clinically prevalent NNRTI-resistant HIV-1 strains. Boehringer Ingelheim has presented data on BILR355BS, a dipyridodiazepinone NNRTI compound, with potent antiviral activity (EC$_{50}$ < 10 nM) against a wide range of NNRTI-resistant viruses but terminated drug development during the Phase II clinical trial [43]. GSK2248761, belonging to the family of 3-phosphindoles, was developed by ViiV Healthcare and completed Phase II studies, but the FDA put further development on hold due to significant adverse events (seizures). It is unclear if or when development will continue. RDEA806, a new family of triazole NNRTIs, entered Phase IIb clinical trials by Ardea Biosciences in 2009. Lersivirine, developed by Pfizer, belongs to the pyrazole family and completed Phase IIb studies in 2010. The resistance profile for compounds in development is similar to that of other next generation NNRTIs. ImQuest Pharmaceuticals has recently reported a pyrimidinedione NNRTI with highly potent anti-HIV-1 activity and a dual mechanism of action which also involves the inhibition of virus entry [18]. Their lead compound (IQP-0528) is expected to soon enter human clinical trials for both therapeutic and topical microbicide use.

Inhibition of the virus-encoded reverse transcriptase can be evaluated in both cell-based and biochemical assays. High MOI and time of drug addition anti-HIV-1 assays are often useful in demonstrating that RT inhibitors must be present prior to 8 hours postinfection of target cells in order to yield antiviral activity. In cell-based assays, PCR amplification of early, intermediate, and late RT products may be analyzed in treated, HIV-1-infected cells to determine inhibition of enzymatic activity compared to an untreated, infected cell culture. A biochemical assay utilizing purified, recombinant RT enzymes can also be used to identify inhibitors of wild type and drug-resistant HIV-1 reverse transcriptase [19].

5. HIV-1 RNase H Inhibitors

The ribonuclease H (RNase H) function of the C terminus of reverse transcriptase is required for successful production of a DNA copy of the HIV-1 genome. RNase H is required for processing the tRNA primer used to begin minus-strand DNA synthesis and degradation of the viral RNA during synthesis, followed by preparation of the polypurine tract (PPT) DNA-RNA hybrid, which serves as the primer for positive-strand DNA synthesis. Essential for RNase H activity is a group of three carboxylate-containing amino acid residues, conserved in the class of polynucleotidyl transferases and a fourth conserved in RNase H [44]. RT-RNase H is absolutely essential for HIV-1 replication and is therefore a logical and thus far unexploited target for antiretroviral intervention. Drug discovery efforts focusing on RT-RNase H have lagged behind those for other HIV-1 targets but are ongoing. Nucleoside and nonnucleoside compounds have been reported to inhibit both the polymerase and RNase H activities, though the mechanism of RNase H inhibition is poorly understood. Studies have shown that the NRTI AZT and the NNRTI EFV act in a synergistic fashion (together they inhibit RT function to a greater extent than the sum of their individual inhibitory activities). It has been demonstrated that RT inhibition by EFV may allow the innate RNAse H activity of RT to cleave the RNA template, which, in turn, increases susceptibility to AZT, yielding a synergistic antiviral interaction of the two drugs [45]. AZT incorporates into the growing DNA chain, stopping reverse transcription unless it is excised. In the presence of EFV, RNase H activity of RT is enhanced, leading to destruction of the RNA template before AZT excision can efficiently occur, increasing the apparent activity of AZT [46]. An obstacle to the development of RNase H inhibitors was highlighted in a study of β-thujaplicinol [47] that measured cleavage of RNA by RNAse H. β-thujaplicinol efficiently cleaved RNA strands; however, in the context of reverse transcriptase tightly bound to the RNA substrate, the conformational change during reverse transcription resulted in β-thujaplicinol being unable to inhibit RNase H. This suggested that RNase H inhibitors, such as β-thujaplicinol and dihydroxyl benzoyl naphthyl hydrazine (DHBNH) [47], that bind directly to the RNase H active site within RT might have difficulty accessing this site during transcription when RT is bound to an RNA template. RNase H inhibitors that do not bind in the active site of RNase H within HIV-1 RT, such as the MK3 naphthyridine [48], should be explored, and potential antagonism with other RT inhibitors will need to be addressed. RNase H proteins are native to all forms of life, so building inhibitor specificity toward HIV-1 RNase H without off-target effects will be critical to developing an effective drug.

Inhibitors of RNase H function have been identified using a biochemical polymerase-independent cleavage assay with a $5'$-[^{32}P]tC5U/p12 substrate [20]. The radioactive RNA-DNA chimera is hybridized to its DNA complement, which mimics processing of the HIV-1-1 PPT primer from nascent DNA, following initiation of second-strand synthesis. Capillary electrophoresis is used to illustrate RNase H cleavage at the PPT RNA-U3 DNA junction and at two additional positions.

6. HIV-1 NCp7 Inhibitors

HIV-1 p7 nucleocapsid protein (NCp7), which contains two highly conserved zinc fingers with a nonclassical Cys-Xaa2-Cys-Xaa4-His-Xaa4-Cys (CHHC) sequence, is a maturational proteolytic product of the p55 precursor polyprotein [49]. The zinc fingers function in selection and incorporation of viral RNA into budding virions while being a component of the p55 precursor. Zinc fingers of the NCp7 are required for the initial infection of target cells, promote initiation of transcription, and increase the efficiency of template switching during reverse transcription. Due to the essential and pluripotent roles in both early and late stages of HIV-1 replication, as well as the conserved Cys and His chelating residues, the HIV-1 zinc fingers represent attractive antiviral targets and would appear to be multifunctional inhibitors of HIV-1. Disulfide-substituted benzamides (DIBAs) were identified as anti-HIV-1 inhibitors with the ability to chemically modify and eject zinc from the zinc finger of NCp7. Antiviral activity of the DIBAs resulted in the formation of noninfectious virus or in the complete inhibition of virus production in vitro, similar to HIV-1 protease inhibitors. Azodicarbonamide (ADA; HPH116) is a nucleocapsid inhibitor that electrophilically attacks the sulfur atoms of the zinc-coordinating cysteine residues of the CCHC domain [50]. ADA is directly virucidal by preventing the initiation of reverse transcription and blocking formation of infectious virus by modification of the CCHC domain within Gag precursors. ADA was evaluated in a Phase II study in 2001, but the status of drug development is unknown. S-acyl-2-mercaptobenzamide thioesters (SAMTs) demonstrate potent antiviral activity in vitro as a virucidal agent and in in vivo SIV studies in Cynomolgus macaques [51]. NV038, a N,N'-bis(1,2,2-thiadiazol-5-yl)benzene-1,2-diamine, targets NCp7 by reacting with the sulfhydryl group of cysteine residues. NV038 acts via a different mechanism than other reported zinc ejectors, as its structural features do not allow an acyl transfer to Cys or a thiol-sulfide interchange [52]. Studies performed at ImQuest BioSciences have demonstrated a significant inability to select for drug resistant viruses to the zinc finger inhibitors as well as their highly synergistic interaction with all classes of antiretroviral agents.

NCp7-targeted inhibitors have been shown to be virucidal in vitro. Cell-free virus is treated with compound then washed away prior to incubation with target cells to demonstrate reduction in virus infectivity [51]. In addition, the zinc finger inhibitors reduce virus production from chronically HIV-1-infected cells. Zinc ejection from purified NCp7 protein can also be assessed biochemically in the presence of inhibitors [52]. Specificity of NCp7 inhibition for the retroviral zinc finger should be addressed by evaluating

The Continuing Evolution of HIV-1 Therapy: Identification and Development of Novel Antiretroviral Agents Targeting
Viral and Cellular Targets

101

the interaction of inhibitors with cellular Sp1, GATA, and PARP zinc fingers.

7. HIV-1 Integrase Inhibitors

Integrase is a viral enzyme that integrates retroviral DNA into the host cell genome. HIV-1 integration occurs through a multistep process that includes two catalytic reactions: 3′ endonucleolytic processing of proviral DNA ends and integration of 3′-processed viral DNA into cellular DNA, referred to as strand transfer. The 3′ processing integrase binds to a short sequence located at either end of the long terminal repeat (LTR) of the viral DNA and catalyzes endonucleotide cleavage, resulting in elimination of a dinucleotide from each of the 3′ ends of the LTR. Cleaved DNA is then used as a substrate for integration. Strand transfer occurs simultaneously at both ends of the viral DNA molecule, with an offset of five base pairs between the two opposite points of insertion. Integration is completed by removal of the unpaired dinucleotides, repair of the single-stranded gaps created between the viral and target DNA, and ligation of the host DNA. Divalent metals, Mg^{2+} or Mn^{2+}, are cofactors required for 3′-processing and strand transfer steps. Raltegravir, the first integrase inhibitor developed by Merck Sharp & Dohme Limited, was FDA approved for use in HIV-1-infected patients in 2007. Other HIV-1 integrase inhibitors currently in Phase III clinical trials include elvitegravir, developed by Japan Tobacco, and dolutegravir, developed jointly by ViiV Healthcare and Shinongi. Raltegravir and elvitegravir possess metal-chelating functions and interact with divalent metals within the active site of HIV-1 integrase. The inhibitors compete directly with viral DNA for binding to the integrase active site at the DDE motif, a highly conserved triad of acidic residues consisting of D64, D116, and E152 which mediate binding of the metal cofactors to the active site, in order to block strand transfer [53]. Two structural components are necessary for integrase binding: a hydrophobic benzyl moiety that buries into a highly hydrophobic pocket near the active site and a chelating triad that binds with two Mg^{2+} ions in a hydrophilic region, anchoring the inhibitor onto the protein surface. Identification of the pharmacophore for inhibition of HIV-1 integrase catalysis has proven to be challenging. For optimal integrase inhibition, the pharmacophore requires a region-specific (N-1) diketoacid of specific length [54]; however, a detailed binding model is lacking, so it has been difficult to develop structure-based design of integrase inhibitors. HIV-1 resistance to raltegravir and elvitegravir has been associated with mutations in the loop of amino acid residues 140–149. Raltegravir has limited intestinal absorption, and thus resistance cannot be overcome by prescribing higher doses. The integrase inhibitor dolutegravir is sensitive to HIV-1 variants resistant to raltegravir or elvitegravir, is bioavailable as a single, oral dose without need of a booster, and has been well tolerated by patients in clinical trials. Clinical trials are underway to support the use of dolutegravir in combination with abacavir and lamivudine, in a new fixed dose combination called 572-Trii. GSK1265744 is in Phase IIa human clinical trials as a new generation candidate to dolutegravir.

Merck has developed a second generation integrase inhibitor, MK-2048, with the same mechanism of action as raltegravir with sensitivity to raltegravir-resistant HIV-1 [55]. MK-2048 is being investigated for use as part of preexposure prophylaxis (PrEP) regimen and has been shown to inhibit the integrase enzyme four times longer than raltegravir. BI224436 is in preclinical development by Gilead Sciences following its purchase from Boehringer Ingleheim as a novel noncatalytic site integrase inhibitor that binds to a conserved allosteric pocket of the HIV-1 integrase enzyme [56]. BI224436 has been shown to retain full antiviral activity against viruses encoding resistance mutations to clinically approved drugs targeting HIV-1 integrase. BI224436 has advanced to Phase I clinical trials following ADME evaluations which indicated favorable metabolic stability, low potential for interactions with CYP3A4 and CYP2D6, high permeability, excellent physicochemical properties, and excellent pharmacokinetic profiles in animals.

Structural studies utilizing cocrystallization with prototype foamy virus (PFV) intasome with raltegravir and elvitegravir have been helpful in establishing the binding mode of integrase strand transfer inhibitors. Crystal structures of PFV intasomes containing primary mutations associated with drug resistance, as well secondary amino acid substitutions which may compensate for the impaired viral fitness, revealed conformational rearrangements within the IN active site contributing to raltegravir resistance [21]. Integration of the 2-LTR circular cDNA into the host DNA mediated by the virus-encoded integrase can be evaluated for inhibition in both cell-based and biochemical assays. In a high MOI single-round HIV-1 infection in cells, PCR detection of the provirus in genomic DNA can be assessed. Amersham produces an HIV-1 integrase scintillation proximity assay (SPA) enzyme kit for biochemical evaluation of potential integrase inhibitors [23]. An in vitro assay utilizing integrase-mutant HIV-1 molecular clones complemented in trans by Vpr-IN fusion proteins enabled the study of integrase function in replicating viruses [22].

8. HIV-1 Regulatory and Accessory Protein Inhibitors

After integration into the host genome, HIV-1 remains quiescent until basal transcription produces a threshold level of the viral transactivator protein, Tat. Tat increases viral mRNA production several hundredfold by increasing the elongation capacity of RNA polymerase II (Pol II) rather than initiation of transcription. Tat is brought into contact with the transcription machinery after binding the transactivation-responsive (TAR) element, a 59-residue stem loop RNA found at the 5′ end of all HIV-1 transcripts. Tat forms a tight, specific complex with TAR RNA centered on a U-rich region found near the apex of the TAR RNA stem. Interactions between Tat and TAR are absolutely required for the increased processivity of Pol II and the production of full length virus transcripts. Tat binds to the cyclin-dependent kinase 7 (CDK7) and activates the phosphorylation of the carboxy-terminal domain of Pol II by TFIIH and the CDK-activating kinase (CAK) complex [57]. Studies suggest the interaction between Tat and its cellular counterpart is critical for the function of Tat and the increased processivity of Pol II. Oligonucleotides have been investigated for inhibition of Tat binding to this recognition site in biochemical assays, but they failed to disrupt HIV-1 replication in acute infection of primary lymphocytes [58]. Natural 4-phenylcoumarins isolated from *Marila pluricostata* were identified as Tat antagonists and were able to inhibit HIV-1 replication in cell-based assays [24]. Based on the beta-turn motif present in HIV-1 Tat, a series of novel benzodiazepine analogs were designed as biological mimetics. Preliminary biological evaluation exhibited inhibitory activity on HIV-1 Tat-mediated LTR transcription [59]. BPRHIV001, a coumarin derivative, has been identified as an HIV-1 Tat transactivation inhibitor (EC_{50} of 1.3 nM) with synergistic effects in combination with currently used reverse transcriptase inhibitors [60].

The Rev protein is an essential factor for HIV-1 replication and promotes the export of unspliced or partially spliced mRNA responsible for the production of the viral structural proteins. Within the N-terminal of Rev is the arginine-rich motif (ARM) which comprises both the nuclear localization signal (NLS) to mediate the nuclear and nucleolar localization of Rev and the RNA-binding domain to mediate binding of Rev to the Rev-Responsive Element (RRE), a 240-base region of complex RNA secondary structure. Flanking the ARM are sequences involved in mediating Rev multimerization that appears to be critical for its biological role. Polymerized Rev that interacts with host cellular factors is a prerequisite for RNA binding. The interaction between the HIV-1 Rev protein and the RRE RNA is an attractive target for antiviral therapy due to its role in facilitating the nuclear export of incompletely processed viral transcripts and its necessity for viral replication. For HIV-1, targeting the host cell factors might elicit fewer drug-resistant viruses. Screening for Rev inhibitors is in the early preclinical drug development stage, and various researchers have targeted the nuclear export factor CRM1, interference with the Rev-RRE interaction, Rev protein itself, and other cellular factors involved in HIV-1 transcription [26]. Leptomycin B (LMB), a *Streptomyces* cytotoxin discovered as a potent antifungal antibiotic that blocks the eukaryotic cell cycle, binds CRM1 and disrupts NES-mediated nuclear transport [61]. Variability in LMB production lots in *Streptomyces* cultures that vary greatly in toxicity has hampered the use of LMB. PKF050-638 is also capable of blocking Rev function by binding to CRM1 at position Cys-539 but its cellular toxicity resulted in the failure to pursue its potential as a therapeutic [62]. Neomycin B is capable of interfering with the Rev-RRE interaction, but poor efficacy (EC_{50} of 2.5 mM), toxicity, and poor oral absorption have prevented its development as a useful antiviral drug [63]. Diphenylfuran cations have also been shown to interfere with the Rev-RRE interaction *in vitro* at $0.1\,\mu M$ concentrations. These aromatic cationic compounds bind tightly to the minor groove of the IIB Rev motif with pronounced selectivity [64]. Antisense oligonucleotides which interact with RRE-IIB have also been investigated and found to bind with specificity and high affinity with apparent dissociation constants in the nanomolar range [65].

Thiabendazole, chlorpropham, and a series of related analogs which inhibit HIV-1 at a late stage, postintegration step of virus replication were identified by The Proctor & Gamble Company and are being investigated by ImQuest BioSciences. The compounds were identified as inhibitors of HIV-1 replication from chronically HIV-1-infected cells with the ability to suppress constitutive virus production in the long term. Mechanistic studies indicate the treatment of infected cells with these compounds results in an accumulation of multiply spliced viral RNA, with a corresponding decrease in the quantity of singly spliced and unspliced viral RNA, suggesting the compounds may inhibit Rev function.

A novel mechanism of antiviral action recently exploited by Trana Discovery involves human transfer RNA (tRNA) as a therapeutic target. The role of $tRNA_{SUU}^{Lys3}$ is essential for the replication and survival of HIV-1 at both reverse transcription as a primer and virus assembly, thereby providing a dual point of intervention by tRNA inhibitors. Efforts to inhibit the $tRNA_{SUU}^{Lys3}$ have centered on mimicking the anticodon stem loop (ASL) of tRNA to prevent binding of viral RNA [66].

Nef is a multifunctional accessory protein of HIV-1 which is critical for high virus replication and disease progression in infected patients. The lack of disease progression in patients infected with *nef*-deleted HIV-1, such as the Sydney Blood Bank Cohort comprised of eight individuals infected with an attenuated, nef/LTR-deleted strain of HIV-1 from a single donor, defines Nef as a pathogenic factor [67]. Developing inhibitors of Nef in order to reduce the severity of HIV-1 disease has been difficult due to the complexity of Nef's multiple functions. Nef is a small protein devoid of enzymatic activity that serves as an adaptor protein to divert host cell proteins to aberrant functions that amplify viral replication. Investigation of Nef function has led to the possibility of developing new anti-HIV-1 drugs targeting Nef's ability to induce CD4 downmodulation, major histocompatibility complex I and II (MHCI/MHCII) downmodulation, Pak2 activation, inhibition of p53 and ASK-1 involved in apoptosis, and enhancement of virion infectivity. Nef-induced CD4 downmodulation involves the internalization of surface CD4 followed by degradation via the endosomal/lysosomal pathway. Inhibition of lysosomal acidification blocks Nef-induced CD4 degradation, without restoring CD4 surface expression. The clathrin-associated adaptor protein 2 (AP2) is a key molecular mediator of Nef-induced CD4 downmodulation, suggesting this interaction is a possible target for antiviral therapy [68]. Another well-conserved property of Nef is its ability to downmodulate MHC class I molecules that enables the infected cell to evade destruction by the immune system during active viral replication. A ternary complex between the cytoplasmic tail of MHC and AP1, with Nef acting as a facilitator, may activate a tyrosine sorting signal in the MHC which diverts newly synthesized MHC molecules from their transit to the plasma membrane to an internal compartment. This ternary complex engages Nef in a novel interaction and could be a potential target for an antiviral compound. Nef may regulate cellular activation through several kinases, such as Pak2 and Hck. Nef binding

The Continuing Evolution of HIV-1 Therapy: Identification and Development of Novel Antiretroviral Agents Targeting
Viral and Cellular Targets

103

with Pak2 has been demonstrated to activate Pak2 in multiple HIV-1 subtypes. However, the structural fluidity of Nef's Pak2 interaction surface could make this Nef interaction difficult to target with antiviral compounds. Structure-function analyses identified an SH3 domain interaction of Hck that interacts with Nef. A series of small Nef interacting proteins composed of an SH3 domain fused to a sequence motif of the CD4 cytoplasmic tail and a prenylation signal for membrane association were investigated [25] and identified two hydrophobic pockets on Nef as potent pharmacophore target sites. Nef augments the infectivity of HIV-1 particles and accounts for the slight delay in replication kinetics observed for nef-deficient HIV-1. Triciribine (TCN) is a tricyclic nucleoside that once phosphorylated to its $5'$ monophosphate form by intracellular adenosine kinase is active against a wide range of HIV-1 and HIV-2 isolates. TCN was determined to be a late stage inhibitor of HIV-1 replication, and sequencing of TCN-resistant HIV-1 resulted in five-point mutations in the DNA sequence of nef [27]. Originally developed as an anticancer therapy, clinical trials indicated severe adverse toxicity with TCN such as hepatic toxicity, hyperglycemia, and thrombocytopenia [69]. Despite the attractiveness of a drug that reduces the inherent infectivity of HIV-1 virions, the prospects for inhibiting Nef-mediated enhancement of infectivity are remote. Overall attempts to develop inhibitors of Nef have demonstrated relatively low binding affinity, high cytotoxicity, and interference with only a subset of Nef interactions and functions.

Viral protein U (Vpu) is a type 1 membrane-associated accessory protein encoded by HIV-1 and functions to form a virus ion channel. Vpu contributes to HIV-1-induced CD4 receptor downregulation by mediating the proteosomal degradation of newly synthesized CD4 in the endoplasmic reticulum. Vpu also enhances the release of progeny virions from infected cells by antagonizing tetherin, an interferon-regulated host restriction factor that directly cross-links virions on the host cell surface [70]. BIT225 was developed by Biotron Limited as a small molecule inhibitor of HIV-1 Vpu to specifically target HIV-1 in the monocyte-macrophage reservoir, similar to tetherin-mediated reduction in infectivity [71]. BIT225 is active against multiple drug-resistant strains of HIV-1, and Phase IIb clinical trials are currently in progress.

Vpr is a multifunctional accessory protein critical for efficient viral infection of CD4-positive T cells and macrophages. Vpr mediates nuclear transport of the HIV-1 preintegration complex (PIC), induces G2 cell cycle arrest, modulates T-cell apoptosis, transcriptionally coactivates viral and host genes, and regulates nuclear factor kappa B (NF-κB) activity [72]. The numerous functions of Vpr in the viral life cycle suggest that Vpr would be an attractive target for HIV-1 therapeutics. Di-tryptophan containing hexameric peptides have been reported to overcome Vpr-mediated cell growth arrest and apoptosis by interfering with nuclear translocation [73]. Damnacanthal (Dam), an anthraquinone derivative isolated from the Tahitian noni fruit, has been identified as an inhibitor of Vpr-induced cell growth cessation [74]. Vipirinin, a 3-phenyl coumarin-based compound

in the RIKEN Natural Products Depository, inhibits Vpr-dependent viral infection of human macrophages. The hydrophobic region of residues Glu-25 and Gln-65 was found to be potentially involved in the binding of vipirinin to Vpr [75].

Viral infectivity factor (Vif) is a small, phosphoprotein essential for HIV-1 replication and pathogenesis. Vif neutralizes the host cell antiviral factor, apolipoprotein B mRNA editing enzyme catalytic polypeptide like 3G (APOBEC3G; A3G), which makes the viral particles more infective [76]. RN-18 was identified as an antagonist of Vif function and inhibited HIV-1 replication only in the presence of A3G. RN-18 increases cellular A3G levels in a Vif-dependent manner and increases A3G incorporation into virions without inhibiting general proteasome-mediated protein degradation in order to decrease virus replication [77].

The expression of HIV-1 regulatory proteins occurs early in the infected cell and is critical for appropriate replication of the virus. The ability of an anti-HIV-1 agent to inhibit these regulatory proteins can be evaluated in cell-based reporter assays, analyzed by Northern or Western blot, and by direct biochemical inhibition assays [24–27, 70, 73, 77].

9. Protease Inhibitors

HIV-1 aspartyl protease is a C2-symmetric homodimer that catalyzes the proteolytic cleavage of the polypeptide precursors into mature enzymes and structural proteins. Inhibitors have been designed to mimic the transition state of the protease substrates. A peptide linkage consisting of –NH–CO– is replaced by a hydroxyethylene group, where the protease is unable to cleave. Mutations that confer resistance to HIV-1 protease inhibitors are located primarily in the active site of the enzyme that directly changes the binding of the inhibitor. Nonactive site mutations have been shown to alter dimer stability and conformational flexibility. Over 26 protease inhibitor-specific mutations have been described, of which 15 are primary mutations significant enough to reduce drug efficacy. High-level drug resistance typically requires multiple mutations in the HIV-1 protease. Often, these resistance-associated mutations reduce the catalytic efficiency of the protease, resulting in immature or noninfectious viruses. In addition, mutations develop within Gag cleavage sites, complementing the changes in the resistant protease. Significant associations have been observed between mutations in the nucleocapsid-p1 (NC-p1) and the p1-p6 cleavage sites and various mutations in protease associated with protease inhibitor resistance [78]. Gag A431V or the I437V mutation, within the NC-p1 cleavage site, has been associated with the V82A, I50L, or I84V protease mutations. Gag L449F/P, R452S, P453L mutations within the p1-p6 cleavage site have been associated with I50V or D30N/N88D protease mutations. Cross-resistance is one of the major problems of protease inhibitor treatment. FDA-approved protease inhibitors saquinavir (Hoffman-La Roche), ritonavir (Abbott Laboratories), and indinavir (Merck) are peptidomimetic compounds designed to fit the C2 symmetry in the protease-binding site. Nelfinavir (Agouron Pharmaceuticals) was the first nonpeptidomimetic compound designed to contain a novel 2-methyl-3-hydroxybenzamide group. Amprenavir (GlaxoSmithKline) is an N,N-disubstituted aminosulfonamide nonpeptide inhibitor with enhanced aqueous solubility compared to previous protease inhibitors and was later replaced on the market with its prodrug, Fosamprenavir, which resulted in lower pill burden. Lopinavir (Abbott Laboratories) is a peptidomimetic protease inhibitor designed for activity against drug-resistant HIV-1 containing mutations at the Val82 residue. Atazanavir (Bristol-Myers Squibb) is an azapeptide protease inhibitor designed to fit the C2 symmetry of the enzyme-binding site and is unique to other PIs as it can only be absorbed in an acidic environment. The resistance profile of atazanavir is also better than previous protease inhibitors. Tipranavir (Boehringer Ingelheim), a nonpeptide inhibitor of protease, was developed from a coumarin template and possesses broad antiviral activity against multiple protease inhibitor-resistant HIV-1. Darunavir (Tibotec, Inc.) is a nonpeptide analog of amprenavir with a critical change at the terminal tetrahydrofuran group, allowing for antiviral activity against amprenavir-resistant HIV-1. Research on new protease inhibitors is directed towards the development of compounds that will not be cross-resistant with other PIs, have a favorable metabolic profile, will not require boosting by RTV, and have a low once-daily pill burden. GlaxoSmithKline discontinued Phase II clinical development of brecanavir due to insurmountable issues regarding formulation. In 2009, GlaxoSmithKline and Concert Pharmaceuticals entered into a collaboration to develop deuterium-containing drugs. CTP518, an analog of atazanavir produced by replacing key hydrogen atoms with deuterium, demonstrated slow hepatic metabolism resulting in an increased half-life and entered Phase I studies in 2010 [79]. CTP518 has the potential to eliminate the need to codose with a boosting agent, such as ritonavir. TM310911, developed by Tibotec Therapeutics, is in Phase II clinical trials with a ritonavir booster. SPI-256, developed by Sequoia Pharmaceuticals in 2008, demonstrated significant potency and a high genetic barrier to resistance *in vitro*. A Phase I study demonstrated safety and tolerability in humans, but SPI-256 development was recently discontinued. SPI-452, a PK enhancer in development by Sequoia Pharmaceuticals, has been shown to increase plasma concentrations of atazanavir and darunavir in Phase I studies without the side effects typically seen with ritonavir as a boosting agent [80]. Cobicistat (GS 9350) by Gilead is a pharmacoenhancer based on CYP3A inhibition, and it represents the PK enhancer in the most advanced development phase. Cobicistat tested against ritonavir with atazanavir plus TDF/FTC and Quad in combination with cobicistat and elvitegravir are all currently in larger Phase III studies.

Cell-based and biochemical assays are available to evaluate the ability of a compound to inhibit the enzymatic cleavage of viral polyproteins by HIV-1 protease. An HIV-1 protease fluorescence resonance energy transfer (FRET) assay kits are commercially available for biochemical evaluation of potential protease inhibitors [29]. HIV-1 protease activity can be monitored in human cells based on expression of a precursor protein harboring the viral protease fused to the reporter protein GFP [28]. Western analysis of intracellular and virion protein processing can be utilized as well to evaluate HIV-1 protease inhibition.

10. Myristoylation Inhibitors

HIV-1 Gag is synthesized in the cytosol as a precursor protein, p55, and is targeted to the plasma membrane where particle assembly and packaging of viral genomic RNA occur. Modification of p55 at the N-terminal glycine residue with myristic acid, a saturated 14-carbon fatty acid, is essential for targeting p55 to the plasma membrane for HIV-1 assembly. Gag myristoylation consists of two reactions: activation of myristic acid to myristoyl-CoA by acyl-CoA synthetase and transfer of the myristoyl group from myristoyl-CoA to the N-terminal glycine of p55 by N-myristoyltransferase (NMT). Several studies have considered NMT as a potential drug target for the inhibition of HIV-1 assembly. NMT inhibitors have been shown to prevent both membrane binding of Gag as well as virus assembly [81]; however, NMT inhibitors are expected to affect a broad spectrum of cellular processes that depend on protein N-myristoylation for membrane binding. Heteroatom-substituted myristic acid analogs, such as 12-methoxydodecanoic acid, can be used by NMT as alternative substrates for covalent attachment to proteins. The hydrophilic nature of these compounds inhibits membrane binding and function of the modified HIV-1 Gag [82]. The biochemical characterization of these compounds in relation to their effect on HIV-1 remains poorly understood. Dinucleoside fatty acyl prodrugs are being explored for the ability to inhibit HIV-1 replication as a topical microbicide by two mechanisms of action including inhibition of reverse transcriptase and inhibition of the cellular N-myristoyl transferase (NMT) [83].

The levels of myristoylation in cells infected with HIV-1 in the presence and absence of compound can be analyzed by labeling infected cells with [^{3}H]myristate and analyzing cell lysates for myristate incorporation into gp41 through immunoprecipitation (IP) with anti-gp41 antibody [81]. Cell-based assays with chronically HIV-1 infected cells can also be used to demonstrate the effects of myristoylation inhibition on proteolytic processing and virus production [84].

11. Maturation Inhibitors

Maturation inhibitors interfere with the final stage of HIV-1 replication, when viral proteins are assembled, packaged, and released from the host cell membrane to form new virus particles. Bevirimat, a betulinic acid-like compound isolated from the Chinese herb *Syzygium claviflorum*, was purchased by Myriad Genetics from Panacos in 2009, as an inhibitor of HIV-1 maturation. Bevirimat binds to the Gag protein and prevents the critical cleavage of p25 (CA-SP1) between Gag codons 363 and 364 to p24 (CA) and p2 (SP1), resulting in virus particles that lack functional capsid protein and have structural defects rendering them incapable of infecting other cells [85]. Clinical trial data reported in 2009 indicated bevirimat was well tolerated and showed good antiviral

The Continuing Evolution of HIV-1 Therapy: Identification and Development of Novel Antiretroviral Agents Targeting Viral and Cellular Targets

105

activity against HIV-1 with specific Gag protein variations. *In vitro* studies demonstrated the presence of a number of single nucleotide polymorphisms, including H358Y, L363F/M, A364V, and A366T/V, in the CA/SP1 cleavage site that resulted in resistance to bevirimat [86]. Mutations at these sites were not, however, detected in the Phase I and II clinical trials for bevirimat, even in nonresponders. Instead, mutations in the QVT motif of the SP1 peptide (Gag positions 369 to 371) were the primary predictors of failure of response to bevirimat. The comparable potency to other approved HIV-1 drugs, combined with the benefits of oral administration, low probability of drug interactions, and long plasma half-life made bevirimat appear to be a promising new drug candidate. However, Myriad announced in 2010 that it was stopping the development of the maturation inhibitors bevirimat and vivecon.

Maturation inhibitors can be assessed in cell-based assays to evaluate the RNA content and infectivity of virions produced following treatment of infected cells [87]. Electron microscopy can also be used to visualize virions budding from the infected cells following treatment.

12. Cellular Targets

Cells have evolved a number of barriers to resist invading microorganisms. One mechanism that appears to be particularly important in counteracting HIV-1 infection is a group of type 1 interferon-inducible, innate restriction factors that includes tetherin and APOBEC3G. Knowledge of the mechanisms by which restriction factors interfere with HIV-1 replication and how their effects are avoided by HIV-1 in human cells could allow for novel forms of therapeutic intervention. Tetherin is a host protein expressed by many cell types following interferon induction, including CD4-positive T cells, that acts at a late stage of HIV-1 replication to trap mature virions at the plasma membrane by cross-linking to prevent cell-free virus release [88, 89]. Tetherin-retained virions can be reinternalized into the infected cell and targeted to late endosomes where they are destroyed by lysosomal enzymes. However, cell to-cell transmission of HIV-1 is an important mode of dissemination and the possibility of using interferon-based therapy to upregulate the natural antiviral activity of host cells has proven ineffective [90]. APOBEC3G was identified as an inhibitor of HIV-1 replication in cells nonpermissive for replication of HIV-1 mutants lacking a functional Vif gene [91]. APOBEC3G protein can be incorporated into HIV-1 particles through interactions with packaged RNA and the enzyme catalyzes the deamination of deoxycytidines generating minus-strand DNA containing many deoxyuracil nucleotides whose replication results in plus-strand G to A mutations [92]. Hypermutation of HIV-1 DNA can be lethal through deposition of many inactivating missense and nonsense mutations in protein-coding sequences. As previously discussed, RN-18 identified by University of Massachusetts Medical School by high throughput screening of a compound library has been reported to inhibit Vif function by increasing ABOBEC3G concentration within the target cells [93].

Lens epithelial-derived growth factor (LEDGF/p75) is a host protein that binds to HIV-1 integrase and is crucial for viral replication [94]. The mechanism of action is not precisely known but evidence suggests that LEDGF/p75 guides integrase to insert viral DNA into transcriptionally active sites of the host genome. Inhibitors being developed are likely to be highly target specific and less prone to the development of resistance.

Tumor susceptibility gene 101 (TSG101) has been reported to be an essential cellular factor for HIV-1 budding [95]. Inhibiting TSG101 engagement by Gag induces a block of budding virus due to the lipid envelope of nascent particles remaining continuous with the host cell membrane. Monoclonal antibodies and cyclic peptides have been investigated as inhibitors of TSG101 interactions with Gag.

Second generation nonnucleoside rhodanine derivatives have been reported to have improved inhibition of the human DEAD-box RNA helicase DDX3 leading to anti-HIV-1 activity [96]. DEAD-box proteins have nucleic acid-dependent ATPase activity and are involved in ATP-dependent RNA unwinding. DDX3 has been shown to possess relaxed nucleotide substrate specificity, being able to accept ribo- and deoxynucleoside triphosphates as well as nucleoside analogs. DDX3 incorporates into the nucleocapsids and is an essential cofactor for HIV-1 replication. Studies indicate DDX3 is dispensable for host cell metabolism and would therefore provide an excellent antiviral target with predicted low levels of drug resistance without leading to toxicity from interference of a cellular pathway [97].

Antithrombin III has been reported to activate two host cell interactomes dependent on the NFκB transcription factor, extracellular signal-regulated kinases (ERK), mitogen-activated protein kinase (MAPK), and prostaglandin-synthetase 2 (PTGS2) nodules which have anti-HIV-1 effects [98]. Acceleration Biopharmaceuticals is investigating protein interactomes to identify nodules with host cell factors and pathways for viral inhibition.

Antimicrobial peptides derived from cathelicidins, selected on criteria of length, charge, and lack of Cys since defensins have already been reported to demonstrate anti-HIV-1 effects, are being investigated by the University of Nebraska as potential microbicides [98]. A wide variety of other antimicrobial peptides which have been identified to possess anti-HIV-1 activity are cataloged in the Antimicrobial Peptide Database (APD) maintained by The University of Nebraska [99].

The investigation of host cell factors involved in HIV-1 replication involves profiling of well-characterized signal transduction pathways and antiviral immune responses in HIV-1-infected and uninfected cells treated with test material then building an interactome to identify nodules whose blockage might inhibit viral replication. RT-PCR-based gene arrays are used to determine if cellular gene expression is altered by infected cells treated with a potential inhibitor [14, 100, 101]. Data analysis requires a large data base to define potential nodules responsible for the gene expression alterations. Knockout experiments with siRNA can be used in cell-based assays to confirm the inhibition of HIV-1 replication is due to a particular host cell factor.

13. Immunotherapy

Another approach to treating HIV-1 infection is to strengthen the immune response of infected patients. Immune stimulators are designed to improve overall immune function and include preclinical research on Alferon, human leukocyte-derived interferon alfa-n3 developed by Hemisperx Biopharma, that is currently in Phase III clinical trials [102]. Such approaches like Proleukin, developed by Novartis as recombinant human interleukin-2, have failed in the past to demonstrate stimulation of CD4-positive T cell production in HIV-1-infected patients enrolled in the studies [103]. CYT107, recombinant human interleukin-7 developed by Cytheris, is in Phase II clinical trials with raltegravir and maraviroc with the hope of improving T cell counts in patients classified as immunological nonresponders on antiviral therapy [104]. As a growth factor and cytokine physiologically produced by marrow or thymic stromal cells and other epithelia, IL-7 has a crucial stimulating effect on T lymphocyte development and on homeostatic expansion of peripheral T-cells. Tarix Pharmaceuticals is developing TXA127, angiotensin 1–7, to stimulate bone marrow production of progenitor cells [105]. TXA127 is currently in Phase I studies.

Immunomodulatory compounds are designed to signal immune cells to respond to infection in specific ways and may have direct or indirect antiviral activity. Many immunomodulatory molecules have been shown to reduce cell surface antigen expression using flow cytometry, which resulted in inhibition of virus replication through an entry blocking mechanism. Many immunomodulatory compounds will either inhibit or induce cellular proliferation of specific cell types. In many cases, *in vitro* cell-based assays are not possible, and efficacy will need to be demonstrated using relevant animal models.

14. Gene Therapy

Several gene therapy strategies are being studied in order to construct CD4 cells resistant to HIV-1 infection by a population of anti-HIV-1 antisense RNA producing lymphocytes. Enzo Biochem completed a Phase II clinical trial for HGTV43, a retrovirus vector used to deliver three genes encoding U1/anti-HIV-1 antisense RNA targeting TAR and two separate sites of tat/rev region [106], with results indicating antisense RNA was produced from CD4-positive lymphocytes throughout the 24-month observance but no recent news on the status of HGTV43 could be found. A Phase II clinical trial of VRX496, developed by VIRxSYS, was completed in 2010. VRX496 gene therapy is derived from a lentivirus vector and appears to sustain expression of the delivered genes of interest for a longer period of time compared to previous gene therapies and does not appear to elicit an inflammatory immune response. VIRxSYS is attempting to develop a therapy that will allow HIV-1 patients with undetectable viral loads on HAART to discontinue the antiretroviral treatment and still control their viral load. The VRX496 Phase II study demonstrated a decrease in viral load for 88% of the enrolled HIV-1-infected patients with suppression of HIV-1 viremia for more than 14 weeks in some patients in the absence of HAART [107].

Gene therapies are investigated *in vitro* using cell-based anti-HIV-1 assays that measure reduced virus replication, and effects on specific proteins or RNA can be analyzed by Western or Northern blots [108].

15. The Problem of Latent Reservoirs

HIV-1 is known to establish latent reservoirs where the virus is maintained for long periods of time in an essentially quiescent state. Low-rate viral replication also comes from anatomical sites, such as the brain, where drug penetration is limited and only suboptimal drug concentration can be achieved [109]. Studies employing HAART intensification strategies have failed to demonstrate any appreciable reduction in virus load in patients, suggesting the inability to further reduce virus production from these latently infected cells [110]. In recent years considerable interest in the ability to eradicate these latent virus reservoirs and cure HIV-1 infection has evolved. In addition to the HAART intensification studies, efforts have been directed at activating virus production from the latently infected cells to target them for destruction by antiretroviral agents or the host immune system. Compounds developed for this purpose primarily include histone deacetylase (HDAC) inhibitors such as valproic acid, vorinostat, givinostat, and belinostat [111] and nontumor promoting phorbol esters such as prostratin [112]. Compounds which target cellular factors and or regulatory/accessory proteins might also be utilized to target and further reduce virus replication in latently infected cells, such as the transcriptional inhibitors being investigated at ImQuest BioSciences.

The identification and evaluation of compounds which specifically target latently infected cells has primarily utilized the latently infected U1 or ACH-2 cell lines. Primary resting CD4-positive T cells provide the optimal intracellular milieu for establishing latency but are inefficiently infected *in vitro*, since HIV is impaired during reverse transcription and integration. Most primary cell models use one or more rounds of cellular stimulation to remove these blocks, followed by HIV infection during the return to a resting state. Unfortunately, although latently infected nondividing T cells are generated, the process often takes several weeks or months of continuous culture. Investigation into direct infection of resting CD4 T cells by spinoculation has resulted in postintegration latency in these spinoculated cells within 72 h in all CD4 T-cell subsets, including both naive and memory T cells [113]. Cells are sorted by FACS analysis, latent proviruses are activated after additional 72 h of cellular stimulation, and latency can be established and reactivation assessed within 6 days. Using novel reporter viruses, an improved version of this primary CD4 T-cell model has been utilized to study latency in all subsets of CD4 T cells [114]. The ability to target virus in latent reservoirs also requires evaluation in animal models of HIV-1 infection where these reservoirs are established and can be appropriately evaluated.

The Continuing Evolution of HIV-1 Therapy: Identification and Development of Novel Antiretroviral Agents Targeting Viral and Cellular Targets

107

16. Summary

Three decades of HIV-1 research have greatly contributed to the knowledge scientists and clinicians have available regarding HIV-1 replication, pathogenesis, and therapeutic strategies. Though great strides have been made in the development of anti-HIV-1 inhibitors targeting various viral enzymes and cellular host factors involved in the virus life cycle, we have learned that multi-drug combinations are necessary for the suppression of viremia and the delayed emergence of drug resistance. More drugs targeting essential virus-specific and/or cellular components of the viral replication pathway and virus transmission are needed to treat and prevent HIV-1 infection. The increasing prevalence of drug-resistant virus strains in patient populations, the increasing incidence of transmission of drug-resistant virus during primary HIV-1 infection, the toxicity of the currently approved therapeutic regimens, and the sometimes difficult regimens that must be followed assure that continued HIV-1 drug development will occur in the foreseeable future. Additionally, development issues must include the ability to safely use drugs in pediatric and pregnant individuals and to specifically target virus in latently infected reservoirs. Finally, increasing emphasis on the eradication of HIV from latent reservoirs in infected individuals will require the development of new and novel treatment strategies. The algorithms available for guiding the screening, identification, characterization, and development of these new compounds have been refined over the years as many thousands of compounds have been evaluated and compounds have been approved for use in humans. Current drug development programs must not only prove the efficacy and safety of the new drug candidates but must also show superiority over existing drugs in the same or similar classes. Thus, drug development must be directed at establishing new and novel drug targets, increasing the potency of existing classes of molecules, decreasing the toxicity or pill burden of existing therapies, or adding new drugs to the HAART regimen with superior combination therapy potential or reduced susceptibility to resistant viruses, including drugs designed specifically to attack existing drug-resistant virus strains. Drug development algorithms in the HIV-1 area must be customizable and highly flexible to assure the ability to characterize novel compounds and therapeutic strategies.

References

[1] Food and Drug Administration, http://www.fda.gov/For-Consumers/byAudience/ForPatientAdvocates/HIVandAIDS-Activities/ucm118915.htm.

[2] E. De Clercq, "HIV inhibitors targeted at the reverse transcriptase," *AIDS Research and Human Retroviruses*, vol. 8, no. 2, pp. 119–134, 1992.

[3] A. Molla, G. Richard Granneman, E. Sun, and D. J. Kempf, "Recent developments in HIV protease inhibitor therapy," *Antiviral Research*, vol. 39, no. 1, pp. 1–23, 1998.

[4] H. J. P. Ryser and R. Flückiger, "Keynote review: progress in targeting HIV-1 entry," *Drug Discovery Today*, vol. 10, no. 16, pp. 1085–1094, 2005.

[5] B. A. Larder, "Viral resistance and the selection of antiretroviral combinations," *Journal of Acquired Immune Deficiency Syndromes and Human Retrovirology*, vol. 10, supplement 1, pp. S28–S33, 1995.

[6] D. L. Mayers, "Drug-resistant HIV-1: the virus strikes back," *Journal of the American Medical Association*, vol. 279, no. 24, pp. 2000–2002, 1998.

[7] D. Boden, A. Hurley, L. Zhang et al., "HIV-1 drug resistance in newly infected individuals," *Journal of the American Medical Association*, vol. 282, no. 12, pp. 1135–1141, 1999.

[8] R. L. Murphy, "Defining the toxicity profile of nevirapine and other antiretroviral drugs," *Journal of Acquired Immune Deficiency Syndromes*, vol. 34, no. 1, pp. S15–S20, 2003.

[9] B. P. Sabundayo, J. H. McArthur, S. J. Langan, J. E. Gallant, and J. B. Margolick, "High frequency of highly active antiretroviral therapy modifications in patients with acute or early human immunodeficiency virus infection," *Pharmacotherapy*, vol. 26, no. 5, pp. 674–681, 2006.

[10] J. E. Gallant, E. Dejesus, J. R. Arribas et al., "Tenofovir DF, emtricitabine, and efavirenz vs. zidovudine, lamivudine, and efavirenz for HIV," *The New England Journal of Medicine*, vol. 354, no. 3, pp. 251–260, 2006.

[11] P. D. Ghys, T. Saidel, H. T. Vu et al., "Growing in silence: selected regions and countries with expanding HIV/AIDS epidemics," *AIDS*, vol. 17, supplement 4, pp. S45–50, 2003.

[12] S. L. Lard-Whiteford, D. Matecka, J. J. O'Rear, I. S. Yuen, C. Litterst, and P. Reichelderfer, "Recommendations for the nonclinical development of topical microbicides for prevention of HIV transmission: an update," *Journal of Acquired Immune Deficiency Syndromes*, vol. 36, no. 1, pp. 541–552, 2004.

[13] M. Oette, R. Kaiser, M. Däumer et al., "Primary HIV drug resistance and efficacy of first-line antiretroviral therapy guided by resistance testing," *Journal of Acquired Immune Deficiency Syndromes*, vol. 41, no. 5, pp. 573–581, 2006.

[14] F. D. Bushman, N. Malani, J. Fernandes et al., "Host cell factors in HIV replication: meta-analysis of genome-wide studies," *PLoS Pathogens*, vol. 5, no. 5, Article ID e1000437, 2009.

[15] F. Hladik and M. J. McElrath, "Setting the stage: host invasion by HIV," *Nature Reviews Immunology*, vol. 8, no. 6, pp. 447–457, 2008.

[16] J. Xu, L. Lecanu, M. Tan, W. Yao, J. Greeson, and V. Papadopoulos, "The benzamide derivative N-[1-(7-tert-Butyl-1H-indol-3-ylmethyl)-2-(4- cyclopropanecarbonyl-3-methyl-piperazin-1-yl)-2-oxo-ethyl]-4-nitro-benzamide (SP-10) reduces HIV-1 infectivity in vitro by modifying actin dynamics," *Antiviral Chemistry and Chemotherapy*, vol. 17, no. 6, pp. 331–342, 2006.

[17] C. Lackman-Smith, C. Osterling, K. Luckenbaugh et al., "Development of a comprehensive human immunodeficiency virus type 1 screening algorithm for discovery and preclinical testing of topical microbicides," *Antimicrobial Agents and Chemotherapy*, vol. 52, no. 5, pp. 1768–1781, 2008.

[18] F. Huang, M. Koenen-Bergmann, T. R. MacGregor, A. Ring, S. Hattox, and P. Robinson, "Pharmacokinetic and safety evaluation of BILR 355, a second-generation nonnucleoside reverse transcriptase inhibitor, in healthy volunteers," *Antimicrobial Agents and Chemotherapy*, vol. 52, no. 12, pp. 4300–4307, 2008.

[19] A. Mahalingam, A. P. Simmons, S. R. Ugaonkar et al., "Vaginal microbicide gel for delivery of IQP-0528, a pyrimidinedione analog with a dual mechanism of action against

HIV-1," *Antimicrobial Agents and Chemotherapy*, vol. 55, no. 4, pp. 1650–1660, 2011.

[20] D. M. Himmel, S. G. Sarafianos, S. Dharmasena et al., "HIV-1 reverse transcriptase structure with RNase H inhibitor dihydroxy benzoyl naphthyl hydrazone bound at a novel site," *ACS Chemical Biology*, vol. 1, no. 11, pp. 702–712, 2006.

[21] J. Levin, "BI224436, a non-catalytic site integrase inhibitor, is a potent inhibitor of the replication of treatment-naïve and raltegravir-resistant clinical isolates of HIV-1," in *Proceedings of the 51th ICAAC Interscience Conference on Antimicrobial Agents and Chemotherapy*, Chicago, Ill, USA, September 2011.

[22] E. P. Garvey, B. A. Johns, M. J. Gartland et al., "The naphthyridinone GSK364735 is a novel, potent human immunodeficiency virus type 1 integrase inhibitor and antiretroviral," *Antimicrobial Agents and Chemotherapy*, vol. 52, no. 3, article 901, 2008.

[23] S. Hare, A. M. Vos, R. F. Clayton, J. W. Thuring, M. D. Cummings, and P. Cherepanov, "Molecular mechanisms of retroviral integrase inhibition and the evolution of viral resistance," *Proceedings of the National Academy of Sciences of the United States of America*, vol. 107, no. 46, pp. 20057–20062, 2010.

[24] F. Hamy, E. R. Felder, G. Heizmann et al., "An inhibitor of the tat/TAR RNA interaction that effectively suppresses HIV-1 replication," *Proceedings of the National Academy of Sciences of the United States of America*, vol. 94, no. 8, pp. 3548–3553, 1997.

[25] O. W. Lindwasser, W. J. Smith, R. Chaudhuri, P. Yang, J. H. Hurley, and J. S. Bonifacino, "A diacidic motif in human immunodeficiency virus type 1 Nef is a novel determinant of binding to AP-2," *Journal of Virology*, vol. 82, no. 3, pp. 1166–1174, 2008.

[26] P.-H. Lin, Y.-Y. Ke, C.-T. Su et al., "Inhibition of HIV-1 Tat-mediated transcription by a coumarin derivative, BPRHIV001, through the Akt pathway," *Journal of Virology*, vol. 85, no. 17, pp. 9114–9126, 2011.

[27] S. Breuer, S. I. Schievink, A. Schulte, W. Blankenfeldt, O. T. Fackler, and M. Geyer, "Molecular design, functional characterization and structural basis of a protein inhibitor against the HIV-1 pathogenicity factor Nef," *PLoS ONE*, vol. 6, no. 5, Article ID e20033, 2011.

[28] D. Lu, Y. Y. Sham, and R. Vince, "Design, asymmetric synthesis, and evaluation of pseudosymmetric sulfoximine inhibitors against HIV-1 protease," *Bioorganic and Medicinal Chemistry*, vol. 18, no. 5, pp. 2037–2048, 2010.

[29] S. Gulnik, M. Eissenstat, and E. Afonina, "Preclinical and early clinical evaluation of SPI-452, a new pharmacokinetic enhancer," in *Proceedings of the 16th CROI Conference on Retroviruses and Opportunistic Infections*, Montreal, Canada, February 2009.

[30] R. Klein, "New class of medications approved for advance HIV," *FDA Consumer*, vol. 37, no. 3, p. 5, 2003.

[31] R. Carmona, L. Pérez-Alvarez, M. Muñoz et al., "Natural resistance-associated mutations to Enfuvirtide (T20) and polymorphisms in the gp41 region of different HIV-1 genetic forms from T20 naive patients," *Journal of Clinical Virology*, vol. 32, no. 3, pp. 248–253, 2005.

[32] L. Krauskof, "Pfizer wins U.S. approval for new HIV drug," Reuters, 2007, http://www.reuters.com/article/2007/08/06/businesspro-pfizer-hiv-dc-idUSN0642522320070806.

[33] W. D. Hardy, R. M. Gulick, H. Mayer et al., "Two-year safety and virologic efficacy of maraviroc in treatment- experienced patients with CCR5-tropic HIV-1 infection: 96-week combined analysis of MOTIVATE 1 and 2," *Journal of Acquired Immune Deficiency Syndromes*, vol. 55, no. 5, pp. 558–564, 2010.

[34] J. M. Jacobson, D. R. Kuritzkes, E. Godofsky et al., "Safety, pharmacokinetics, and antiretroviral activity of multiple doses of ibalizumab (formerly TNX-355), an anti-CD4 monoclonal antibody, in human immunodeficiency virus type 1-infected adults," *Antimicrobial Agents and Chemotherapy*, vol. 53, no. 2, pp. 450–457, 2009.

[35] W. Popik, T. M. Alce, and W. C. Au, "Human immunodeficiency virus type 1 uses lipid raft-colocalized CD4 and chemokine receptors for productive entry into CD4⁺ T cells," *Journal of Virology*, vol. 76, no. 10, pp. 4709–4722, 2002.

[36] J. P. Moore and R. F. Jarrett, "Sensitive ELISA for the gp120 and gp160 surface glycoproteins of HIV-1," *AIDS Research and Human Retroviruses*, vol. 4, no. 5, pp. 369–379, 1988.

[37] R. W. Shafer, A. K. N. Iversen, M. A. Winters, E. Aguiniga, D. A. Katzenstein, and T. C. Merigan, "Drug resistance and heterogeneous long-term virologic responses of human immunodeficiency virus type 1-infected subjects to zidovudine and didanosine combination therapy," *Journal of Infectious Diseases*, vol. 172, no. 1, pp. 70–78, 1995.

[38] A. Hachiya, E. N. Kodama, M. M. Schuckmann et al., "K70Q adds high-level tenofovir resistance to "Q151M complex" HIV reverse transcriptase through the enhanced discrimination mechanism," *PLoS ONE*, vol. 6, no. 1, Article ID e16242, 2011.

[39] K. Das, R. P. Bandwar, K. L. White et al., "Structural basis for the role of the K65R mutation in HIV-1 reverse transcriptase polymerization, excision antagonism, and tenofovir resistance," *Journal of Biological Chemistry*, vol. 284, no. 50, pp. 35092–35100, 2009.

[40] S. G. Sarafianos, S. H. Hughes, and E. Arnold, "Designing anti-AIDS drugs targeting the major mechanism of HIV-1 RT resistance to nucleoside analog drugs," *International Journal of Biochemistry and Cell Biology*, vol. 36, no. 9, pp. 1706–1715, 2004.

[41] E. R. Lanier, R. G. Ptak, B. M. Lampert et al., "Development of hexadecyloxypropyl tenofovir (CMX157) for treatment of infection caused by wild-type and nucleoside/nucleotide-resistant HIV," *Antimicrobial Agents and Chemotherapy*, vol. 54, no. 7, pp. 2901–2909, 2010.

[42] M. Markowitz, "GS-7340 demonstrates greater declines in HIV-1 RNA than TDF during 14 days of monotherapy in HIV-1-infected subjects," in *Proceedings of the 18th Conference on Retroviruses and Opportunistic Infections*, March 2011.

[43] C. Chu, "Unique antiviral activity of dioxolane-thymine (DOT) against HIV drug resistant mutants," in *Proceedings of the 4th IAS Conference on HIV Pathogenesis, Treatment and Prevention*, 2007.

[44] P. L. Boyer, M. J. Currens, J. B. McMahon, M. R. Boyd, and S. H. Hughes, "Analysis of nonnucleoside drug-resistant variants of human immunodeficiency virus type 1 reverse transcriptase," *Journal of Virology*, vol. 67, no. 4, pp. 2412–2420, 1993.

[45] J. Radzio and N. Sluis-Cremer, "Efavirenz accelerates HIV-1 reverse transcriptase ribonuclease H cleavage, leading to diminished zidovudine excision," *Molecular Pharmacology*, vol. 73, no. 2, pp. 601–606, 2008.

[46] G. N. Nikolenko, S. Palmer, F. Maldarelli, J. W. Mellors, J. M. Coffin, and V. K. Pathak, "Mechanism for nucleoside analog-mediated abrogation of HIV-1 replication: balance between

The Continuing Evolution of HIV-1 Therapy: Identification and Development of Novel Antiretroviral Agents Targeting Viral and Cellular Targets

109

RNase H activity and nucleotide excision," *Proceedings of the National Academy of Sciences of the United States of America*, vol. 102, no. 6, pp. 2093–2098, 2005.

[47] W. Yang and T. A. Steitz, "Recombining the structures of HIV integrase, RuvC and RNase H," *Structure*, vol. 3, no. 2, pp. 131–134, 1995.

[48] M. Wendeler, H. F. Lee, A. Bermingham et al., "Vinylogous ureas as a novel class of inhibitors of reverse transcriptase-associated ribonuclease H activity," *ACS Chemical Biology*, vol. 3, no. 10, pp. 635–644, 2008.

[49] C. A. Shaw-Reid, V. Munshi, P. Graham et al., "Inhibition of HIV-1 ribonuclease H by a novel diketo acid, 4-[5-(benzoylamino)thien-2-yl]-2,4-dioxobutanoic acid," *Journal of Biological Chemistry*, vol. 278, no. 5, pp. 2777–2780, 2003.

[50] J. A. Turpin, S. J. Terpening, C. A. Schaeffer et al., "Inhibitors of human immunodeficiency virus type 1 zinc fingers prevent normal processing of gag precursors and result in the release of noninfectious virus particles," *Journal of Virology*, vol. 70, no. 9, pp. 6180–6189, 1996.

[51] W. G. Rice, J. A. Turpin, M. Huang et al., "Azodicarbonamide inhibits HIV-1 replication by targeting the nucleocapsid protein," *Nature Medicine*, vol. 3, no. 3, pp. 341–345, 1997.

[52] M. L. Schito, A. C. Soloff, D. Slovitz et al., "Preclinical evaluation of a zinc finger inhibitor targeting lentivirus nucleocapsid protein in SIV-infected monkeys," *Current HIV Research*, vol. 4, no. 3, pp. 379–386, 2006.

[53] C. Pannecouque, B. Szafarowicz, N. Volkova et al., "Inhibition of HIV-1 replication by a bis-thiadiazolbenzene-1,2-diamine that chelates zinc ions from retroviral nucleocapsid zinc fingers," *Antimicrobial Agents and Chemotherapy*, vol. 54, no. 4, pp. 1461–1468, 2010.

[54] J. A. Grobler, K. Stillmock, B. Hu et al., "Diketo acid inhibitor mechanism and HIV-1 integrase: implications for metal binding in the active site of phosphotransferase enzymes," *Proceedings of the National Academy of Sciences of the United States of America*, vol. 99, no. 10, pp. 6661–6666, 2002.

[55] Z. Wang, J. Tang, C. E. Salomon, C. D. Dreis, and R. Vince, "Pharmacophore and structure-activity relationships of integrase inhibition within a dual inhibitor scaffold of HIV reverse transcriptase and integrase," *Bioorganic and Medicinal Chemistry*, vol. 18, no. 12, pp. 4202–4211, 2010.

[56] O. Goethals, A. Vos, M. Van Ginderen et al., "Primary mutations selected in vitro with raltegravir confer large fold changes in susceptibility to first-generation integrase inhibitors, but minor fold changes to inhibitors with second-generation resistance profiles," *Virology*, vol. 402, no. 2, pp. 338–346, 2010.

[57] T. M. Fletcher, M. A. Soares, S. McPhearson et al., "Complementation of integrase function in HIV-1 virions," *EMBO Journal*, vol. 16, no. 16, pp. 5123–5138, 1997.

[58] T. P. Cujec, H. Okamoto, K. Fujinaga et al., "The HIV transactivator TAT binds to the CDK-activating kinase and activates the phosphorylation of the carboxy-terminal domain of RNA polymerase II," *Genes and Development*, vol. 11, no. 20, pp. 2645–2657, 1997.

[59] L. M. Bedoya, M. Beltrán, R. Sancho et al., "4-Phenylcoumarins as HIV transcription inhibitors," *Bioorganic and Medicinal Chemistry Letters*, vol. 15, no. 20, pp. 4447–4450, 2005.

[60] Y. B. Tang, C. M. Zhang, C. Fang et al., "Design, synthesis and evaluation of novel 2H-1, 4-benzodiazepine-2-ones as inhibitors of HIV-1 transcription," *Yaoxue Xuebao*, vol. 46, no. 6, pp. 688–694, 2011.

[61] Y. Cao, X. Liu, and E. De Clercq, "Cessation of HIV-1 transcription by inhibiting regulatory protein Rev-mediated RNA transport," *Current HIV Research*, vol. 7, no. 1, pp. 101–108, 2009.

[62] B. Wolff, J. J. Sanglier, and Y. Wang, "Leptomycin B is an inhibitor of nuclear export: inhibition of nucleo-cytoplasmic translocation of the human immunodeficiency virus type 1 (HIV-1) Rev protein and Rev-dependent mRNA," *Chemistry and Biology*, vol. 4, no. 2, pp. 139–147, 1997.

[63] A. Cochrane, "Controlling HIV-1 rev function," *Current Drug Targets*, vol. 4, no. 4, pp. 287–295, 2004.

[64] M. Baba, "Inhibitors of HIV-1 gene expression and transcription," *Current Topics in Medicinal Chemistry*, vol. 4, no. 9, pp. 871–882, 2004.

[65] J. R. Thomas and P. J. Hergenrother, "Targeting RNA with small molecules," *Chemical Reviews*, vol. 108, no. 4, pp. 1171–1224, 2008.

[66] C. E. Prater, A. D. Saleh, M. P. Wear, and P. S. Miller, "Allosteric inhibition of the HIV-1 Rev/RRE interaction by a 3′-methylphosphonate modified antisense oligo-2′-O-methylribonucleotide," *Oligonucleotides*, vol. 17, no. 3, pp. 275–290, 2007.

[67] TRANA Discovery, http://www.tranadiscovery.com/.

[68] J. Zaunders, W. B. Dyer, and M. Churchill, "The Sydney Blood Bank Cohort: implications for viral fitness as a cause of elite control," *Current Opinion in HIV and AIDS*, vol. 6, no. 3, pp. 151–156, 2011.

[69] R. G. Ptak, B. G. Gentry, T. L. Hartman et al., "Inhibition of human immunodeficiency virus type 1 by triciribine involves the accessory protein nef," *Antimicrobial Agents and Chemotherapy*, vol. 54, no. 4, pp. 1512–1519, 2010.

[70] L. G. Feun, N. Savaraj, and G. P. Bodey, "Phase I study of tricyclic nucleoside phosphate using a five-day continuous infusion schedule," *Cancer Research*, vol. 44, no. 8, pp. 3608–3612, 1984.

[71] M. Dubé, M. G. Bego, C. Paquay, and É. A. Cohen, "Modulation of HIV-1-host interaction: Role of the Vpu accessory protein," *Retrovirology*, vol. 7, article 144, 2010.

[72] B. D. Kuhl, V. Cheng, D. A. Donahue et al., "The HIV-1 Vpu viroporin inhibitor BIT225 does not affect Vpu-mediated tetherin antagonism," *PLoS ONE*, vol. 6, no. 11, Article ID e27660, 2011.

[73] M. Kogan and J. Rappaport, "HIV-1 Accessory Protein Vpr: relevance in the pathogenesis of HIV and potential for therapeutic intervention," *Retrovirology*, vol. 8, article 25, 2011.

[74] X. J. Yao, J. Lemay, N. Rougeau et al., "Genetic selection of peptide inhibitors of human immunodeficiency virus type 1 Vpr," *Journal of Biological Chemistry*, vol. 277, no. 50, pp. 48816–48826, 2002.

[75] E. B. B. Ong, N. Watanabe, A. Saito et al., "Vipirinin, a coumarin-based HIV-1 Vpr inhibitor, interacts with a hydrophobic region of Vpr," *Journal of Biological Chemistry*, vol. 286, no. 16, pp. 14049–14056, 2011.

[76] M. Kamata, R. P. Wu, D. S. An et al., "Cell-based chemical genetic screen identifies damnacanthal as an inhibitor of HIV-1 Vpr induced cell death," *Biochemical and Biophysical Research Communications*, vol. 351, no. 3, p. 791, 2006.

[77] Z. Y. Li, P. Zhan, and X. Y. Liu, "Progress in the study of HIV-1 Vif and related inhibitors," *Yaoxue Xuebao*, vol. 45, no. 6, pp. 684–693, 2010.

[78] H. Côté, Z. Brumme, and P. Harrigan, "Human Immunodeficiency Virus Type 1 protease cleavage site mutations associated with protease inhibitor cross-resistance selected by

Indinavir, Ritonavir, and/or Saquinavir," *Journal of Virology*, vol. 75, no. 2, pp. 589–594, 2001.

[79] M. Kolli, E. Stawiski, C. Chappey, and C. A. Schiffer, "Human immunodeficiency virus type 1 protease-correlated cleavage site mutations enhance inhibitor resistance," *Journal of Virology*, vol. 83, no. 21, pp. 11027–11042, 2009.

[80] R. Tung, "The development of deuterium-containing drugs," *Innovations in Pharmaceutical Technology*, no. 32, pp. 24–28, 2010.

[81] K. Lindsten, T. Uhlíková, J. Konvalinka, M. G. Massuci, and N. P. Dantuma, "Cell-based fluorescence assay for human immunodeficiency virus type 1 protease activity," *Antimicrobial Agents and Chemotherapy*, vol. 45, no. 9, pp. 2616–2622, 2001.

[82] M. Bryant and L. Ratner, "Myristoylation-dependent replication and assembly of human immunodeficiency virus 1," *Proceedings of the National Academy of Sciences of the United States of America*, vol. 87, no. 2, pp. 523–527, 1990.

[83] G. B. Dreyer, B. W. Metcalf, T. A. Tomaszek et al., "Inhibition of human immunodeficiency virus 1 protease in vitro: rational design of substrate analogue inhibitors," *Proceedings of the National Academy of Sciences of the United States of America*, vol. 86, no. 24, pp. 9752–9756, 1989.

[84] O. W. Lindwasser and M. D. Resh, "Myristoylation as a target for inhibiting HIV assembly: unsaturated fatty acids block viral budding," *Proceedings of the National Academy of Sciences of the United States of America*, vol. 99, no. 20, pp. 13037–13042, 2002.

[85] M. L. Bryant, R. O. Heuckeroth, J. T. Kimata, L. Ratner, and J. I. Gordon, "Replication of human immunodeficiency virus 1 and Moloney murine leukemia virus is inhibited by different heteroatom-containing analogs of myristic acid," *Proceedings of the National Academy of Sciences of the United States of America*, vol. 86, no. 22, pp. 8655–8659, 1989.

[86] A. T. Nguyen, C. L. Feasley, K. W. Jackson et al., "The prototype HIV-1 maturation inhibitor, bevirimat, binds to the CA-SP1 cleavage site in immature Gag particles," *Retrovirology*, Article ID 8, p. 101, 2011.

[87] F. Li, R. Goila-Gaur, K. Salzwedel et al., "PA-457: a potent HIV inhibitor that disrupts core condensation by targeting a late step in Gag processing," *Proceedings of the National Academy of Sciences of the United States of America*, vol. 100, no. 23, pp. 13555–13560, 2003.

[88] C. Jolly, N. J. Booth, and S. J. D. Neil, "Cell-cell spread of human immunodeficiency virus type 1 overcomes tetherin/BST-2-mediated restriction in T cells," *Journal of Virology*, vol. 84, no. 23, pp. 12185–12199, 2010.

[89] S. J. D. Neil, T. Zang, and P. D. Bieniasz, "Tetherin inhibits retrovirus release and is antagonized by HIV-1 Vpu," *Nature*, vol. 451, no. 7177, pp. 425–430, 2008.

[90] S. Neil and P. Bieniasz, "Human immunodeficiency virus, restriction factors, and interferon," *Journal of Interferon and Cytokine Research*, vol. 29, no. 9, pp. 569–580, 2009.

[91] A. M. Sheehy, N. C. Gaddis, J. D. Choi, and M. H. Malim, "Isolation of a human gene that inhibits HIV-1 infection and is suppressed by the viral Vif protein," *Nature*, vol. 418, no. 6898, pp. 646–650, 2002.

[92] R. S. Harris, K. N. Bishop, A. M. Sheehy et al., "DNA deamination mediates innate immunity to retroviral infection," *Cell*, vol. 113, no. 6, pp. 803–809, 2003.

[93] R. Nathans, H. Cao, N. Sharova et al., "Small-molecule inhibitionof HIV-1 Vif," *Nature Biotechnology*, vol. 26, no. 10, pp. 1187–1192, 2008.

[94] G. Maertens, P. Cherepanov, W. Pluymers et al., "LEDGF/p75 is essential for nuclear and chromosomal targeting of HIV-1 integrase in human cells," *Journal of Biological Chemistry*, vol. 278, no. 35, pp. 33528–33539, 2003.

[95] J. E. Garrus, U. K. Von Schwedler, O. W. Pornillos et al., "Tsg101 and the vacuolar protein sorting pathway are essential for HIV-1 budding," *Cell*, vol. 107, no. 1, pp. 55–65, 2001.

[96] G. Maga, F. Falchi, M. Radi et al., "Toward the discovery of novel anti-HIV drugs. second-generation inhibitors of the cellular ATPase DDX3 with improved anti-HIV activity: synthesis, structure-activity relationship analysis, cytotoxicity studies, and target validation," *ChemMedChem*, vol. 6, no. 8, pp. 1371–1389, 2011.

[97] A. Garbelli, S. Beermann, G. Di Cicco, U. Dietrich, and G. Maga, "A motif unique to the human dead-box protein DDX3 is important for nucleic acid binding, ATP hydrolysis, RNA/DNA unwinding and HIV-1 replication," *PLoS ONE*, vol. 6, no. 5, Article ID e19810, 2011.

[98] J. B. Whitney, M. Asmal, and R. Geiben-Lynn, "Serpin induced antiviral activity of prostaglandin synthetase-2 against HIV-1 replication," *PLoS ONE*, vol. 6, no. 4, Article ID e18589, 2011.

[99] G. Wang, K. M. Watson, and R. W. Buckheit Jr., "Anti-human immunodeficiency virus type 1 activities of antimicrobial peptides derived from human and bovine cathelicidins," *Antimicrobial Agents and Chemotherapy*, vol. 52, no. 9, pp. 3438–3440, 2008.

[100] Z. Wang and G. Wang, "APD: the antimicrobial peptide database," *Nucleic Acids Research*, vol. 32, pp. D590–D592, 2004.

[101] T. Murali, M. D. Dyer, D. Badger, B. M. Tyler, and M. G. Katze, "Network-based prediction and analysis of HIV dependency factors," *PLoS Computational Biology*, vol. 7, no. 9, Article ID e1002164, 2011.

[102] R. G. Ptak, W. Fu, B. E. Sanders-Beer et al., "Cataloguing the HIV type 1 human protein interaction network," *AIDS Research and Human Retroviruses*, vol. 24, no. 12, pp. 1497–1502, 2008.

[103] B. Alston, J. H. Ellenberg, H. C. Standiford et al., "A multicenter, randomized, controlled trial of three preparations of low-dose oral α-interferon in HIV-infected patients with CD4[+] counts between 50 and 350 cells/mm[3] ," *Journal of Acquired Immune Deficiency Syndromes and Human Retrovirology*, vol. 22, no. 4, pp. 348–357, 1999.

[104] J. A. Tavel, A. Babiker, C. Carey et al., "Effects of intermittent IL-2 alone or with peri-cycle antiretroviral therapy in early HIV infection: the STALWART study," *PLoS ONE*, vol. 5, no. 2, Article ID e9334, 2010.

[105] Moore et al., "CYT107 enters phase II clinical trial in HIV-infected patients," *Immunotherapy*, vol. 2, no. 6, pp. 753–755, 2010.

[106] S. Heringer-Walther, K. Eckert, S. M. Schumacher et al., "Angiotensin-(1–7) stimulates hematopoietic progenitor cells in vitro and in vivo," *Haematologica*, vol. 94, no. 6, pp. 857–860, 2009.

[107] D. Liu, "Engraftment and development of HGTV43-transduced CD[34+] PBSC in HIV-1 seropositive individuals," in *Proceedings of the 14th International Conference on AIDS*, September 2011.

[108] C. June, "Gene modification at clinical scale: engineering resistance to HIV infection via targeted disruption of the HIV co-receptor CCR5 gene in CD4+ T cells using modified zinc finger protein nucleases," in *Proceedings of the 11th Annual Meeting ofthe American Society of Gene Therapy*, Boston, Mass, USA, May 2008.

The Continuing Evolution of HIV-1 Therapy: Identification and Development of Novel Antiretroviral Agents Targeting
Viral and Cellular Targets

111

[109] M. Tuomela, I. Stanescu, and K. Krohn, "Validation overview of bio-analytical methods," *Gene Therapy*, vol. 12, no. 1, pp. S131–S138, 2005.

[110] J. Jones et al., "No decrease in residual viremia during raltegravir intensification in patients on standard ART," in *Proceedings of the 16th Conference on Retroviruses and Opportunistic Infections (CROI)*, Montreal, Canada, February 2009.

[111] T. W. Chun and A. S. Fauci, "Latent reservoirs of HIV: obstacles to the eradication of virus," *Proceedings of the National Academy of Sciences of the United States of America*, vol. 96, no. 20, pp. 10958–10961, 1999.

[112] S. Matalon, T. A. Rasmussen, and C. A. Dinarello, "Histone deacetylase inhibitors for purging HIV-1 from the latent reservoir," *Molecular Medicine*, vol. 17, no. 5-6, pp. 466–472, 2011.

[113] J. Kulkosky, D. M. Culnan, J. Roman et al., "Prostratin: activation of latent HIV-1 expression suggests a potential inductive adjuvant therapy for HAART," *Blood*, vol. 98, no. 10, pp. 3006–3015, 2001.

[114] M. J. Pace, L. Agosto, E. H. Graf, and U. O'Doherty, "HIV reservoirs and latency models," *Virology*, vol. 411, no. 2, pp. 344–354, 2011.

12

Restriction of Retroviral Replication by Tetherin/BST-2

Jason Hammonds, Jaang-Jiun Wang, and Paul Spearman

Department of Pediatrics, Emory University and Children's Healthcare of Atlanta, 2015 Uppergate Drive, Atlanta, GA 30322, USA

Correspondence should be addressed to Paul Spearman, paul.spearman@emory.edu

Academic Editor: Abraham Brass

Tetherin/BST-2 is an important host restriction factor that limits the replication of HIV and other enveloped viruses. Tetherin is a type II membrane glycoprotein with a very unusual domain structure that allows it to engage budding virions and retain them on the plasma membrane of infected cells. Following the initial report identifying tetherin as the host cell factor targeted by the HIV-1 Vpu gene, knowledge of the molecular, structural, and cellular biology of tetherin has rapidly advanced. This paper summarizes the discovery and impact of tetherin biology on the HIV field, with a focus on recent advances in understanding its structure and function. The relevance of tetherin to replication and spread of other retroviruses is also reviewed. Tetherin is a unique host restriction factor that is likely to continue to provide new insights into host-virus interactions and illustrates well the varied ways by which host organisms defend against viral pathogens.

1. Introduction

Viruses and their host organisms engage in a series of conflicts in which viruses can be thought of as leading the offense, placing the host on defense. Host defenses against retroviral replication have arisen in a wide variety of forms. Classical cellular and humoral immune responses may limit retroviral replication and may be sufficient to prevent adverse outcomes in some host-virus interactions. However, throughout the evolution of mammals a series of alternative host defense factors have arisen whose apparent primary function is to counteract retroviruses in ways that lie outside of classical innate or adaptive immunity. These intrinsic host defense mechanisms against viruses have come to light largely through comparative studies of inhibition or "restriction" of replication of HIV or SIV in cells from different origins and are collectively referred to as host restriction factors. APOBEC3G, TRIM5alpha, and tetherin are the most prominent of a series of host restriction factors to be identified in recent years that limit HIV replication. This paper focuses on the discovery and subsequent characterization of tetherin, with an emphasis on recent work aimed at elucidating how its structure leads to retention of particles on the plasma membrane and on how Vpu acts to overcome tetherin-mediated restriction.

2. Identification of Tetherin as an Antiviral Host Restriction Factor

The discovery of tetherin is intimately linked to studies of the effects of the HIV accessory gene Vpu. Vpu is a small integral membrane protein encoded by HIV-1 and a limited subset of SIV species. Early studies utilizing HIV proviruses deficient for Vpu expression revealed that fewer particles were released from infected cells despite apparently normal production of all other viral proteins [1, 2]. Furthermore, electron microscopic analysis revealed striking accumulations of particles at the cell surface and within intracellular compartments of infected cells, revealing a defect at a late stage of particle release [3]. Subsequent work revealed that one of two important functions of Vpu was the downregulation of CD4 through interactions with cellular proteasomal degradation pathways [4–9]. Vpu was found to bind both CD4 and the human beta transducing-repeat containing protein (β-TrCP) [10, 11], connecting CD4 to the ubiquitin-proteasome machinery and inducing its degradation in the endoplasmic reticulum. Casein kinase phosphorylation sites on the Vpu cytoplasmic tail at residues 52 and 56 were found to be critical for β-TrCP interactions and for CD4 downregulation [10, 12]. This line of investigation along with other investigations into Vpu function prior to the discovery of tetherin is reviewed in [13]. However, the ability of Vpu to enhance particle release in human cells was not explained by downregulation of CD4 and remained a mystery for many years.

Experiments leading to the discovery of the function of the HIV Vif protein and its host restriction factor APOBEC3G [14, 15] provided a potential clue to the particle release function of Vpu. Like the infectivity conferred by Vif, the particle release function of Vpu proved to be cell type specific, suggesting that it might be overcoming a cellular factor involved in limiting particle release [16, 17]. A key experiment demonstrated that heterokaryons between restrictive, Vpu-responsive HeLa cells and permissive, Vpu-unresponsive Cos-7 cells were restricted in particle release, suggesting that a negative (restricting) factor was dominant [18]. Vpu was able to enhance particle release in the heterokaryons, demonstrating that the factor from human cells restricting particle release could be overcome by Vpu [18].

Several cellular factors were described as potential targets of Vpu prior to or concomitant with the identification of tetherin, including TASK-1 [19] and CAML [20]. However, neither of these factors has subsequently proven to be the restriction factor targeted by Vpu. Instead, a series of key findings led by Stuart Neil in the Bieniasz laboratory resulted in the ultimate identification of tetherin as the restriction factor targeted by Vpu. First, these investigators demonstrated clearly that the effect of Vpu was on particle release rather than other steps in virus assembly, while retention of virions and subsequent endocytosis occurred in the absence of Vpu [21]. The specific particle retention activity was found to be prominent in HeLa cells as before, while a subset of human cells such as HOS or 293T cells lacked this activity. The next key observation was that the restricting activity could be induced by type I interferons. Neil and colleagues demonstrated that retention of Vpu-deficient HIV-1 particles at the plasma membrane could be induced in 293T or HOS cells and that treatment with the protease subtilisin released the particles from the cell surface [22]. Furthermore, the restricting activity extended to additional virus genera, as Ebola VP40 release was similarly deficient in an IFN-induced manner and its release could be enhanced by Vpu. These results suggested that an interferon-inducible, proteinaceous tether was responsible for retaining enveloped viruses at the cell surface. In 2008 this factor was identified by the same group as BST-2/CD317 and renamed tetherin because of this prominent biological function [23].

BST-2 had first been cloned as a membrane antigen present on bone marrow stromal cells and synovial cells that was thought to be involved in pre-B-cell growth [24]. The same protein had been identified as a membrane antigen termed HM1.24, present on terminally differentiated B cells, and was thought to be a potential anticancer target for multiple myeloma [25]. The terminology for the HM1.24 antigen was later changed to CD317 [26]. BST-2 was later shown to be an interferon-inducible antigen and identical to plasmacytoid dendritic cell antigen-1 (PDCA1) in mice [27]. CD317/BST-2 is a highly unusual type II integral membrane protein, with a transmembrane domain near its N-terminus and a C-terminal glycosyl-phosphatidylinositol (GPI) anchor (Figure 1). The protein localizes to lipid rafts on the plasma membrane and to the trans-Golgi network (TGN) and is endocytosed from the plasma membrane through a clathrin-dependent pathway [28]. Remarkably, a membrane proteomic screen examining the effects of the K5 protein of KSHV revealed a marked downregulation of CD317/BST-2 and even showed almost as an afterthought that HIV-1 Vpu downregulated the protein [29]. This published observation led the Guatelli group to examine CD317/BST-2 as a candidate restriction factor targeted by Vpu, and their findings were published soon after the identification of tetherin by the Bieniasz group [30]. For the purpose of this paper, BST-2/CD317/tetherin will be hereafter referred to simply as tetherin.

3. Structural Biology of Tetherin and Functional Implications

One of the most fascinating aspects of tetherin biology is how its structure allows for retention of enveloped virions through protein-lipid and protein-protein interactions occurring at the particle budding site. As already mentioned, tetherin's basic domain structure is highly unusual. Tetherin is a type II membrane protein bearing a small N-terminal cytoplasmic domain, a transmembrane region, an ectodomain forming a coiled-coil in tetherin dimers, and a C-terminal GPI anchor (Figure 1) [31]. The double-membrane anchor plays a key role in the ability of tetherin to restrict enveloped virus particle release, presumably because one anchor is present on the plasma membrane of the cell and the second is inserted into the viral membrane [23] (Figure 2). Three cysteines in the N-terminal ectodomain of tetherin (C53, C63, C91) are capable of forming disulfide-linked dimers [32, 33], and mutation of all three abolished dimer formation and greatly reduced the ability of tetherin to restrict Vpu-deficient HIV release [34]. Two N-linked glycosylation sites (N65 and N92) lead to some variability of migration on SDS-PAGE analysis and appear to play a role in correct folding and transport of tetherin to the cell surface in one report [34], while another group found that alteration of N-linked glycosylation sites had no effect on virus restriction or cell surface levels [33].

Four reports of the tetherin ectodomain structure have been published [35–38]. The ectodomain forms a long extended rod-like conformation in a loose or imperfect coiled-coil parallel dimer [35, 38], suggesting that there is some conformational flexibility in the C-terminal portion of the ectodomain that may be required to accommodate dynamic changes in membrane deformation at the particle budding site. Disulfide bonds stabilize the dimeric N-terminal region, which cannot stably dimerize in their absence [38]. Unexpectedly, tetrameric forms of tetherin were also detected in crystallization studies [36, 38]. The biological function of tetherin tetramers remains uncertain and mutations designed to disrupt the tetramer did not prevent tetherin-mediated particle restriction [36, 38]. The crystal structure of murine BST-2/tetherin ectodomain

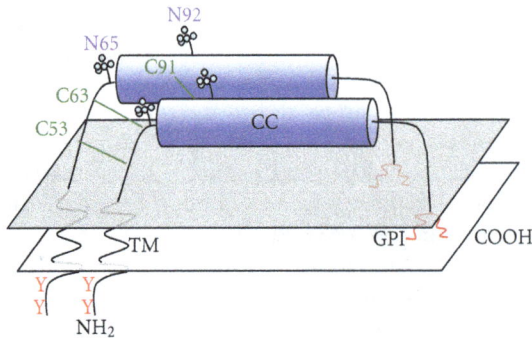

FIGURE 1: Schematic representation of tetherin domain structure. Tetherin is depicted as a parallel dimer with both transmembrane (TM) and glycophosphatidylinositol (GPI) membrane anchors in the same membrane. Disulfide linkages are depicted in green, and N-linked glycosylation sites pictured. CC: coiled coil; Y: tyrosine residues critical for endocytic motif.

revealed similar ectodomain architecture, and suggested that tetrameric assemblies may form a curved assembly that functions as a sensor of membrane curvature, analogous to BAR domains [37]. The authors of this paper suggest that tetrameric assemblies may facilitate the clustering of tetherin around the neck of a budding virus as has been seen in immunoelectron microscopic analysis [39, 40]. At the current time, the significance of the tetrameric assemblies remains unclear but quite intriguing.

While tetherin is thought to be a raft-associated protein through its C-terminal GPI anchor, a recent report questioned this and suggested that instead the C-terminus of tetherin acts as a second transmembrane domain [41]. This unexpected result is intriguing and awaits further verification.

4. Tetherin Clustering in Membrane Microdomains and Role of the Actin Cytoskeleton

The functional significance of tetherin's unusual structure and topology to its mechanism of restriction of viral budding have not yet been entirely delineated. However, there is significant biochemical and microscopic evidence that tetherin functions as a physical tether connecting virions to the plasma membrane. Immunoelectron microscopic analysis has shown clear evidence of clustering of tetherin on discrete cell surface microdomains and sometimes on filopodia or at the location of coated pits, in the absence of viral infection [39, 40]. In infected cells, immunogold beads are most often observed at the neck of the budding particle and at the site of connections between particle membranes [39, 40] (Figure 2(a)) Tetherin is enriched on the particle membrane itself [39, 40, 42], as well as on filamentous connections that sometimes are present linking particles to one another [40]. Microdomain clustering of tetherin can also be readily observed by superresolution light microscopic techniques [43, 44]. We recently described a tetherin ectodomain mutant with four substitutions in the coiled-coil region (4S)

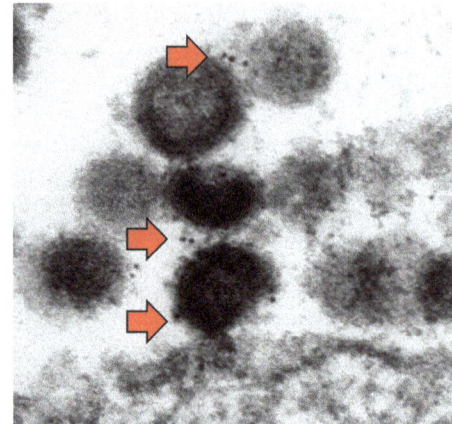

Immunogold label = tetherin

(a)

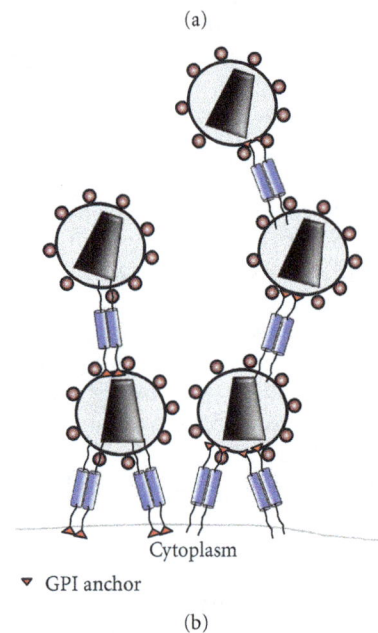

Cytoplasm

▼ GPI anchor

(b)

FIGURE 2: (a) Tetherin on the cell surface of A3.01 T cells infected with NLUdel virus, treated with indinavir to preserve particle morphology for preparation. Arrows indicate immunogold beads; primary antibody was rabbit anti-tetherin polyclonal antisera. (b) Schematic depiction of parallel homodimers of tetherin retaining HIV particles on the plasma membrane; tetherin is not to scale in this diagram.

that was expressed well on the cell surface, yet lost the ability to cluster in plasma membrane microdomains and was unable to restrict release of viral particles [43]. The loss of discrete puncta formation of the 4S mutant was associated with an increase in lateral mobility as measured by fluorescence recovery after photobleaching (FRAP), while wildtype, restrictive tetherin was constrained in lateral mobility when compared with classical GPI-anchored proteins [43]. These findings imply that tetherin's restriction of particle release requires localization in discrete microdomains that help to form or are in the immediate vicinity of the developing particle bud. In other words, tetherin's presence on the plasma membrane globally may not be as important as its discrete localization at the site of particle budding. While clustering appears to be associated with restriction, relief of restriction by Vpu is not achieved through removal of tetherin from lipid rafts as measured by partitioning into detergent-resistant membranes [45, 46]. The lack of mobility of tetherin in clustered plasma membrane sites is potentially regulated through interactions not only with lipid microdomains but also with the underlying cytoskeleton.

The potential for regulation of tetherin clustering through interactions with the underlying actin cytoskeleton is supported by the report from Rollason and colleagues of a direct interaction between tetherin and the RhoGAP protein RICH2 [47]. RICH2 contains both an N-terminal BAR domain and a Rho/Rac/cdc42 GAP domain [48, 49]. The presence of a BAR domain capable of inducing membrane tubulation is curious, given the previously mentioned modeling of tetherin tetramers as a BAR domain [37]. The potential for tetherin to act as a link to the regulation of Rac and Rho through the GAP activity of RICH2 is also intriguing. Perhaps more directly relevant to peripheral clustering of tetherin is the known interaction of RICH2 with EBP50 (ERM-binding phosphoprotein 50) through its C-terminal ESTAL domain [50, 51]. EBP50 acts as a linker between ERM proteins and the cytoplasmic tails of integral membrane proteins, in this case tetherin. This suggests that tetherin is connected indirectly to the underlying cortical actin cytoskeleton through a RICH2-EBP50-ezrin complex. Because RICH2 interacts with the same region of the tetherin cytoplasmic tail that binds $\mu 1$ and $\mu 2$ and directs its clathrin-mediated endocytosis [28], the interaction with RICH2 and the actin cytoskeleton might be predicted to stabilize tetherin on the plasma membrane and prevent its endocytosis. Much remains to be learned about the functional role of tetherin's interaction with RICH2 and connection to actin, as well as with the potential modulation of Rho family GTPases. One pressing question that has not yet been addressed is whether this cytoskeletal anchoring plays a role in restriction of particle release and in the punctate clustering of tetherin on the cell surface.

A counterargument against the role of additional cellular factors in tetherin-mediated restriction may be made in light of evidence from the Bieniasz laboratory demonstrating that an artificial tetherin-like molecule pieced together from domains of three distinct proteins (art-tetherin) can restrict particle release [34]. This strategy employed stitching together the cytoplasmic tail and transmembrane domain of the transferrin receptor, the helical coiled-coil domain of DMPK (dystrophia myotonica protein kinase), and the C-terminus of uPAR that includes a GPI anchor. The investigators in effect recreated the domain architecture of tetherin from sequence-unrelated proteins and quite strikingly were able to inhibit HIV particle release through overexpression of art-tetherin [34]. Despite the ability of this artificial construct to restrict particle release, cellular interactors of wild-type tetherin in relevant human cells clearly play a role in its endocytosis and recycling, and the potential for functional significance of the RICH2-EBP50-ezrin-actin linkage remains.

5. Counteraction of Tetherin-Mediated Restriction of Particle Release by Vpu

Following the identification of tetherin as the restriction factor responsible for retention of HIV particles, attention turned to understanding the molecular and cellular mechanisms underlying the relief of tetherin-mediated restriction by Vpu. Comparison of the effects of Vpu on tetherin molecules from nonhuman primates helped to identify critical domains involved in tetherin-Vpu interactions and provided important clues to the evolution of tetherin and of viral countermeasures designed to overcome restriction. Counteraction of tetherin-mediated restriction was mapped to specific interactions between the transmembrane domain of Vpu and the transmembrane domain of tetherin [34, 52–55]. Coimmunoprecipitation studies performed by several groups confirmed a physical interaction between tetherin and Vpu, and the interaction required residues within the TM domains of both Vpu and tetherin as suggested by genetic studies [54, 56–58]. A single-residue alteration in human tetherin to one found in tetherin from the Tantalus monkey (T45I) rendered it Vpu insensitive, yet still able to restrict HIV-1 [55]. Tetherin variants from rhesus macaques and mice were similarly able to restrict HIV-1 release and yet were insensitive to Vpu, and transfer of the corresponding TM region between tetherin molecules from different species conferred sensitivity or resistance [52]. Furthermore, there is strong evidence of positive selection among primate tetherin molecules, and the selected changes were enriched in the N-terminal and TM regions of tetherin, suggesting frequent episodes of evolution under selection pressure to evade viral countermeasures [52, 55]. The discovery that SIV Nef proteins downregulate tetherin from rhesus macaque, sooty mangabey, and African green monkey but are inactive against human tetherin provided evidence that primate lentiviruses have targeted tetherin in different ways over evolutionary history [56, 59]. The Vpu proteins from SIVgsn, SIVmus, and SIVmon are able to downregulate both CD4 and tetherin in cells from their cognate primate species, while Vpu from SIVcpz, the precursor virus of HIV-1, is unable to downregulate chimpanzee tetherin and instead utilizes Nef for this function [60]. The Vpu protein of HIV-1 group M, but not group O or group N, is able to downregulate both tetherin and CD4, and the presence of this fully functional Vpu has been proposed as a reason for the worldwide spread of group M versus the nonpandemic HIV-1 strains [60, 61]. Thus, species-specific differences in tetherin and in lentiviral countermeasures against tetherin have played a major role in cross-species transmission and subsequent spread of lentiviruses and have likely been an important contributor to the current HIV-1 pandemic. While these species-specific differences are the rule, there are exceptions. Shingai and colleagues demonstrated that some HIV-1 Vpu proteins are able to antagonize rhesus tetherin, indicating that some HIV-1 isolates encode a Vpu protein with a broader host range [62].

Tetherin cell surface levels are downregulated by Vpu, and degradation of tetherin by Vpu has been observed in a wide variety of cell types [30, 54, 63, 64]. The logical hypothesis suggested by this association was that Vpu overcomes restriction by removing tetherin from plasma membrane viral assembly sites and targeting tetherin for degradation, as has been well established for CD4. The downregulation of CD4 by Vpu requires the phosphorylation of serines 52 and 56 on the Vpu cytoplasmic tail, interaction with β-TrCP, and degradation of CD4 through the ubiquitin-proteasome pathway [10–12, 65]. The mechanism and importance of downregulation of tetherin by Vpu, however, have not yet been as clearly worked out. Several groups have reported that relief of tetherin-mediated restriction of particle release can occur in the absence of degradation of tetherin [57, 66, 67], indicating that degradation is not the essential step in the action of Vpu that leads to relief of restriction. Goffinet and colleagues generated a series of tetherin cytoplasmic tail mutants including lysine mutants that were not degraded upon expression of Vpu. The mutants remained competent for restriction of particle release, and despite their lack of degradation Vpu potently relieved the restriction to particle release [66]. The involvement of β-TrCP in Vpu-mediated targeted degradation of tetherin has been supported by a number of investigators [54, 63, 64, 68], which would seem to suggest that a proteasomal pathway of degradation similar to that involved in the Vpu-β-TrCP-CD4 pathway is essential. Proteasomal degradation of tetherin has indeed been supported in some studies [63, 64] but is not universally accepted as the major pathway. Instead, a β-TrCP-dependent endolysosomal pathway for tetherin degradation has been reported [54, 58, 68]. According to this model, Vpu still acts as an adaptor molecule linking tetherin to β-TrCP, but does not connect tetherin to the ER-associated protein degradation (ERAD) pathway. Instead, interactions in the TGN or early endosome compartments direct tetherin to degradation in lysosomal compartments. There still is work to be done to clarify this pathway and to derive a clearer understanding of the role of β-TrCP and of the degradation of tetherin that is initiated or facilitated by Vpu.

The site of interaction of Vpu with tetherin is not known with certainty. Expression of Vpu alters the intracellular pattern of tetherin, with decreased cell surface of tetherin and prominent colocalization of tetherin and Vpu in the TGN [23, 43, 57, 68]. Mutants of Vpu that are unable to interact with tetherin fail to redistribute tetherin to the TGN, suggesting that tetherin may be retained in the TGN through TM-TM interactions with Vpu [57]. The rate of tetherin endocytosis from the plasma membrane is not significantly altered by Vpu [43, 57, 69]. These data suggest that Vpu may alter delivery of newly synthesized tetherin to the plasma membrane and/or disrupt outward tetherin recycling from the endosomal recycling compartment. Taken together with the data described above regarding endolysosomal degradation, a consistent model would posit that Vpu interacts with and traps tetherin in the TGN or other post-ER compartments, thereafter shunting tetherin to degradation in lysosomal compartments and preventing newly synthesized tetherin from trafficking to the plasma membrane. Alternatively, Vpu may disrupt outward trafficking of tetherin to the particle assembly microdomain on

the plasma membrane through additional effects on host trafficking factors.

6. Counteraction of Tetherin by Other Viruses

The significance of tetherin as a bona fide host restriction factor is convincingly demonstrated by the fact that diverse families of enveloped viruses have developed distinct mechanisms to overcome its inhibitory effects. One of the earliest factors identified that enhanced the release of vpu-deficient HIV-1 and produced efficient release of HIV-2 in restrictive cell types was the envelope glycoprotein of certain strains of HIV-2, in particular ROD10 Env [70–72]. Although the effect of HIV-2 Env on particle release was described well before the identification of tetherin as the target of Vpu, it is now clear that it does so through acting as a tetherin antagonist. HIV-2 Env appears to exclude tetherin from the site of viral budding through direct interaction with tetherin leading to sequestration within the TGN [73]. Determinants of tetherin antagonism by HIV-2 Env include a highly conserved endocytic-sorting motif (GYXXθ) in the cytoplasmic tail of gp41 [73, 74]. This sorting motif binds clathrin in an AP-2-dependent manner and is responsible for the redistribution of tetherin from the plasma membrane and concentration within endosomal compartments, in particular the TGN [73, 75, 76]. Interestingly, the gp41 ectodomain of HIV-2 Env has also been implicated in tetherin antagonism [73, 77]. The exact region required for physical tetherin interaction remains unclear due to the inability to differentiate those areas responsible for interaction and those residues involved in maintenance of tertiary Env structure. Additionally, proteolytic Env cleavage into gp120/gp41 subunits is required, as the unprocessed form is incompetent for virion egress and tetherin sequestration [5, 64]. It is interesting to note that, while Vpu expression leads to reduced cellular levels of tetherin, HIV-2 Env reduces cell surface levels but not total cellular levels of tetherin [73]. Finally, the ability of HIV-2 Env to counteract restriction is dependent on conservation of the tetherin ectodomain sequence [78]. Together, these data strongly suggest an interaction between the tetherin and mature HIV-2 Env ectodomains that leads to intracellular trapping of tetherin and abrogates restriction of particle release.

The K5 protein of KSHV (Human Herpesvirus 8; HHV-8) was the first viral component shown to specifically target tetherin prior to its identification as a viral restriction factor [29]. The K5 protein is a RICH-CH (MARCH) family of cellular transmembrane E3 ubiquitin ligases. This family of proteins facilitates the ubiquitination and subsequent degradation of transmembrane proteins. K5 exhibits potent immunomodulatory function resulting in the degradation of major histocompatibility complex (MHC) proteins (MHC), adhesion molecules, and NK receptor ligands while also promoting the degradation of tetherin through ubiquitination of lysine residues in the tetherin cytoplasmic tail [79, 80]. K5-mediated tetherin degradation is ESCRT-dependent, and ubiquitination of K18 in the CT of tetherin by K5 is critical for the efficient release of KSHV [79, 80]. In the case of K5, it is clear that ubiquitination in a post-ER compartment targets tetherin for degradation via ubiquitin-dependent endolysosomal pathways [80].

Ebola virus overcomes tetherin-mediated restriction through the activity of its surface glycoprotein (GP) [81]. The Ebola virus GP has a broad species specificity comprising an ability to antagonize both human and murine tetherin. The Ebola GP mechanism of action appears to be novel, as it relieves restriction without reducing tetherin cell surface concentration and can even relieve the restriction conferred by a wholly artificial tetherin molecule [82]. It was recently reported that the GP2 subunit of Ebola interacts with tetherin, and another filovirus GP (Marburg virus GP) was shown to have anti-tetherin activity [83]. The mechanism of action of Ebola GP is perhaps the least clear of the tetherin antagonists that have been described to date.

7. *In Vivo* Significance of Tetherin for Viral Spread and Pathogenesis

The importance of tetherin for restricting viral replication is strongly supported by the multiple mechanisms described above by which viruses can overcome its tethering function and by the evidence of positive selection of tetherin in the primate lineage. The assumption would logically be that tetherin inhibits release of free virus, preventing infection of additional cells and limiting overall replication (and potentially pathogenesis) within an organism. However, whether or not tetherin restricts cell-cell spread remains to be definitively established. Casartelli and coworkers demonstrated that the formation of virologic synapses was not prevented by tetherin, but that tetherin did limit cell-cell transmission of virus [84]. Another group found similarly that cell-cell transmission was inhibited by tetherin in a flow-cytometry-based assay [85]. In contrast, Jolly and colleagues demonstrated that depletion of tetherin diminished virologic synapse formation and cell-cell spread and suggested that under some circumstances tetherin may actually enhance cell-cell transmission [86]. Depletion of tetherin in mature dendritic cells was not associated with a significant enhancement of transmission to CD4+ T cells in another report, although modest enhancement or inhibition of cell-cell transmission was seen that differed with the stimulus utilized for maturation of dendritic cells [87]. Currently there is a need for further investigation into this question, as there is not a clear consensus in the field.

Tetherin knockout mice have provided additional weight to the argument that this protein has evolved as an interferon-induced host defense mechanism to limit viral replication *in vivo*. Liberatore and Bieniasz used poly(I : C) to enhance tetherin expression in wild-type mice and found that replication of Moloney murine leukemia virus (Mo-MLV) in these mice was significantly attenuated as compared with tetherin-deficient mice [88]. Using a murine leukemia virus strain that induces a strong interferon response, they then demonstrated that tetherin-deficient mice developed both higher levels of MLV viremia and enhanced pathology [88]. A different strategy utilizing a naturally occurring polymorphism in tetherin in NZW mice allowed Barrett and colleagues to study Friend virus replication in mice

homozygous for enhanced versus normal tetherin cell surface expression. These investigators demonstrated that enhanced cell surface tetherin *in vivo* correlated with diminished replication of Friend virus and improved outcomes [89]. Together these reports provide solid evidence that tetherin acts as an antiretroviral host restriction factor *in vivo*. A modest inhibitory effect of tetherin on Mo-MLV replication was also reported by Swiecki and colleagues, consistent with the effects seen by Liberatore and Bieniasz in the absence of IFN induction [90]. Surprisingly, however, these authors observed lower viral titers and enhanced virus-specific CD8+ T-cell responses in tetherin-deficient mice infected with vesicular stomatitis virus or influenza virus. Thus, while tetherin's antiretroviral effects are clear, there may be more complexity in how tetherin alters antigen processing and affects the replication of other enveloped viruses *in vivo*.

8. Summary

Tetherin is an unusual host protein that restricts enveloped particle release at the very latest stage of the viral life-cycle through physically tethering virions to the plasma membrane. A number of unrelated viruses have developed the means to overcome restriction by tetherin and have done so through different mechanisms. The acquisition of Vpu by primate lentiviruses and its ability to counteract restriction by human tetherin is thought to be an important factor in cross-species transmission and potentially in the magnitude of the HIV-1 pandemic itself. The flurry of recent studies examining tetherin and its antagonists emphasizes the significance of this potent antiviral host restriction factor. Future studies should shed light not only on the mechanism of action of Vpu, but will likely identify additional enveloped viruses that have developed the means to antagonize tetherin. Studies examining the cellular interactions of tetherin are also poised to provide new insights into the nature of the particle assembly site, trafficking of membrane glycoproteins to the particle assembly site, and the role of the cortical actin cytoskeleton in particle release.

Acknowledgments

This work was supported by NIH AI058828 and by funds from Children's Healthcare of Atlanta. The work was partly supported by the Emory Center for AIDS Research (P30 AI050409) and by the Robert P. Apkarian Integrated Electron Microscopy Core Laboratory of Emory University.

References

[1] K. Strebel, T. Klimkait, and M. A. Martin, "A novel gene of HIV-1, vpu, and its 16-kilodalton product," *Science*, vol. 241, no. 4870, pp. 1221–1223, 1988.

[2] E. F. Terwilliger, B. Godin, J. G. Sodroski, and W. A. Haseltine, "Construction and use of a replication-competent human immunodeficiency virus (HIV-1) that expresses the chloramphenicol acetyltransferase enzyme," *Proceedings of the National Academy of Sciences of the United States of America*, vol. 86, no. 10, pp. 3857–3861, 1989.

[3] T. Klimkait, K. Strebel, M. D. Hoggan, M. A. Martin, and J. M. Orenstein, "The human immunodeficiency virus type 1-specific protein vpu is required for efficient virus maturation and release," *Journal of Virology*, vol. 64, no. 2, pp. 621–629, 1990.

[4] R. L. Willey, A. Buckler-White, and K. Strebel, "Sequences present in the cytoplasmic domain of CD4 are necessary and sufficient to confer sensitivity to the human immunodeficiency virus type 1 Vpu protein," *Journal of Virology*, vol. 68, no. 2, pp. 1207–1212, 1994.

[5] M. E. Lenburg and N. R. Landau, "Vpu-induced degradation of CD4: requirement for specific amino acid residues in the cytoplasmic domain of CD4," *Journal of Virology*, vol. 67, no. 12, pp. 7238–7245, 1993.

[6] M. J. Vincent, N. U. Raja, and M. A. Jabbar, "Human immunodeficiency virus type 1 Vpu protein induces degradation of chimeric envelope glycoproteins bearing the cytoplasmic and anchor domains of CD4: role of the cytoplasmic domain in Vpu-induced degradation in the endoplasmic reticulum," *Journal of Virology*, vol. 67, no. 9, pp. 5538–5549, 1993.

[7] R. J. Geraghty and A. T. Panganiban, "Human immunodeficiency virus type 1 Vpu has a CD4- and an envelope glycoprotein-independent function," *Journal of Virology*, vol. 67, no. 7, pp. 4190–4194, 1993.

[8] M. Y. Chen, F. Maldarelli, M. K. Karczewski, R. L. Willey, and K. Strebel, "Human immunodeficiency virus type 1 Vpu protein induces degradation of CD4 in vitro: the cytoplasmic domain of CD4 contributes to Vpu sensitivity," *Journal of Virology*, vol. 67, no. 7, pp. 3877–3884, 1993.

[9] R. L. Willey, F. Maldarelli, M. A. Martin, and K. Strebel, "Human immunodeficiency virus type 1 Vpu protein induces rapid degradation of CD4," *Journal of Virology*, vol. 66, no. 12, pp. 7193–7200, 1992.

[10] F. Margottin, S. P. Bour, H. Durand et al., "A novel human WD protein, h-βTrCP, that interacts with HIV-1 Vpu connects CD4 to the ER degradation pathway through an F-box motif," *Molecular Cell*, vol. 1, no. 4, pp. 565–574, 1998.

[11] U. Schubert, L. C. Antón, I. Bačík et al., "CD4 glycoprotein degradation induced by human immunodeficiency virus type 1 Vpu protein requires the function of proteasomes and the ubiquitin- conjugating pathway," *Journal of Virology*, vol. 72, no. 3, pp. 2280–2288, 1998.

[12] M. Paul and M. A. Jabbar, "Phosphorylation of both phosphoacceptor sites in the HIV-1 Vpu cytoplasmic domain is essential for Vpu-mediated ER degradation of CD4," *Virology*, vol. 232, no. 1, pp. 207–216, 1997.

[13] S. Bour and K. Strebel, "The HIV-1 Vpu protein: a multifunctional enhancer of viral particle release," *Microbes and Infection*, vol. 5, no. 11, pp. 1029–1039, 2003.

[14] A. M. Sheehy, N. C. Gaddis, J. D. Choi, and M. H. Malim, "Isolation of a human gene that inhibits HIV-1 infection and is suppressed by the viral Vif protein," *Nature*, vol. 418, no. 6898, pp. 646–650, 2002.

[15] J. H. M. Simon, D. L. Miller, R. A. M. Fouchier, M. A. Soares, K. W. C. Peden, and M. H. Malim, "The regulation of primate immunodeficiency virus infectivity by Vif is cell species restricted: a role for Vif in determining virus host range and cross-species transmission," *The EMBO Journal*, vol. 17, no. 5, pp. 1259–1267, 1998.

[16] R. J. Geraghty, K. J. Talbot, M. Callahan, W. Harper, and A. T. Panganiban, "Cell type-dependence for Vpu function," *Journal of Medical Primatology*, vol. 23, no. 2-3, pp. 146–150, 1994.

[17] H. Sakai, K. Tokunaga, M. Kawamura, and A. Adachi, "Function of human immunodeficiency virus type 1 Vpu protein in various cell types," *Journal of General Virology*, vol. 76, part 11, pp. 2717–2722, 1995.

[18] V. Varthakavi, R. M. Smith, S. P. Bour, K. Strebel, and P. Spearman, "Viral protein U counteracts a human host cell restriction that inhibits HIV-1 particle production," *Proceedings of the National Academy of Sciences of the United States of America*, vol. 100, no. 25, pp. 15154–15159, 2003.

[19] K. Hsu, J. Seharaseyon, P. Dong, S. Bour, and E. Marbán, "Mutual functional destruction of HIV-1 Vpu and host TASK-1 channel," *Molecular Cell*, vol. 14, no. 2, pp. 259–267, 2004.

[20] V. Varthakavi, E. Heimann-Nichols, R. M. Smith et al., "Identification of calcium-modulating cyclophilin ligand as a human host restriction to HIV-1 release overcome by Vpu," *Nature Medicine*, vol. 14, no. 6, pp. 641–647, 2008.

[21] S. J. Neil, S. W. Eastman, N. Jouvenet, and P. D. Bieniasz, "HIV-1 Vpu promotes release and prevents endocytosis of nascent retrovirus particles from the plasma membrane," *PLoS Pathogens*, vol. 2, no. 5, article e39, 2006.

[22] S. J. D. Neil, V. Sandrin, W. I. Sundquist, and P. D. Bieniasz, "An interferon-alpha-induced tethering mechanism inhibits HIV-1 and Ebola virus particle release but is counteracted by the HIV-1 Vpu protein," *Cell Host and Microbe*, vol. 2, no. 3, pp. 193–203, 2007.

[23] S. J. D. Neil, T. Zang, and P. D. Bieniasz, "Tetherin inhibits retrovirus release and is antagonized by HIV-1 Vpu," *Nature*, vol. 451, no. 7177, pp. 425–430, 2008.

[24] J. Ishikawa, T. Kaisho, H. Tomizawa et al., "Molecular cloning and chromosomal mapping of a bone marrow stromal cell surface gene, BST2, that may be involved in pre-B-cell growth," *Genomics*, vol. 26, no. 3, pp. 527–534, 1995.

[25] T. Goto, S. J. Kennel, M. Abe et al., "A novel membrane antigen selectively expressed on terminally differentiated human B cells," *Blood*, vol. 84, no. 6, pp. 1922–1930, 1994.

[26] M. Vidal-Laliena, X. Romero, S. March, V. Requena, J. Petriz, and P. Engel, "Characterization of antibodies submitted to the B cell section of the 8th Human Leukocyte Differentiation Antigens Workshop by flow cytometry and immunohistochemistry," *Cellular Immunology*, vol. 236, no. 1-2, pp. 6–16, 2005.

[27] A. L. Blasius, E. Giurisato, M. Cella, R. D. Schreiber, A. S. Shaw, and M. Colonna, "Bone marrow stromal cell antigen 2 is a specific marker of type I IFN-producing cells in the naive mouse, but a promiscuous cell surface antigen following IFN stimulation," *Journal of Immunology*, vol. 177, no. 5, pp. 3260–3265, 2006.

[28] R. Rollason, V. Korolchuk, C. Hamilton, P. Schu, and G. Banting, "Clathrin-mediated endocytosis of a lipid-raft-associated protein is mediated through a dual tyrosine motif," *Journal of Cell Science*, vol. 120, no. 21, pp. 3850–3858, 2007.

[29] E. Bartee, A. McCormack, and K. Früh, "Quantitative membrane proteomics reveals new cellular targets of viral immune modulators," *PLoS Pathogens*, vol. 2, no. 10, article e107, 2006.

[30] N. Van Damme, D. Goff, C. Katsura et al., "The interferon-induced protein BST-2 restricts HIV-1 release and is downregulated from the cell surface by the viral Vpu protein," *Cell Host and Microbe*, vol. 3, no. 4, pp. 245–252, 2008.

[31] S. Kupzig, V. Korolchuk, R. Rollason, A. Sugden, A. Wilde, and G. Banting, "Bst-2/HM1.24 is a raft-associated apical membrane protein with an unusual topology," *Traffic*, vol. 4, no. 10, pp. 694–709, 2003.

[32] T. Ohtomo, Y. Sugamata, Y. Ozaki et al., "Molecular cloning and characterization of a surface antigen preferentially overexpressed on multiple myeloma cells," *Biochemical and Biophysical Research Communications*, vol. 258, no. 3, pp. 583–591, 1999.

[33] A. J. Andrew, E. Miyagi, S. Kao, and K. Strebel, "The formation of cysteine-linked dimers of BST-2/tetherin is important for inhibition of HIV-1 virus release but not for sensitivity to Vpu," *Retrovirology*, vol. 6, article 80, 2009.

[34] D. Perez-Caballero, T. Zang, A. Ebrahimi et al., "Tetherin inhibits HIV-1 release by directly tethering virions to cells," *Cell*, vol. 139, no. 3, pp. 499–511, 2009.

[35] A. Hinz, N. Miguet, G. Natrajan et al., "Structural basis of HIV-1 tethering to membranes by the BST-2/tetherin ectodomain," *Cell Host and Microbe*, vol. 7, no. 4, pp. 314–323, 2010.

[36] H. L. Schubert, Q. Zhai, V. Sandrin et al., "Structural and functional studies on the extracellular domain of BST2/tetherin in reduced and oxidized conformations," *Proceedings of the National Academy of Sciences of the United States of America*, vol. 107, no. 42, pp. 17951–17956, 2010.

[37] M. Swiecki, S. M. Scheaffer, M. Allaire, D. H. Fremont, M. Colonna, and T. J. Brett, "Structural and biophysical analysis of BST-2/tetherin ectodomains reveals an evolutionary conserved design to inhibit virus release," *The Journal of Biological Chemistry*, vol. 286, no. 4, pp. 2987–2997, 2011.

[38] H. Yang, J. Wang, X. Jia et al., "Structural insight into the mechanisms of enveloped virus tethering by tetherin," *Proceedings of the National Academy of Sciences of the United States of America*, vol. 107, no. 43, pp. 18428–18432, 2010.

[39] K. Fitzpatrick, M. Skasko, T. J. Deerinck, J. Crum, M. H. Ellisman, and J. Guatelli, "Direct restriction of virus release and incorporation of the interferon-induced protein BST-2 into HIV-1 particles," *PLoS Pathogens*, vol. 6, no. 3, Article ID e1000701, 2010.

[40] J. Hammonds, J. J. Wang, H. Yi, and P. Spearman, "Immuno-electron microscopic evidence for tetherin/BST2 as the physical bridge between HIV-1 virions and the plasma membrane," *PLoS Pathogens*, vol. 6, no. 2, Article ID e1000749, 2010.

[41] A. J. Andrew, S. Kao, and K. Strebel, "C-terminal hydrophobic region in human bone marrow stromal cell antigen 2 (BST-2)/tetherin protein functions as second transmembrane motif," *The Journal of Biological Chemistry*, vol. 286, no. 46, pp. 39967–39981, 2011.

[42] A. Habermann, J. Krijnse-Locker, H. Oberwinkler et al., "CD317/tetherin is enriched in the HIV-1 envelope and down-regulated from the plasma membrane upon virus infection," *Journal of Virology*, vol. 84, no. 9, pp. 4646–4658, 2010.

[43] J. Hammonds, L. Ding, H. Chu et al., "The tetherin/BST-2 coiled-coil ectodomain mediates plasma membrane microdomain localization and restriction of particle release," *Journal of Virology*, vol. 86, no. 4, pp. 2259–2272, 2012.

[44] M. Lehmann, S. Rocha, B. Mangeat et al., "Quantitative multicolor super-resolution microscopy reveals tetherin HIV-1 interaction," *PLoS Pathogens*, vol. 7, no. 12, Article ID e1002456, 2011.

[45] L. A. Lopez, S. J. Yang, C. M. Exline, S. Rengarajan, K. G. Haworth, and P. M. Cannon, "Anti-tetherin activities of HIV-1 Vpu and ebola virus glycoprotein do not involve removal of tetherin from lipid rafts," *Journal of Virology*, vol. 86, no. 10, pp. 5467–5480, 2012.

[46] J. V. Fritz, N. Tibroni, O. T. Keppler, and O. T. Fackler, "HIV-1 Vpu's lipid raft association is dispensable for counteraction of the particle release restriction imposed by CD317/Tetherin," *Virology*, vol. 424, no. 1, pp. 33–44, 2012.

[47] R. Rollason, V. Korolchuk, C. Hamilton, M. Jepson, and G. Banting, "A CD317/tetherin-RICH2 complex plays a critical role in the organization of the subapical actin cytoskeleton in polarized epithelial cells," *The Journal of Cell Biology*, vol. 184, no. 5, pp. 721–736, 2009.

[48] Y. Katoh and M. Katoh, "Identification and characterization of ARHGAP27 gene in silico," *International Journal of Molecular Medicine*, vol. 14, no. 5, pp. 943–947, 2004.

[49] N. Richnau and P. Aspenström, "Rich, a rho GTPase-activating protein domain-containing protein involved in signaling by Cdc42 and Rac1," *The Journal of Biological Chemistry*, vol. 276, no. 37, pp. 35060–35070, 2001.

[50] D. Reczek and A. Bretscher, "Identification of EPI64, a TBC/rabGAP domain-containing microvillar protein that binds to the first PDZ domain of EBP50 and E3KARP," *Journal of Cell Biology*, vol. 153, no. 1, pp. 191–206, 2001.

[51] Z. Songyang, S. E. Shoelson, M. Chaudhuri et al., "SH2 domains recognize specific phosphopeptide sequences," *Cell*, vol. 72, no. 5, pp. 767–778, 1993.

[52] M. W. McNatt, T. Zang, T. Hatziioannou et al., "Species-specific activity of HIV-1 Vpu and positive selection of tetherin transmembrane domain variants," *PLoS Pathogens*, vol. 5, no. 2, Article ID e1000300, 2009.

[53] L. Rong, J. Zhang, J. Lu et al., "The transmembrane domain of BST-2 determines its sensitivity to down-modulation by human immunodeficiency virus type 1 Vpu," *Journal of Virology*, vol. 83, no. 15, pp. 7536–7546, 2009.

[54] J. L. Douglas, K. Viswanathan, M. N. McCarroll, J. K. Gustin, K. Früh, and A. V. Moses, "Vpu directs the degradation of the human immunodeficiency virus restriction factor BST-2/tetherin via a βTrCP-dependent mechanism," *Journal of Virology*, vol. 83, no. 16, pp. 7931–7947, 2009.

[55] R. K. Gupta, S. Hué, T. Schaller, E. Verschoor, D. Pillay, and G. J. Towers, "Mutation of a single residue renders human tetherin resistant to HIV-1 Vpu-mediated depletion," *PLoS Pathogens*, vol. 5, no. 5, Article ID e1000443, 2009.

[56] B. Jia, R. Serra-Moreno, W. Neidermyer et al., "Species-specific activity of SIV Nef and HIV-1 Vpu in overcoming restriction by tetherin/BST2," *PLoS Pathogens*, vol. 5, no. 5, Article ID e1000429, 2009.

[57] M. Dubé, B. B. Roy, P. Guiot-Guillain et al., "Antagonism of tetherin restriction of HIV-1 release by Vpu involves binding and sequestration of the restriction factor in a perinuclear compartment," *PLoS Pathogens*, vol. 6, no. 4, Article ID e1000856, 2010.

[58] Y. Iwabu, H. Fujita, M. Kinomoto et al., "HIV-1 accessory protein Vpu internalizes cell-surface BST-2/tetherin through transmembrane interactions leading to lysosomes," *The Journal of Biological Chemistry*, vol. 284, no. 50, pp. 35060–35072, 2009.

[59] F. Zhang, S. J. Wilson, W. C. Landford et al., "Nef proteins from simian immunodeficiency viruses are tetherin antagonists," *Cell Host and Microbe*, vol. 6, no. 1, pp. 54–67, 2009.

[60] D. Sauter, M. Schindler, A. Specht et al., "Tetherin-driven adaptation of Vpu and Nef function and the evolution of pandemic and nonpandemic HIV-1 strains," *Cell Host and Microbe*, vol. 6, no. 5, pp. 409–421, 2009.

[61] R. K. Gupta and G. J. Towers, "A tail of Tetherin: how pandemic HIV-1 conquered the world," *Cell Host and Microbe*, vol. 6, no. 5, pp. 393–395, 2009.

[62] M. Shingai, T. Yoshida, M. A. Martin, and K. Strebel, "Some human immunodeficiency virus type 1 Vpu proteins are able to antagonize macaque BST-2 In Vitro and In vivo: Vpu-Negative simian-human immunodeficiency viruses are

attenuated In vivo," *Journal of Virology*, vol. 85, no. 19, pp. 9708–9715, 2011.

[63] C. Goffinet, I. Allespach, S. Homann et al., "HIV-1 antagonism of CD317 is species specific and involves Vpu-mediated proteasomal degradation of the restriction factor," *Cell Host and Microbe*, vol. 5, no. 3, pp. 285–297, 2009.

[64] B. Mangeat, G. Gers-Huber, M. Lehmann, M. Zufferey, J. Luban, and V. Piguet, "HIV-1 Vpu neutralizes the antiviral factor tetherin/BST-2 by binding it and directing its beta-TrCP2-dependent degradation," *PLoS Pathogens*, vol. 5, no. 9, Article ID e1000574, 2009.

[65] F. Margottin, S. Benichou, H. Durand et al., "Interaction between the cytoplasmic domains of HIV-1 Vpu and CD4: role of Vpu residues involved in CD4 interaction and in vitro CD4 degradation," *Virology*, vol. 223, no. 2, pp. 381–386, 1996.

[66] C. Goffinet, S. Homann, I. Ambiel et al., "Antagonism of CD317 restriction of human immunodeficiency virus type 1 (HIV-1) particle release and depletion of CD317 are separable activities of HIV-1 Vpu," *Journal of Virology*, vol. 84, no. 8, pp. 4089–4094, 2010.

[67] E. Miyagi, A. J. Andrew, S. Kao, and K. Strebe, "Vpu enhances HIV-1 virus release in the absence of Bst-2 cell surface down-modulation and intracellular depletion," *Proceedings of the National Academy of Sciences of the United States of America*, vol. 106, no. 8, pp. 2868–2873, 2009.

[68] R. S. Mitchell, C. Katsura, M. A. Skasko et al., "Vpu antagonizes BST-2-mediated restriction of HIV-1 release via β-TrCP and endo-lysosomal trafficking," *PLoS Pathogens*, vol. 5, no. 5, Article ID e1000450, 2009.

[69] A. J. Andrew, E. Miyagi, and K. Strebel, "Differential effects of human immunodeficiency virus type 1 Vpu on the stability of BST-2/tetherin," *Journal of Virology*, vol. 85, no. 6, pp. 2611–2619, 2011.

[70] S. Bour, U. Schubert, K. Peden, and K. Strebel, "The envelope glycoprotein of human immunodeficiency virus type 2 enhances viral particle release: a Vpu-like factor?" *Journal of Virology*, vol. 70, no. 2, pp. 820–829, 1996.

[71] S. Bour and K. Strebel, "The human immunodeficiency virus (HIV) type 2 envelope protein is a functional complement to HIV type 1 Vpu that enhances particle release of heterologous retroviruses," *Journal of Virology*, vol. 70, no. 12, pp. 8285–8300, 1996.

[72] G. D. Ritter, G. Yamshchikov, S. J. Cohen, and M. J. Mulligan, "Human immunodeficiency virus type 2 glycoprotein enhancement of particle budding: role of the cytoplasmic domain," *Journal of Virology*, vol. 70, no. 4, pp. 2669–2673, 1996.

[73] A. Le Tortorec and S. J. D. Neil, "Antagonism to and intracellular sequestration of human tetherin by the human immunodeficiency virus type 2 envelope glycoprotein," *Journal of Virology*, vol. 83, no. 22, pp. 11966–11978, 2009.

[74] P. Abada, B. Noble, and P. M. Cannon, "Functional domains within the human immunodeficiency virus type 2 envelope protein required to enhance virus production," *Journal of Virology*, vol. 79, no. 6, pp. 3627–3638, 2005.

[75] H. Hauser, L. A. Lopez, S. J. Yang et al., "HIV-1 Vpu and HIV-2 Env counteract BST-2/tetherin by sequestration in a perinuclear compartment," *Retrovirology*, vol. 7, article 51, 2010.

[76] B. Noble, P. Abada, J. Nunez-Iglesias, and P. M. Cannon, "Recruitment of the adaptor protein 2 complex by the human immunodeficiency virus type 2 envelope protein is necessary for high levels of virus release," *Journal of Virology*, vol. 80, no. 6, pp. 2924–2932, 2006.

[77] S. Bour, H. Akari, E. Miyagi, and K. Strebel, "Naturally occurring amino acid substitutions in the HIV-2 ROD envelope glycoprotein regulate its ability to augment viral particle release," *Virology*, vol. 309, no. 1, pp. 85–98, 2003.

[78] R. K. Gupta, P. Mlcochova, A. Pelchen-Matthews et al., "Simian immunodeficiency virus envelope glycoprotein counteracts tetherin/BST-2/CD317 by intracellular sequestration," *Proceedings of the National Academy of Sciences of the United States of America*, vol. 106, no. 49, pp. 20889–20894, 2009.

[79] M. Mansouri, K. Viswanathan, J. L. Douglas et al., "Molecular mechanism of BST2/tetherin downregulation by K5/MIR2 of Kaposi's sarcoma-associated herpesvirus," *Journal of Virology*, vol. 83, no. 19, pp. 9672–9681, 2009.

[80] C. Pardieu, R. Vigan, S. J. Wilson et al., "The RING-CH ligase K5 antagonizes restriction of KSHV and HIV-1 particle release by mediating ubiquitin-dependent endosomal degradation of tetherin," *PLoS Pathogens*, vol. 6, no. 4, Article ID e1000843, 2010.

[81] P. Bates, R. L. Kaletsky, J. R. Francica, and C. Agrawal-Gamse, "Tetherin-mediated restriction of filovirus budding is antagonized by the Ebola glycoprotein," *Proceedings of the National Academy of Sciences of the United States of America*, vol. 106, no. 8, pp. 2886–2891, 2009.

[82] L. A. Lopez, S. J. Yang, H. Hauser et al., "Ebola virus glycoprotein counteracts BST-2/tetherin restriction in a sequence-independent manner that does not require tetherin surface removal," *Journal of Virology*, vol. 84, no. 14, pp. 7243–7255, 2010.

[83] A. Kühl, C. Banning, A. Marzi et al., "The Ebola virus glycoprotein and HIV-1 VPU employ different strategies to counteract the antiviral factor tetherin," *Journal of Infectious Diseases*, vol. 204, supplement 3, pp. S850–S860, 2011.

[84] N. Casartelli, M. Sourisseau, J. Feldmann et al., "Tetherin restricts productive HIV-1 cell-to-cell transmission," *PLoS Pathogens*, vol. 6, no. 6, Article ID e1000955, 2010.

[85] B. D. Kuhl, R. D. Sloan, D. A. Donahue, T. Bar-Magen, C. Liang, and M. A. Wainberg, "Tetherin restricts direct cell-to-cell infection of HIV-1," *Retrovirology*, vol. 7, article 115, 2010.

[86] C. Jolly, N. J. Booth, and S. J. D. Neil, "Cell-cell spread of human immunodeficiency virus type 1 overcomes tetherin/BST-2-mediated restriction in T cells," *Journal of Virology*, vol. 84, no. 23, pp. 12185–12199, 2010.

[87] C. M. Coleman, P. Spearman, and L. Wu, "Tetherin does not significantly restrict dendritic cell-mediated HIV-1 transmission and its expression is upregulated by newly synthesized HIV-1 Nef," *Retrovirology*, vol. 8, article 26, 2011.

[88] R. A. Liberatore and P. D. Bieniasz, "Tetherin is a key effector of the antiretroviral activity of type I interferon in vitro and in vivo," *Proceedings of the National Academy of Sciences of the United States of America*, vol. 108, no. 44, pp. 18097–18101, 2011.

[89] B. S. Barrett, D. S. Smith, S. X. Li, K. Guo, K. J. Hasenkrug, and M. L. Santiago, "A single nucleotide polymorphism in tetherin promotes retrovirus restriction in vivo," *PLoS Pathogens*, vol. 8, no. 3, Article ID e1002596, 2012.

[90] M. Swiecki, Y. Wang, S. Gilfillan, D. J. Lenschow, and M. Colonna, "Cutting edge: paradoxical roles of BST2/tetherin in promoting type I IFN response and viral infection," *Journal of Immunology*, vol. 188, no. 6, pp. 2488–2492, 2012.

13

TRIM5 and the Regulation of HIV-1 Infectivity

Jeremy Luban

Department of Microbiology and Molecular Medicine, University of Geneva, 1211 Geneva, Switzerland

Correspondence should be addressed to Jeremy Luban, jeremy.luban@unige.ch

Academic Editor: Abraham Brass

The past ten years have seen an explosion of information concerning host restriction factors that inhibit the replication of HIV-1 and other retroviruses. Among these factors is TRIM5, an innate immune signaling molecule that recognizes the capsid lattice as soon as the retrovirion core is released into the cytoplasm of otherwise susceptible target cells. Recognition of the capsid lattice has several consequences that include multimerization of TRIM5 into a complementary lattice, premature uncoating of the virion core, and activation of TRIM5 E3 ubiquitin ligase activity. Unattached, K63-linked ubiquitin chains are generated that activate the TAK1 kinase complex and downstream inflammatory mediators. Polymorphisms in the capsid recognition domain of TRIM5 explain the observed species-specific differences among orthologues and the relatively weak anti-HIV-1 activity of human TRIM5. Better understanding of the complex interaction between TRIM5 and the retrovirus capsid lattice may someday lead to exploitation of this interaction for the development of potent HIV-1 inhibitors.

1. Introduction

HIV-1 was identified only two years after the first report of AIDS in 1981 [1]. The HIV-1 genome was cloned and sequenced, ORFs were identified, and functions of the gene products pinpointed. At a time when few antivirals were in clinical use, HIV-1 proteins were isolated, their activities were described, their structures were determined, and inhibitors were identified [2–5]. The first anti-HIV-1 drug, AZT, was approved for patients in 1987, and effective combinations of anti-HIV-1 drugs were in the clinic by the mid-1990s. Thanks to these anti-HIV-1 drugs, the number of AIDS cases plummeted in countries like the United States. HIV-1 infection became an outpatient disease. Yet, despite the impact of basic science on disease in individuals with HIV-1 infection, the AIDS pandemic has not gone away.

2. Ongoing Pandemic and the Need for More Basic Research

Failure to control the AIDS pandemic may be attributable to a number of factors, including the need for improvement in drugs and more ready access to those drugs that already exist.

Aside from one extraordinary case of a person who underwent bone marrow transplantation with cells from a CCR5-defective donor [6], there has been no documented cure of HIV-1 infection. Aside from a small effect in one vaccination trial [7], there is no evidence for prevention of HIV-1 infection in people by a vaccine. Without prospects for curative drugs or a preventive vaccine, the cost of HIV-1 infection to individuals and to society will remain high. In New York City there are currently ~110,000 people living with HIV-1 and ~1,600 HIV-related deaths annually (NYC Dept of Health). The toll of AIDS is much greater in medically underserved regions of the world, despite improved distribution of anti-HIV-1 drugs in these places. According to the UNAIDS report concluding in 2010 (http://www.unaids.org/en/), 34 million people were living with HIV infection, and in that year alone there were 2.7 million new infections.

3. Host Factors and HIV-1 Infectivity

Much remains to be learned about the function of each of the HIV-1 gene products and the optimization of drugs that inhibit their function. In recent years the focus of much HIV-1 molecular biology research has shifted to host factors that regulate HIV-1 infection. Initially these studies involved searches for host factors that physically interact with individual viral proteins. The cellular proteins cyclophilin A and LEDGF, for example, were found to interact with HIV-1 capsid (CA) and HIV-1 integrase (IN), respectively, [8, 9]. Both of these protein-protein interactions have been studied extensively and have offered novel approaches to HIV-1 inhibition and potential new anti-HIV-1 drug candidates [9–12].

Functional screens have also yielded information concerning host factors that regulate infection by HIV-1 and other retroviruses [13–16]. More recently, several groups have reported human genome-wide RNAi screens to identify factors that regulate HIV-1 infectivity [17–21]. Among host factors identified in these screens are host proteins such as TNPO3 that play critical roles in the poorly understood early events of HIV-1 infection that culminate in establishment of the provirus [15, 22–25]. Ultimately, information springing from the study of any one of these host factors has the potential to be exploited towards the development of drugs that disrupt HIV-1 in people.

4. Restriction Factors

Over the past 10 years, in addition to the identification of host factors that promote HIV-1 infectivity, several host factors have been discovered that block HIV-1 infection [26]. Comparative analysis of the genes encoding these proteins, which have been called restriction factors, indicates that some of them have evolved in response to challenge with pathogenic retroviruses [27, 28]. Study of these factors has offered a wealth of information concerning requirements for HIV-1 replication, novel ways that HIV-1 might be targeted therapeutically, potential paths to cure HIV-1 infection, and ways in which innate immune detection of HIV-1 might be amplified to improve vaccination protocols.

5. Fv1 and Capsid-Specific Restriction

When HIV-1 and other retroviruses undergo membrane fusion with susceptible target cells, the virion core is released into the target cell cytoplasm. The core of the virion consists of a capsid-protein lattice, within which there are two copies of the viral genome, along with the reverse transcriptase and IN proteins. An extraordinary series of experiments spanning several decades demonstrated that the retroviral CA protein lattice is the viral determinant of sensitivity to a murine-specific restriction factor called Fv1 [29, 30]. Curiously, *Fv1* encodes a retroviral Gag polyprotein [29]. The mechanism of Fv1 restriction is still unknown, but these studies established the concept of retrovirus CA-specific restriction and inspired the search for similar factors targeting HIV-1 CA.

6. Cyclophilin A and Capsid-Specific Restriction

Cyclophilin A was the first HIV-1 CA-specific host factor that was identified [9, 31]. Though cyclophilin A is not a restriction factor itself, it controls the accessibility of CA to other host factors that inhibit reverse transcription and other processes essential to the early steps of the infection cycle

[32]. One apparent effect of these host factors is to influence these early steps via effects on stability of the HIV-1 virion core [15, 32–36]. The identity of these cyclophilin-regulated host factors is unknown. Additional screens have identified CPSF6 as a conditional regulator of HIV-1 infection, that acts in a capsid-specific manner [15, 37]. CPSF6 is a possible candidate for one such cyclophilin A-regulated restriction factor.

Cyclophilin A cDNAs have retrotransposed many times in evolution, in several cases creating new genes that regulate HIV-1 infectivity in a capsid-specific manner. The first of the cyclophilin A-targeted restriction factors to be identified was the TRIM5-cyclophilin A fusion protein found in South American owl monkeys [38]. A similar, though independently derived, TRIM5-cyclophilin A fusion gene that acts as a capsid-specific restriction factor was created in Asian macaques [39–42]. Nup358/RanBP2, a nuclear pore protein that possesses a cyclophilin A domain also plays a role in HIV-1 infectivity [15, 17, 19, 43].

7. The Discovery of TRIM5 as an HIV-1 CA-Specific Restriction Factor

Early studies with HIV-1 showed that infection of cells from nonhuman primates is too inefficient to establish spreading infection [44–48]. It was then shown that dominant-acting inhibitors were present in these species, and that the viral capsid was the main determinant for sensitivity [49–51]. In 2004, two groups independently identified TRIM5 orthologues as being responsible for these species-specific, capsid-specific blocks [38, 52]. The owl monkey orthologue (known as TRIM5-Cyp) targets HIV-1 capsid via its carboxy-terminal cyclophilin A domain [38, 53], and the rhesus macaque orthologue (the alpha isoform) targets HIV-1 capsid via its carboxy-terminal PRY-SPRY domain [52]. Human TRIM5alpha potently restricts EIAV and N-tropic MLV, but it only weakly inhibits HIV-1 lab strains. Differences in specificity between human and macaque TRIM5alpha map to a small block of residues in the PRY-SPRY domain [28, 32, 54, 55]. Though standard HIV-1 lab strains are only weakly inhibited by human TRIM5alpha, some primary HIV-1 isolates are much more sensitive [56, 57].

8. The Problem of CA Recognition

One of the biggest ongoing challenges for researchers studying TRIM5 is to understand the structural basis for CA recognition. TRIM5 is a multimer, and CA recognition does not occur via a simple protein-protein interaction. Rather, TRIM5 recognizes a complex surface involving the CA lattice [58, 59]. In fact, TRIM5 spontaneously forms a hexameric protein lattice, and this propensity to form a lattice is greatly stimulated in the presence of the CA lattice [60] (Figure 1). This explains why a simple binding assay has not been developed. Extensive efforts have been made by several groups to develop soluble subdomains of the CA lattice that might be used in binding studies [61, 62]. The soluble hexamer unit, for example, seems not to bind to TRIM5 [63, 64]. In contrast, promising results have been obtained

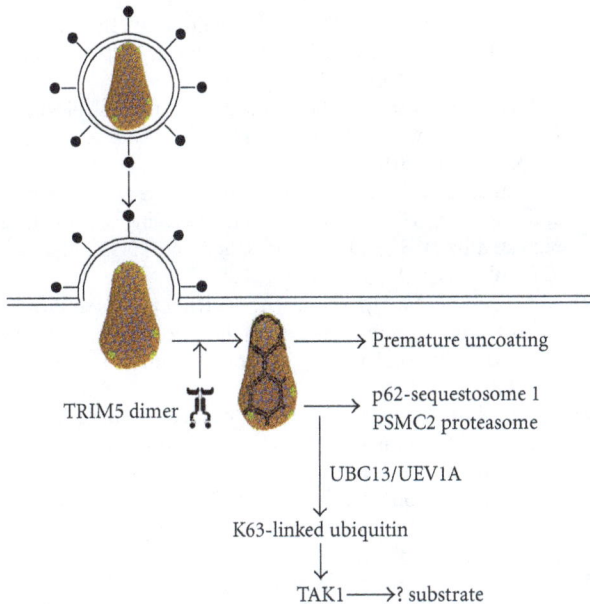

FIGURE 1: Schematic diagram showing current models of TRIM5-mediated restriction. Free TRIM5 probably exists as a dimer in the target cell cytoplasm. Upon interaction with the capsid of a restriction-sensitive retrovirus, the propensity of TRIM5 to form a complementary hexameric lattice is stimulated. This increases its intrinsic E3 ubiquitin ligase activity. If avidity for the retrovirus capsid is sufficient, the virion core prematurely uncoats and reverse transcription is blocked. Depending upon the proximity of particular cellular E2 enzymes, TRIM5 will either autoubiquitinate and traffic towards proteasomes, or it will activate the TAK1 kinase and downstream signaling molecules.

with a CA trimer [64]. A requirement for additional host factors such as SUMO-1 may complicate the situation with CA recognition even further [65].

9. TRIM5 and E3 Ubiquitin Ligase Activity

At latest count, the human TRIM family comprises ∼100 genes [66]. Like other members of this large family, TRIM5 possesses an N-terminal RING domain, a B-box domain, and a coiled-coil domain. The B box and coiled-coil domains promote multimerization of TRIM5 required for restriction activity [67, 68]. The TRIM5 RING domain confers E3 ubiquitin ligase activity, and, in cooperation with certain E2 enzymes, TRIM5 is autocatalytic, covalently attaching ubiquitin to itself [69]. Mutations on the putative E2-interacting face which disrupt this autocatalytic activity block restriction activity [70]. Ubiquitination of TRIM5 contributes to the short half-life of this protein [71], and challenge of cells with viruses bearing restriction-sensitive capsids promotes the proteasome-dependent degradation of TRIM5 [72]. Though TRIM5-stimulated ubiquitination of viral proteins has not been detected, TRIM5 may contribute to the restriction mechanism by recruiting viral components to the proteasome for degradation (Figure 1). TRIM5 interacts biochemically with the proteasome component PSMC2 and colocalizes with proteasomes in infected cells [73]. TRIM5

also associates with the proteasomal adaptor protein p62 [74] though p62 seems to stabilize TRIM5 protein levels.

In certain experimental conditions, restriction activity has been reported in the absence of the RING domain or in the absence of ubiquitination. There are several possible explanations for these discrepancies. One possibility is that, when avidity for a particular CA is great enough, TRIM5 binding to the CA is sufficient to disassemble the virion core prior to reverse transcription [59] (Figure 1). Another possible explanation stems from the fact that TRIM5 blocks multiple steps in the restriction pathway [75]. Disruption of the RING domain rescues the TRIM5-mediated block to reverse transcription and premature uncoating but not subsequent blocks in the infection cycle that lead up to integration [76, 77].

10. TRIM5, TAK1, and Inflammation

In combination with the heterodimeric E2, UBC13/UEV1A, TRIM5 catalyzes the synthesis of unattached, K63-linked ubiquitin chains that multimerize and activate the TAK1 kinase complex [63]. These K63-linked ubiquitin chains are not generated by TRIM5 when other E2 enzymes are substituted for UBC13/UEV1A. Disruption of TAK1 or of UBC13/UEV1A prevents restriction activity. Taken together, these observations suggest that the activated TAK1 complex contributes to TRIM5-mediated restriction activity via phosphorylation of a critical cofactor (Figure 1). The identity of this putative cofactor is not known, and direct phosphorylation of CA by TAK1 has not been detected.

Coming at it from another direction, the synthesis of K63-linked ubiquitin chains that activate TAK1 is stimulated by TRIM5 interaction with a restricted capsid lattice [63]. TAK1 activation leads to NFκB and AP-1 signaling which activate inflammatory cytokine transcription. In other words, TRIM5 functions as a pattern recognition receptor specific for the retrovirus capsid lattice. The consequence of TRIM5-mediated signaling for HIV-1-associated inflammation and pathology is only now being considered.

11. Future Directions of TRIM5 Research

If a robust assay was developed for TRIM5 interaction with the retrovirus capsid lattice, it would inform attempts to influence HIV-1 CA recognition by TRIM5, and perhaps to develop HIV-1 inhibitors that increase the avidity of this specific interaction. If the avidity of human TRIM5 for the HIV-1 capsid lattice could be increased experimentally, the resulting increase in capsid-stimulated signaling might also be exploited as an adjuvant for anti-HIV-1 immunization.

Recent publicity concerning the apparent cure from HIV-1 infection of a leukemia patient in Berlin with transplantation of cells from a CCR5-mutant donor [6, 78] has generated excitement concerning prospects for curing HIV-1 infection. This case has also renewed interest in basic research concerning gene therapy against HIV-1 and the regulation of HIV-1 latency in people who are already infected with HIV-1. Concerning gene therapy, the most promising approaches at this point involve either disruption of CCR5 [79] or transduction of hematopoietic stem cells with potent HIV-1 restriction factors such as engineered, human TRIM5-cyclophilin A fusion proteins [80].

Acknowledgments

This work was supported by NIH Grant RO1AI59159 and Swiss National Science Foundation Grant 3100A0-128655.

References

[1] F. Barré-Sinoussi, J. C. Chermann, F. Rey et al., "Isolation of a T-lymphotropic retrovirus from a patient at risk for acquired immune deficiency syndrome (AIDS)," *Science*, vol. 220, pp. 868–871, 1983.

[2] J. C.-H. Chen, J. Krucinski, L. J. W. Miercke et al., "Crystal structure of the HIV-1 integrase catalytic core and C-terminal domains: a model for viral DNA binding," *Proceedings of the National Academy of Sciences of the United States of America*, vol. 97, no. 15, pp. 8233–8238, 2000.

[3] E. E. Kim, C. T. Baker, M. D. Dwyer et al., "Crystal structure of HIV-1 protease in complex with VX-478, a potent and orally bioavailable inhibitor of the enzyme," *Journal of the American Chemical Society*, vol. 117, no. 3, pp. 1181–1182, 1995.

[4] S. G. Sarafianos, K. Das, C. Tantillo et al., "Crystal structure of HIV-1 reverse transcriptase in complex with a polypurine tract RNA:DNA," *EMBO Journal*, vol. 20, no. 6, pp. 1449–1461, 2001.

[5] B. G. Turner and M. F. Summers, "Structural biology of HIV," *Journal of Molecular Biology*, vol. 285, no. 1, pp. 1–32, 1999.

[6] G. Hütter, D. Nowak, M. Mossner et al., "Long-term control of HIV by CCR5 delta32/delta32 stem-cell transplantation," *The New England Journal of Medicine*, vol. 360, no. 7, pp. 692–698, 2009.

[7] S. Rerks-Ngarm, P. Pitisuttithum, S. Nitayaphan et al., "Vaccination with ALVAC and AIDSVAX to prevent HIV-1 infection in Thailand," *The New England Journal of Medicine*, vol. 361, no. 23, pp. 2209–2220, 2009.

[8] P. Cherepanov, G. Maertens, P. Proost et al., "HIV-1 integrase forms stable tetramers and associates with LEDGF/p75 protein in human cells," *Journal of Biological Chemistry*, vol. 278, no. 1, pp. 372–381, 2003.

[9] J. Luban, K. L. Bossolt, E. K. Franke, G. V. Kalpana, and S. P. Goff, "Human immunodeficiency virus type 1 Gag protein binds to cyclophilins A and B," *Cell*, vol. 73, no. 6, pp. 1067–1078, 1993.

[10] F. Christ, A. Voet, A. Marchand et al., "Rational design of small-molecule inhibitors of the LEDGF/p75-integrase interaction and HIV replication," *Nature Chemical Biology*, vol. 6, no. 6, pp. 442–448, 2010.

[11] E. K. Franke and J. Luban, "Inhibition of HIV-1 replication by cyclosporine A or related compounds correlates with the ability to disrupt the Gag-cyclophilin A interaction," *Virology*, vol. 222, no. 1, pp. 279–282, 1996.

[12] M. Thali, A. Bukovsky, E. Kondo et al., "Functional association of cyclophilin A with HIV-1 virions," *Nature*, vol. 372, no. 6504, pp. 363–365, 1994.

[13] G. Gao and S. P. Goff, "Somatic cell mutants resistant to retrovirus replication: intracellular blocks during the early stages of infection," *Molecular Biology of the Cell*, vol. 10, no. 6, pp. 1705–1717, 1999.

[14] G. Gao, X. Guo, and S. P. Goff, "Inhibition of retroviral RNA production by ZAP, a CCCH-type zinc finger protein," *Science*, vol. 297, no. 5587, pp. 1703–1706, 2002.

[15] K. Lee, Z. Ambrose, T. D. Martin et al., "Flexible Use of Nuclear Import Pathways by HIV-1," *Cell Host and Microbe*, vol. 7, no. 3, pp. 221–233, 2010.

[16] S. T. Valente, G. M. Gilmartin, K. Venkatarama, G. Arriagada, and S. P. Goff, "HIV-1 mRNA 3′ end processing is distinctively regulated by eIF3f, CDK11, and splice factor 9G8," *Molecular Cell*, vol. 36, no. 2, pp. 279–289, 2009.

[17] A. L. Brass, D. M. Dykxhoorn, Y. Benita et al., "Identification of host proteins required for HIV infection through a functional genomic screen," *Science*, vol. 319, no. 5865, pp. 921–926, 2008.

[18] F. D. Bushman, N. Malani, J. Fernandes et al., "Host cell factors in HIV replication: meta-analysis of genome-wide studies," *PLoS Pathogens*, vol. 5, no. 5, Article ID e1000437, 2009.

[19] R. König, Y. Zhou, D. Elleder et al., "Global analysis of host-pathogen interactions that regulate early-stage HIV-1 replication," *Cell*, vol. 135, no. 1, pp. 49–60, 2008.

[20] M. L. Yeung, L. Houzet, V. S. R. K. Yedavalli, and K.-T. Jeang, "A genome-wide short hairpin RNA screening of Jurkat T-cells for human proteins contributing to productive HIV-1 replication," *Journal of Biological Chemistry*, vol. 284, no. 29, pp. 19463–19473, 2009.

[21] H. Zhou, M. Xu, Q. Huang et al., "Genome-scale RNAi screen for host factors required for HIV replication," *Cell Host and Microbe*, vol. 4, no. 5, pp. 495–504, 2008.

[22] F. Christ, W. Thys, J. De Rijck et al., "Transportin-SR2 Imports HIV into the nucleus," *Current Biology*, vol. 18, no. 16, pp. 1192–1202, 2008.

[23] A. De Iaco and J. Luban, "Inhibition of HIV-1 infection by TNPO3 depletion is determined by capsid and detectable after viral cDNA enters the nucleus," *Retrovirology*, vol. 8, article 98, 2011.

[24] L. Krishnan, K. A. Matreyek, I. Oztop et al., "The requirement for cellular transportin 3 (TNPO3 or TRN-SR2) during infection maps to human immunodeficiency virus type 1 capsid and not integrase," *Journal of Virology*, vol. 84, no. 1, pp. 397–406, 2010.

[25] L. Zhou, E. Sokolskaja, C. Jolly, W. James, S. A. Cowley, and A. Fassati, "Transportin 3 promotes a nuclear maturation step required for efficient HIV-1 integration," *PLoS Pathogens*, vol. 7, Article ID e1002194.

[26] K. Strebel, J. Luban, and K.-T. Jeang, "Human cellular restriction factors that target HIV-1 replication," *BMC Medicine*, vol. 7, article 48, 2009.

[27] S. L. Sawyer, M. Emerman, and H. S. Malik, "Ancient adaptive evolution of the primate antiviral DNA-editing enzyme APOBEC3G," *PLoS Biology*, vol. 2, no. 9, Article ID E275, 2004.

[28] S. L. Sawyer, L. I. Wu, M. Emerman, and H. S. Malik, "Positive selection of primate TRIM5α identifies a critical species-specific retroviral restriction domain," *Proceedings of the National Academy of Sciences of the United States of America*, vol. 102, no. 8, pp. 2832–2837, 2005.

[29] S. Best, P. L. Tissier, G. Towers, and J. P. Stoye, "Positional cloning of the mouse retrovirus restriction gene Fv1," *Nature*, vol. 382, no. 6594, pp. 826–829, 1996.

[30] T. Pincus, W. P. Rowe, and F. Lilly, "A major genetic locus affecting resistance to infection with murine leukemia viruses. II. Apparent identity to a major locus described for resistance to friend murine leukemia virus," *Journal of Experimental Medicine*, vol. 133, no. 6, pp. 1234–1241, 1971.

[31] E. K. Franke, H. E. H. Yuan, and J. Luban, "Specific incorporation of cyclophilin A into HIV-1 virions," *Nature*, vol. 372, no. 6504, pp. 359–362, 1994.

[32] J. Luban, "Cyclophilin A, TRIM5, and resistance to human immunodeficiency virus type 1 infection," *Journal of Virology*, vol. 81, no. 3, pp. 1054–1061, 2007.

[33] L. Yuan, A. K. Kar, and J. Sodroski, "Target cell type-dependent modulation of human immunodeficiency virus type 1 capsid disassembly by cyclophilin A," *Journal of Virology*, vol. 83, no. 21, pp. 10951–10962, 2009.

[34] J. Luban, "Absconding with the chaperone: essential cyclophilin-gag interaction in HIV-1 virions," *Cell*, vol. 87, no. 7, pp. 1157–1159, 1996.

[35] M. Qi, R. Yang, and C. Aiken, "Cyclophilin A-dependent restriction of human immunodeficiency virus type 1 capsid mutants for infection of nondividing cells," *Journal of Virology*, vol. 82, no. 24, pp. 12001–12008, 2008.

[36] C. Song and C. Aiken, "Analysis of human cell heterokaryons demonstrates that target cell restriction of cyclosporine-resistant human immunodeficiency virus type 1 mutants is genetically dominant," *Journal of Virology*, vol. 81, no. 21, pp. 11946–11956, 2007.

[37] K. Lee, A. Mulky, W. Yuen et al., "HIV-1 capsid targeting domain of cleavage and polyadenylation specificity factor 6," *Journal of Virology*, vol. 86, no. 7, pp. 3851–3860, 2012.

[38] D. M. Sayah, E. Sokolskaja, L. Berthoux, and J. Luban, "Cyclophilin A retrotransposition into TRIM5 explains owl monkey resistance to HIV-1," *Nature*, vol. 430, no. 6999, pp. 569–573, 2004.

[39] G. Brennan, Y. Kozyrev, and S.-L. Hu, "TRIMCyp expression in old world primates macaca nemestrina and macaca fascicularis," *Proceedings of the National Academy of Sciences of the United States of America*, vol. 105, no. 9, pp. 3569–3574, 2008.

[40] R. M. Newman, L. Hall, A. Kirmaier et al., "Evolution of a TRIM5-CypA splice isoform in old world monkeys," *PLoS Pathogens*, vol. 4, no. 2, Article ID e1000003, 2008.

[41] C. A. Virgen, Z. Kratovac, P. D. Bieniasz, and T. Hatziioannou, "Independent genesis of chimeric TRIM5-cyclophilin proteins in two primate species," *Proceedings of the National Academy of Sciences of the United States of America*, vol. 105, no. 9, pp. 3563–3568, 2008.

[42] S. J. Wilson, B. L. J. Webb, L. M. J. Ylinen, E. Verschoor, J. L. Heeney, and G. J. Towers, "Independent evolution of an antiviral TRIMCyp in rhesus macaques," *Proceedings of the National Academy of Sciences of the United States of America*, vol. 105, no. 9, pp. 3557–3562, 2008.

[43] T. Schaller, K. E. Ocwieja, J. Rasaiyaah et al., "HIV-1 capsid-cyclophilin interactions determine nuclear import pathway, integration targeting and replication efficiency," *PLoS Pathogens*, vol. 7, Article ID e1002439, 2011.

[44] J. Balzarini, E. De Clercq, and K. Uberla, "SIV/HIV-1 hybrid virus expressing the reverse transcriptase gene of HIV-1 remains sensitive to HIV-1-specific reverse transcriptase inhibitors after passage in rhesus macaques," *Journal of Acquired Immune Deficiency Syndromes and Human Retrovirology*, vol. 15, no. 1, pp. 1–4, 1997.

[45] S. Himathongkham and P. A. Luciw, "Restriction of HIV-1 (subtype B) replication at the entry step in rhesus macaque cells," *Virology*, vol. 219, no. 2, pp. 485–488, 1996.

[46] W. Hofmann, D. Schubert, J. LaBonte et al., "Species-specific, postentry barriers to primate immunodeficiency virus infection," *Journal of Virology*, vol. 73, no. 12, pp. 10020–10028, 1999.

[47] J. Li, C. I. Lord, W. Haseltine, N. L. Letvin, and J. Sodroski, "Infection of cynomolgus monkeys with a chimeric HIV-1/SIV(mac) virus that expresses the HIV-1 envelope glycoproteins," *Journal of Acquired Immune Deficiency Syndromes*, vol. 5, no. 7, pp. 639–646, 1992.

[48] R. Shibata, M. Kawamura, H. Sakai, M. Hayami, A. Ishimoto, and A. Adachi, "Generation of a chimeric human and simian immunodeficiency virus infectious to monkey peripheral blood mononuclear cells," *Journal of Virology*, vol. 65, no. 7, pp. 3514–3520, 1991.

[49] C. Besnier, Y. Takeuchi, and G. Towers, "Restriction of lentivirus in monkeys," *Proceedings of the National Academy of Sciences of the United States of America*, vol. 99, no. 18, pp. 11920–11925, 2002.

[50] S. Cowan, T. Hatziioannou, T. Cunningham, M. A. Muesing, H. G. Gottlinger, and P. D. Bieniasz, "Cellular inhibitors with Fv1-like activity restrict human and simian immunodeficiency virus tropism," *Proceedings of the National Academy of Sciences of the United States of America*, vol. 99, no. 18, pp. 11914–11919, 2002.

[51] C. Münk, S. M. Brandt, G. Lucero, and N. R. Landau, "A dominant block to HIV-1 replication at reverse transcription in simian cells," *Proceedings of the National Academy of Sciences of the United States of America*, vol. 99, no. 21, pp. 13843–13848, 2002.

[52] M. Stremlau, C. M. Owens, M. J. Perron, M. Kiessling, P. Autissier, and J. Sodroski, "The cytoplasmic body component TRIM5α restricts HIV-1 infection in old world monkeys," *Nature*, vol. 427, no. 6977, pp. 848–853, 2004.

[53] S. Nisole, C. Lynch, J. P. Stoye, and M. W. Yap, "A Trim5-cyclophilin A fusion protein found in owl monkey kidney cells can restrict HIV-1," *Proceedings of the National Academy of Sciences of the United States of America*, vol. 101, no. 36, pp. 13324–13328, 2004.

[54] M. Stremlau, M. Perron, S. Welikala, and J. Sodroski, "Species-specific variation in the B30.2(SPRY) domain of TRIM5α determines the potency of human immunodeficiency virus restriction," *Journal of Virology*, vol. 79, no. 5, pp. 3139–3145, 2005.

[55] M. W. Yap, S. Nisole, and J. P. Stoye, "A single amino acid change in the SPRY domain of human Trim5α leads to HIV-1 restriction," *Current Biology*, vol. 15, no. 1, pp. 73–78, 2005.

[56] E. Battivelli, D. Lecossier, S. Matsuoka, J. Migraine, F. Clavel, and A. J. Hance, "Strain-specific differences in the impact of human TRIM5α, different TRIM5α alleles, and the inhibition of capsid-cyclophilin a interactions on the infectivity of HIV-1," *Journal of Virology*, vol. 84, no. 21, pp. 11010–11019, 2010.

[57] E. Battivelli, J. Migraine, D. Lecossier, P. Yeni, F. Clavel, and A. J. Hance, "Gag cytotoxic T lymphocyte escape mutations can increase sensitivity of HIV-1 to human TRIM5alpha, linking intrinsic and acquired immunity," *Journal of Virology*, vol. 85, pp. 11846–11854, 2011.

[58] S. Sebastian and J. Luban, "TRIM5α selectively binds a restriction-sensitive retroviral capsid," *Retrovirology*, vol. 2, article 40, 2005.

[59] M. Stremlau, M. Perron, M. Lee et al., "Specific recognition and accelerated uncoating of retroviral capsids by the TRIM5α restriction factor," *Proceedings of the National Academy of Sciences of the United States of America*, vol. 103, no. 14, pp. 5514–5519, 2006.

[60] B. K. Ganser-Pornillos, V. Chandrasekaran, O. Pornillos, J. G. Sodroski, W. I. Sundquist, and M. Yeager, "Hexagonal assembly of a restricting TRIM5alpha protein," *Proceedings of the

National Academy of Sciences of the United States of America, vol. 108, no. 2, pp. 534–539, 2011.

[61] I.-J. L. Byeon, X. Meng, J. Jung et al., "Structural convergence between Cryo-EM and NMR reveals intersubunit interactions critical for HIV-1 capsid function," *Cell,* vol. 139, no. 4, pp. 780–790, 2009.

[62] O. Pornillos, B. K. Ganser-Pornillos, B. N. Kelly et al., "X-ray structures of the hexameric building block of the hiv capsid," *Cell,* vol. 137, no. 7, pp. 1282–1292, 2009.

[63] T. Pertel, S. Hausmann, D. Morger et al., "TRIM5 is an innate immune sensor for the retrovirus capsid lattice," *Nature,* vol. 472, no. 7343, pp. 361–365, 2011.

[64] G. Zhao, D. Ke, T. Vu et al., "Rhesus TRIM5α disrupts the HIV-1 capsid at the inter-hexamer interfaces," *PLoS Pathogens,* vol. 7, no. 3, Article ID e1002009, 2011.

[65] G. Arriagada, L. N. Muntean, and S. P. Goff, "SUMO-interacting motifs of human TRIM5α are important for antiviral activity," *PLoS Pathogens,* vol. 7, no. 4, Article ID e1002019, 2011.

[66] K. Han, D. I. Lou, and S. L. Sawyer, "Identification of a genomic reservoir for new trim genes in primate genomes," *PLoS Genetics,* vol. 7, Article ID e1002388.

[67] F. Diaz-Griffero, X.-R. Qin, F. Hayashi et al., "A B-box 2 surface patch important for TRIM5α self-association, capsid binding avidity, and retrovirus restriction," *Journal of Virology,* vol. 83, no. 20, pp. 10737–10751, 2009.

[68] X. Li and J. Sodroski, "The TRIM5α B-box 2 domain promotes cooperative binding to the retroviral capsid by mediating higher-order self-association," *Journal of Virology,* vol. 82, no. 23, pp. 11495–11502, 2008.

[69] L. Xu, L. Yang, P. K. Moitra et al., "BTBD1 and BTBD2 colocalize to cytoplasmic bodies with the RBCC/tripartite motif protein, TRIM5δ," *Experimental Cell Research,* vol. 288, no. 1, pp. 84–93, 2003.

[70] M. Lienlaf, F. Hayashi, F. Di Nunzio et al., "Contribution of E3-ubiquitin ligase activity to HIV-1 restriction by TRIM5alpha(rh): structure of the RING domain of TRIM5alpha," *Journal of Virology,* vol. 85, pp. 8725–8737, 2011.

[71] F. Diaz-Griffero, X. Li, H. Javanbakht et al., "Rapid turnover and polyubiquitylation of the retroviral restriction factor TRIM5," *Virology,* vol. 349, no. 2, pp. 300–315, 2006.

[72] C. J. Rold and C. Aiken, "Proteasomal degradation of TRIM5α during retrovirus restriction," *PLoS Pathogens,* vol. 4, no. 5, Article ID e1000074, 2008.

[73] Z. Lukic, S. Hausmann, S. Sebastian et al., "TRIM5alpha associates with proteasomal subunits in cells while in complex with HIV-1 virions," *Retrovirology,* vol. 8, article 93, 2011.

[74] C. O'Connor, T. Pertel, S. Gray et al., "p62/sequestosome-1 associates with and sustains the expression of retroviral restriction factor TRIM5α," *Journal of Virology,* vol. 84, no. 12, pp. 5997–6006, 2010.

[75] L. Berthoux, S. Sebastian, E. Sokolskaja, and J. Luban, "Lv1 inhibition of human immunodeficiency virus type 1 is counteracted by factors that stimulate synthesis or nuclear translocation of viral cDNA," *Journal of Virology,* vol. 78, no. 21, pp. 11739–11750, 2004.

[76] A. Roa, F. Hayashi, Y. Yang et al., "Ring domain mutations uncouple TRIM5α restriction of HIV-1 from inhibition of reverse transcription and acceleration of uncoating," *Journal of Virology,* vol. 86, pp. 1717–1727, 2012.

[77] X. Wu, J. L. Anderson, E. M. Campbell, A. M. Joseph, and T. J. Hope, "Proteasome inhibitors uncouple rhesus TRIM5α restriction of HIV-1 reverse transcription and infection," *Proceedings of the National Academy of Sciences of the United States of America,* vol. 103, no. 19, pp. 7465–7470, 2006.

[78] K. Allers, G. Hütter, J. Hofmann et al., "Evidence for the cure of HIV infection by CCR5Δ32/Δ32 stem cell transplantation," *Blood,* vol. 117, no. 10, pp. 2791–2799, 2011.

[79] N. Holt, J. Wang, K. Kim et al., "Human hematopoietic stem/progenitor cells modified by zinc-finger nucleases targeted to CCR5 control HIV-1 *in vivo,*" *Nature Biotechnology,* vol. 28, no. 8, pp. 839–847, 2010.

[80] M. R. Neagu, P. Ziegler, T. Pertel et al., "Potent inhibition of HIV-1 by TRIM5-cyclophilin fusion proteins engineered from human components," *Journal of Clinical Investigation,* vol. 119, no. 10, pp. 3035–3047, 2009.

14

Probing Retroviral and Retrotransposon Genome Structures: The "SHAPE" of Things to Come

Joanna Sztuba-Solinska and Stuart F. J. Le Grice

RT Biochemistry Section, HIV Drug Resistance Program, National Cancer Institute, Fredrick, MD 21702-1201, USA

Correspondence should be addressed to Stuart F. J. Le Grice, legrices@mail.nih.gov

Academic Editor: Abdul A. Waheed

Understanding the nuances of RNA structure as they pertain to biological function remains a formidable challenge for retrovirus research and development of RNA-based therapeutics, an area of particular importance with respect to combating HIV infection. Although a variety of chemical and enzymatic RNA probing techniques have been successfully employed for more than 30 years, they primarily interrogate small (100–500 nt) RNAs that have been removed from their biological context, potentially eliminating long-range tertiary interactions (such as kissing loops and pseudoknots) that may play a critical regulatory role. Selective 2′ hydroxyl acylation analyzed by primer extension (SHAPE), pioneered recently by Merino and colleagues, represents a facile, user-friendly technology capable of interrogating RNA structure with a single reagent and, combined with automated capillary electrophoresis, can analyze an entire 10,000-nucleotide RNA genome in a matter of weeks. Despite these obvious advantages, SHAPE essentially provides a nucleotide "connectivity map," conversion of which into a 3-D structure requires a variety of complementary approaches. This paper summarizes contributions from SHAPE towards our understanding of the structure of retroviral genomes, modifications to which technology that have been developed to address some of its limitations, and future challenges.

1. Introduction

Cis-acting sequences within the (+) strand RNA genomes of retroviruses and long terminal repeat (LTR) containing retrotransposons control several critical events in their life cycle, including transcription [1], translation [2], dimerization [3], packaging [4], RNA export [5], and DNA synthesis [6]. Development of novel RNA-based strategies to ameliorate human immunodeficiency virus (HIV) pathogenesis would therefore benefit from an improved understanding of RNA structure and how this mediates interactions with both host and viral proteins. Historically, deciphering higher-order RNA structure has taken advantage of base- and structure-specific nucleases (e.g., RNases A, T1, T2 [7] and nuclease S1 [8]) or chemicals (e.g., dimethyl sulfate, diethyl pyrocarbonate [9, 10], and Pb^{2+} [11]). While these approaches have produced seminal advances in elucidating features of the HIV-1 and HIV-2 genomes [12–23], the necessity in most cases for multiple reaction conditions can be considered a limitation. Moreover, in almost all instances, enzymatic and chemical RNA footprinting has been performed on short RNAs prepared by *in vitro* transcription and labeled with ^{32}P, eliminating any positional context, that is, regulatory roles that might be mediated by long-range, tertiary interactions. Although this challenge has in part been addressed by Paillart et al. via *ex vivo* footprinting of virion-associated RNA with dimethyl sulfate [24], a more "user-friendly" approach capable of providing information on RNA structure both *in vitro* and *ex virio*, and with fewer base-specific reagents, would clearly be advantageous.

Selective 2′ hydroxyl acylation analyzed by primer extension (SHAPE), reported in 2005 by Merino and colleagues [25], has emerged as a facile technique that addresses many of these concerns. Since the target of the probing agent (N-methyl isatoic anhydride (NMIA) [25] or 1-methyl-7-nitroisatoic anhydride (1M7) [26]) is the ribose 2′ hydroxyl, all four RNA bases are simultaneously probed with a single reagent. Secondly, when combined with fluorimetric detection, multiplexing and automated capillary electrophoresis, SHAPE profiles of complete, 10,000 nt retroviral genomes can be generated in a matter of weeks [27]. By comparing reactivity profiles obtained *in vitro* and *ex vivo*, these studies have also provided important information on HIV genome organization and the role played by chaperone proteins. Finally, the recent advent of the self-inactivating electrophile benzoyl cyanide (BCN) [28] opens the possibility of time-resolved SHAPE, which promises to provide important glimpses into RNA conformational dynamics.

Despite these benefits, it should be borne in mind that SHAPE effectively provides a secondary structure nucleotide "connectivity" profile; that is, it does not report directly on long-distance tertiary interactions such as kissing loops and pseudoknots and is best used in conjunction with other solution techniques, such as X-ray crystallography, NMR spectroscopy, and small angle X-ray scattering in order to generate an accurate 3-D model. Where possible, combining structural data with a genetic analysis, via construction of disruptive and complementary mutations, should be seen as an important complement. In this communication, we have reviewed the basic SHAPE methodology and its application to understanding the structure of regulatory elements of both retroviral and retrotransposon genomes. Modifications to the probing technology which have allowed us to (i) investigate tertiary interactions important for regulating nucleocytoplasmic RNA transport and (ii) combine chemical modification with tandem mass spectrometry to understand conformational dynamics of RNA/DNA hybrids containing polypurine tract (PPT) primers of (+) strand DNA synthesis, are presented. Finally, future challenges of SHAPE, including increasing sensitivity where the amount of biological material is limiting, and studying interconverting RNA structures, are also discussed.

2. SHAPE Methodology

A brief outline of SHAPE methodology is presented in Figure 1. As originally conceived, this chemoenzymatic strategy assesses local flexibility in RNA via accessibility of the ribose 2′-OH group to acylation by the electrophilic reagent NMIA. In flexible regions (such as loops, bulges, and junctions), RNA adopts conformations that will promote formation of a nucleophilic 2′-oxyanion which reacts with NMIA to form a bulky 2′-O-adduct [25] (Figure 1(a)). Recent modifications to the strategy have taken advantage of 1M7 [26] and BCN [28], which are more labile towards hydrolysis and self-inactivation, making them particularly advantageous for performing time-resolved footprinting. Modified RNAs are subsequently evaluated by primer extension with an RNase H-deficient reverse transcriptase, creating a cDNA library corresponding to stops at sites of adduct formation in the RNA when analyzed by high resolution gel electrophoresis (Figure 1(b)). End-labeling with ^{32}P allows primer extension products of 50–300 nt to be fractionated by conventional denaturing polyacrylamide gel electrophoresis, while

multiplexing with fluorescently-labeled primers and automated capillary electrophoresis permits resolution of 500–750 nt in a single electropherogram (Figure 1(c)). Finally, autoradiograms or electropherograms are quantified and computationally deconvoluted in order to obtain the energy-minimized RNA structure (Figure 1(d)).

In contrast to the many benefits of SHAPE, analyzing sites of adduct formation by primer extension has limitations for structural studies aimed at very short RNAs. Since SHAPE information is tabulated indirectly through the length and frequency of a given cDNA, information on ~ 50 nt at the 3′ terminus of the RNA molecule is lost as a consequence of both primer binding and reduced processivity of the retroviral reverse transcriptase used for cDNA synthesis. In an attempt to address this shortcoming, Steen et al. [33] recently combined chemical acylation with sensitivity to exonucleolytic degradation, based on the observation that RNase R exonucleases processively cleave RNA in a 3′ → 5′ direction. Screening several sources of RNases R identified an enzyme from *Mycoplasma genitalium* capable of processively degrading RNA, including through base-paired regions, but not beyond sites of adduct formation. The approach of RNase-directed SHAPE provides a facile and important complement to examine structural features at the termini of important regulatory RNAs. Although there is currently no commercial source for *Mycoplasma genitalium* RNase R, methods for purifying this enzyme from recombinant *E. coli* have been published [34].

3. SHA-MS Combines Chemical Acylation with Mass Spectrometry

As originally conceived, SHAPE was designed to interrogate structural features of RNA molecules ranging in size from several hundred to several thousand nucleotides. A critical feature of retrovirus and retrotransposon replication is initiation of (+) strand, DNA-dependent DNA synthesis from the polypurine tract (PPT) RNA primer. Although we have gleaned important information on PPT function using mutants of HIV RT [35–37] and targeted insertion of nucleoside analogs at, and in the vicinity of, the PPT-U3 junction [38–42], the structural basis for PPT primer recognition remains elusive. Since our nucleoside analog strategy has mandated analysis of short RNA/DNA hybrids (25–30 bp), identifying structural anomalies by SHAPE becomes impractical. However, since RNA 2′-OH acylation results in a mass increment of 133 Da, we reasoned that adduct formation could be evaluated by electrospray ionization (ESI) mass spectrometry (MS). As illustrated in Figure 2(a), discrete PPT RNAs containing between one and four NMIA adducts could be detected by nanospray ESI-MS, while the DNA complement, as predicted, was insensitive to modification. Tandem mass spectrometry was subsequently used to define the positions of adduct formation indicating that, in addition to terminal ribonucleotides, which might be predicted to "fray," ribonucleotides-11 and -12 of the wild type PPT (defining position -1 as the ribonucleotide 5′ of

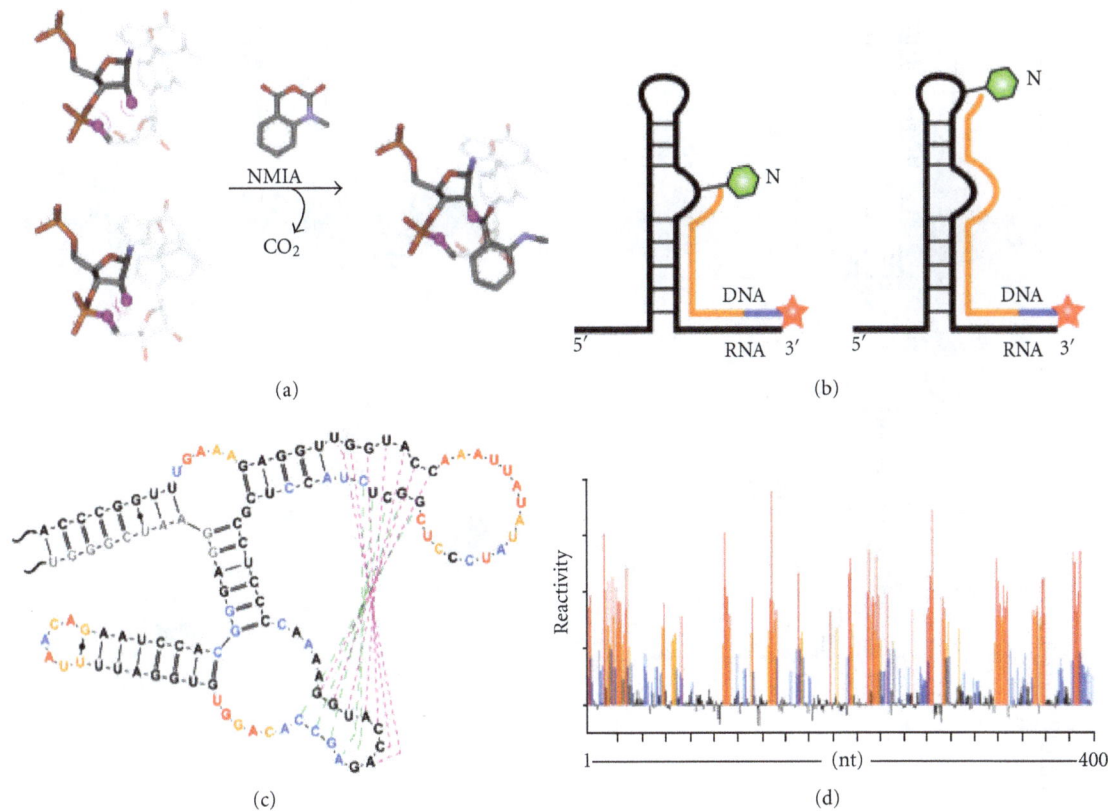

FIGURE 1: Overview of SHAPE technology. (a) Ribose 2′ OH of RNA at flexible, or unpaired nucleotides is selectively modified by NMIA. (b) Positions of adduct formation result in impaired primer extension during subsequent cDNA synthesis. (c) Radiolabeled or fluorescently-labeled primer extension products are resolved by high resolution polyacrylamide gel electrophoresis or automated capillary electrophoresis. (d) Electropherograms are computationally deconvoluted to obtain normalized NMIA reactivities, from which a secondary structure model is constructed.

the PPT/U3 junction) were sensitive to acylation. These positions, corresponding to bases of the mispaired or "unzipped" component of the PPT observed crystallographically [43], suggest that either mispairing alters the geometry of the ribose 2′-OH or that the unzipped region of the PPT is transiently unpaired.

The utility of our approach [29], designated selective 2′ hydroxyl acylation analyzed by mass spectrometry (SHA-MS [29]), was perhaps better demonstrated by analyzing nucleoside analog-substituted PPTs. As might be predicted, substituting template thymine −13T with the nonhydrogen bonding pyrimidine isostere 2,4-difluorotoluene (dF [41]) expanded the NMIA sensitivity profile to include ribonucleotides -11, -12, and -13. However, replacing template nucleotide-8T with dF rendered not only primer nucleotides -11 and -12 insensitive to acylation, but also the complementary primer nucleotide -8, possibly indicating a local difference in base stacking that masks the ribose 2′-OH. Surprisingly, while the PPT RNA primer of the *Saccaromyces cerevisiae* LTR-retrotransposon Ty3 was insensitive to NMIA, acylation of ribonucleotide +1G was observed. These results were in agreement with NMR data [44], suggesting that a unique geometry at the Ty3 PPT/U3 junction may contribute towards recognition specificity. When complemented with

KMnO4 footprinting, which differentiates between thymines in a single-stranded and duplex configuration [45], SHA-MS provides a valuable, high resolution approach to interrogate the geometry of short, purine-rich RNA/DNA hybrids where conventional probing strategies are impractical.

4. Antisense (AI)-Interfered SHAPE: Deciphering Tertiary Interactions

Originally defined as an intermolecular interactions that mediate HIV-1 RNA genome dimerization [46], kissing loops have also been identified in the genomes of hepatitis C virus [47], chrysanthemum chlorotic mottle viroid [48], and a group C enterovirus [49]. Furthermore, pseudoknots, (tertiary interactions containing at least two stem-loop structures wherein a portion of one stem is intercalated between two halves of the other) are associated with translational control via internal ribosome entry sites [50], ribosomal frameshifting [51], and tRNA mimicry [52, 53]. Analysis of the RNA transport element of the murine retrotransposon *MusD* (MTE) revealed a complex structure containing a combination of a kissing loop and a pseudoknot [30]. Such tertiary interactions are particularly challenging for SHAPE

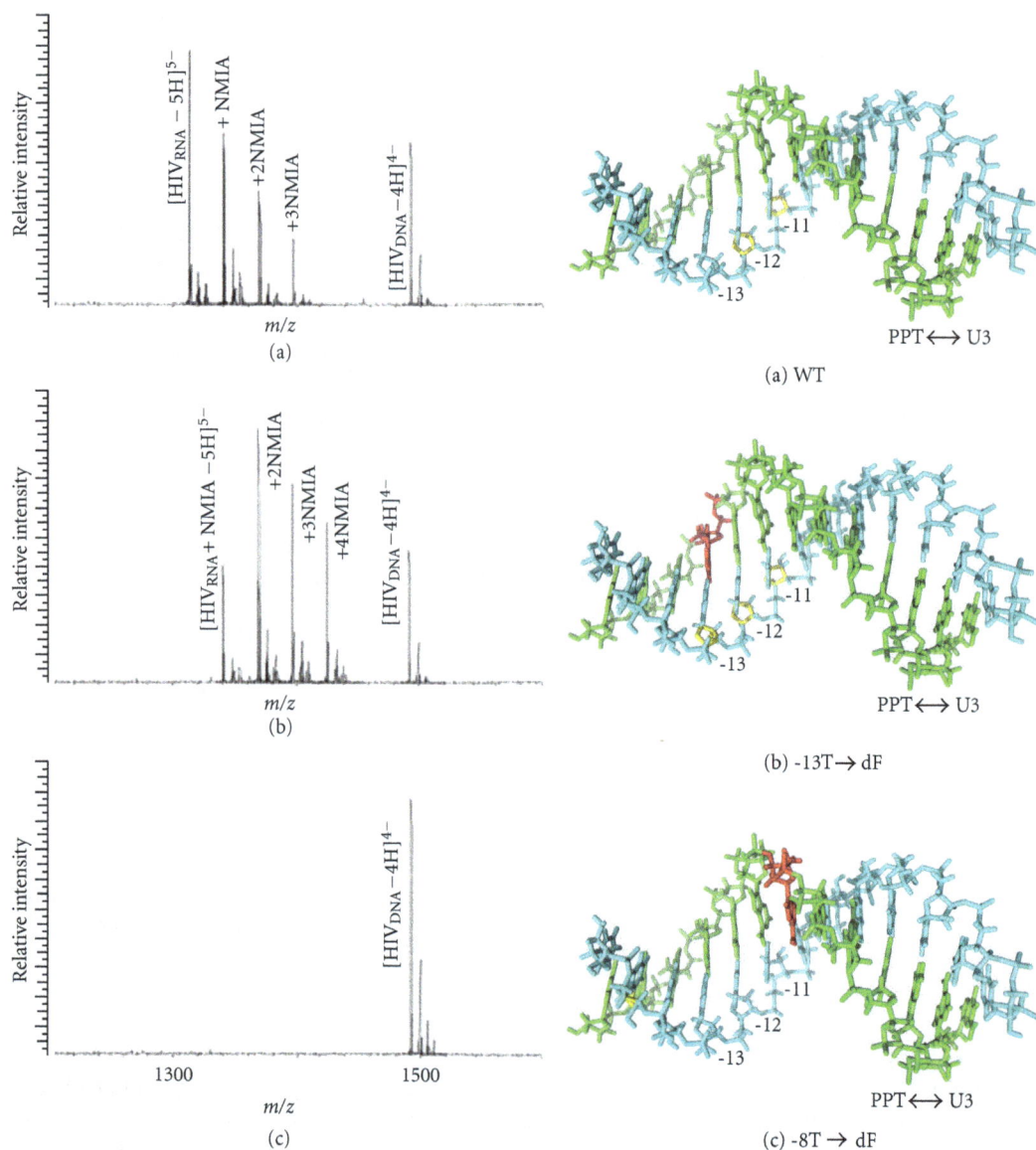

FIGURE 2: Examining RNA/DNA structural dynamics by combining chemical acylation with mass spectrometry. *Left*, Nano-ESI mass spectra of a model HIV-1 PPT RNA/DNA hybrid following treatment with a 10-fold (a), 50-fold (b), and 100-fold NMIA excess (c). At limiting NMIA concentrations (a) and (b), the majority of the PPT RNA is unmodified, and RNAs containing one, two, three, or four NMIA adducts can be observed, while excess acylation (c) results in overmodification of the entire RNA strand. In all cases, however, the PPT DNA complement is not modified by NMIA owing to the absence of a ribose 2′-OH group. *Right*, NMIA sensitivity of the wild type (a) and dF-modified (b) and (c) HIV-1 PPT RNA/DNA hybrids. In all cases, DNA and RNA nucleotides are represented in green and blue, respectively. NMIA-sensitive ribonucleotides are in yellow and positions of dF substitution in red. The position of the PPT/U3 junction has been indicated. Adapted from [29].

and in the first instance require manual identification. In order to verify the identity of these structures, we developed an oligonucleotides-based interfering strategy designated antisense (ai)-interfered SHAPE, the basis of which is illustrated in Figure 3(a).

This strategy involves hybridization of short (5–10 nts) oligonucleotides to the proposed RNA duplex and determining whether this induces enhanced NMIA reactivity of the displaced strand. In view of their length, antisense oligonucleotides were constructed containing 2′-O-methyl

and locked nucleic acid substitutions, both of which have been shown to improve duplex stability. Such interfering oligonucleotides are invasive inasmuch that they will hybridize to their partner sequence in an RNA that has already adopted its 3D structure. When applied to the MusD MTE, an interfering octanucleotide hybridized to internal loop 8 (IL8) stimulated NMIA reactivity at several positions in its kissing partner, loop 3 (L3, Figure 3(b)). Importantly, and as suggested earlier, the L3/IL8 kissing interaction suggested by ai-SHAPE was confirmed genetically *in vivo*,

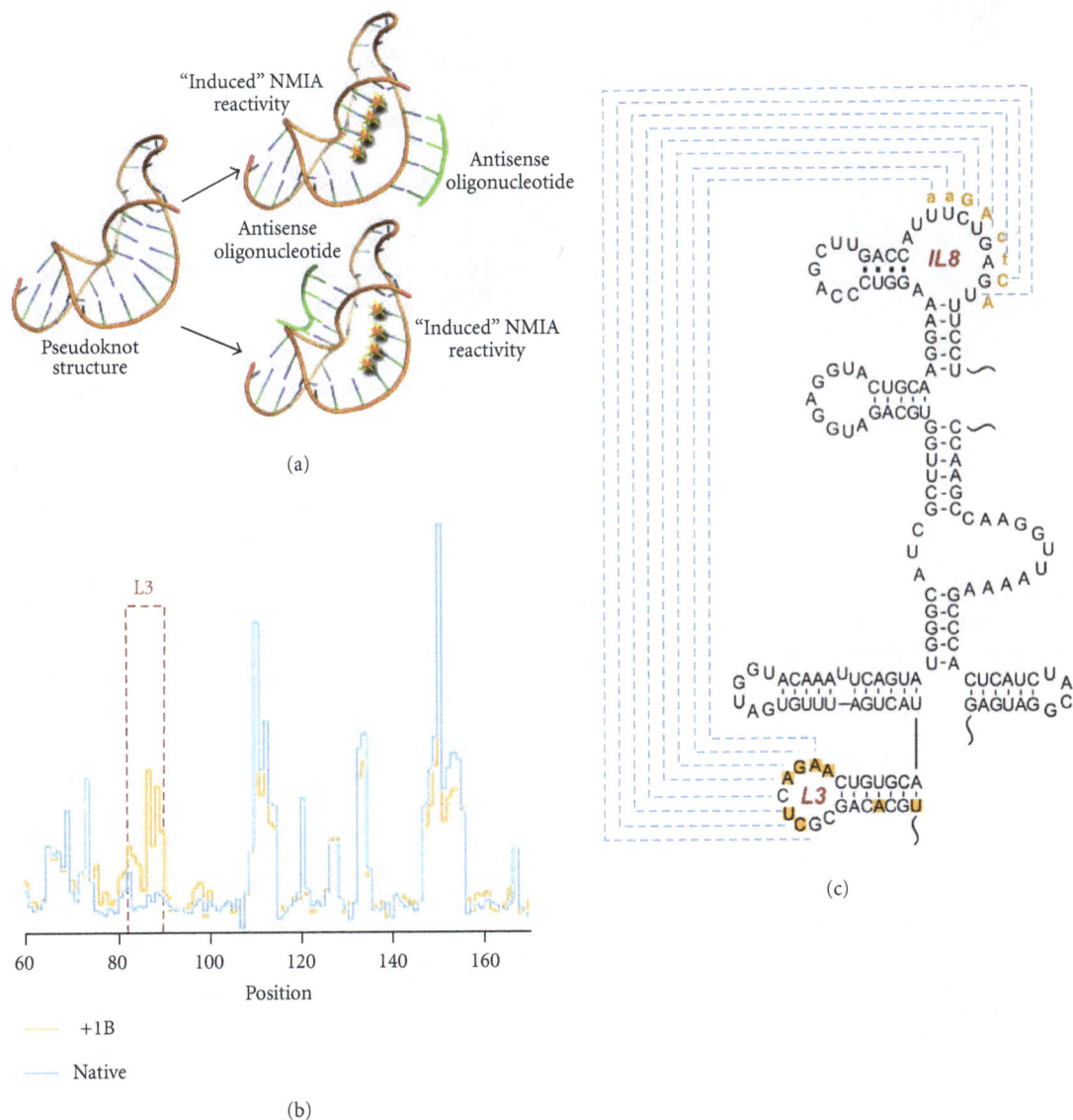

(a)

(b)

(c)

FIGURE 3: Examining RNA tertiary interactions by ai-SHAPE. (a) ai-SHAPE principal, that is, hybridization of an interfering oligonucleotide (green) to one partner of the proposed RNA duplex increases acylation sensitivity of its base-paired counterpart. (b) Electropherogram of NMIA reactivity of MTE nucleotides 60–170 in the absence (blue trace) and presence of the interfering oligonucleotide 1B (yellow trace). Loop L3 has been highlighted by the red box. (c) Secondary structure map for a portion of the *MusD* RNA transport element MTE, illustrating the L3/IL8 kissing interaction. The sequence of the interfering oligonucleotide hybridized to IL8 is indicated in orange, while nucleotides of loop L3 and the neighboring helix that exhibited enhanced NMIA reactivity are depicted within orange boxes adapted from [30].

where MusD-dependent nucleocytoplasmic RNA transport was abrogated and restored by disruptive and compensatory kissing loop mutations, respectively. The structure of the MusD pseudoknot was likewise confirmed by ai-SHAPE, while a genetic analysis indicated that the ability to assume a pseudoknot configuration was a more critical determinant of function than absolute nucleotide sequence.

5. Interconverting RNAs: Choosing between Dimerization and Protein Synthesis

Riboswitches, located in the noncoding region of several mRNAs, have been demonstrated to regulate RNA stability,

protein synthesis, and splicing via a conformational change mediated by binding of a high-affinity ligand [54–56]. The highly-structured 5′ untranslated regions of many retroviruses can be considered formally analogous to a riboswitch, inasmuch as overlapping sequences have been proposed to mediate both genome dimerization/packaging and translation [57, 58]. An inconclusive acylation pattern in our recent SHAPE study of the 5′ UTR of the feline immunodeficiency virus (FIV) genome [31] led us to postulation that certain regions were metastable, allowing them to adopt alternative structures, a notion strengthened by the observation of two closely-migrating RNA species following fractionation by nondenaturing polyacrylamide gel electrophoresis. The

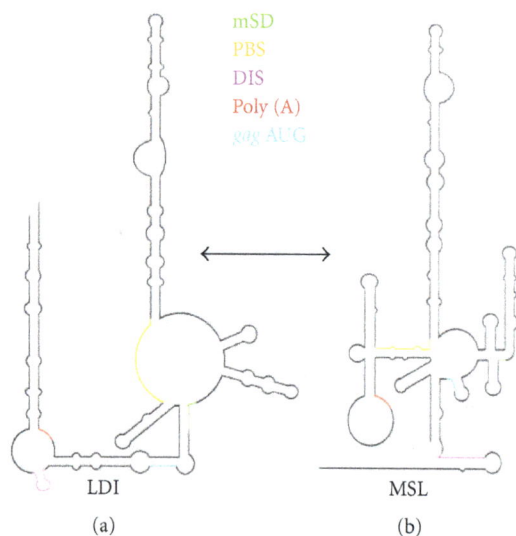

FIGURE 4: Proposed interconverting structures of the FIV 5′ leader RNA controlling genome dimerization/packaging and translation. For both the long-distance interaction (LDI) (a) and multiple stem-loops (MSL) structures (b), important regulatory sequences have been color-coded. mSD, major splice donor sequence; PBS, tRNA$_{Lys,3}$ primer binding site; DIS, dimer initiation site; Poly(A), poly (A) hairpin; gag AUG, gag initiator methionine codon. See text for fuller details adapted from [31].

hypothesis that best unified our experimental data is illustrated in Figure 4, suggesting that alternate structures for the FIV 5′ UTR mediate different events in the retrovirus life cycle.

The long range interaction (LRI) model, originally proposed by Kenyon et al. [59], exposes the putative FIV dimer initiation sequence (DIS), while the gag initiation codon is embedded within a short helix, and free energy calculations suggest this model would support genome dimerization and packaging. The LRI structure also exposes the tRNA primer binding site from which reverse transcription is initiated following infection. The alternative, multiple stem-loop (MSL) structure occludes the DIS, while the gag initiation codon is positioned within a short stem-loop, the stability of which would facilitate translation over dimerization and packaging. In the MSL, the tRNA primer binding site is also inaccessible. Support for interconverting structures of the 5′ UTR was provided by the observation that the FIV mutant AN14, demonstrated in vivo to have impaired packaging [60], exhibited impaired dimerization in vitro, while dimerization was enhanced when the RNA was stabilized in the LRI form [31]. Though a later section will address future SHAPE strategies, our study of the FIV leader RNA provides another good example of combining chemical probing with functional studies, while at the same time highlighting one of its challenges, namely, how to deal with interconverting RNAs. One potential solution might be to perform nondenaturing electrophoretic separation immediately following chemical acylation. Since SHAPE relies on single-hit kinetics, modified RNAs should still resolve as discrete species. Polymerizing gels with disruptable crosslinking

agents such as N,N′-bisacryloylcystamine (BAC) or N,N-diallyltartardiamide (DATD) would allow solubilization and recovery of nucleic acid for subsequent cDNA synthesis. Should ribose acylation alter RNA conformation, in-gel probing directly following fractionation by nondenaturing electrophoresis is an alternative strategy.

6. Investigating RNA Tertiary Structure with "Threading Intercalators"

Understanding RNA structure-function relationship requires accurate three-dimensional structure modeling methods. At present, there is a substantial gap in obtaining high-throughput 3D information for RNA molecules larger than 150 nts. The techniques frequently used to obtain atomic resolution of RNAs, such as NMR spectroscopy and X-ray crystallography, have restrictions that preclude structural analysis. In NMR spectroscopy, the excited signal from individual atomic nuclei becomes congested and difficult to analyze with the increasing size of RNA molecule. Even though X-ray crystallography does not suffer from size limitations, obtaining crystals for flexible and diverse RNA structures represents a great challenge. These difficulties however, are now being addressed by combining SHAPE with methidiumpropyl-EDTA- (MPE-) directed through-space hydroxyl radical cleavage, as outlined schematically in Figure 5. In the past, MPE has been successfully applied as a tool for footprinting binding sites of small molecules on heterogeneous DNA [61], RNA folding analysis [61, 62] and examining RNA-binding properties of phospho- and dephospho-RNA-dependent protein kinase [63]. Recently, Gherghe et al. successfully combined SHAPE with MPE-directed hydroxyl radical cleavage to study tRNA^{Asp} tertiary structure [64].

MPE is a methidium intercalator moiety tethered to EDTA that preferentially intercalates at G-C rich helices in RNA at sites adjacent to a single nucleotide bulge. The intercalated MPE occupies roughly the same space as a single base pair and is oriented in the motif such that the EDTA moiety points toward the bulge. Upon addition of Fe(II) and a reducing agent, ferrous ion binds the EDTA and generates short-lived hydroxyl radicals that cleave proximal regions of the RNA backbone [65]. The MPE binding site can be placed at RNA helical motifs by replacing four consecutive base pairs with CGAG/C(C/U)G motif [64]. Provided that this replacement is compatible with the native structure of RNA, cleavage at positions proximal in space to the unique location of the bound MPE affords information about the nucleotides neighboring the intercalating ligand. Cleavage intensity at each position can be calculated as a ratio relative to the mean value for all intensities, after subtracting background cleavage observed for the native RNA sequence that does not contain an MPE binding site. Subsequently, MPE-directed through-space cleavage experiments yield high quality, long range constrains that refine nucleotide positions in RNA to atomic resolution of 4 Å rmsd [64]. As a result, the combined experimental and computational approach has the potential to yield native-like models for functionally crucial RNA

(a)

(b)

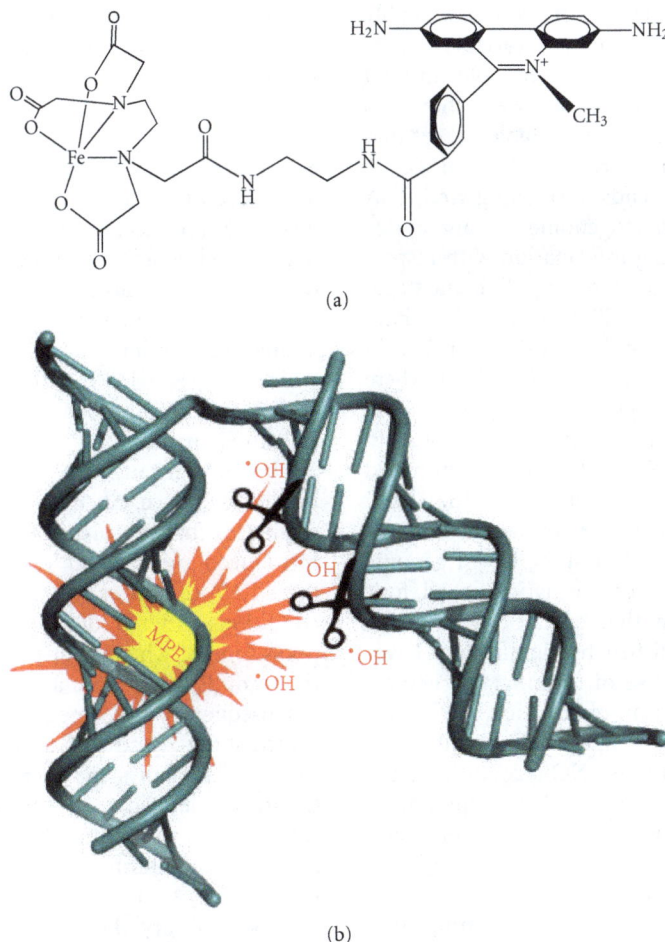

FIGURE 5: (a) Structure of the threading intercalator, MPE. (b) Examining RNA tertiary interactions by through-space hydroxyl radical cleavage (–OH) with the threading intercalator methidiumpropyl EDTA (MPE). Once a SHAPE profile for the RNA under investigation is determined, an MPE intercalation site is introduced by replacing four consecutive nucleotides with the CGAG/C(C/U)G recognition motif. SHAPE is then repeated to determine that sequence changes are nonperturbing, after which site-directed hydroxyl radical cleavage is performed to identify neighboring sites in the RNA. Repeating this process with independent RNAs containing unique MPE intercalation sites cumulatively provides information on tertiary interactions.

molecules. Currently, MPE is not commercially available, and its application to through-space cleavage has only been demonstrated with a well-characterized RNA (yeast tRNA^{Asp}). However, synthesis of MPE has been reported, and this strategy opens the intriguing possibility of developing "molecular rulers" by introducing linkers of different length between the intercalating and hydroxyl radical generating moieties.

7. Bringing It All Together-Determining Full Genome Structures by SHAPE

Most structural analyses have historically targeted small RNA motifs (<500 nt) in artificial contexts and, in the absence of complementary genetic and phylogenetic data, may not accurately relate their structures to the biology of the larger RNAs from which they were derived. In contrast, SHAPE provides an unprecedented opportunity to view an entire RNA molecule, giving the researcher the opportunity to connect simple elements to the components of larger RNA motifs. This concept has recently been exemplified through the application of SHAPE to decode the structure of the entire HIV-1 genome (~9750 nucleotides) at single-nucleotide resolution [27]. This seminal study determined that, although the HIV-1 genome is less structured than ribosomal RNA, it nonetheless contains independent RNA folding domains. Some functionally significant RNA motifs were shown to belong to the larger elements, an example of which is the *gag-pol* ribosomal frameshift signal, which constituted one component of a three-helix structure (P1-P2-P3). The slippery sequence forms one of the three helices (P2), while two others (P1 and P3) are stabilized by an anchoring stem with two bulges. Additional RNA elements were identified in protein-coding regions of the genome, from which it has been tentatively postulated that RNA structure constitutes an additional organizational level of the genetic code. Since many proteins appear to fold co-translationally, highly structured RNA might induce pausing of the translational machinery, promoting protein folding in a more native-like conformation. In contrast, highly unstructured regions were observed in hypervariable regions of the HIV-1 genome, which have important roles in viral host evasion. These unstructured regions were shown as separated from the rest of the genome by stable helices that have been proposed to function as structural "insulators."

The versatility of SHAPE extends to studying viral RNA not only in the context of the intact genome, but also at different biological states, providing information with respect to RNA conformational changes underlying different stages of viral life cycle. As an example, Wilkinson et al. [66] have provided structural information on the HIV-1 leader RNA in four biological states, namely (i) in vivo, (ii) ex vivo, where genomic RNA had been gently deproteinized, (iii) in vivo, but where important interactions between the nucleocapsid protein (NC) and genomic RNA had been compromised by covalent modification with aldrithiol-2 (AT-2 [67]), and (iv) genomic RNA prepared by in vitro transcription. This study concluded that the first 1000 nt of the HIV-1 genome exists in a single, predominant conformation in all four states. RNA of noncoding regions that regulate different steps of viral life cycle was distinguished by significantly lower sensitivity to acylation (predictive of secondary structure) than coding regions. A comparison of acylation profiles for the in vivo state with those following covalent modification by AT-2 defined several high affinity NC recognition sites, consistent with the role of this critical RNA chaperone in governing packaging of viral RNA. All NC binding sites were characterized by a G-rich single-stranded sequence flanked by stable helices. Additionally, RNA motifs where NC increases local flexibility were also identified, comprising single-stranded A/U-rich motifs adjacent to a duplex in which the first base pair includes a guanosine nucleotide. Collectively, this genome-probing approach suggests that local protein interactions can be organized by the long-range architecture of RNA. Although a limited region of the genome of the formerly known gammaretrovirus xenotropic murine leukemia virus related virus (XMRV) was examined using this strategy, it yielded similar conclusions on high affinity NC binding sites [68]. Future studies directed towards whole-genome structural analysis would, however, benefit from development of methods that enhanced SHAPE sensitivity, thereby reducing the culture volumes of potentially biohazardous material required. Efforts in this direction are discussed in the following section.

8. Increasing SHAPE Sensitivity for In Vivo Structure Analysis

In most instances, RNA structural analysis is performed on material either made synthetically or via in vitro transcription, where the amount of starting material is not a major consideration. Although in vivo and ex vivo analysis of the entire HIV-1 genome has been reported [27], this has required virus isolation from substantial culture volumes and is not readily adaptable to routine laboratory procedures. Thus, in circumstances where the amounts of biological material may be both biohazardous and limiting, methods of

increasing SHAPE sensitivity that have broader applicability would be a major advantage. Efforts in this direction are summarized below.

8.1. (i) SHAPE-Seq. Approximately 1–3 pmol of RNA is usually needed to accurately map a reactivity spectrum for any given RNA molecule [69]. This limits the application of SHAPE to biological samples for which significant amounts of RNA are available. The recently-described SHAPE-Seq technology provides a means of signal intensification to address this limitation [32]. This innovative methodology, which merges SHAPE with a multiplexed hierarchical bar coding and deep sequencing strategy, is outlined schematically in Figure 6.

Initially, input RNA templates are bar-coded with a unique sequence. Such barcodes comprise tetranucleotide sequences that are placed in the 3' structural cassette and introduced prior to in vitro transcription. Subsequently, these RNA templates are mixed and refolded under desired conditions. After folding, the mixture is divided into two pools, one of which is treated with modifying agent, while the second treated with a control solvent. Primer extension is subsequently performed with an end-labeled DNA primer tagged at the 5' end with tetranucleotide "handle" sequence. This handle allows the user to distinguish between cDNA fragments derived from the positive or control reactions. Additionally, the 5' tail of the reverse transcription primer contains an Illumina adapter necessary for paired-end sequencing. As a result, reverse transcription generates a bar-coded library of uniquely-sized cDNAs corresponding to stops at sites of adduct formation in the target RNA. The process is followed by hydrolysis of RNA and single-stranded (ss) cDNA ligation to incorporate the second Illumina adapter. Single-stranded cDNA ligation is achieved using a thermostable ligase (circLigase, Epicentre Biotechnologies, Madison, WI) and a blocking group on the 3' end of the adapter to prevent concatemerization [70]. Finally, 9 to 12 cycles of PCR, employing primers that bind to the Illumina adapter sequences, amplify the cDNA library before multiplex paired-end deep sequencing of primer extension products. Since the RNA modification position and the identity barcode are on opposite ends of the cDNA fragments, only 50 nucleotides need to be read on each terminus. After sequencing, the reads are separated first by handle sequence, then barcode, and subsequently aligned to probed RNAs.

When compared to conventional SHAPE, SHAPE-Seq permits rapid, fully-automated analysis and eliminates the necessity for manual, time-consuming data manipulations associated with quantification of fluorescently-labeled cDNAs by capillary electrophoresis. By ligating single-stranded cDNA products with 5' adapters followed by PCR-amplification, with minute amounts of RNA needed to generate the reactivity spectrum of a given RNA, SHAPE-Seq represents a more generally-applicable and sensitive technique studying RNA samples that are limiting, from a biohazardous source, or both. For example, it was shown for the RNase P specificity domain that with as little as 0.1 pmol

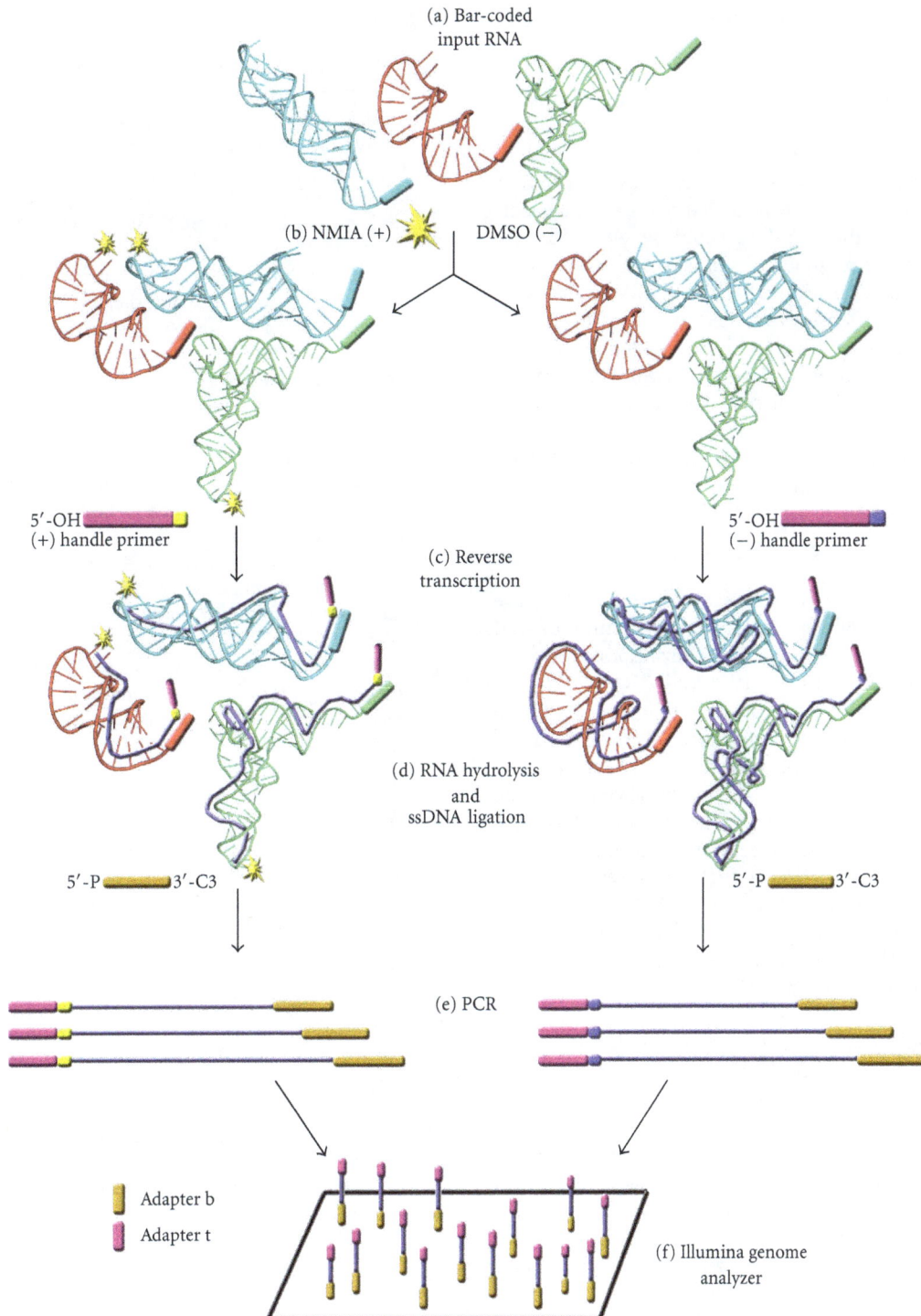

FIGURE 6: Summary of SHAPE-Seq methodology. (a) Input RNAs are bar-coded during *in vitro* transcription, followed by refolding under desired conditions and modification with SHAPE reagent (NMIA, 1M7). (b) The mixture is split into NMIA-treated and control pools. (c) Reverse transcription is performed with end-labeled primer containing a "handle" at the 5′ end and an Illumina adapter t. (d) The process is followed by hydrolysis of RNA and single-stranded (ss) cDNA ligation to incorporate the second Illumina adapter b. (e) After 9 to 12 cycles of PCR amplification, the cDNA library is analyzed by multiplex paired-end deep sequencing (f) adapted from [32].

of input RNA, SHAPE-Seq reactivities of over 800 bar-coded RNA species could be inferred [32]. SHAPE-Seq has the additional advantage of being able to simultaneously determine structural retformation from many RNAs through direct sequencing of the 3′ RNA bar codes. Although the

additional steps of SHAPE-Seq, (adapter ligation, PCR amplification, sequencing) might result in decreased sensitivity to some structural effects, as has been observed for the UUCG tetraloop of RNase P, this is offset with the ability of this technique to study structural changes involving interaction of various species within a population of RNA molecules.

8.2. (ii) Femtomole SHAPE. Using a two-color automated capillary electrophoresis with subfemtomole sensitivity, Grohman et al. [68] have recently reported *in vivo* analysis of a short portion of the formerly known XMRV genome. In contrast to earlier *in vivo* studies that required 1–3 pmole of input RNA, acylation profiles could be obtained with as little as 50 fmole aliquots of genomic RNA. As might be predicted, structural features of the XMRV leader RNA were similar to the extensively-studied counterpart Moloney murine leukemia virus, although binding sites unique to the XMRV nucleocapsid protein were proposed. More importantly, this study, which required in-house construction of a dedicated two-color capillary electrophoresis instrument, opens the exciting prospect of future functional studies on low abundance RNAs of clinical significance.

9. Future Perspectives

Rather than giving an exhaustive review of projects that have made use of SHAPE, which have included structures of wild type and mutant variants of the HIV-1 Rev response element [71], NC binding sites of the HIV-2 leader RNA [72], and RNA control of foamy virus protease activity [73], we have attempted here to highlight variations in this novel technology which facilitate interrogation of retroviral RNAs varying in size from 25–30 nt to intact, 9.5 kb retroviral genomes. The unequivocal benefit of this strategy is its ability to interrogate all four RNA bases with a single reagent, requiring thereafter simply fractionation of cDNA products. However, we should stress that SHAPE, while predictive of RNA structure, is best used with complementary genetic, phylogenetic, chemical modification (Pb^{2+} cleavage, ai-SHAPE and threading intercalators) and biophysical approaches (X-ray crystallography, NMR spectroscopy and small angle X-ray scattering). The benefits of capillary electrophoresis-based high throughput SHAPE must also be balanced by the demand this makers on the number of fluorescent oligonucleotide primers required for multiplexing, and the necessity for expensive instrumentation, features that also hold for femtomole SHAPE and SHAPE-Seq. Moreover, Kladwang and coworkers [74] compared SHAPE and crystallographic data for six RNAs and demonstrated significantly high (~20%) false negative and discovery rates, as well as several helix prediction errors, concluding that helix-by-helix confidence estimates may be critical for interpreting results from this powerful methodology. These issues notwithstanding, SHAPE should be seen as the beginning, and not the end, of an exciting path towards understanding the architecture of retroviral RNA genomes and the contribution this makes to biological function.

Acknowledgment

S. F. J. Le Grice and J. Sztuba-Solinska are supported by the Intramural Research Program of the National Cancer Institute, National Institutes of Health, USA.

References

[1] B. Berkhout, "Structural features in TAR RNA of human and simian immunodeficiency viruses: a phylogenetic analysis," *Nucleic Acids Research*, vol. 20, no. 1, pp. 27–31, 1992.

[2] W. Wilson, M. Braddock, S. E. Adams, P. D. Rathjen, S. M. Kingsman, and A. J. Kingsman, "HIV expression strategies: ribosomal frameshifting is directed by a short sequence in both mammalian and yeast systems," *Cell*, vol. 55, no. 6, pp. 1159–1169, 1988.

[3] E. Skripkin, J. C. Paillart, R. Marquet, B. Ehresmann, and C. Ehresmann, "Identification of the primary site of the human immunodeficiency virus type 1 RNA dimerization in vitro," *Proceedings of the National Academy of Sciences of the United States of America*, vol. 91, no. 11, pp. 4945–4949, 1994.

[4] A. Lever, H. Gottlinger, W. Haseltine, and J. Sodroski, "Identification of a sequence required for efficient packaging of human immunodeficiency virus type 1 RNA into virions," *Journal of Virology*, vol. 63, no. 9, pp. 4085–4087, 1989.

[5] B. R. Cullen, "Human immunodeficiency virus: nuclear RNA export unwound," *Nature*, vol. 433, no. 7021, pp. 26–27, 2005.

[6] S. F. J. Le Grice, " In the Beginning': initiation of minus strand DNA synthesis in retroviruses and LTR-containing retrotransposons," *Biochemistry*, vol. 42, no. 49, pp. 14349–14355, 2003.

[7] H. Donis Keller, A. M. Maxam, and W. Gilbert, "Mapping adenines, guanines, and pyrimidines in RNA," *Nucleic Acids Research*, vol. 4, no. 8, pp. 2527–2538, 1977.

[8] R. M. Wurst, J. N. Vournakis, and A. M. Maxam, "Structure mapping of 5'-32P-labeled RNA with S1 nuclease," *Biochemistry*, vol. 17, no. 21, pp. 4493–4499, 1978.

[9] D. A. Peattie, "Direct chemical method for sequencing RNA," *Proceedings of the National Academy of Sciences of the United States of America*, vol. 76, no. 4, pp. 1760–1764, 1979.

[10] D. A. Peattie and W. Gilbert, "Chemical probes for higher-order structure in RNA.," *Proceedings of the National Academy of Sciences of the United States of America*, vol. 77, no. 8, pp. 4679–4682, 1980.

[11] W. J. Krzyzosiak, T. Marciniec, M. Wiewiorowski, P. Romby, J. P. Ebel, and R. Giegé, "Characterization of the lead(II)-induced cleavages in tRNAs in solution and effect of the Y-base removal in yeast tRNAphe," *Biochemistry*, vol. 27, no. 15, pp. 5771–5777, 1988.

[12] J. Kjems, M. Brown, D. D. Chang, and P. A. Sharp, "Structural analysis of the interaction between the human immunodeficiency virus Rev protein and the Rev response element," *Proceedings of the National Academy of Sciences of the United States of America*, vol. 88, no. 3, pp. 683–687, 1991.

[13] B. Berkhout and L. Schoneveld, "Secondary structure of the HIV-2 leader RNA comprising the tRNA-primer binding site," *Nucleic Acids Research*, vol. 21, no. 5, pp. 1171–1178, 1993.

[14] J. C. Paillar, R. Marquet, E. Skripkin, B. Ehresmann, and C. Ehresmann, "Mutational analysis of the bipartite dimer linkage structure of human immunodeficiency virus type 1 genomic RNA," *Journal of Biological Chemistry*, vol. 269, no. 44, pp. 27486–27493, 1994.

[15] B. Berkhout, B. Klaver, and A. T. Das, "A conserved hairpin structure predicted for the poly(A) signal of human and simian immunodeficiency viruses," *Virology*, vol. 207, no. 1, pp. 276–281, 1995.

[16] G. Isel, C. Ehresmann, G. Keith, B. Ehresmann, and R. Marquet, "Initiation of reverse transcription of HIV-1: secondary structure of the HIV-1 RNA/tRNA3(Lys) (Template/Primer)

complex," *Journal of Molecular Biology*, vol. 247, no. 2, pp. 236–250, 1995.

[17] B. Berkhout, "Structure and function of the human immunodeficiency virus leader RNA," *Progress in Nucleic Acid Research and Molecular Biology*, vol. 54, pp. 1–34, 1996.

[18] B. Berkhout, "The primer binding site on the RNA genome of human and simian immunodeficiency viruses is flanked by an upstream hairpin structure," *Nucleic Acids Research*, vol. 25, no. 20, pp. 4013–4017, 1997.

[19] A. T. Das, B. Klaver, and B. Berkhout, "A hairpin structure in the R region of the human immunodeficiency virus type 1 RNA genome is instrumental in polyadenylation site selection," *Journal of Virology*, vol. 73, no. 1, pp. 81–91, 1999.

[20] F. Jossinet, J. C. Paillart, E. Westhof et al., "Dimerization of HIV-1 genomic RNA of subtypes A and B: RNA loop structure and magnesium binding," *RNA*, vol. 5, no. 9, pp. 1222–1234, 1999.

[21] B. Berkhout and J. L. B. Van Wamel, "The leader of the HIV-1 RNA genome forms a compactly folded tertiary structure," *RNA*, vol. 6, no. 2, pp. 282–295, 2000.

[22] J. S. Lodmell, C. Ehresmann, B. Ehresmann, and R. Marquet, "Structure and dimerization of HIV-1 kissing loop aptamers," *Journal of Molecular Biology*, vol. 311, no. 3, pp. 475–490, 2001.

[23] H. Huthoff and B. Berkhout, "Multiple secondary structure rearrangements during HIV-1 RNA dimerization," *Biochemistry*, vol. 41, no. 33, pp. 10439–10445, 2002.

[24] J. C. Paillart, M. Dettenhofer, X. F. Yu, C. Ehresmann, B. Ehresmann, and R. Marquet, "First snapshots of the HIV-1 RNA structure in infected cells and in virions," *Journal of Biological Chemistry*, vol. 279, no. 46, pp. 48397–48403, 2004.

[25] E. J. Merino, K. A. Wilkinson, J. L. Coughlan, and K. M. Weeks, "RNA structure analysis at single nucleotide resolution by Selective 2'-Hydroxyl Acylation and Primer Extension (SHAPE)," *Journal of the American Chemical Society*, vol. 127, no. 12, pp. 4223–4231, 2005.

[26] S. A. Mortimer and K. M. Weeks, "A fast-acting reagent for accurate analysis of RNA secondary and tertiary structure by SHAPE chemistry," *Journal of the American Chemical Society*, vol. 129, no. 14, pp. 4144–4145, 2007.

[27] J. M. Watts, K. K. Dang, R. J. Gorelick et al., "Architecture and secondary structure of an entire HIV-1 RNA genome," *Nature*, vol. 460, no. 7256, pp. 711–716, 2009.

[28] S. A. Mortimer and K. M. Weeks, "Time-resolved RNA SHAPE chemistry: quantitative RNA structure analysis in one-second snapshots and at single-nucleotide resolution," *Nature Protocols*, vol. 4, no. 10, pp. 1413–1421, 2009.

[29] K. B. Turner, Y. Y. B. Hye, R. G. Brinson, J. P. Marino, D. Fabris, and S. F. J. Le Grice, "SHAMS: combining chemical modification of RNA with mass spectrometry to examine polypurine tract-containing RNA/DNA hybrids," *RNA*, vol. 15, no. 8, pp. 1605–1613, 2009.

[30] M. Legiewicz, A. S. Zolotukhin, G. R. Pilkington et al., "The RNA transport element of the murine musD retrotransposon requires long-range intramolecular interactions for function," *Journal of Biological Chemistry*, vol. 285, no. 53, pp. 42097–42104, 2010.

[31] J. C. Kenyon et al., "SHAPE analysis of the FIV Leader RNA reveals a structural switch potentially controlling viral packaging and genome dimerization," *Nucleic Acids Research*, vol. 39, no. 15, pp. 6692–6704, 2011.

[32] J. B. Lucks, S. A. Mortimer, C. Trapnell et al., "Multiplexed RNA structure characterization with selective 2'-hydroxyl acylation analyzed by primer extension sequencing (SHAPE-Seq)," *Proceedings of the National Academy of Sciences of the*

United States of America, vol. 108, no. 27, pp. 11063–11068, 2011.

[33] K. A. Steen, A. Malhotra, and K. M. Weeks, "Selective 2'-hydroxyl acylation analyzed by protection from exoribonuclease," *Journal of the American Chemical Society*, vol. 132, no. 29, pp. 9940–9943, 2010.

[34] M. S. Lalonde, Y. Zuo, J. Zhang et al., "Exoribonuclease R in Mycoplasma genitalium can carry out both RNA processing and degradative functions and is sensitive to RNA ribose methylation," *RNA*, vol. 13, no. 11, pp. 1957–1968, 2007.

[35] J. W. Rausch, D. Lener, J. T. Miller, J. G. Julias, S. H. Hughes, and S. F. J. Le Grice, "Altering the RNase H primer grip of human immunodeficiency virus reverse transcriptase modifies cleavage specificity," *Biochemistry*, vol. 41, no. 15, pp. 4856–4865, 2002.

[36] M. D. Powell et al., "Residues in the alphaH and alphaI helices of the HIV-1 reverse transcriptase thumb subdomain required for the specificity of RNase H-catalyzed removal of the polypurine tract primer," *Journal of Biological Chemistry*, vol. 274, no. 28, pp. 19885–19893, 1999.

[37] J. W. Rausch and S. F. J. Le Grice, "Substituting a conserved residue of the ribonuclease H domain alters substrate hydrolysis by retroviral reverse transcriptase," *Journal of Biological Chemistry*, vol. 272, no. 13, pp. 8602–8610, 1997.

[38] J. W. Rausch and S. F. J. Le Grice, "Purine analog substitution of the HIV-1 polypurine tract primer defines regions controlling initiation of plus-strand DNA synthesis," *Nucleic Acids Research*, vol. 35, no. 1, pp. 256–268, 2007.

[39] H. Y. Yi-Brunozzi and S. F. J. Le Grice, "Investigating HIV-1 polypurine tract geometry via targeted insertion of abasic lesions in the (-)-DNA template and (+)-RNA primer," *Journal of Biological Chemistry*, vol. 280, no. 20, pp. 20154–20162, 2005.

[40] C. Dash, J. W. Rausch, and S. F. J. Le Grice, "Using pyrrolodeoxycytosine to probe RNA/DNA hybrids containing the human immunodeficiency virus type-1 3' polypurine tract," *Nucleic Acids Research*, vol. 32, no. 4, pp. 1539–1547, 2004.

[41] J. W. Rausch, J. Qu, H. Y. Yi-Brunozzi, E. T. Kool, and S. F. J. Le Grice, "Hydrolysis of RNA/DNA hybrids containing nonpolar pyrimidine isosteres defines regions essential for HIV type 1 polypurine tract selection," *Proceedings of the National Academy of Sciences of the United States of America*, vol. 100, no. 20, pp. 11279–11284, 2003.

[42] D. Lener, M. Kvaratskhelia, and S. F. J. Le Grice, "Nonpolar thymine isosteres in the Ty3 polypurine tract DNA template modulate processing and provide a model for its recognition by Ty3 reverse transcriptase," *Journal of Biological Chemistry*, vol. 278, no. 29, pp. 26526–26532, 2003.

[43] S. G. Sarafianos, K. Das, C. Tantillo et al., "Crystal structure of HIV-1 reverse transcriptase in complex with a polypurine tract RNA:DNA," *EMBO Journal*, vol. 20, no. 6, pp. 1449–1461, 2001.

[44] H. Y. Yi-Brunozzi, D. M. Brabazon, D. Lener, S. F. J. Le Grice, and J. P. Marino, "A ribose sugar conformational switch in the LTR-retrotransposon Ty3 polypurine tract-containing RNA/DNA hybrid," *Journal of the American Chemical Society*, vol. 127, no. 47, pp. 16344–16345, 2005.

[45] M. Kvaratskhelia, S. R. Budihas, and S. F. J. Le Grice, "Preexisting distortions in nucleic acid structure aid polypurine tract selection by HIV-1 reverse transcriptase," *Journal of Biological Chemistry*, vol. 277, no. 19, pp. 16689–16696, 2002.

[46] M. Haddrick, A. L. Lear, A. J. Cann, and S. Heaphy, "Evidence that a kissing loop structure facilitates genomic

RNA dimerisation in HIV-1," *Journal of Molecular Biology*, vol. 259, no. 1, pp. 58–68, 1996.

[47] Y. Song, P. Friebe, E. Tzima, C. Jünemann, R. Bartenschlager, and M. Niepmann, "The hepatitis C virus RNA 3′-untranslated region strongly enhances translation directed by the internal ribosome entry site," *Journal of Virology*, vol. 80, no. 23, pp. 11579–11588, 2006.

[48] S. Gago, M. De La Peña, and R. Flores, "A kissing-loop interaction in a hammerhead viroid RNA critical for its in vitro folding and in vivo viability," *RNA*, vol. 11, no. 7, pp. 1073–1083, 2005.

[49] H. L. Townsend, B. K. Jha, R. H. Silverman, and D. J. Barton, "A putative loop E motif and an H-H kissing loop interaction are conserved and functional features in a group C enterovirus RNA that inhibits ribonuclease L," *RNA Biology*, vol. 5, no. 4, pp. 263–272, 2008.

[50] S. K. Jang, H. G. Krausslich, M. J. H. Nicklin, G. M. Duke, A. C. Palmenberg, and E. Wimmer, "A segment of the 5′ nontranslated region of encephalomyocarditis virus RNA directs internal entry of ribosomes during in vitro translation," *Journal of Virology*, vol. 62, no. 8, pp. 2636–2643, 1988.

[51] A. S. Zolotukhin, H. Uranishi, S. Lindtner, J. Bear, G. N. Pavlakis, and B. K. Felber, "Nuclear export factor RBM15 facilitates the access of DBP5 to mRNA," *Nucleic Acids Research*, vol. 37, no. 21, pp. 7151–7162, 2009.

[52] E. Hiriart, H. Gruffat, M. Buisson et al., "Interaction of the Epstein-Barr virus mRNA export factor EB2 with human Spen proteins SHARP, OTT1, and a novel member of the family, OTT3, links Spen proteins with splicing regulation and mRNA export," *Journal of Biological Chemistry*, vol. 280, no. 44, pp. 36935–36945, 2005.

[53] I. Tretyakova, A. S. Zolotukhin, W. Tan et al., "Nuclear export factor family protein participates in cytoplasmic mRNA trafficking," *Journal of Biological Chemistry*, vol. 280, no. 36, pp. 31981–31990, 2005.

[54] C. Lu, F. Ding, A. Chowdhury et al., "SAM recognition and conformational switching mechanism in the Bacillus subtilis yitJ S box/SAM-I riboswitch," *Journal of Molecular Biology*, vol. 404, no. 5, pp. 803–818, 2010.

[55] A. Haller, M. F. Souliere, and R. Micura, "The dynamic nature of RNA as key to understanding riboswitch mechanisms," *Accounts of Chemical Research*, vol. 44, no. 12, pp. 1339–1348, 2011.

[56] Q. Vicens, E. Mondragon, and R. T. Batey, "Molecular sensing by the aptamer domain of the FMN riboswitch: a general model for ligand binding by conformational selection," *Nucleic Acids Research*, vol. 39, no. 19, pp. 8586–8598, 2011.

[57] T. E. M. Abbink and B. Berkhout, "A novel long distance base-pairing interaction in human immunodeficiency virus type 1 rna occludes the gag start codon," *Journal of Biological Chemistry*, vol. 278, no. 13, pp. 11601–11611, 2003.

[58] T. E. M. Abbink, M. Ooms, P. C. J. Haasnoot, and B. Berkhout, "The HIV-1 leader RNA conformational switch regulates RNA dimerization but does not regulate mRNA translation," *Biochemistry*, vol. 44, no. 25, pp. 9058–9066, 2005.

[59] J. C. Kenyon, A. Ghazawi, W. K. S. Cheung, P. S. Phillip, T. A. Rizvi, and A. M. L. Lever, "The secondary structure of the 52 end of the FIV genome reveals a long-range interaction between R/U5 and gag sequences, and a large, stable stem-loop," *RNA*, vol. 14, no. 12, pp. 2597–2608, 2008.

[60] T. A. Rizvi, J. C. Kenyon, J. Ali et al., "Optimal packaging of FIV genomic RNA depends upon a conserved long-range interaction and a palindromic sequence within gag," *Journal of Molecular Biology*, vol. 403, no. 1, pp. 103–119, 2010.

[61] M. W. Dyke and P. B. Dervan, "Methidiumpropyl-EDTA-Fe(II) and DNase I footprinting report different small molecule binding site sizes on DNA," *Nucleic Acids Research*, vol. 11, no. 16, pp. 5555–5567, 1983.

[62] J. M. Kean, S. A. White, and D. E. Draper, "Detection of high-affinity intercalator sites in a ribosomal RNA fragment by the affinity cleavage intercalator methidiumpropyl-EDTA-iron(II)," *Biochemistry*, vol. 24, no. 19, pp. 5062–5070, 1985.

[63] N. V. Jammi and P. A. Beal, "Phosphorylation of the RNA-dependent protein kinase regulates its RNA-binding activity," *Nucleic Acids Research*, vol. 29, no. 14, pp. 3020–3029, 2001.

[64] C. M. Gherghe, C. W. Leonard, F. Ding, N. V. Dokholyan, and K. M. Weeks, "Native-like RNA tertiary structures using a sequence-encoded cleavage agent and refinement by discrete molecular dynamics," *Journal of the American Chemical Society*, vol. 131, no. 7, pp. 2541–2546, 2009.

[65] R. P. Hertzberg and P. B. Dervan, "Cleavage of DNA with methidiumpropyl-EDTA-iron(II): reaction conditions and product analyses," *Biochemistry*, vol. 23, no. 17, pp. 3934–3945, 1984.

[66] K. A. Wilkinson, R. J. Gorelick, S. M. Vasa et al., "High-throughput SHAPE analysis reveals structures in HIV-1 genomic RNA strongly conserved across distinct biological states," *PLoS Biology*, vol. 6, no. 4, article no. e96, 2008.

[67] J. L. Rossio, M. T. Esser, K. Suryanarayana et al., "Inactivation of human immunodeficiency virus type 1 infectivity with preservation of conformational and functional integrity of virion surface proteins," *Journal of Virology*, vol. 72, no. 10, pp. 7992–8001, 1998.

[68] J. K. Grohman, S. Kottegoda, R. J. Gorelick, N. L. Allbritton, and K. M. Weeks, "Femtomole SHAPE reveals regulatory structures in the authentic XMRV RNA genome," *Journal of the American Chemical Society*, vol. 133, no. 50, pp. 20326–20334, 2011.

[69] K. A. Wilkinson, E. J. Merino, and K. M. Weeks, "Selective 2′-hydroxyl acylation analyzed by primer extension (SHAPE): Quantitative RNA structure analysis at single nucleotide resolution," *Nature Protocols*, vol. 1, no. 3, pp. 1610–1616, 2006.

[70] T. W. Li and K. M. Weeks, "Structure-independent and quantitative ligation of single-stranded DNA," *Analytical Biochemistry*, vol. 349, no. 2, pp. 242–246, 2006.

[71] M. Legiewicz, C. S. Badorrek, K. B. Turner et al., "Resistance to RevM10 inhibition reflects a conformational switch in the HIV-1 Rev response element," *Proceedings of the National Academy of Sciences of the United States of America*, vol. 105, no. 38, pp. 14365–14370, 2008.

[72] K. J. Purzycka, K. Pachulska-Wieczorek, and R. W. Adamiak, "The in vitro loose dimer structure and rearrangements of the HIV-2 leader RNA," *Nucleic Acids Research*, vol. 39, no. 16, pp. 7234–7248, 2011.

[73] M. J. Hartl, J. Bodem, F. Jochheim, A. Rethwilm, P. Rosch, and B. M. Wöhrl, "Regulation of foamy virus protease activity by viral RNA: a novel and unique mechanism among retroviruses," *Journal of Virology*, vol. 85, no. 9, pp. 4462–4469, 2011.

[74] W. Kladwang, C. C. VanLang, P. Cordero, and R. Das, "Understanding the errors of SHAPE-directed RNA structure modeling," *Biochemistry*, vol. 50, no. 37, pp. 8049–8056, 2011.

15

Modulator of Apoptosis 1: A Highly Regulated RASSF1A-Interacting BH3-Like Protein

Jennifer Law,[1] Victor C. Yu,[2] and Shairaz Baksh[1]

[1] Department of Pediatrics, Faculty of Medicine and Dentistry, University of Alberta, 3055 Katz Group Centre for Pharmacy and Health Research, 113 Street 87 Avenue, Edmonton, AB, Canada T6G 2E1
[2] Department of Pharmacy, Faculty of Science, National University of Singapore, 18 Science Drive 4, Singapore 117543

Correspondence should be addressed to Shairaz Baksh, sbaksh@ualberta.ca

Academic Editor: Dae-Sik Lim

Modulator of apoptosis 1 (MOAP-1) is a BH3-like protein that plays key roles in both the intrinsic and extrinsic modes of cell death or apoptosis. MOAP-1 is part of the Ras association domain family 1A (RASSF1A)/MOAP-1 pro-apoptotic extrinsic signaling pathway that regulates apoptosis by utilizing death receptors such as tumor necrosis factor α (TNFα) or TNF-related apoptosis-inducing ligand (TRAIL) to inhibit abnormal growth. RASSF1A is a bona fide tumor suppressor gene that is epigenetically silenced by promoter-specific methylation in numerous human cancers. MOAP-1 is a downstream effector of RASSF1A that promotes Bax activation and cell death and is highly regulated during apoptosis. We speculate that MOAP-1 and RASSF1A are important elements of an "apoptotic checkpoint" that directly influences the outcome of cell death. The failure to regulate this pro-apoptotic pathway may result in the appearance of cancer and possibly other disorders. Although loss of RASSF1A expression is frequently observed in human cancers, it is currently unknown if MOAP-1 expression may also be affected during carcinogenesis to result in uncontrolled malignant growth. In this article, we will summarize what is known about the biological role(s) of MOAP-1 and how it functions as a downstream effector to RASSF1A.

1. Introduction

Cancer is a disease of uncontrolled cell proliferation and is the third leading causing of death worldwide following cardiovascular and infectious diseases [2]. The abnormal proliferation of cells during cancer development results from a multistep process involving the deregulation of genes that promote cell growth (oncogenes) and those that normally function to restrain growth (tumor suppressors). Interestingly, approximately 90% of the genes that are associated with cancer development have now been identified as being tumor suppressors [3]. Moreover, many of these growth inhibitory genes encode proteins that are involved in cell death. RASSF1A has multiple biological functions including the regulation of Bax-mediated cell death [4–6]. MOAP-1, a highly regulated pro-apoptotic protein, serves a critical role during mitochondrial-dependent apoptosis by influencing and sustaining Bax activation [7, 8]. In this review, we will discuss how MOAP-1 is regulated and how it serves as a pivotal RASSF1A effector protein to regulate cell death.

2. Apoptosis: A Regulated Biological Process to Modulate Growth

A well-known mechanism of tumor suppression is the elimination of unwanted cells through a sequence of events known as *apoptosis* [9]. The significance of apoptosis in metazoan biology is highlighted by the number of diseases that are associated with its deregulation [10]. Apoptosis plays a critical role during the development of multicellular organisms and adult tissue homeostasis and is vital to the removal of damaged or dangerous cells. It can be initiated through two main pathways in response to intracellular or extracellular signals of cell death [11]. The intrinsic apoptotic signaling pathway is activated in response to a diverse set of signals originating from within cells due to cellular stresses such as DNA damage, hypoxia, toxins, or starvation [12]. In contrast, the extrinsic pathway of cell death is activated by the binding of death-inducing ligands to death receptors.

Activation of the extrinsic apoptotic signaling pathway occurs through cell surface death receptor/ligand combinations that include TNF-R1/TNFα, Fas receptor (R) (CD95/ APO-1)/Fas ligand, as well as TRAIL-R (1/2)/TRAIL [13]. Activated death receptors trigger a series of events resulting in the formation of trimeric receptor complexes and the death-inducing signaling complex (DISC) [14]. DISC assembly and subsequent activation of initiator caspases (mainly caspase-8) convey signals to the mitochondria to promote the release of small molecules (such as cytochrome c) from the mitochondrial matrix into the cytosol and the assembly of the apoptosome complex to activate downstream effector caspases (such as caspase-3) [15]. Intrinsic pathway stimulation can also lead to cytochrome c release and activation of effector caspases. Once activated, effector caspases cleave several nuclear proteins [such as lamin B and poly(ADP-ribose) polymerase] and activate specific DNA endonucleases. These events result in many of the biochemical and morphological changes observed during apoptosis, including nuclear and cytoplasmic breakdown.

Mitochondria play an important role in the induction of apoptosis through the release of proteins that promote caspase activation and the breakdown of cellular components [16]. Regulation of the mitochondrial events during apoptosis is controlled by proteins of the B-cell lymphoma-2 (Bcl-2) family and is composed of three different subgroups known as the anti-apoptotic, multidomain pro-apoptotic and BH3-only proteins [12, 17]. The anti-apoptotic and BH3-only proteins are involved in inhibiting or promoting the function of multi-domain pro-apoptotic molecules, respectively. In contrast, it is members of the multidomain subgroup that are directly responsible for the mitochondrial outer membrane permeabilization that occurs during apoptosis [18, 19]. Two members from this group, Bax and Bak, are required for apoptosis to occur [20]. Although the functions of Bax and Bak are closely regulated by its Bcl-2 family members, it is now known that, for at least Bax activation, other proteins may also be involved in its modulation. One of these molecules is the RASSF1A-binding protein, MOAP-1. RASSF1A functions to "open" MOAP-1 to allow for MOAP-1-induced Bax conformational change by exposing the epitope, [12]GPTSSEQIMKTGA[24], and allowing for the subsequent insertion of Bax into the mitochondrial membrane. Once inserted, Bax can cooperatively drive cell death in association with Bak [21].

3. Ras Association Domain Family

RASSF1A is a bona fide tumor suppressor molecule that serves as the founding member of the RASSF group of proteins [22]. Currently, the RASSF protein family is comprised of ten different members known as RASSF1–10 that each share the presence of a Ras association (RA) domain within its primary amino acid sequence [23–26]. Of this protein family, RASSF1 is the most thoroughly characterized and studied thus far. A loss or decrease in RASSF1A expression is frequently observed in a wide range of human cancers due to epigenetic transcriptional silencing [27–30].

The tumor suppressor functions of RASSF1A include the ability to regulate microtubule dynamics [31–33], mitosis [32, 34–37], and apoptosis [5, 6, 38–41]. Due to the

particular focus of this paper, we will only discuss in detail what is known about RASSF1A-dependent cell death involving MOAP-1. It is now known that several pro-apoptotic pathways can be modulated by RASSF1A. One such pathway for the induction of RASSF1A-mediated apoptosis involves protein interactions with the Hippo signaling components, serine/threonine kinases mammalian Ste20-like (MST) 1 and 2 (reviewed separately in this issue). The Hippo pathway is a conserved signaling pathway essential for organ growth regulation in *Drosophila* and vertebrates [42]. Currently, there is evidence to support the role for RASSF1A in modulating the kinase activity of MST1/2 and thus MST1/2-mediated cell death [38, 39]. RASSF1A can also induce apoptosis through an MST2-specific pathway by releasing MST2 from its inhibitor, Raf1, and allowing for large tumor suppressor homology (*Drosophila*) (LATS)1-mediated activation of the transcriptional regulator Yes-associated protein (YAP)1 [41]. In turn, YAP1 can translocate to the nucleus and associate with the p73 transcription factor in order to induce the transcription of pro-apoptotic gene p53-upregulated modulator of apoptosis (*PUMA*) to aid in Hippo-mediated cell death.

A second pathway involves MOAP-1. In response to death receptor signaling involving TNFα or TRAIL, RASSF1A can associate with MOAP-1 in order to promote Bax conformational change, translocation and integration into the mitochondrial membrane to perturb mitochondrial permeability [5, 6]. This is followed by the release of cytochrome c to activate downstream caspases and to promote nuclear and cytoplasmic breakdown. Furthermore, we speculate that MOAP-1 may cooperate with RASSF1A to promote tumor suppression. RASSF1A has been extensively reviewed in the literature. In contrast, there are currently no reviews that specifically address what is known about the biology of MOAP-1. Indeed, MOAP-1 remains separate from the canonical group of Bax-regulatory molecules and therefore has not garnered as much attention as the proteins of the Bcl-2 family. In the remainder of this review, we will document what is currently known about MOAP-1 and will discuss evidence providing insight into the complexities of this protein and its biological function(s).

4. Modulator of Apoptosis 1: A Brief History

MOAP-1 was first reported as a mitochondria-enriched 39.5 kDa molecule that was first identified as a novel Bax-associating protein in a yeast two-hybrid screen [7]. Located at genetic locus 14q32 (Figure 1), MOAP-1 is a negatively charged protein that contains 351 amino acid residues in humans and an isoelectric point (pI) of 4.939 at pH 7.0 (Ensembl protein ID: ENST00000298894). *MOAP-1* is highly conserved in chimpanzee (*Pan troglodytes*), rat (*Rattus norvegicus*) and mouse (*Mus musculus*), and its coding sequence is contained within a single exon in both mouse and humans (Figure 2). Since its discovery in 2001, research has established a central role for MOAP-1 in both mitochondrial and death receptor-mediated apoptosis [5, 8]. When overexpressed in mammalian cells, MOAP-1 induces caspase-dependent apoptosis whereas MOAP-1

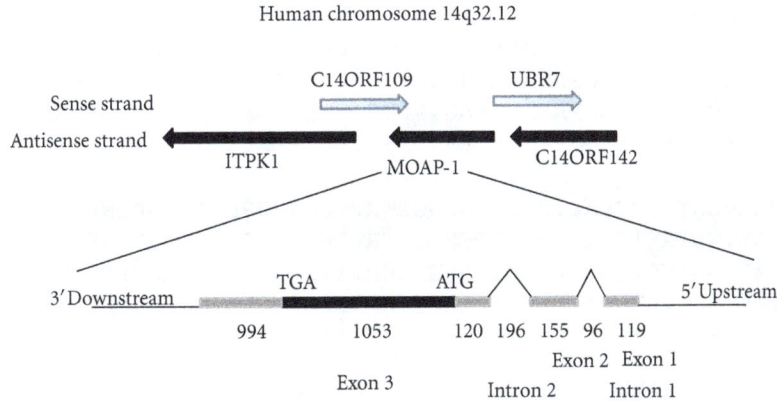

FIGURE 1: Gene structure of human MOAP-1. The entire protein coding sequence of MOAP-1 is contained within exon 3 and is located on the anti-sense strand of chromosome 14. Genbank accession: NM_022151.4. More information can be found at http://www.ncbi.nlm.nih.gov/gene/64112. Numbers below schematic denote the size of the intron or exon.

knockdown cells are resistant to a variety of apoptotic stimuli including staurosporine, serum withdrawal, UV irradiation, TNFα, and TRAIL [8]. Altogether, these results demonstrate the importance of MOAP-1 in apoptosis and functions as a key effector of Bax conformational change and activation.

5. MOAP-1 Expression in Normal and Cancer Cells

MOAP-1 is a ubiquitously expressed protein that is present at moderate levels under normal cellular conditions and is constitutively degraded by the ubiquitin-proteasome system [7, 43]. Given that RASSF1A expression is frequently lost during carcinogenesis and Bax is mutated in a large percentage of gastrointestinal and colorectal cancers, it is plausible that MOAP-1 expression and/or function may also be regulated during cancer development [29, 44, 45]. Indeed, immunohistochemical analysis of MOAP-1 performed over a wide range of human cancer tissues demonstrates either a negative or a weak staining pattern for this protein (Table 1 and please see site http://www.proteinatlas.org/search/moap1 under "moap1 or pnma4" for immunohistochemical pictures of MOAP-1 staining in numerous cancer cells). In support of this immunohistochemical data, we have also found a loss or reduction of MOAP-1 expression in an extensive panel of cancer cell lines ranging from breast, brain, lung, skin and blood cancers [Law et al., unpublished observations]. Furthermore, in a classical xenograft assay, both RASSF1A and MOAP-1 can suppress tumor formation in HCT116 colon cancer cells suggesting tumor suppressor function (Figure 3) and functional importance for both genes in growth inhibition in normal cells.

Currently, the mechanism responsible for the loss of MOAP-1 expression in cancer cells remains unknown. It is possible that expression changes in MOAP-1 may arise by promoter specific epigenetic methylation, by miRNA/siRNA regulation of the mRNA, and/or by alterations in MOAP-1 protein stability due to ubiquitin-directed proteolysis. The *MOAP-1* promoter displays 17 potential CpG islands that

TABLE 1: Summary of MOAP-1 staining patterns in human malignant tissues. Data source was The Human Protein Atlas (http://www.proteinatlas.org/search/moap1). Antibody used for all MOAP-1 immunohistochemistry: Sigma-Aldrich HPA000939.

Cancer tissue type	MOAP-1 staining pattern
Colorectal cancer	Weak
Breast cancer	Negative
Prostate cancer	Negative
Ovarian cancer	Negative
Cervical cancer	Negative
Endometrial cancer	Negative
Malignant carcinoid	Negative
Head and neck cancer	Negative
Thyroid cancer	Negative
Malignant glioma	Weak
Malignant lymphoma	Negative
Lung cancer	Weak
Malignant melanoma	Negative
Skin cancer	Negative
Testis cancer	Moderate
Urothelial cancer	Negative
Renal cancer	Negative
Stomach cancer	Weak
Pancreatic cancer	Negative
Liver cancer	Negative

may be epigenetically modified to result in loss of gene expression, as suggested using MethPrimer online software [46]. To date, no miRNA or siRNA has been identified for MOAP-1 although we suspect that specific miRNA(s) may exist to reduce or shut down MOAP-1 expression. The last potential mechanism regulating MOAP-1 expression is posttranslational modification by ubiquitination and degradation by the proteasomal degradation machinery [43]. Future investigations will be required in order to understand

```
hMOAP-1   MTLRLLEDWCRGMDMNPRKALLIAGISQSCSVAEIEEALQAGLAPLGEYRLLGRMFRRDE   60
cMOAP-1   MTLRLLEDWCRGMDMNPRKALLIAGISQSCSVAEIEEALQAGLAPLGEYRLLGRMFRRDE   60
mMOAP-1   MTLRLLEDWCRGMDMNPRKALLVAGIPPTCGVADIEEALQAGLAPLGEHRLLGRMFRRDE   60
rMOAP-1   MTLRLLEDWCRGMDMNPRKALLVAGIPPTCGVADIEEALQVGLAPLGEHRLLGRMFRRDE   60
          *********************:***.  :*.**:*****.*******:**********

hMOAP-1   NRKVALVGLTAETSHALVPKEIPGKGGIWRVIFKPPDPDNTFLSRLNEFLAGEGMTVGEL   120
cMOAP-1   NRKVALVGLTAETSHALVPKEIPGKGGIWRVIFKPPDPDNTFLSRLNEFLAGEGMTVGEL   120
mMOAP-1   NKNVALIGLTVETGSALVPKEIPAKGGVWRVIFKPPDTDSDFLCRLNEFLKGEGMTMGEL   120
rMOAP-1   NKNVALVGLTVETGSALVPKEIPAKGGVWRVIFKPPDADSDFLCRLNEFLKGEGMTMGEL   120
          *::***:***.**. ********.***:*********.*. **.****** *****:***

hMOAP-1   SRALGHENGSLDPEQG-MIPEMWAPMLAQAL-EALQPALQCLKYKKLRVFSGRESPEPGE   178
cMOAP-1   TRALGHENGSLDPEQG-MVPEMWAPMLAQAL-EALQPALQCLKYKKLRVFSGRESPEPGE   178
mMOAP-1   TRVLGNRNDPLGLDPGIMIPEIRAPMLAQALNEALKPTLQYLRYKKLSVFSGRDPPGPGE   180
rMOAP-1   TRVLGNRNDPLCLDQNVMIPEIRAPMLAQALDEALKPTLQYLRYKKLSVFSGRDPPGPGE   180
          :*.**:.*..*   : .  *:**: ******* ***:*:** *:**** *****:..* ***

hMOAP-1   EEFGRWMFHTTQMIKAWQVPDVEKRRRLLESLRGPALDVIRVLKINNPLITVDECLQALE   238
cMOAP-1   EEFGRWMFHTTQMIKAWQVPDVEKRRRLLESLRGPALDVIRVLKINNPLITVDECLQALE   238
mMOAP-1   EEFESWMFHTSQVMKTWQVSDVEKRRRLIESLRGPAFEIIRVLKINNPFITVAECLKTLE   240
rMOAP-1   EEFESWMFHTSQVMKTWQVSDVEKRRRLMESLRGPAFEIIRVLKINNPFITVAECLKTLE   240
          ***    *****:*::*:***.********.*******::.:*********:*** ***::**

hMOAP-1   EVFGVTDNPRELQVKYLTTYQKDEEKLSAYVLRLEPLLQKLVQRGAIERDAVNQARLDQV   298
cMOAP-1   EVFGVTDNPRELQVKYLTTYQKDEEKLSAYVLRLEPLLQKLVQRGAIERDAVNQARLDQV   298
mMOAP-1   TIFGIIDNPRALQVKYLTTYQKTDEKLSAYVLRLEPLLQKLVQKGAIEKEVVNQARLDQV   300
rMOAP-1   TIFGIIDNPRALQVRYLTTYQKSGEKLSAYVLRLEPLLQKLVQKGAIEKEVVNQARLDQV   300
          :**: **** ***:******* ****************:****:. .*********

hMOAP-1   IAGAVHKTIRRELNLPEDGPAPGFLQLLVLIKDYEAAEEEEALLQAILEGNFT   351
cMOAP-1   IAGAVHKTIRRELNLPEDGPAPGFLQLLVLIKDYEAAEEEEALLQAVLEGNFT   351
mMOAP-1   IAGAVHKSVRRELGLPEGSPAPGLLQLLTLIKDKEAEEEE-VLLQAELEGYCT   352
rMOAP-1   IAGAIHKSVRRELGLPEGSPAPGLLQLLTLIKDKEAEEEE-VLLQAELEGHFT   352
          ****:**::****.***..****:****.****.**.***.:****.***. *
```

(a)

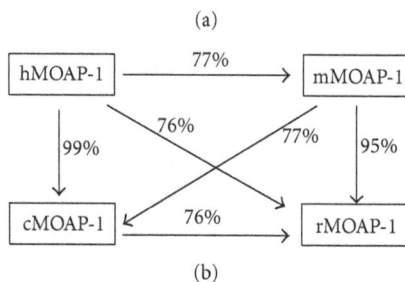

(b)

FIGURE 2: A comparison of MOAP-1 orthologs. (a) Multiple sequence alignments of MOAP-1 orthologs present in human (h), mouse (m), rat (r) and chimpanzee (c). Sequence alignments were performed using ClustalW2. NCBI reference sequences (mRNA and protein): NM_022151.4 and NP_071434.2 (human); NM_022323.7 and NP_071718.1 (mouse); NM_001013101.1 and NP_001013119.1 (rat); XM_510137.3 and XP_510137 (chimpanzee). (b) Percent amino acid identity between MOAP-1 orthologs calculated based on sequence alignments in (2a). Analysis carried out using ClustalW2.

the ubiquitination of MOAP-1 and the biological outcome of these ubiquitination events.

Like most disease-associated genes, polymorphisms may exist to result in the loss of the encoded protein function.

Single nucleotide polymorphisms (SNPs) of MOAP-1 have been documented in two databases suggesting disease-associated changes [47, 48]. Although the population distribution has not been determined as of yet, two somatically

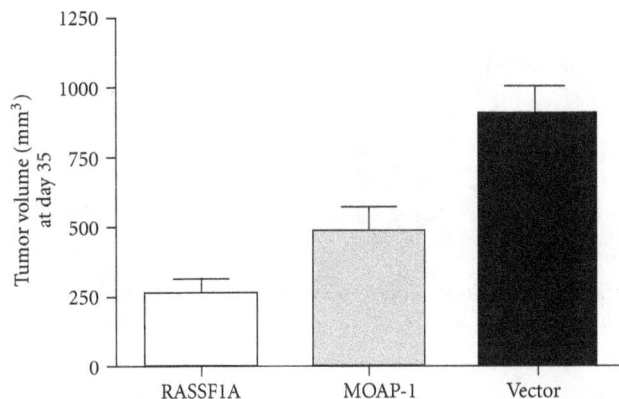

FIGURE 3: Tumor inhibiting potential of the RASSF1A/MOAP-1 tumor suppressor pathway. A classical xenograft assay was carried out. Male athymic nude mice were injected subcutaneously with 1×10^6 transiently transfected HCT116 cells mixed with matrigel mix into the right and left flank areas. Tumor volumes were measured until day 35 and plotted. P values for MOAP-1 versus vector (0.019); RASSF1A versus vector (0.0001); MOAP-1 versus RASSF1A (0.02), $n = 12$–14. Statistical analysis was evaluated by Student's t-test (two-tailed). Protein expression at the time of subcutaneous injection was confirmed by immunoblotting (data not shown). Protein expression in HCT116 cells can be detected up to 10 days post-transfection. However, at the end of experiment, we could not detect protein expression of HA-RASSF1A or Myc-MOAP-1 in the resulting tumors. We argue that the growth properties of HCT116 cells containing the indicated expression constructs were programmed within the first 7–10 days and continued on that program even though expression detection of the indicated genes was not possible. Please refer to [1] for more details on this issue.

derived SNPs (resulting in a predicted amino acid change) have been observed in melanoma patients—a proline to serine change at amino acid 79 (P79S with a nucleotide change of CCT → TCT) and an alanine to aspartic acid change at position 335 (A335D with a nucleotide change of GCT → GAT) [49]. Interestingly, the P79S polymorphism may suggest the creation of a potentially novel serine phosphorylation site to affect the cell death properties of MOAP-1, whereas A335D amino acid change would affect the TNF-R1-binding site on MOAP-1 (please see Figure 4). Further verification of these SNPs is warranted with respect to penetrance within the normal and disease groups, origin of these potential polymorphic changes, and their biological significance. Regardless of how MOAP-1 may lose expression and/or function, we speculate that the combined loss of both MOAP-1 and RASSF1A expression may be a common event occurring during carcinogenesis to result in the functional loss of the MOAP-1/RASSF1A cell death pathway and enhanced proliferation of malignant cells. Furthermore, the absence of MOAP-1 in cancer cells would also impact to some extent on the intrinsic apoptotic pathway(s) where MOAP-1 has been shown to play a role [8] and which is also the target of many chemotherapeutic drugs. Future investigations will be required in order to determine the cause(s) underlying MOAP-1 expression changes in human cancer.

Evidence from the literature indicates downregulation of *MOAP-1* expression in macrophage cells upon overexpression of the transcription factor MafB [50]. Upregulation of MafB is commonly observed in alveolar macrophages that have been exposed to cigarette smoke, and, incidentally, these cells also display increased viability [50, 51]. It has been proposed that MafB may promote macrophage survival through inhibition of apoptosis, which may be achieved through downregulation of pro-apoptotic molecules such as MOAP-1 [50]. In addition, analysis of the promoter region of MOAP-1 for transcription factor binding sites identified several interesting sites for NFκB (CCCTGGTCCC CAAGGAAATA CCT GCAAAAG) and c-Rel (ATCGGAATGA CCCTCTCGGC) and three sites for STAT1 (CTTGCTCCCT TAGGGGAACA) using the online, publicly available Transcription Factor Search (TFSEARCH) software. It remains to be determined if these are functional transcription factor binding sites but does provide hints to the complexity of MOAP-1 expression and reaffirms its importance in both cell death and growth control.

6. Interaction of MOAP-1 with Bcl-2 Family Members

As a pro-apoptotic molecule, MOAP-1 selectively interacts with members of the Bcl-2 protein family. In particular, its association with Bax requires the presence of a Bcl-2 homology 3 (BH3)-like domain within amino acids 120–127 and the same domain is also essential for mediating apoptosis [7]. Interestingly, the association of MOAP-1 to Bax requires all three BH (BH1, BH2, BH3) domains of the latter protein and is thus in contrast to other known Bax-associating partners (Figure 4). Additionally, it is speculated that MOAP-1 may associate at the hydrophobic cleft of Bax since critical point mutations in any of the three BH domains in Bax result in a loss of MOAP-1 association. The interaction between MOAP-1 and Bax occurs upon induction of apoptosis in response to activators of both the intrinsic and extrinsic cell death pathways and facilitates the release of cytochrome c from the mitochondria [8].

In addition to Bax, MOAP-1 also associates with the pro-survival anti-apoptotic proteins Bcl-2 and Bcl-X_L but not additional Bcl-2 family members Bid, BimL, Bak, Bad or Bcl-w under the same experimental conditions [7]. Evidence suggests that its interactions with Bcl-2 and Bcl-X_L may function to restrain the pro-apoptotic activity of MOAP-1 since overexpression of Bcl-X_L is sufficient to block MOAP-1-mediated cell death. Therefore, it appears that MOAP-1 may function similar to the canonical BH3-only proteins of the Bcl-2 family that are known to promote Bax activation and which are also inhibited by its anti-apoptotic family members.

7. Cooperation of MOAP-1 with RASSF1A in Death Receptor-Mediated Apoptosis

MOAP-1 is required for execution of both the intrinsic and extrinsic pathways of apoptosis where it is required for Bax

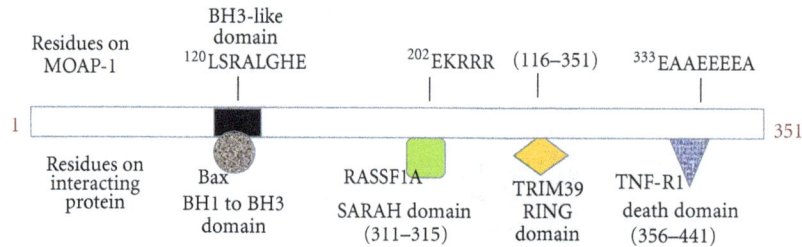

FIGURE 4: A schematic of MOAP-1 with indicated areas of contact with other proteins documented above schematic. Residues empirically determined to be required for protein interactions are indicated below each region.

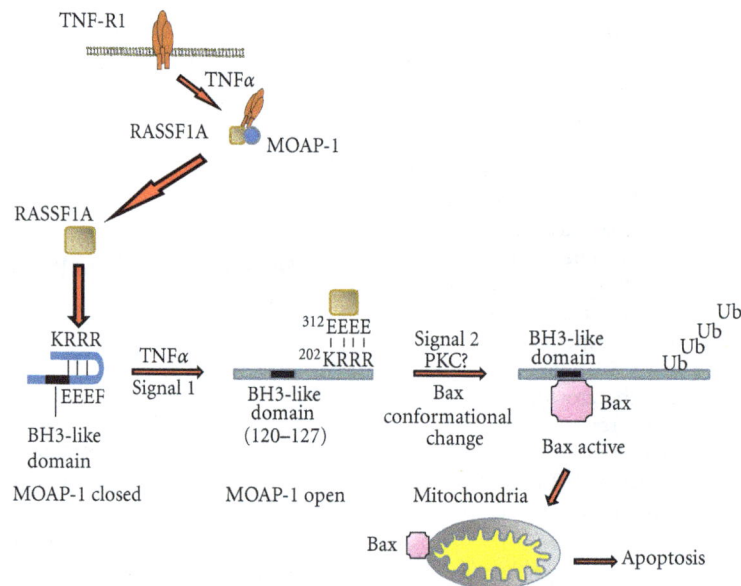

FIGURE 5: MOAP-1 cooperates with RASSF1A during death receptor-dependent apoptosis and promotes Bax activation. In response to death receptor stimulation, MOAP-1 is first recruited to the receptor and then followed by RASSF1A association at the MOAP-1/receptor complex. The association of MOAP-1 to RASSF1A promotes a conformational change in MOAP-1 that exposes its BH3-like domain required for Bax association. The subsequent interaction between MOAP-1 and Bax induces a conformational change in Bax that enables its translocation, and insertion into the mitochondrial outer membrane resulting in the release of cytochrome c and other apoptogenic factors, leading to apoptosis.

conformational change and translocation from the cytosol to the mitochondria prior to the release of apoptogenic factors [5, 8]. Although the mechanistic details of its role in the intrinsic pathway are currently unknown, the death receptor-dependent pathway involving MOAP-1 has been delineated to a great extent [5, 6] (Figure 5).

Under nonstimulated conditions, MOAP-1 is normally held in a "closed" conformation through an intraelectrostatic interaction involving regions [178]EEEF and [202]KRRR [6]. However, stimulation of cells with TNFα or TRAIL results in the recruitment of MOAP-1 to the receptor via a basic sequence ([336]EEEEA) at its C-terminal end (Figures 4 and 5). Prior to death receptor association, RASSF1A is released from association with 14-3-3 and loses its ability to homodimerize [52]. Upon binding to the receptor through its N-terminal cysteine-rich (C1) domain, RASSF1A induces a conformational change in MOAP-1 to a more "open" state

(Figure 5, Signal 1, TNFα) that exposes its BH3-like domain and allows it to bind and promote the activation of Bax [6].

The association of MOAP-1 with RASSF1A involves the sequence [202]KRRR in the former protein and [312]EEEE in the latter. Although activated K-Ras has been reported to be required for stabilization of the MOAP-1/RASSF1A protein complex [53], we are is able to consistently detect robust associations between MOAP-1 and RASSF1A in experiments that do not require the presence of overexpressed active K-Ras [5, 6]. Therefore, we are currently unable to explain or support the results of Vos and colleagues. Nonetheless, MOAP-1-induced Bax conformational change enables Bax to translocate from the cytosol to the mitochondria where it can insert into the mitochondrial membrane and promote the release of cytochrome c as well as other apoptosis-inducing factors, resulting in cell death. Therefore, MOAP-1 functions alongside RASSF1A as a key component linking death receptor signaling to Bax activation and mitochondria-associated cell death. The MOAP-1/RASSF1A pathway exists as a separate, parallel signaling cascade that links the extrinsic and intrinsic pathways of apoptosis independent of tBid and caspase 8 [5].

In addition to RASSF1A, MOAP-1 has also demonstrated the ability to associate with a second RASSF family member, RASSF6 [54]. The interaction between RASSF6 and MOAP-1 is enhanced by the presence of activated K-Ras, and, furthermore, RASSF6 is also able to promote apoptosis. Therefore, it has been speculated that Ras may activate the pro-apoptotic function of RASSF6 and that RASSF6 may cooperate with MOAP-1 in a pathway similar to RASSF1A in order to induce cell death. However, this hypothesis still needs to be verified.

8. Regulation of MOAP-1 Stability by Apoptotic Signals

Under nonstimulated conditions, MOAP-1 is constitutively degraded by the ubiquitin-proteasome system and is normally a short-lived protein with a half-life of approximately 25 minutes [43]. However, evidence suggests that targeting of MOAP-1 to the proteasome may involve an unconventional mechanism given that no specific lysine residue can be identified as the site of polyubiquitination [43]. Indeed, mutation of any individual lysine residue or combination of residues fails to abolish MOAP-1 ubiquitination. Thus, the process involved in controlling MOAP-1 turnover remains to be determined.

In addition to regulation of basal MOAP-1 expression, MOAP-1 is also rapidly upregulated in response to multiple apoptotic stimuli including serum withdrawal, etoposide, TRAIL, and the endoplasmic reticulum stress inducer thapsigargin [43]. The increase in MOAP-1 protein arises through inhibition of its polyubiquitination and subsequent proteasomal degradation. Research findings demonstrate that elevation of MOAP-1 levels occurs prior to cell commitment to apoptosis and that the stabilization of MOAP-1 helps to sensitize cells to apoptosis by increasing the levels of activated Bax.

Intriguingly, stabilization of MOAP-1 in response to apoptosis employs the RING domain protein tripartite motif containing 39 (TRIM39) [55]. TRIM39 has not yet been functionally characterized but belongs to the tripartite motif (TRIM) family of proteins that are commonly involved in innate immunity [56] and contains three zinc-binding domains including a RING, B box, and coiled-coil region. Although a large number of proteins that contain RING domains also function as E3 ligases [57], TRIM39 associates with MOAP-1 in a manner that promotes its stabilization rather than its polyubiquitination [55]. TRIM39 also sensitizes cells to apoptosis by inhibiting MOAP-1 ubiquitination (through an unknown mechanism) and thus allows for the accumulation of MOAP-1 that can then can activate Bax. Furthermore, it was observed that both TRIM39 and MOAP-1 influence each other's localization to the mitochondria when overexpressed in HEK293 cells [55]. The upregulation of MOAP-1 protein levels can also occur in response to

chemical toxins and clinical drugs reaffirming our speculation that MOAP-1 in cancer cells may be important for patient response to certain chemotherapeutic treatments [58]. Incubation of chronic lymphocytic leukemia cells with the apoptosis-inducing compound 5-aminoimidazole-4-carboxamideriboside or acadesine (AICAR) has been shown to result in a significant increase in *MOAP-1* expression [58]. Although the pathway through which AICAR induces cell death remains unknown, it is achieved through a mechanism that is independent of both AMPK and p53. In a second example, the addition of the novel immunosuppressant 2-amino-2[2-(4-octylphenyl) ethyl]-1,3-propane-diol hydrochloride (FTY720) to Jurkat cells results in a greater than tenfold upregulation of MOAP-1 mRNA levels [59]. It is believed that the potent immunosuppressive function of FTY720 may be attributed to its ability to induce lymphocyte apoptosis [60]. However, FTY720 has also been shown to induce apoptosis in a variety of different cancer cell types and to prevent breast cancer metastasis in mouse models [61–64]. Thus, it is plausible that the immunosuppressive and/or antitumorigenic effects of FTY720 may be partially mediated by MOAP-1.

We have evidence for a nondegradative ubiquitination of MOAP-1. This post-translational modification proceeds through a mechanism that is responsive to death receptor stimulation and a novel protein kinase C (PKC) dependent event [Law et al, unpublished observations] that may allow MOAP-1 to associate with and promote Bax activation (Figure 5, Signal 2). Interestingly, MOAP-1 has two potential binding sites for TRAF2, an E3-ubiquitin ligase important for TNF-R1-dependent signaling. These sites are at [178]EPGEEFGRW AND [331]DYEAAEEEAL with the underlined residues forming the core of the TRAF2 association site [64]. The first potential site is part of the intraelectrostatic pair that overlaps with the BH3-domain of MOAP-1. We are currently investigating the possible involvement of TRAF2 in MOAP-1 ubiquitination and the functional importance of several potential lysine residues for ubiquitin-dependent modification. We speculate that the ubiquitination of MOAP-1 may influence MOAP-1-mediated growth suppression and/or MOAP-1-directed apoptosis. This form of MOAP-1 ubiquitination adds to the complexity of MOAP-1 stability by a degradative-dependent ubiquitination to modulate the biological functions of MOAP-1.

9. MOAP-1: A Paraneoplastic Antigen

In addition to its role as a pro-apoptotic molecule, MOAP-1 is also the fourth member of the paraneoplastic Ma antigen (PNMA) family and is consequently also known as PNMA4. Paraneoplastic antigens (also termed "onconeural antigens") are proteins that are restricted in expression to immune-privileged sites within the body (such as the brain) and are therefore recognized as foreign molecules by the immune system when aberrantly expressed at other sites [65, 66]. Remarkably, these foreign proteins are expressed by systemic tumors in a subset of cancer patients which subsequently trigger an immune-mediated antitumor response. In some patients, this immune response is not only directed against the tumor itself but also towards the sites within the body that ordinarily express the protein. In the case of the brain, this immune response results in neuronal degeneration and the development of an autoimmune neurologic disease known as a paraneoplastic neurological disorder (PND).

The PNMA family consists of six members (PNMAs 1–6) that, with the exceptions of PNMAs 4, 5, and 6, were originally identified through screening of complementary DNA libraries using antibody-containing sera from patients with PNDs [67]. Although MOAP-1/PNMA4 is ubiquitously expressed with higher levels in the heart and brain [7], each of the other family members are more restricted in expression [67–71]. The detection of antibodies to PNMAs 1–3 in PND patients is associated with disorders affecting the limbic system, brain, stem and cerebellum but is not indicative of any particular cancer type [68–70, 72]. In contrast, MOAP-1 has a well-established role in apoptosis and—similar to PNMA5 and PNMA6—is not associated with the development of PNDs to date. MOAP-1 displays the greatest amino acid sequence homology with PNMA1 (58%) which functions as a neuronal-specific pro-apoptotic molecule [73]. PNMA1 contains both a BH3-like domain and a conserved RASSF1A association site similar to that found on MOAP-1 (Figure 4). However, PNMA1 does not associate with either Bax or RASSF1A [73], and, therefore, although unknown, the mechanism by which it induces cell death presumably differs from MOAP-1. It remains to be determined how, and if, MOAP-1 may impinge on the pathogenesis of paraneoplastic syndromes.

10. Concluding Remarks

MOAP-1 is a highly regulated pro-apoptotic molecule that demonstrates multiple potential properties of a candidate tumor suppressor protein. Given that MOAP-1 regulates RASSF1A pro-apoptotic function and RASSF1A is also epigenetically silenced in a large number of human cancers, it is possible that the combined loss of MOAP-1 and RASSF1A during carcinogenesis may result in the inhibition of extrinsically activated cell death signaling pathways in cancer cells. RASSF1A has now been demonstrated to influence several other biological processes such as cell cycle, microtubule dynamics, and cell migration. Therefore, it will be interesting to explore which of these biological processes MOAP-1 may also be involved in and that may be important for it to behave as a potential tumor suppressor protein.

Abbreviations

AICAR:	5-Aminoimidazole-4-carboxamideribosideor acadesine
Bcl-2:	B-cell lymphoma-2
BH3:	Bcl-2 homology 3
FTY720:	2-Amino-2[2-(4-octylphenyl) ethyl]-1,3-propane-diol hydrochloride
LATS1:	Large tumor suppressor, homolog 1
MOAP-1:	Modulator of apoptosis
MST:	Mammalian sterile 20-like
pI:	Isoelectric point
PND:	Paraneoplastic neurological disorder
PNMA:	Paraneoplastic Ma antigen
RAS:	Rat sarcoma
RASSF:	Ras association domain family
RA:	Ras association domain
SNP:	Single nucleotide polymorphism
TNFα:	Tumor necrosis factor α
TNF-R1:	Tumor necrosis factor receptor 1
TRAIL:	TNF-related apoptosis-inducing ligand
TRIM39:	Tripartite motif containing 39
YAP1:	Yes-associated protein 1.

Acknowledgments

The authors would like to thank all the members of the Baksh laboratories, past and present, for their helpful discussions and dedication to understanding the biological role of RASSF1A. They are grateful to the excellent technical help of Christina Onyskiw and for the support of the Department of Pediatrics, especially the division of Hematology/Oncology/Palliative Care/Epidemiology under the guidance of Dr. Paul Grundy, and currently, Dr. David Eisenstat. Investigation of the biological role of RASSF1A has been supported by grants from the Canadian Breast Cancer Foundation (Prairies/NWT), Canadian Institutes of Health Research, Women and Children's Health Research Institute, Alberta Heritage Foundation for Medical Research, Canadian Foundation for Innovation/Alberta Small Equipment Grants Program, and The Stollery Children's Foundation/Hair Massacure Grant generously donated by the MacDonald family. V. C. Yu is supported by National University of Singapore (R-148-000-121-133) and National Medical Research Council of Singapore (NMRC-IRG11-076). He is also a Minjiang Scholar Chair Professor at School of Life Sciences, Xiamen University, China). Please note that excerpts of this review have been extracted from the M.Sc. thesis of Jennifer Law entitled "MOAP-1: a candidate tumor suppressor protein" (Department of Biochemistry, Faculty of Medicine and Dentistry, University of Alberta, 2011).

References

[1] M. El-Kalla, C. Onyskiw, and S. Baksh, "Functional importance of RASSF1A microtubule localization and polymorphisms," *Oncogene*, vol. 29, no. 42, pp. 5729–5740, 2010.

[2] World Health Organization, *The Global Burden of Disease: 2004 Update*, 2008.

[3] B. Vogelstein, *The Cancer Genome*, American Association for Cancer Research, Washington, DC, USA, 2010.

[4] H. Donninger, M. D. Vos, and G. J. Clark, "The RASSF1A tumor suppressor," *Journal of Cell Science*, vol. 120, no. 18, pp. 3163–3172, 2007.

[5] S. Baksh, S. Tommasi, S. Fenton et al., "The tumor suppressor RASSF1A and MAP-1 link death receptor signaling to bax conformational change and cell death," *Molecular Cell*, vol. 18, no. 6, pp. 637–650, 2005.

[6] C. J. Foley, H. Freedman, S. L. Choo et al., "Dynamics of RASSF1A/MOAP-1 association with death receptors," *Molecular and Cellular Biology*, vol. 28, no. 14, pp. 4520–4535, 2008.

[7] K. O. Tan, K. M. L. Tan, S. L. Chan et al., "MAP-1, a novel proapoptotic protein containing a BH3-like motif that associates with bax through Its Bcl-2 homology domains," *Journal of Biological Chemistry*, vol. 276, no. 4, pp. 2802–2807, 2001.

[8] K. O. Tan, N. Y. Fu, S. K. Sukumaran et al., "MAP-1 is a mitochondrial effector of Bax," *Proceedings of the National Academy of Sciences of the United States of America*, vol. 102, no. 41, pp. 14623–14628, 2005.

[9] C. Mondello and A. I. Scovassi, "Apoptosis: a way to maintain healthy individuals," *Sub-cellular biochemistry*, vol. 50, pp. 307–323, 2010.

[10] D. A. Carson and J. M. Ribeiro, "Apoptosis and disease," *The Lancet*, vol. 341, no. 8855, pp. 1251–1254, 1993.

[11] A. Strasser, A. W. Harris, D. C. S. Huang, P. H. Krammer, and S. Cory, "Bcl-2 and Fas/APO-1 regulate distinct pathways to lymphocyte apoptosis," *The EMBO Journal*, vol. 14, no. 24, pp. 6136–6147, 1995.

[12] S. Elmore, "Apoptosis: a review of programmed cell death," *Toxicologic Pathology*, vol. 35, no. 4, pp. 495–516, 2007.

[13] S. J. Baker and E. P. Reddy, "Modulation of life and death by the TNF receptor superfamily," *Oncogene*, vol. 17, no. 25, pp. 3261–3270, 1998.

[14] A. Thorburn, "Death receptor-induced cell killing," *Cellular Signalling*, vol. 16, no. 2, pp. 139–144, 2004.

[15] S. W. G. Tait and D. R. Green, "Mitochondria and cell death: outer membrane permeabilization and beyond," *Nature Reviews Molecular Cell Biology*, vol. 11, no. 9, pp. 621–632, 2010.

[16] C. Wang and R. J. Youle, "The role of mitochondria in apoptosis," *Annual Review of Genetics*, vol. 43, pp. 95–118, 2009.

[17] S. L. Chan and V. C. Yu, "Proteins of the Bcl-2 family in apoptosis signalling: from mechanistic insights to therapeutic opportunities," *Clinical and Experimental Pharmacology and Physiology*, vol. 31, no. 3, pp. 119–128, 2004.

[18] S. Cory and J. M. Adams, "The BCL2 family: regulators of the cellular life-or-death switch," *Nature Reviews Cancer*, vol. 2, no. 9, pp. 647–656, 2002.

[19] R. J. Youle and A. Strasser, "The BCL-2 protein family: opposing activities that mediate cell death," *Nature Reviews Molecular Cell Biology*, vol. 9, no. 1, pp. 47–59, 2008.

[20] T. Lindsten, A. J. Ross, A. King et al., "The combined functions of proapoptotic Bcl-2 family members Bak and Bax are essential for normal development of multiple tissues," *Molecular Cell*, vol. 6, no. 6, pp. 1389–1399, 2000.

[21] M. Giam, D. C. S. Huang, and P. Bouillet, "BH3-only proteins and their roles in programmed cell death," *Oncogene*, vol. 27, no. 1, pp. S128–S136, 2008.

[22] R. Dammann, C. Li, J. H. Yoon, P. L. Chin, S. Bates, and G. P. Pfeifer, "Epigenetic inactivation of a RAS association domain family protein from the lung tumour suppressor locus 3p21.3," *Nature Genetics*, vol. 25, no. 3, pp. 315–319, 2000.

[23] A. M. Richter, G. P. Pfeifer, and R. H. Dammann, "The RASSF proteins in cancer; from epigenetic silencing to functional characterization," *Biochimica et Biophysica Acta*, vol. 1796, no. 2, pp. 114–128, 2009.

[24] V. Sherwood, A. Recino, A. Jeffries, A. Ward, and A. D. Chalmers, "The N-terminal RASSF family: a new group of Ras-association-domain-containing proteins, with emerging links to cancer formation," *Biochemical Journal*, vol. 425, no. 2, pp. 303–311, 2010.

[25] N. Underhill-Day, V. Hill, and F. Latif, "N-terminal RASSF family (RASSF7-RASSF10): a mini review," *Epigenetics*, vol. 6, no. 3, pp. 284–292, 2011.

[26] L. van der Weyden and D. J. Adams, "The Ras-association domain family (RASSF) members and their role in human tumourigenesis," *Biochimica et Biophysica Acta*, vol. 1776, no. 1, pp. 58–85, 2007.

[27] G. P. Pfeifer, J. H. Yoon, L. Liu, S. Tommasi, S. P. Wilczynski, and R. Dammann, "Methylation of the RASSF1A gene in human cancers," *Biological Chemistry*, vol. 383, no. 6, pp. 907–914, 2002.

[28] A. Agathanggelou, W. N. Cooper, and F. Latif, "Role of the Ras-association domain family 1 tumor suppressor gene in human cancers," *Cancer Research*, vol. 65, no. 9, pp. 3497–3508, 2005.

[29] L. B. Hesson, W. N. Cooper, and F. Latif, "The role of RASSF1A methylation in cancer," *Disease Markers*, vol. 23, no. 1-2, pp. 73–87, 2007.

[30] R. Dammann, U. Schagdarsurengin, C. Seidel et al., "The tumor suppressor RASSF1A in human carcinogenesis: an update," *Histology and Histopathology*, vol. 20, no. 2, pp. 645–663, 2005.

[31] A. Dallol, A. Agathanggelou, S. Tommasi, G. P. Pfeifer, E. R. Maher, and F. Latif, "Involvement of the RASSF1A tumor suppressor gene in controlling cell migration," *Cancer Research*, vol. 65, no. 17, pp. 7653–7659, 2005.

[32] M. D. Vos, A. Martinez, C. Elam et al., "A role for the RASSF1A tumor suppressor in the regulation of tubulin polymerization and genomic stability," *Cancer Research*, vol. 64, no. 12, pp. 4244–4250, 2004.

[33] L. Liu, A. Vo, and W. L. McKeehan, "Specificity of the methylation-suppressed A isoform of candidate tumor suppressor RASSF1 for microtubule hyperstabilization is determined by cell death inducer C19ORF5," *Cancer Research*, vol. 65, no. 5, pp. 1830–1838, 2005.

[34] M. S. Song, J. S. Chang, S. J. Song, T. H. Yang, H. Lee, and D. S. Lim, "The centrosomal protein RAS association domain family protein 1A (RASSF1A)-binding protein 1 regulates mitotic progression by recruiting RASSF1A to spindle poles," *Journal of Biological Chemistry*, vol. 280, no. 5, pp. 3920–3927, 2005.

[35] S. L. Fenton, A. Dallol, A. Agathanggelou et al., "Identification of the E1A-regulated transcription factor p120 E4F as an interacting partner of the RASSF1A candidate tumor suppressor gene," *Cancer Research*, vol. 64, no. 1, pp. 102–107, 2004.

[36] J. Ahmed-Choudhury, A. Agathanggelou, S. L. Fenton et al., "Transcriptional regulation of cyclin A2 by RASSF1A through the enhanced binding of p120E4F to the cyclin A2 promoter," *Cancer Research*, vol. 65, no. 7, pp. 2690–2697, 2005.

[37] L. Shivakumar, J. Minna, T. Sakamaki, R. Pestell, and M. A. White, "The RASSF1A tumor suppressor blocks cell cycle progression and inhibits cyclin D1 accumulation," *Molecular and Cellular Biology*, vol. 22, no. 12, pp. 4309–4318, 2002.

[38] H. J. Oh, K. K. Lee, S. J. Song et al., "Role of the tumor suppressor RASSF1A in Mst1-mediated apoptosis," *Cancer Research*, vol. 66, no. 5, pp. 2562–2569, 2006.

[39] M. Praskova, A. Khoklatchev, S. Ortiz-Vega, and J. Avruch, "Regulation of the MST1 kinase by autophosphorylation, by the growth inhibitory proteins, RASSF1 and NORE1, and by Ras," *Biochemical Journal*, vol. 381, no. 2, pp. 453–462, 2004.

[40] A. Khokhlatchev, S. Rabizadeh, R. Xavier et al., "Identification of a novel Ras-regulated proapoptotic pathway," *Current Biology*, vol. 12, no. 4, pp. 253–265, 2002.

[41] D. Matallanas, D. Romano, K. Yee et al., "RASSF1A elicits apoptosis through an MST2 pathway directing proapoptotic transcription by the p73 tumor suppressor protein," *Molecular Cell*, vol. 27, no. 6, pp. 962–975, 2007.

[42] G. Halder and R. L. Johnson, "Hippo signaling: growth control and beyond," *Development*, vol. 138, no. 1, pp. 9–22, 2011.

[43] N. Y. Fu, S. K. Sukumaran, and V. C. Yu, "Inhibition of ubiquitin-mediated degradation of MOAP-1 by apoptotic stimuli promotes Bax function in mitochondria," *Proceedings of the National Academy of Sciences of the United States of America*, vol. 104, no. 24, pp. 10051–10056, 2007.

[44] N. Rampino, H. Yamamoto, Y. Ionov et al., "Somatic frameshift mutations in the BAX gene in colon cancers of the microsatellite mutator phenotype," *Science*, vol. 275, no. 5302, pp. 967–969, 1997.

[45] H. Yamamoto, F. Itoh, H. Fukushima et al., "Frequent bax frameshift mutations in gastric cancer with high but not low microsatellite instability," *Journal of Experimental and Clinical Cancer Research*, vol. 18, no. 1, pp. 103–106, 1999.

[46] L. C. Li and R. Dahiya, "MethPrimer: designing primers for methylation PCRs," *Bioinformatics*, vol. 18, no. 11, pp. 1427–1431, 2002.

[47] "dbSNP Short genetic variations," NCBI, 2012, http://www.ncbi.nlm.nih.gov/projects/SNP/snp_ref.cgi?showRare=on&chooseRs=coding&go=Go&locusId=64112.

[48] "Transcript: MOAP1-001 (ENST00000298894)," Ensembl, 2012, http://www.ensembl.org/Homo_sapiens/Transcript/ProtVariations?g=ENSG00000165943;peptide=ENSP000002-98894;r=14:93648541-93651273;t=ENST00000298894.

[49] X. Wei, V. Walia, J. C. Lin et al., "Exome sequencing identifies GRIN2A as frequently mutated in melanoma," *Nature Genetics*, vol. 43, no. 5, pp. 442–448, 2011.

[50] J. I. Machiya, Y. Shibata, K. Yamauchi et al., "Enhanced expression of MafB inhibits macrophage apoptosis induced by cigarette smoke exposure," *American Journal of Respiratory Cell and Molecular Biology*, vol. 36, no. 4, pp. 418–426, 2007.

[51] K. Tomita, G. Caramori, S. Lim et al., "Increased p21CIP1/WAF1 and B cell lymphoma leukemia-xL expression and reduced apoptosis in alveolar macrophages from smokers," *American Journal of Respiratory and Critical Care Medicine*, vol. 166, no. 5, pp. 724–731, 2002.

[52] H. A. Ghazaleh, R. S. Chow, S. L. Choo et al., "14-3-3 Mediated regulation of the tumor suppressor protein, RASSF1A," *Apoptosis*, vol. 15, no. 2, pp. 117–127, 2010.

[53] M. D. Vos, A. Dallol, K. Eckfeld et al., "The RASSF1A tumor suppressor activates bax via MOAP-1," *Journal of Biological Chemistry*, vol. 281, no. 8, pp. 4557–4563, 2006.

[54] N. P. C. Allen, H. Donninger, M. D. Vos et al., "RASSF6 is a novel member of the RASSF family of tumor suppressors," *Oncogene*, vol. 26, no. 42, pp. 6203–6211, 2007.

[55] S. S. Lee, N. Y. Fu, S. K. Sukumaran, K. F. Wan, Q. Wan, and V. C. Yu, "TRIM39 is a MOAP-1-binding protein that stabilizes MOAP-1 through inhibition of its poly-ubiquitination process," *Experimental Cell Research*, vol. 315, no. 7, pp. 1313–1325, 2009.

[56] K. Ozato, D. M. Shin, T. H. Chang, and H. C. Morse, "TRIM family proteins and their emerging roles in innate immunity," *Nature Reviews Immunology*, vol. 8, no. 11, pp. 849–860, 2008.

[57] R. J. Deshaies and C. A. P. Joazeiro, "RING domain E3 ubiquitin ligases," *Annual Review of Biochemistry*, vol. 78, pp. 399–434, 2009.

[58] A. F. Santidrián, D. M. González-Girones, D. Iglesias-Serret et al., "AICAR induces apoptosis independently of AMPK and p53 through up-regulation of the BH3-only proteins BIM and NOXAin chronic lymphocytic leukemia cells," *Blood*, vol. 116, no. 16, pp. 3023–3032, 2010.

[59] F. Wang, W. Tan, D. Guo, X. Zhu, K. Qian, and S. He, "Altered expression of signaling genes in jurkat cells upon FTY720 induced apoptosis," *International Journal of Molecular Sciences*, vol. 11, no. 9, pp. 3087–3105, 2010.

[60] S. Suzuki, X. K. Li, S. Enosawa, and T. Shinomiya, "A new immunosuppressant, FTY720, induces bcl-2-associated apoptotic cell death in human lymphocytes," *Immunology*, vol. 89, no. 4, pp. 518–523, 1996.

[61] H. Azuma, S. Takahara, N. Ichimaru et al., "Marked prevention of tumor growth and metastasis by a novel immunosuppressive agent, FTY720, in mouse breast cancer models," *Cancer Research*, vol. 62, no. 5, pp. 1410–1419, 2002.

[62] J. D. Wang, S. Takahara, N. Nonomura et al., "Early induction of apoptosis in androgen-independent prostate cancer cell line by FTY720 requires caspase-3 activation," *Prostate*, vol. 1, pp. 50–55, 1999.

[63] T. Shinomiya, X. K. Li, H. Amemiya, and S. Suzuki, "An immunosuppressive agent, FTY720, increases intracellular concentration of calcium ion and induces apoptosis in HL-60," *Immunology*, vol. 91, no. 4, pp. 594–600, 1997.

[64] T. Matsuda, H. Nakajima, I. Fujiwara, N. Mizuta, and T. Oka, "Caspase requirement for the apoptotic death of WR19L-induced by FTY720," *Transplantation Proceedings*, vol. 30, no. 5, pp. 2355–2357, 1998.

[65] K. Musunuru and R. B. Darnell, "Paraneoplastic neurologic disease antigens: RNA-binding proteins and signaling proteins in neuronal degeneration," *Annual Review of Neuroscience*, vol. 24, pp. 239–262, 2001.

[66] W. K. Roberts and R. B. Darnell, "Neuroimmunology of the paraneoplastic neurological degenerations," *Current Opinion in Immunology*, vol. 16, no. 5, pp. 616–622, 2004.

[67] M. Schüller, D. Jenne, and R. Voltz, "The human PNMA family: novel neuronal proteins implicated in paraneoplastic neurological disease," *Journal of Neuroimmunology*, vol. 169, no. 1-2, pp. 172–176, 2005.

[68] J. Dalmau, S. H. Gultekin, R. Voltz et al., "Ma1, a novel neuron- and testis-specific protein, is recognized by the serum of patients with paraneoplastic neurological disorders," *Brain*, vol. 122, no. 1, pp. 27–39, 1999.

[69] R. Voltz, S. H. Gultekin, M. R. Rosenfeld et al., "A serologic marker of paraneoplastic limbic and brain-stem encephalitis in patients with testicular cancer," *The New England Journal of Medicine*, vol. 340, no. 23, pp. 1788–1795, 1999.

[70] M. R. Rosenfeld, J. G. Eichen, D. F. Wade, J. B. Posner, and J. Dalmau, "Molecular and clinical diversity in paraneoplastic immunity to Ma proteins," *Annals of Neurology*, vol. 50, no. 3, pp. 339–348, 2001.

[71] M. Takaji, Y. Komatsu, A. Watakabe, T. Hashikawa, and T. Yamamori, "Paraneoplastic antigen-like 5 gene (PNMA5) is preferentially expressed in the association areas in a primate specific manner," *Cerebral Cortex*, vol. 19, no. 12, pp. 2865–2879, 2009.

[72] L. A. Hoffmann, S. Jarius, H. L. Pellkofer et al., "Anti-Ma and anti-Ta associated paraneoplastic neurological syndromes: 22 newly diagnosed patients and review of previous cases," *Journal of Neurology, Neurosurgery and Psychiatry*, vol. 79, no. 7, pp. 767–773, 2008.

[73] H. L. Chen and S. R. D'Mello, "Induction of neuronal cell death by paraneoplastic Ma1 antigen," *Journal of Neuroscience Research*, vol. 88, no. 16, pp. 3508–3519, 2010.

16

Three-Dimensional Molecular Modeling of a Diverse Range of SC Clan Serine Proteases

Aparna Laskar,[1] Aniruddha Chatterjee,[2, 3] Somnath Chatterjee,[1] and Euan J. Rodger[2]

[1] *Infectious Diseases and Immunology Division, CSIR-Indian Institute of Chemical Biology, West Bengal, Kolkata 700032, India*
[2] *Department of Pathology, Dunedin School of Medicine, University of Otago, P.O. Box 913, Dunedin 9054, New Zealand*
[3] *National Research Centre for Growth and Development, University of Auckland, Auckland 1142, New Zealand*

Correspondence should be addressed to Euan J. Rodger, euan.rodger@otago.ac.nz

Academic Editor: Alessandro Desideri

Serine proteases are involved in a variety of biological processes and are classified into clans sharing structural homology. Although various three-dimensional structures of SC clan proteases have been experimentally determined, they are mostly bacterial and animal proteases, with some from archaea, plants, and fungi, and as yet no structures have been determined for protozoa. To bridge this gap, we have used molecular modeling techniques to investigate the structural properties of different SC clan serine proteases from a diverse range of taxa. Either SWISS-MODEL was used for homology-based structure prediction or the LOOPP server was used for threading-based structure prediction. The predicted models were refined using Insight II and SCRWL and validated against experimental structures. Investigation of secondary structures and electrostatic surface potential was performed using MOLMOL. The structural geometry of the catalytic core shows clear deviations between taxa, but the relative positions of the catalytic triad residues were conserved. Evolutionary divergence was also exhibited by large variation in secondary structure features outside the core, differences in overall amino acid distribution, and unique surface electrostatic potential patterns between species. Encompassing a wide range of taxa, our structural analysis provides an evolutionary perspective on SC clan serine proteases.

1. Introduction

Serine proteases account for over a third of all known proteolytic enzymes and are involved in a range of physiological processes including digestion, immunity, blood clotting, fibrinolysis, reproduction, and protein folding [1]. The proteolytic mechanism of these proteases involves nucleophilic attack of the carbonyl atom of the substrate peptide bond by a catalytic serine (Ser) residue in the active site of the enzyme. In addition to the nucleophilic Ser residue, this reaction is dependent on other critical amino acids in the catalytic site such as an Aspartate (Asp) and a Histidine (His) that together form what is referred to as the catalytic triad (or a dyad in some cases) [2]. The presence of this catalytic triad in at least four distinct protein folds indicates the same mechanism evolved four separate times during evolution [3].

The MEROPS classification system (http://merops .sanger.ac.uk/) has grouped proteases into families according to statistically significant similarities in the amino acid sequence. These protease families are further grouped into clans that have dissimilar amino acid sequences but typically have structural homology and/or the same linear order of catalytic triad residues [4]. The SC clan of serine proteases is widely distributed across all taxa, and in contrast to other clans, it includes both endopeptidases and exopeptidases. At the core of all SC clan proteases is an α/β hydrolase fold, which typically consists of an eight-stranded β-sheet flanked by two or more α-helices. The α-helices contribute to substrate specificity, and the curvature of the β-sheet may also affect interactions with the substrate [1]. The α/β hydrolase fold is a common hydrolytic enzyme structure and is found in many other enzymes such as lipases, peroxidases, and esterases [5]. The SC clan has the same classical serine protease catalytic triad residue formation as clans SB and PA, but with the amino acid sequence order of Ser, Asp, and His. Typically, these residues are confined to the C-terminal region within about 130 residues. The proteolytic mechanism is initiated by the nucleophilic Ser158 (standard serine carboxypeptidase 2 *CBP2* numbering) hydroxyl group transferring a proton to the carbonyl of the peptide substrate. This reaction is catalyzed by the His413 acting as a general base, which is thought to be supported by a hydrogen bond to Asp361. The resulting tetrahedral intermediate breaks down to an acylenzyme intermediate, followed by the formation of a second tetrahedral intermediate. With the protonation of Ser158 by His413, the second tetrahedral intermediate breaks down and the cleaved substrate is released [2].

The five main families in this clan have distinct specificities and have different peptidase activities as represented by the archetypes prolyl oligopeptidase (S9 family), carboxypeptidase Y (S10 family), Xaa-Pro dipeptidyl-peptidase (S15 family), lysosomal Pro-Xaa carboxypeptidase (S28 family), and prolyl aminopeptidase (S33 family).

Because of their abundance and biological significance, the S9 serine protease family has been the most intensively studied. The proteases of this family are up to three times larger than their classic serine protease counterparts, trypsin and subtilisin (25–30 kDa). Many members hydrolyze the peptide bond on the C-terminal side of proline, but the exceptions include oligopeptidase B, which recognizes arginine or lysine, and acylaminoacyl peptidase, which is a cytoplasmic omega-peptidase that releases an N-acylated amino acid [6]. Notably, the central tunnel of an unusual N-terminal β-propeller domain covers the catalytic site and selectively restricts access to oligopeptides of approximately 30 amino acids in length [7]. The S9 family appears to be important in the processing and degradation of peptide hormones, and, therefore, these proteases are important targets of drug design [8]. In humans, prolyl oligopeptidase is involved in several neurological conditions and control of blood pressure [9–11], dipeptidyl peptidase 4 in type 2 diabetes and cancer [12, 13], and acylaminoacyl peptidase in small-cell lung and renal cancer [14, 15]. Both prolyl oligopeptidase and oligopeptidase B seem to facilitate the virulence of protozoan parasites such as *Trypanosoma cruzi* and *Trypanosoma brucei*, which result in the trypanosome infections Chagas disease and sleeping sickness, respectively [16, 17]. Dipeptidyl peptidase 4 contributes to the pathogenicity of *Porphyromonas gingivalis*, the gram-negative bacteria associated with periodontitis [18].

Proteases in the S10 family are serine carboxypeptidases, which cleave C-terminal peptide bonds. They generally prefer hydrophobic amino acids but exhibit broad substrate specificity. In contrast with most other serine proteases, which are typically active at neutral/alkaline pH, family S10 proteases maintain catalytic activity in an acidic environment [19]. This family mostly contributes to proteolytic degradation and protein processing within specific cellular compartments such as vacuoles in fungi and plants (carboxypeptidase Y) and lysosomes in animals (serine carboxypeptidase A) [20, 21]. Members of the S15 family selectively cleave Xaa-Pro, in which Xaa is an N-terminal amino acid. In *Lactobacillus helveticus*, which is used for commercial cheese-making, Xaa-Pro dipeptidyl-peptidase is involved in the casein-degradation pathway, providing essential amino acids for the bacteria [22]. The S28 proteases are a distinct family

TABLE 1: Experimental structures and predicted structures of SC serine proteases across different taxa.

Species	Structure	MEROPS ID
Bacteria		
Xanthomonas campestris	PDB: 1AZW	MER002678
Myxococcus xanthus	PDB: 2BKL	MER005694
Streptomyces lividans	PDB: 1A88	MER026339
Archaea		
Thermoplasma acidophilum	PDB: 1MU0	MER003537
Aeropyrum pernix	PDB: 1VE7	MER005807
Fungi		
Saccharomyces cerevisiae	PDB: 1AC5	MER000413
Saccharomyces cerevisiae	PDB: 1WPX	MER002010
Animalia		
Homo sapiens	PDB: 1N1M	MER000401
	PDB: 1QFS	MER000392
Sus scrofa	PDB: 1ORW	MER028372
	PDB: 2BUC	MER028372
Plantae		
Arabidopsis thaliana	**PMDB: PM0078228**	MER045469
Protozoa		
Plasmodium falciparum	**PMDB: PM0078229**	MER035185

of eukaryotic carboxypeptidases that selectively cleave a Pro-Xaa bond, in which Xaa is a C-terminal amino acid. The human lysosomal Pro-Xaa carboxypeptidase (*PRCP*) is thought to be involved in regulating blood pressure by inactivating angiotensin II [23]. Dipeptidyl peptidase 2, which has a similar substrate specificity to dipeptidyl peptidase 4 of the S9 family, is essential for maintaining lymphocytes and fibroblasts in a quiescent state [24]. Proteases in the S33 family are prolyl aminopeptidases, which preferentially cleave an N-terminal proline residue peptide bond. Many of the bacteria and fungi that produce prolyl aminopeptidases are pathogenic and have therefore been proposed as a viable drug target [25].

2. Material and Methods

Structural data for 3 bacterial, 2 archaeal, 2 fungal, and 4 animal SC clan serine protease structures (Table 1) were obtained from the Protein Data Bank (PDB, http://www.rcsb.org/pdb/). Our inhouse modeling software package MODELYN [26] was developed to perform customized molecular editing and *in silico* structural analysis. It has a set of powerful menus for batch processing commands leading to automated implementation of complicated tasks, including complete model building based on sequence homology and batch processing of replacement mutations. ANALYN [26] is an ancillary protein sequence analysis program that assists MODELYN by analyzing homologous sequences and formulating the strategy for model building. In addition to the experimental structures, amino acid sequences of SC serine proteases (Table 1) for 1 plant (*Arabidopsis thaliana*) and 1 protozoan (*Plasmodium falciparum*) were obtained from the MEROPS protease database (http://merops.sanger.ac.uk/) in

TABLE 2: SWISS-MODEL homology results of *Arabidopsis thaliana* SC serine protease target sequence with known PDB structures.

PDB ID	Resolution (Å)	R-value	Score (bits)	Expect value	AA identity (%)
2BKLB	1.50	0.161	47.4	9×10^{-7}	30
2BKLA	1.50	0.161	47.4	9×10^{-7}	30
1YR2A	1.80	0.162	40.8	9×10^{-6}	34
1QFSA	2.00	0.201	39.7	9×10^{-5}	30
1VZ2A	2.20	0.165	39.7	2×10^{-2}	30

FASTA format [27]. These sequences were initially submitted to SWISS-MODEL for homology-based structure prediction [28]. If a sequence had less than 25% sequence similarity with known experimental structures, these sequences were then submitted to the LOOPP server [29] for threading-based structure prediction as previously described [30, 31]. This analysis reported a ranked list of possible structure predictions for each of the protease sequences, including match scores, sequence identity (%), and the extent of sequence coverage (%). Predicted structures were superposed with respect to a selected set of Cα atoms on the structure with the highest match score, and a suitable starting scaffold was determined using MODELYN. Root mean square deviation (RMSD) values helped to identify the common segments, corresponding to the structurally conserved regions. The initial structures were refined using the DISCOVER and ANALYSIS modules within the software package Insight II [32] through energy minimization and molecular dynamics. The side chains were regenerated using SCRWL [33], and the overall structure was energy minimized. The SCWRL software package was used for prediction of protein side-chains of a fixed backbone, using graph theory to solve the combinatorial problem (details of the structure refinement are given in the Supplementary Material available online at doi:10.1155/2012/580965). PROCHECK was used to check the distribution of φ-ψ dihedral angles and identify Ramachandran outliers [34]. The CHARMM module within Insight II was used to apply dihedral constraints in these segments. MOLPROBITY [35] and MODELYN were used to validate the structural models against experimental structure data. MOLPROBITY provides all-atom contact analysis and gives quantitative information on the steric interactions (H-bond and van der Waals contacts) at the interfaces between components. This program is widely used for quality validation of three-dimensional (3D) protein structures by measuring deviations of bond lengths, bond angles from standard values, overall atom clashscores, and rotamer outliers. MODELYN was used to analyze other structural parameters, including the distance between Cα atoms of the catalytic triad. Verify3D [36], ProSA [37], and ERRAT [38] were also used to further assess the quality of the protease models. Verify3D analyzes the compatibility of the model against its own amino acid sequence. The Verify3D score (the sum of scores for individual residues using a 21-residue sliding window) is normalized to the length of the sequence: $\log_2(\text{Verify3D score}/L^2)$ [39]. ProSA calculates an overall quality score (Z score) of a model in comparison to a range of characteristics expected for native protein structures. ERRAT analyzes the statistics of nonbonded interactions between different atom types (9-residue sliding window) and provides an overall quality factor that is expressed as the percentage of the protein for which the calculated error value falls below the 95% threshold. The ribbon structure and electrostatic potential surface of the structures were determined by MOLMOL [40]. To determine sequence conservation between species, ClustalW [41] was used for multiple sequence alignment. For each sequence, PEPSTATS [42] was used to determine the molar percentage of each amino acid physicochemical class. A flowchart of the modeling and structure refinement strategy has been included as Supplementary Figure S1.

3. Results

3.1. Modeling of Protease Structures. The plant protease from *A. thaliana* had significant homology with proteases of known experimental structure for successful structure prediction using SWISS-MODEL. The homology model was essentially built on the structures 2BKL, 1YR2, 1VZ2 (prolyl oligopeptidases from *Myxococcus xanthus*, *Novosphingobium capsulatum*, and *Sus scrofa* resp.), and 1QGS (an spsA glycosyltransferase from *Bacillus subtilis*), with sequence identity ranging from 30% to 34% (Table 2). Homology-based structure prediction for the *P. falciparum* protease was unsuccessful due to insufficient sequence similarity with known experimental structures. The amino acid sequence was then submitted to the LOOPP server for threading-based structure prediction, which yielded a list of 14 different PDB experimental structures that matched the protease sequence. The matching structures showed good confidence scores ranging from 2.7 to 3.5, sequence identity ranging from 13% to 19%, with best length coverage between 86% and 100% (Table 3). The matched structures were superposed with respect to a selected set of *P. falciparum* protease Cα atoms (43% superposition), with the structure 1U2E (an MhpC C–C bond hydrolase from *Escherichia coli*) having the best score of 3.5 (RMSD values were between 0.332 and 0.564 Å, which helped to identify common segments corresponding to structurally conserved regions). From these superposed structures, the variable loop regions were identified on the starting scaffold derived from 1U2E. Structural refinement of the two models using Insight II and SCRWL is provided in detail as Supplementary Material (additional file 1). The overall backbone conformations of the predicted structures were measured, and Ramachandran outliers were corrected for by applying dihedral constraints in these segments (Table 4). The general structural parameters and the overall quality of the final refined model were compared to experimental structure data (Table 5). The physical parameters were

TABLE 3: LOOPP server results for secondary structure matches of *Plasmodium falciparum* SC serine protease target sequence with known PDB structures.

| PDB ID | Secondary structure | | | Score | Sequence identity (%) | Length (%) |
	Helical structure (%)	Extended (%)	Loops/other (%)			
Target	36.20	19.00	44.80	—	—	—
1U2E	39.44	18.31	42.25	3.547	19.46	100.00
1J1I	41.80	19.53	38.67	3.484	14.88	97.29
1UKS	40.15	17.84	42.01	3.353	18.35	98.64
1C4X	40.71	16.43	42.86	3.294	14.68	98.64
1CQW	32.98	18.85	48.17	3.207	15.67	85.97
1BN6	32.46	19.37	48.17	3.205	15.67	85.97
1MJS	32.53	18.69	48.79	3.158	18.64	99.55
1BN7	32.98	18.47	43.55	3.042	17.73	99.55
1B6G	33.44	14.75	51.80	3.036	15.84	100.00
1FJ2	29.65	19.91	50.44	2.999	19.71	94.12
1JJF	32.40	20.40	47.20	2.997	16.29	100.00
1EHY	38.43	17.08	44.48	2.937	14.55	99.55
1IMJ	30.73	21.95	47.32	2.933	12.79	92.31
1A85	42.28	15.81	41.91	2.688	17.06	95.48

TABLE 4: Backbone refinement of the modeled SC proteases from *Arabidopsis thaliana* and *Plasmodium falciparum*.

| Structural model | φ-ψ distribution in the regions of Ramachandran plot Number of residues (percentage) | | | |
	Most favoured	Additionally allowed	Generously allowed	Disallowed
A. thaliana				
Before backbone refinement	190 (78.8%)	43 (17.8%)	4 (1.7%)	4 (1.7%)
After backbone refinement	194 (80.8%)	46 (19.2%)	0 (0.0%)	0 (0.0%)
P. falciparum				
Before backbone refinement	132 (65.7%)	53 (26.4%)	10 (5.0%)	6 (3.0%)
After backbone refinement	129 (65.5%)	68 (34.5%)	0 (0.0%)	0 (0.0%)

TABLE 5: Structural validation of the modeled SC proteases from *Arabidopsis thaliana* and *Plasmodium falciparum*.

Structural model	All-atom clashscore (No/1000 atoms)	Rotamer outliers (%)	RMSD of bond length (Å)	RMSD of bond angle (degree)
X-ray structure (1YR2)	3.76	2.40	0.240	2.76
Homology model of *A. thaliana* protease	5.37	2.14	0.270	3.70
X-ray structure (1U2E)	23.50	9.62	0.013	2.27
Threading model of *P. falciparum* protease	11.30	4.50	0.028	3.26

	Average Verify3D-1D score	Normalized 3D profile score ($\log_2$(Verify3D/L^2))	ProSA Z-score	ERRAT quality factor (%)
X-ray structure (1YR2)	0.46	−10.53	−9.68	95.2
Homology model of *A. thaliana* protease	0.34	−9.66	−7.55	88.9
X-ray structure (1U2E)	0.42	−10.86	−9.88	93.7
Threading model of *P. falciparum* protease	0.29	−9.58	−4.53	75.5

TABLE 6: Structural parameters of experimentally determined and predicted 3D structures of SC serine proteases.

ID	Taxa	Species	Superposed of AA %	RMSD Å	Distances between the catalytic triad Å		
					(D-H)	(H-S)	(S-D)
1AZW	Bacteria	*X. campestris*	10.20	1.330	4.4	7.8	10.5
2BKL	Bacteria	*M. xanthus*	27.60	0.726	4.6	8.3	10.6
1A88	Bacteria	*S. lividans*	12.30	1.080	4.6	7.9	10.6
1VE7	Archaea	*A. pernix*	15.76	1.083	4.7	8.4	10.4
1MU0	Archaea	*T. acidophilum*	17.00	1.013	4.3	7.8	10.1
1AC5	Fungi	*S. cerevisiae*	17.86	1.441	4.8	7.7	10.8
1WPX	Fungi	*S. cerevisiae*	12.30	1.168	4.6	7.4	10.4
1N1M	Animalia	*H. sapiens*	16.29	1.163	4.8	7.7	10.7
1QFS	Animalia	*S. scrofa*	100	0.000	4.6	8.1	10.5
1ORW	Animalia	*S. scrofa*	18.29	1.092	4.8	7.7	10.7
2BUC	Animalia	*S. scrofa*	17.15	1.108	4.8	7.7	10.7
Mean ± SD of the Cα distances between the triad residues					4.6 ± 0.03	7.9 ± 0.06	10.5 ± 0.04
PM0078228	Plantae	*A. thaliana*	16.7	1.021	4.5	8.2	11.2
PM0078229	Protozoa	*P. falciparum*	69.3	0.686	4.6	8.1	10.5
Mean ± SD of the Cα distances between the triad residues					4.6 ± 0.01	8.2 ± 0.01	10.9 ± 0.25

comparable between the experimental and predicted structures. The good scores provided by Verify3D, ProSA, and ERRAT further validated the overall quality of the refined models from *A. thaliana* (PMDB: PM0078228) and *P. falciparum* (PMDB: PM0078229).

3.2. Catalytic Core Geometry. Superposition of the *A. thaliana* and *P. falciparum* proteases on the representative 1U2E protease structure found that 17% to 69% of the Cα atoms superposed with an RMSD below 1.1 Å (Table 6). In comparison, X-ray protease structures had 10% to 28% of the Cα atoms superposed with an RMSD below 1.5 Å (Table 6). The superposed structures have a common core structure with large variation in loops outside the core (Figure 1). The Cα atom distances of Asp to His, His to Ser, and Asp to Ser averaged over the experimentally determined structures were 4.6 ± 0.03, 7.9 ± 0.06, and 10.5 ± 0.04 Å, respectively (Table 6). The small standard deviations (SDs) indicated that the structural environment around the catalytic triad was highly conserved. Averaged over the predicted structures, the Cα atom distances between the catalytic triad residues were 4.6 ± 0.01, 8.2 ± 0.01, and 10.9 ± 0.25 Å, respectively, in good agreement with the values averaged over the experimental structures. Multiple sequence alignment (Figure 2) confirmed sequence conservation of the catalytic triad residues at Ser158 Asp361 His413 (serine carboxypeptidase 2 numbering). Among the sequences analyzed, the highly conserved amino acids Gly156 and Gly160 had the occupancy percentage of 70% and 77%, respectively, which has been previously described [43]. In addition, Gly161, Asp315, Val317, and Gly343 were all highly conserved with an occupancy percentage of 75% in the S9 family member sequences analyzed. As confirmed in other serine proteases, such residues may confer stabilization of the catalytic site via a hydrogen-bonding interaction

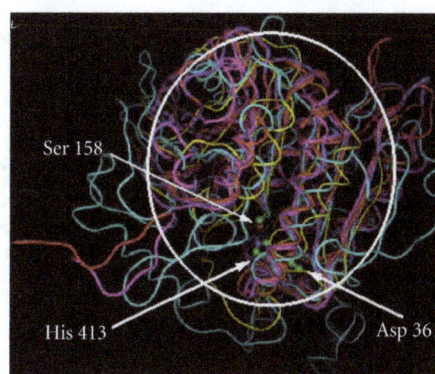

FIGURE 1: Superposed structures of X-ray and modeled structures of the selected proteases of the SC clan. Structures of the plant (PM0078228, *Arabidopsis thaliana*, purple) and protozoan (PM0078229, *Plasmodium falciparum*, yellow) SC proteases were superposed with the animal (1QFS, *Sus scrofa*, red), archaeon (1VE7, *Aeropyrum pernix*, magenta), fungal (1WPX, *Saccharomyces cerevisiae*, cyan), and bacterial (2BKL, *Myxococcus xanthus*, orange) X-ray structures. The catalytic triad residues (Ser, Asp, His; *CBP2* residue numbering used as a standard reference) are shown in ball and stick models, and the core regions of the structures are indicated by the white circle.

[44, 45]. By incorporating an evolutionarily diverse range of SC serine proteases, our analysis indicates that although the core structures deviated considerably during evolution, the relative positions of the catalytic triad Cα atoms maintained very close relative distances and were potentially stabilized by other highly conserved residues.

3.3. Structural Analysis. The catalytic core of all SC clan proteases bears an α/β hydrolase fold, which typically consists

```
                                                    158
T. aestivum (CBP2)        ----------ERFPHYKYRDFYIAGESYAGHYVPELSQLVH----------------------  192
X. campestris (1AZW)      -------IERLRTHLGVDRWQVPFGGSWGSTLALAYAQTHPQQVTELVLRGIFLLRRFELE  144
T. acidophilum (1MU0)     -------EALRSKLFGNEKVFLMGSSYGGALALAYAVKYQDHLKGLIVSGGLSSVPLTVK   139
A. pernix (1VE7)          ------LRRSIAGSRLVWVESFDGSRVPTYVLESGRAPTPGPTVVLVHGGPFAEDSDSW   370
S. cerevisiae (1WPX)      --------GEPSVLPSEECSAMEDSLERCLGLIESCYDSQSVVWSCVPATIYCNNAQLAP  247
S. cerevisiae (1AC5)      -------NFKHTNAHENCQNLINSASTDEAAHFSYQECENILNLLLSYTRESSQKGTAD-  283
H. sapiens (1QFS)         --------------------------NGGSNGGLLVATCANQRPDLFGCVIAQVGVMDMLKFHK  588
M. xanthus (2BKL)         --------------------------YGGSNGGLLVGAAMTQRPELYGAVVCAVPLLDMVRYHL  566
S. scrofa (1 ORW)         --------------------------WGWSYGGYVTSMVLGAGSGVFKCGIAVAPVSKWEYYDS  622
S. scrofa (2BUC)          --------------------------WGWSYGGYVTSMVLGAGSGVFKCGIAVAPVSKWEYYDS  622
H. sapiens (1N1M)         --------------------------WGWSYGGYVTSMVLGSGSGVFKCGIAVAPVSRWEYYDS  261
S. lividans (1A88)        --------------------------ATVSQGLIDHWWLQGMMGAANAHYECIAAFSETD----  194
A. thaliana (MER045469)   --------------------------VGSSAGAPIAGSAVEQVEQVVGYVSLGYPFGLMAS---  154
P. falciparum (MER035185) --------------------------YGRSLGSAASVHIATKRDLLGLVL-------------  144
                                                    • *
```

```
                                                    361
T. aestivum (CBP2)        --------------------------------WVFSGDTDAVVPLTATRYSIGALG--  380
X. campestris (1AZW)      FFEVED--QLLRDAHRIADIPG-----------VIVHGRYDVVCPLQSAWDLHKAWPKA-  284
T. acidophilum (1MU0)     ---------ITDKISAIKIPT-----------LITVGEYDEVTPNVARVIHEKIAGS--  261
A. pernix (1VE7)          EQLTGGSREIMRSRSPINHVDR------IKEPLALIHPQNESRTPLKPLLRLMGELLAR-  534
S. cerevisiae (1WPX)      -DKDFICNWLGNKAWTDVLPWKYDEEFASQKVRNWTASITDEVAGEVKSY---------  384
S. cerevisiae (1AC5)      GDKDLICNNKGVLDTIDNLKWGGIK----------GFSDDAVSFWIHKSKSTDDSEEFS  435
H. sapiens (1QFS)         CSDSKQHFEWLIKYSPLHNVKLPEADDIQYPSMLLLTADHDRVVPLHSLKFIATLQYIV  660
M. xanthus (2BKL)         TAEKPEDFKTLHAYSPYHHVRP----DVRYPALLMMAADHDRVDPMHARKFVAAVQNSP  634
S. scrofa (1 ORW)         LPTPEDNLDYYRNSTVMSRAEN------FKQVEYLLIHGTADNVHFQQSAQLSKALVDA-  688
S. scrofa (2BUC)          LPTPEDNLDYYRNSTVMSRAEN------FKQVEYLLIHGTADNVHFQQSAQLSKALVDA-  688
H. sapiens (1N1M)         LPTPEDNLDHYRNSTVMSRAEN------FKQVEYLLIHGTADNVHFQQSAQISKALVDV-  327
S. lividans (1A88)        ---------FTDDLKRIDVPV-----------LVAHGTDDQVVPYADAAPKSAELLAN  244
A. thaliana (MER045469)   ------------------------------FVMGTQDGFTSVSQLKKKLKSAVGR-  195
P. falciparum (MER035185) ------------------------------FIHGKKDKLLSYH------------  207
                                                    • *
```

```
                                                    413
T. aestivum (CBP2)        --------KGLTLVSVRGAGHEVPLHRPRQALVLFQYF-------------------  453
X. campestris (1AZW)      ----------QLQISPASGHSAFEPENVDALVRATDGFA---------------  318
T. acidophilum (1MU0)     ----------ELHVFRDCSHLTMWEDREGYNKLLSDFILKHL-------------  298
A. pernix (1VE7)          ------GKTFEAHIIPDAGHAINTMEDAVKILLPAVFFLATQRER---------  578
S. cerevisiae (1WPX)      K-------HFTYLRVFNGGHMVPFDVPENALSMVNEWIHGGFSL----------  426
S. cerevisiae (1AC5)      GYVKYDR-NLTFVSVYNASHMVPFDKSLVSRGIVDIYSNDVMIIDNNGKNVMITT----  488
H. sapiens (1QFS)         GRSRKQNNPLLIHVDTKAGHGAGKPTAKVIEEVSDMFAFIARCLNIDWIP---------  715
M. xanthus (2BKL)         GNP-----ATALLRIEANAGHGGADQVAKAIESSVDLYSFLFQVLDVQ----------  682
S. scrofa (1 ORW)         ------GVDFQTMWYTDEDHGIASNMAHQHIYTHMSHFLKQCFSLP-------------  733
S. scrofa (2BUC)          ------GVDFQTMWYTDEDHGIASNMAHQHIYTHMSHFLKQCFSLP-------------  733
H. sapiens (1N1M)         ------GVDFQAMWYTDEDHGIASSTAHQHIYTHMSHFIKQCFS-------------  370
S. lividans (1A88)        ---------ATLKSYEGLPHGMLSTHPEVLNPDLLAFVKS-------------  280
A. thaliana (MER045469)   --------RTETHLIEGVSHFQMEG-------------------------------  215
P. falciparum (MER035185) --------KEQTPVFLNKKHVRKETNNKLYNVLHNNITFHNFNDFL-----------  675
                                                    *
```

FIGURE 2: Multiple amino acid sequence alignment of SC serine proteases. ClustalW was used to align amino acid sequences of SC serine proteases for which their structures were determined experimentally or predicted computationally (highlighted in yellow). Wheat serine carboxypeptidase 2 (CBP2, highlighted in magenta) is used as a standard reference for residue numbering. Only the regions showing the conserved catalytic residues Ser (S), Asp (D), and His (H) are shown (highlighted in blue). Other residues showing medium (•) conservation are highlighted in gray.

of a β-sheet flanked by two or more α-helices. Figures 3(a) and 3(b) are representative X-ray structures of an animal SC protease (1QFS, prolyl oligopeptidase from *Sus scrofa*), comprising 8 β-sheets and 12 α-helices. Figures 3(c) and 3(d) are representative X-ray structures of an archaeon SC protease (1VE7, acylpeptidase hydrolase from *Aeropyrum pernix*), comprising 30 β-sheets and 14 α-helices. The *A. thaliana* SC protease model had 8 β-sheets and 12 α-helices, with Ser120, Asp176, and His198 in separate turn/coil structures (Figure 3(e)). The electrostatic potentials around the Asp and His catalytic residues were mostly electronegative, and there was a patch of electropositive potential around the Ser residue of the catalytic triad (Figure 3(f)). The electronegative region in the catalytic site of the modeled protease could facilitate specificity by favoring positively charged C-terminal amino acid side chains at specific sites within the binding pocket. The *A. thaliana* protease had a higher proportion (>SD of the mean) of aliphatic residues

(32%, molar percentage), compared to other species (see Table S1), which could influence stability of the enzyme at a wide range of temperatures [46]. According to MEROPS annotation (MER045469), this protease has been assigned to the S9 family, but it has an unknown function. Our homology model was essentially built on the structures 2BKL, 1YR2, and 1VZ2, which are prolyl oligopeptidases (S9 family) from *Myxococcus xanthus*, *Novosphingobium capsulatum*, and *Sus scrofa*, respectively. There have been 23 genes encoding prolyl oligopeptidase-like proteins identified in *A. thaliana* [47]. Although the function of most of these is unknown, there is some evidence that prolyl oligopeptidase is involved in seed development [48]. *A. thaliana* is a highly studied model organism, and mutational analysis of this protease would be useful to explore these features.

The protease model from *P. falciparum* had 7 β-sheets and 7 α-helices, with Ser124, Asp188, and His217 in separate turn/coil structures (Figure 3(g)). The surface electrostatic

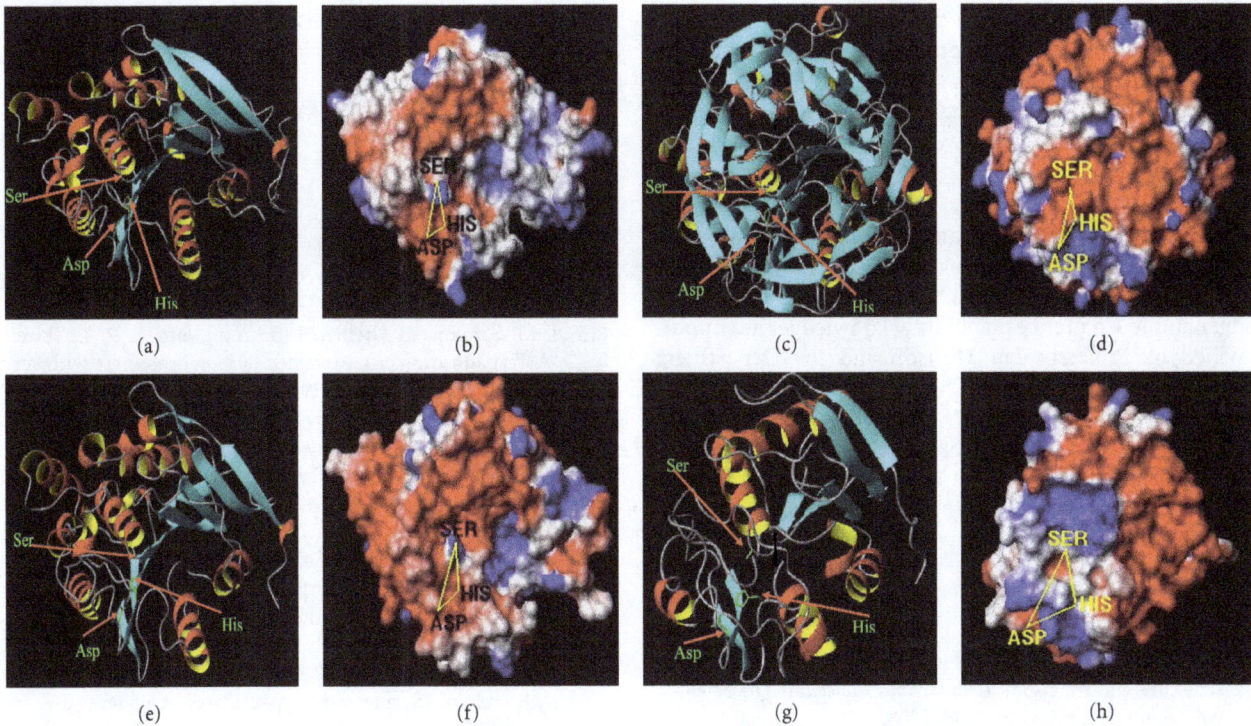

FIGURE 3: Representative X-ray SC protease structures and modeled SC protease structures from *Arabidopsis thaliana* and *Plasmodium falciparum*. Ribbon models of *S. scrofa*, 1QFS (a), *A. pernix*, 1VE7 (c), *A. thaliana* (e), and *P. falciparum* (g) SC protease structures show β-sheets with an arrow directed to the C-terminus (light blue), α-helices (red and yellow), turn/loops (gray), and catalytic triad residue side chains (green sticks). Surface electrostatic potential models of *S. scrofa*, 1QFS (b), *A. pernix*, 1VE7 (d), *A. thaliana*, PM0078228 (f), and *P. falciparum*, PM0078229 (h) SC protease structures show electronegative (red), electropositive (blue), and electroneutral (white) amino acid side chains. Electrostatic potential thresholds: $-1.4 \, kT/e < 0.0 \, kT/e < +1.4 \, kT/e$ (red → white → blue).

potentials around the catalytic site were very different to those of other clan members studied, with large patches of electropositive and electroneutral regions around the catalytic triad residues (Figure 3(h)). The largely electropositive catalytic site of this modeled protease suggests it favors a negatively charged substrate. The largely electroneutral regions possibly relax the stringency of the substrate binding, allowing for a number of different protein substrates. In comparison with the other species analyzed (see Table S1), the *P. falciparum* protease had a higher proportion (>SD of the mean) of polar residues (60%, molar percentage) and basic amino acids (16%), which indicates it could favor a more hydrophilic environment. Like the modeled protease from *A. thaliana*, this protease (MER035185) has also been assigned to the S9 family. Although the function of this protease is not known, it is of interest that both prolyl oligopeptidase and oligopeptidase B of the S9 family appear to facilitate the virulence of other protozoan parasites such as *Trypanosoma cruzi* and *Trypanosoma brucei* [16, 17]. Further investigation of substrate specificity and other properties contributing to it would be beneficial for functional analysis of this protease, as it could be a potential target for rational antimalarial drug design.

The following predicted structures are available in the Protein Model Database (PMDB) (http://mi.caspur.it/PMDB/):

(1) SC serine protease from *Arabidopsis thaliana* (PMDB ID: PM0078228),

(2) SC serine protease from *Plasmodium falciparum* (PMDB ID: PM0078229).

4. Conclusion

In conjunction with 11 experimentally determined 3D protein structures, our analysis of predicted structures from a plant and a protozoan encompassed an evolutionarily diverse range of SC clan proteases. The structural geometry of the catalytic core clearly deviated considerably during evolution, but the relative positions of the catalytic triad residues were conserved, and other highly conserved residues possibly provide stabilization of the core. Evolutionary divergence was also exhibited by large variation in secondary structure features outside the core, differences in overall amino acid distribution, and unique surface electrostatic potential patterns between species. These features are probably associated with environmental adaptation, subcellular localisation, and the diverse functions of the different protease orthologs. The modeled proteases from *A. thaliana* and *P. falciparum* appear to be prolyl oligopeptidases of the S9 family. Evidence indicates that prolyl oligopeptidase is involved in plant seed development [48] and facilitates the virulence of protozoan parasites [16, 17]. Further structural investigation of these proteases would be useful for protein engineering strategies and for rational drug design in the case of the *P. falciparum* protease.

Acknowledgments

A. Laskar and S. Chatterjee are grateful for the funding and infrastructural support provided by the Indian Institute of Chemical Biology, Kolkata, West Bengal, India. E. J. Rodger and A. Chatterjee gratefully acknowledge the support provided by Professor Ian Morison and the Department of Pathology, University of Otago, Dunedin, the Health Research Council (E. J. Rodger), and the National Research Centre for Growth and Development (A. Chatterjee), New Zealand. These entities did not have a role in study design, collection, analysis, or interpretation of data, writing the paper, or decision to submit paper for publication.

References

[1] M. J. Page and E. Di Cera, "Serine peptidases: classification, structure and function," *Cellular and Molecular Life Sciences*, vol. 65, no. 7-8, pp. 1220–1236, 2008.

[2] L. Hedstrom, "Serine protease mechanism and specificity," *Chemical Reviews*, vol. 102, no. 12, pp. 4501–4523, 2002.

[3] M. J. Page and E. Di Cera, "Evolution of peptidase diversity," *Journal of Biological Chemistry*, vol. 283, no. 44, pp. 30010–30014, 2008.

[4] N. D. Rawlings, A. J. Barrett, and A. Bateman, "MEROPS: the database of proteolytic enzymes, their substrates and inhibitors," *Nucleic Acids Research*, vol. 40, no. D1, pp. D343–D350, 2012.

[5] M. Holmquist, "Alpha/Beta-hydrolase fold enzymes: structures, functions and mechanisms," *Current Protein and Peptide Science*, vol. 1, no. 2, pp. 209–235, 2000.

[6] D. Rea and V. Fülöp, "Structure-function properties of prolyl oligopeptidase family enzymes," *Cell Biochemistry and Biophysics*, vol. 44, no. 3, pp. 349–365, 2006.

[7] V. Fülöp, Z. Böcskei, and L. Polgár, "Prolyl oligopeptidase: an unusual β-propeller domain regulates proteolysis," *Cell*, vol. 94, no. 2, pp. 161–170, 1998.

[8] L. Polgár, "The prolyl oligopeptidase family," *Cellular and Molecular Life Sciences*, vol. 59, no. 2, pp. 349–362, 2002.

[9] M. Maes, "Alterations in plasma prolyl endopeptidase activity in depression, mania, and schizophrenia: effects of antidepressants, mood stabilizers, and antipsychotic drugs," *Psychiatry Research*, vol. 58, no. 3, pp. 217–225, 1995.

[10] M. Maes, P. Monteleone, R. Bencivenga et al., "Lower serum activity of prolyl endopeptidase in anorexia and bulimia nervosa," *Psychoneuroendocrinology*, vol. 26, no. 1, pp. 17–26, 2001.

[11] W. R. Welches, K. B. Brosnihan, and C. M. Ferrario, "A comparison of the properties and enzymatic activities of three angiotensin processing enzymes: angiotensin converting enzyme, prolyl endopeptidase and neutral endopeptidase 24.11," *Life Sciences*, vol. 52, no. 18, pp. 1461–1480, 1993.

[12] T. Vilsb ll, T. Krarup, C. F. Deacon, S. Madsbad, and J. J. Holst, "Reduced postprandial concentrations of intact biologically active glucagon-like peptide 1 in type 2 diabetic patients," *Diabetes*, vol. 50, no. 3, pp. 609–613, 2001.

[13] B. Pro and N. H. Dang, "CD26/dipeptidyl peptidase IV and its role in cancer," *Histology and Histopathology*, vol. 19, no. 4, pp. 1345–1351, 2004.

[14] S. L. Naylor, A. Marshall, C. Hensel, P. F. Martinez, B. Holley, and A. Y. Sakaguchi, "The DNF15S2 locus at 3p21 is transcribed in normal lung and small cell lung cancer," *Genomics*, vol. 4, no. 3, pp. 355–361, 1989.

[15] R. Erlandsson, F. Boldog, B. Persson et al., "The gene from the short arm of chromosome 3, at D3F15S2, frequently deleted in renal cell carcinoma, encodes acylpeptide hydrolase," *Oncogene*, vol. 6, no. 7, pp. 1293–1295, 1991.

[16] J. M. Santana, P. Grellier, J. Schrével, and A. R. L. Teixeira, "A Trypanosoma cruzi-secreted 80 kDa proteinase with specificity for human collagen types I and IV," *Biochemical Journal*, vol. 325, no. 1, pp. 129–137, 1997.

[17] R. E. Morty, J. D. Lonsdale-Eccles, J. Morehead et al., "Oligopeptidase B from Trypanosoma brucei, a new member of an emerging subgroup of serine oligopeptidases," *Journal of Biological Chemistry*, vol. 274, no. 37, pp. 26149–26156, 1999.

[18] Y. Kumagai, K. Konishi, T. Gomi, H. Yagishita, A. Yajima, and M. Yoshikawa, "Enzymatic properties of dipeptidyl aminopeptidase IV produced by the periodontal pathogen Porphyromonas gingivalis and its participation in virulence," *Infection and Immunity*, vol. 68, no. 2, pp. 716–724, 2000.

[19] K. Breddam, "Serine carboxypeptidases. A review," *Carlsberg Research Communications*, vol. 51, no. 2, pp. 83–128, 1986.

[20] J. A. Endrizzi, "2.8-Å structure of yeast serine carboxypeptidase," *Biochemistry®*, vol. 33, no. 37, pp. 11106–11120, 1994.

[21] D. I. Liao and S. J. Remington, "Structure of wheat serine carboxypeptidase II at 3.5-Å resolution. A new class of serine proteinase," *Journal of Biological Chemistry*, vol. 265, no. 12, pp. 6528–6531, 1990.

[22] G. U. Yüksel and J. L. Steele, "DNA sequence analysis, expression, distribution, and physiological role of the Xaa-prolyldipeptidyl aminopeptidase gene from Lactobacillus helveticus CNRZ32," *Applied Microbiology and Biotechnology*, vol. 44, no. 6, pp. 766–773, 1996.

[23] C. E. Odya, D. V. Marinkovic, and K. J. Hammon, "Purification and properties of prolylcarboxypeptidase (angiotensinase C) from human kidney," *Journal of Biological Chemistry*, vol. 253, no. 17, pp. 5927–5931, 1978.

[24] D. A. Mele, P. Bista, D. V. Baez, and B. T. Huber, "Dipeptidyl peptidase 2 is an essential survival factor in the regulation of cell quiescence," *Cell Cycle*, vol. 8, no. 15, pp. 2425–2434, 2009.

[25] L. Zhang, Y. Jia, L. Wang, and R. Fang, "A proline iminopeptidase gene upregulated in planta by a LuxR homologue is essential for pathogenicity of Xanthomonas campestris pv. campestris," *Molecular Microbiology*, vol. 65, no. 1, pp. 121–136, 2007.

[26] C. Mandal, "MODELYN: a molecular modelling program, version PC-1.0. Indian copyright No. 9/98," *Copyright Office, Government of India*, 1998.

[27] N. D. Rawlings, A. J. Barrett, and A. Bateman, "MEROPS: the peptidase database," *Nucleic Acids Research*, vol. 38, no. 1, Article ID gkp971, pp. D227–D233, 2009.

[28] T. Schwede, J. Kopp, N. Guex, and M. C. Peitsch, "SWISS-MODEL: an automated protein homology-modeling server," *Nucleic Acids Research*, vol. 31, no. 13, pp. 3381–3385, 2003.

[29] J. Meller and R. Elber, "Linear programming optimization and a double statistical filter for protein threading protocols," *Proteins*, vol. 45, no. 3, pp. 241–261, 2001.

[30] A. Laskar, E. Rodger, A. Chatterjee, and C. Mandal, "Modeling and structural analysis of evolutionarily diverse S8 family

serine proteases," *Bioinformation*, vol. 7, no. 5, pp. 239–245, 2011.

[31] A. Laskar, E. Rodger, A. Chatterjee, and C. Mandal, "Modeling and structural analysis of PA clan serine proteases," *BMC Research Notes*, vol. 5, no. 1, p. 256, 2012.

[32] *Insight II Modeling Environment*, Molecular Simulations, San Diego, Calif, USA, 2005.

[33] A. A. Canutescu, A. A. Shelenkov, and R. L. Dunbrack Jr, "A graph-theory algorithm for rapid protein side-chain prediction," *Protein Science*, vol. 12, no. 9, pp. 2001–2014, 2003.

[34] R. A. Laskowski, "PROCHECK: a program to check the stereochemical quality of protein structures," *Journal of Applied Crystallography*, vol. 26, pp. 283–291, 1993.

[35] I. W. Davis, L. W. Murray, J. S. Richardson, and D. C. Richardson, "MolProbity: structure validation and all-atom contact analysis for nucleic acids and their complexes," *Nucleic Acids Research*, vol. 32, pp. W615–W619, 2004.

[36] R. Luthy, J. U. Bowie, and D. Eisenberg, "Assesment of protein models with three-dimensional profiles," *Nature*, vol. 356, no. 6364, pp. 83–85, 1992.

[37] M. Wiederstein and M. J. Sippl, "ProSA-web: interactive web service for the recognition of errors in three-dimensional structures of proteins," *Nucleic acids research*, vol. 35, pp. W407–410, 2007.

[38] C. Colovos and T. O. Yeates, "Verification of protein structures: patterns of nonbonded atomic interactions," *Protein Science*, vol. 2, no. 9, pp. 1511–1519, 1993.

[39] Y. D. Yang, P. Spratt, H. Chen, C. Park, and D. Kihara, "Sub-AQUA: real-value quality assessment of protein structure models," *Protein Engineering, Design and Selection*, vol. 23, no. 8, pp. 617–632, 2010.

[40] R. Koradi, M. Billeter, and K. Wüthrich, "MOLMOL: a program for display and analysis of macromolecular structures," *Journal of Molecular Graphics*, vol. 14, no. 1, pp. 51–55, 1996.

[41] J. D. Thompson, D. G. Higgins, and T. J. Gibson, "CLUSTAL W: improving the sensitivity of progressive multiple sequence alignment through sequence weighting, position-specific gap penalties and weight matrix choice," *Nucleic Acids Research*, vol. 22, no. 22, pp. 4673–4680, 1994.

[42] P. Rice, L. Longden, and A. Bleasby, "EMBOSS: the European molecular biology open software suite," *Trends in Genetics*, vol. 16, no. 6, pp. 276–277, 2000.

[43] M. M. Krem and E. Di Cera, "Molecular markers of serine protease evolution," *EMBO Journal*, vol. 20, no. 12, pp. 3036–3045, 2001.

[44] R. J. Siezen and J. A. M. Leunissen, "Subtilases: the superfamily of subtilisin-like serine proteases," *Protein Science*, vol. 6, no. 3, pp. 501–523, 1997.

[45] L. B. Smillie and B. S. Hartley, "Histidine sequences in the active centres of some 'serine' proteinases," *Biochemical Journal*, vol. 101, no. 1, pp. 232–241, 1966.

[46] S. Roy, N. Maheshwari, R. Chauhan, N. K. Sen, and A. Sharma, "Structure prediction and functional characterization of secondary metabolite proteins of Ocimum," *Bioinformation*, vol. 6, no. 8, pp. 315–319, 2011.

[47] L. P. Tripathi and R. Sowdhamini, "Cross genome comparisons of serine proteases in Arabidopsis and rice," *BMC Genomics*, vol. 7, p. 200, 2006.

[48] L. Gutierrez, M. Castelain, J. L. Verdeil, G. Conejero, and O. Van Wuytswinkel, "A possible role of prolyl oligopeptidase during Linum usitatissimum (flax) seed development," *Plant Biology*, vol. 10, no. 3, pp. 398–402, 2008.

17

HIV-1 Reverse Transcriptase Still Remains a New Drug Target: Structure, Function, Classical Inhibitors, and New Inhibitors with Innovative Mechanisms of Actions

Francesca Esposito, Angela Corona, and Enzo Tramontano

Department of Life and Environmental Sciences, University of Cagliari, Cittadella Universitaria di Monserrato, SS 554, 09042 Monserrato, Italy

Correspondence should be addressed to Enzo Tramontano, tramon@unica.it

Academic Editor: Gilda Tachedjian

During the retrotranscription process, characteristic of all retroviruses, the viral ssRNA genome is converted into integration-competent dsDNA. This process is accomplished by the virus-coded reverse transcriptase (RT) protein, which is a primary target in the current treatments for HIV-1 infection. In particular, in the approved therapeutic regimens two classes of drugs target RT, namely, nucleoside RT inhibitors (NRTIs) and nonnucleoside RT inhibitors (NNRTIs). Both classes inhibit the RT-associated polymerase activity: the NRTIs compete with the natural dNTP substrate and act as chain terminators, while the NNRTIs bind to an allosteric pocket and inhibit polymerization noncompetitively. In addition to these two classes, other RT inhibitors (RTIs) that target RT by distinct mechanisms have been identified and are currently under development. These include translocation-defective RTIs, delayed chain terminators RTIs, lethal mutagenesis RTIs, dinucleotide tetraphosphates, nucleotide-competing RTIs, pyrophosphate analogs, RT-associated RNase H function inhibitors, and dual activities inhibitors. This paper describes the HIV-1 RT function and molecular structure, illustrates the currently approved RTIs, and focuses on the mechanisms of action of the newer classes of RTIs.

1. Introduction

Since the human immunodeficiency virus (HIV) has been established to be the etiological agent of the acquired immunodeficiency syndrome (AIDS) [1, 2], an originally unpredicted number of drugs have been approved for the treatment of the HIV-infected patients [3]. This success in effective drugs identification, certainly unique in the treatment of viral infections, together with the use of such armamentarium in different combination therapeutic regimens, has transformed a highly lethal syndrome into a chronic disease [4]. The management of this disease, however, is still complex and worrisome due to problems such as monitoring of therapy efficacy, chronic administration drug toxicity, poor tolerability, drug resistance development, or therapy adjustment after treatment failures [4]. For all these reasons, the search for new inhibitors, possibly acting with molecular mechanisms different from the ones of the already approved drugs or anyway showing different patterns of drug resistance and, possibly, with diverse drug-associated chronic toxicity, is still a worldwide health care issue.

The success in HIV infection therapy is certainly related to the fact that the HIV life cycle has been intensely dissected; several of its steps have been validated as drug targets, and, subsequently, a number of viral inhibitors have been identified and developed against many of them [3, 4]. Among the HIV proteins which have been deeply characterized as major drug targets is the reverse transcriptase (RT), the virus coded enzyme that converts the ssRNA viral genome into the dsDNA provirus which is consequently imported into the cell host nucleus and integrated into the host chromosome by another virus-coded protein, integrase (IN). The present paper focuses on the RT function within the virus cycle, its molecular structure, the mechanism of action of the

HIV-1 Reverse Transcriptase Still Remains a New Drug Target: Structure, Function, Classical Inhibitors, and New Inhibitors with Innovative Mechanisms of Actions

159

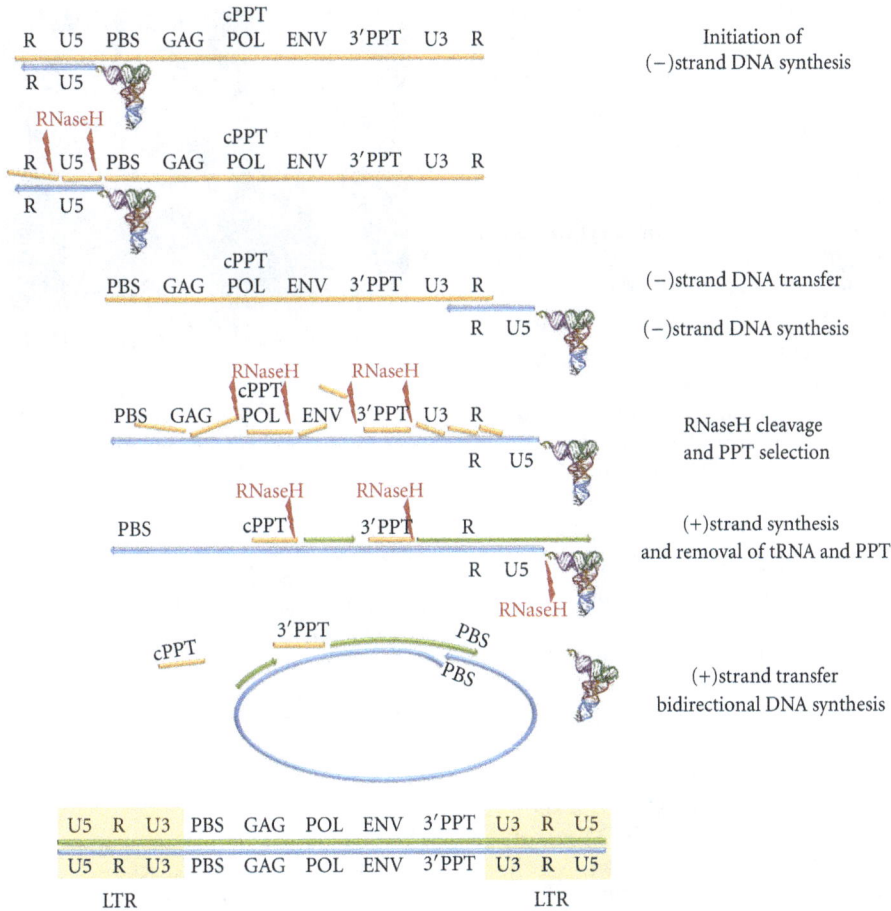

FIGURE 1: HIV-1 reverse transcription process. Step 1: host cell tRNALys3 hybridizes to the PBS near the 5′-end of the (+)strand RNA genome (orange). (−)strand DNA (blue) synthesis starts using host tRNALys3 as a primer. DNA synthesis proceeds up to the 5′-end of the RNA genome. Step 2: RNase H hydrolysis of the RNA portion of the RNA:DNA hybrid product exposes the ssDNA product determining the (−)strand strong stop DNA. Step 3: strand transfer of the (−)strand DNA through its hybridization with the R region at the 3′-end of the ssRNA genome and further elongation of the (−)strand DNA. Step 4: DNA synthesis proceeds, and the RNase H function cleaves the RNA strand of the RNA:DNA at numerous points leaving intact two specific sequences (cPPT, 3′PPT) resistant to the RNase H cleavage. Step 5: (−)strand DNA synthesis (green) initiation using PPTs as primers. Step 6: RNase H hydrolysis of the PPT segments and the junction of the tRNA:DNA hybrid, freeing the PBS sequence of the (+)strand DNA. Step 8: strand transfer of the PBS sequence of the (+)strand DNA that anneals to the PBS on the (−)strand DNA. DNA synthesis then continues with strand displacement synthesis. Step 9: the product is a linear dsDNA with long terminal repeats (LTRs) at both ends.

currently approved RT inhibitors (RTIs), and the newer classes of RTIs and their modes of action.

2. Retrotranscription Process

After the HIV particle fuses with the host cell surface, the viral particle content is released within the host cell cytoplasm where the viral ssRNA genome serves as template to obtain a proviral dsDNA that is integrated into the host genome, becoming a source of mRNAs coding for viral proteins and ssRNA genomes that, together, will form the new viral particles. The conversion of the viral ssRNA genome into integration-competent dsDNA, termed retrotranscription (Figure 1), is characteristic of all retroviruses, and its accomplishment requires viral as well as

cellular elements, among which the most important is the virus-coded RT protein.

Each HIV particle contains two copies of (+)ssRNA genome sequence of 9,7 kb [5] coding for structural and nonstructural proteins and having, in the 5′- and 3′-ends, two identical sequences. Near the 5′-end of the viral genome there is an 18-nucleotides-long segment, termed primer binding site (PBS), which is complementary to the 3′-end 18 nucleotides of the human tRNALys3. When the cellular tRNA is hybridized to the PBS, it serves as an RNA primer, and the RT-associated DNA polymerase function can initiate the first (−)strand DNA synthesis using viral RNA genome as a template (Figure 1). After tRNA elongation until the ssRNA 5′-end, there is a first (−)strand strong-stop DNA. In fact, the (−)strand DNA synthesis generates an RNA:DNA hybrid that is a substrate for the

RT-associated ribonuclease H (RNase H) function which selectively degrades the RNA strand of the RNA:DNA hybrid [6], leaving the nascent (−)strand DNA free to hybridize with the complementary sequence at the 3′-end of one of the two viral genomic ssRNAs. A strand transfer, therefore, occurs from the R region at the 5′-end of the genome to the equivalent R region at the 3′-end (see Figure 1). After this step, termed (−)strand transfer, (−)strand synthesis can continue along the viral RNA starting from its 3′-end. Whilst DNA synthesis proceeds, the RNase H function cleaves the RNA strand of the RNA:DNA at numerous points. Although most of the RNase H cleavages do not appear to be sequence specific, there are two specific purine-rich sequences, known as the polypurine tracts (PPTs), that are resistant to the RNase H cleavage and remain annealed with the nascent (−)strand DNA. These two well-defined sites are located in the central part of the HIV-1 genome. In particular, the 3′-end PPT defines the 5′-end of the viral coding (+)strand DNA synthesis since this PPT serves as primer [7, 8]. The (+)strand DNA synthesis continues to the 5′-end of the (−)strand DNA and uses also the 18-nucleotides PBS sequence of the tRNA as a template. Importantly, the 19th base from the 3′-end of tRNALys3 is a methyl A, and the presence of this modified base blocks the RT, generating a (+)strand strong-stop DNA. Subsequently, the RNase H function cleaves the RNA segment of the tRNA:DNA hybrid, freeing the PBS sequence of the (+)strand DNA and allowing it to anneal to the complementary site near the 3′-end of the extended (−)strand DNA [9]. Then, a bidirectional synthesis occurs to complete a viral dsDNA that has a 90-nucleotides single-stranded flap at the center. This unusual situation is probably solved by host mechanisms, and one candidate for flap removal is the flap endonuclease-1 (FEN-1) [8]. Finally, a specific cleavage removes the PPT primers and exposes the integration sequence to facilitate the insertion of the viral dsDNA into the host chromosome.

3. RT Structure and Functions

As a major target for anti-HIV therapy, RT has been the subject of extensive research through crystal structure determinations, biochemical assays, and single-molecule analyses. RT derives from a virus-coded polyprotein that is processed by the viral protease to give rise to two related subunits of different length, the p66 and the p51, that share a common amino terminus and combine in a stable asymmetric heterodimer [10]. Analysis of the crystal structure of RT reveals that p66 is composed of two spatially distinct domains, polymerase and RNase H domains (Figure 2). The polymerase domain shows a characteristic highly conserved structure that resembles a right hand, consisting of fingers (residues 1–85 and 118–155), palm (residues 86–117 and 156–237), and thumb (residues 238–318) subdomains. The p66 subunit also comprises the connection subdomain (residues 319–426) and RNase H domain (residues 427–560) [11, 12]. The p51 subunit lacks the RNase H domain and has the same four subdomains of the p66 polymerase domain whose relative positions, however, are different. For

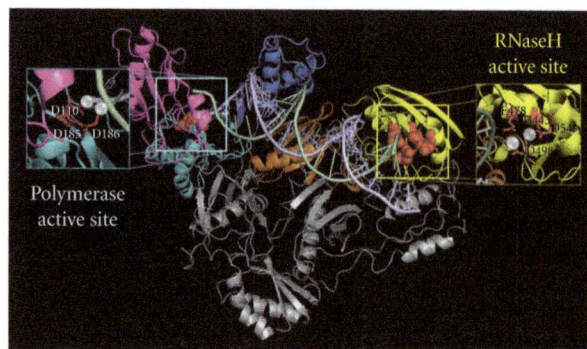

FIGURE 2: Structure of HIV-1 RT. The enzyme has two domains: the p66 (colored) and the p51 (gray). The polymerase domain shows a characteristic highly conserved structure that resembles a right hand, consisting of fingers domain (magenta), palm domain (cyan), thumb domain (blue). The p66 subunit also comprises the connection domain (orange) and RNase H domain (yellow). The polymerase active site is located in the middle of palm, fingers, and thumb subdomains. The three catalytic aspartic acid residues (D110, D185 and D186) located in the palm subdomain of p66 that bind the cofactor divalent ions (Mg^{2+}) are shown (red). The RNase H domain is located at C-terminus of the p66 subunit, 60 Å far from polymerase active site. The RNase H active site contains a DDE motif comprising the carboxylates residues D443, E478, D498, and D549 that can coordinate two divalent Mg^{2+}.

this reason, the p51 subunit folds differently from p66; it does not have enzymatic activities while it serves to anchor the proper folding of the p66 subunit that performs all the catalytic functions.

RT is primarily responsible for several distinct activities that are all indispensable for the retrotranscription process: RNA- and DNA-dependent DNA synthesis, RNase H activity, strand transfer, and strand displacement synthesis [13]. The presence of all these functions in a single protein is facilitated by the highly dynamic RT nature which allows RT to spontaneously slide over long distances of RNA:DNA and DNA:DNA duplexes, to easily target the primer terminus for DNA polymerization, to rapidly access multiple sites, and, hence, to make up for its low processivity [13]. RT sliding does not require energy from nucleotide hydrolysis, and it is supposed to be a thermally driven diffusion process [13]. Noteworthy, it has been recently shown that RT can bind to the nucleic acid substrates in two different orientations, termed "RNase H cleavage competent orientation" and "polymerase competent orientation," and that each of them allows to catalyze one of the two RT-associated enzymatic activities [14]. These two binding modes are in a dynamic equilibrium, and it has been demonstrated that RT can spontaneously and rapidly switch between these orientations without dissociating from the substrate. This flipping can be influenced by the presence of small molecules as nucleotides that stabilize the polymerase competent orientation or inhibitors that, conversely, destabilize it [8]. Together, shuttling and switching give rise to a very complex series of conformational changes that increase enormously the replication efficiency, combining DNA polymerization and RNA cleavage.

HIV-1 Reverse Transcriptase Still Remains a New Drug Target: Structure, Function, Classical Inhibitors, and New Inhibitors with Innovative Mechanisms of Actions

161

3.1. RNA- and DNA-Dependent DNA Synthesis. The DNA synthesis, catalyzed by both RT-associated RNA- and DNA-dependent DNA polymerase activities (RDDP and DDDP, resp.), occurs with a mechanism that is similar to other DNA polymerases [15]. The polymerase active site is located in the middle of the palm, fingers, and thumb subdomains. In particular, the palm subdomain is very important for positioning of the primer terminus in the correct orientation for nucleophilic attack on an incoming dNTP [16]. Three aspartic acids residues (D110, D185, and D186) located in the palm subdomain of p66 bind the divalent ion cofactor (Mg^{2+}) through their catalytic carboxylates group, and are essential for catalysis (Figure 2) [17]. DNA synthesis requires that RT binds to the template:primer on the priming binding site; this interaction is stabilized by a change of the conformation of the p66 thumb (from close to open). Then, the dNTP binds at the nucleotide binding site to form an RT:DNA:dNTP ternary complex [18]. Afterwards, a conformational change of the fingers traps the dNTP, precisely aligning the α-phosphate of the dNTP and the 3′-OH of the primer inside of polymerase active site (this is actually the rate limiting step). Under these conditions, the enzyme catalyzes the formation of a phosphodiester bond between the primer 3′-OH and the dNMP with the release of a pyrophosphate. Then, the pyrophosphate is free to go out of the catalytic site. Finally, translocation of the elongated DNA primer frees the nucleotide-binding site for the next incoming dNTP or, alternatively, RT can dissociate from the complex. Compared to cellular DNA polymerases, RT exhibits a very low processivity, typically dissociating from the substrate after synthesizing only a few to a few hundred nucleotides. This may contribute to the fidelity of RT and results in the accumulation of mutations during reverse transcription.

Importantly, during its DNA polymerase activity RT can run up against several template secondary structures. Particularly, the RNA template can form stable RNA:RNA interactions that can occlude the polymerization site and/or displace the primer terminus. In this case, RT has been shown to realize a strand displacement synthesis, in which the sliding movement can contribute to the reannealing of the primer, displacing the RNA [17].

3.2. DNA-Directed RNA Cleavage. RT is able to degrade selectively the RNA portion of an RNA:DNA hybrid and to remove the priming tRNA and PPT. This RNase H function is essential for virus replication since RNase H-deficient viruses are noninfectious [19]. The RNase H domain is located at C-terminus of the p66 subunit, 60 Å far from polymerase active site (Figure 2) equivalent to 17 nucleotides of a DNA:DNA hybrid and/or 18 nucleotides of a RNA:DNA hybrid [20]. The RNase H active site contains a highly conserved, essential, DDE motif comprising the carboxylates residues D443, E478, D498, and D549, that can coordinate two divalent Mg^{2+} cations, consistently with the proposed phosphoryl transfer geometry [21]. Mutations in any of the D443, D498, and E478 residues abolish enzyme activity [22, 23]. The RNase H domain can catalyze a phosphoryl transfer

through nucleophilic substitution reactions on phosphate ester. This action occurs through the deprotonation of a water molecule, with the production of a nucleophilic hydroxide group that attacks the scissile phosphate group on the RNA previously activated by coordination with the Mg^{2+} cofactor [24]. The reason for the RNase H cleavage specificity for the RNA portion of the RNA:DNA hybrid mainly relies on its particular minor groove width and its interaction with the "primer grip" (an extensive network of contacts between the hybrid phosphate backbone and several residues far ~4–9 bp from the RNase H active site) [16]. The RNA:DNA hybrid has a minor groove width of ~9-10 Å, that is intermediate between the A- and B-form of other double-stranded nucleic acids (dsNA). The HIV-1 RNase H hydrolyzes much less efficiently hybrids with lower widths, such as the PPTs that show a width of 7 Å probably due to the presence of A-tracts [17, 25]. This fact allows the PPT recognition as RNA primers for DNA synthesis and may also represent a further specific viral target.

The RNase H catalysis can occur in a polymerase-dependent or polymerase-independent mode, and it is possible to distinguish three different cleavage types: "DNA 3′-end-directed cleavage," "RNA 5′-end-directed cleavage," and "internal cleavage" [26]. The former acts during (−)strand DNA synthesis, when the RNase H active site cleaves the RNA in a position based on the binding of the polymerase active site to the 3′-end of the new (−)DNA [27]. The second one acts when RT binds to a recessed RNA 5′-end annealed to a longer DNA strand, and the RNase H function cleaves the RNA strand 13–19 nucleotides away from its 5′-end. The internal cleavage occurs since the RNA cleavage is slower than DNA synthesis, and, given that a viral particle contains 50–100 RTs molecules and only two copies of (+)RNA, all the nonpolymerizing RTs can bind to the hybrid and degrade the RNA segment by a polymerase-independent mode [16].

3.3. Strand Transfer. The strand transfer is a critical step during the reverse transcription process in which two complementary ssNAs have to anneal to allow the pursuance of DNA synthesis (Figure 1). In both (−) and (+)strand transfers the ssNA develops secondary structures: the R region consists of a strong-structured motif TAR hairpin and a poly(A) hairpin [28]. Also the PBS sequence at the 3′-end of the (−)strand DNA can form a stable hairpin structure. Therefore, RT is helped in performing this step by the presence of the viral-coded nucleocapsid (NC) protein [29, 30]. The strand transfer process, together with the RT fidelity and the presence of other host factors such as APOBEC [31], helps to explain the high rate of recombination events to allow HIV to evolve rapidly and develop resistance to drugs.

3.4. Pyrophosphorolysis. As most DNA polymerases, RT can catalyze the reversal of the dNTP incorporation that is termed pyrophosphorolysis. RT has the ability to carry out this reverse reaction using a pyrophosphate (PPi) molecule or an NTP, such as ATP, as the acceptor substrate [32–34] giving rise to a dinucleotide tetraphosphate (formed by the excised dNMP and the acceptor ATP substrate) and

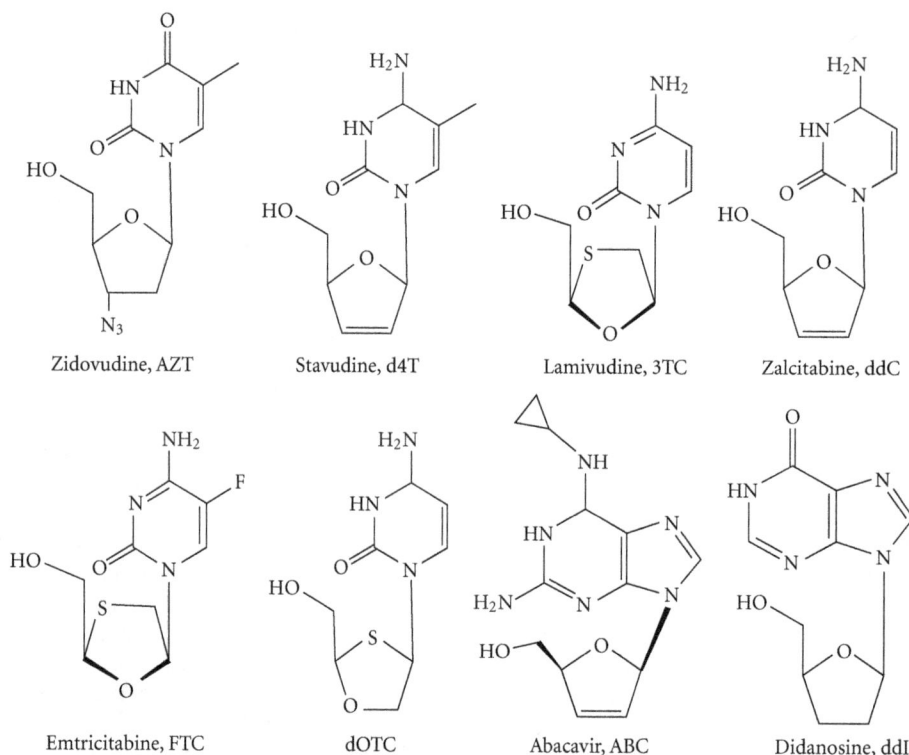

FIGURE 3: Chemical structures of approved NRTIs.

a free 3′-OH end as reaction products. This RT function is particularly important, as discussed later, in some drug resistance mechanisms.

4. Current RTIs: Structure, Mode of Action, and Resistance

The approved combination treatments used for HIV-1 include two classes of RTIs that target the viral enzyme with two different mechanism of action. The first class comprises compounds known as nucleoside/nucleotide RT inhibitors (NRTIs/NtRTIs), while the second class comprises compounds known as nonnucleoside RT inhibitors (NNRTIs).

4.1. Nucleoside RT Inhibitors. There are currently eight NRTIs clinically available, structurally resembling both pyrimidine and purine analogues [3]. Pyrimidine nucleoside analogues include thymidine analogues such as 3′-azido-2′,3′-dideoxythymidine (zidovudine, AZT), and 2′,3′-di-dehydro-2′,3′-dideoxythymidine (stavudine, d4T) and cytosine analogues such as (−)-2′,3′-dideoxy-3′-thiacytidine (lamivudine, 3TC), 2′,3′-dideoxycytidine (zalcitabine, ddC) which, however, is no longer recommended due to peripheral neuropathy [35], (−)-2′,3′-dideoxy-5-fluoro-3′-thiacytidine (emtricitabine, FTC), and [(−)-2′-deoxy-3′-oxa-4′-thiacytidine) (dOTC). Purine nucleoside analogues include (IS-4R)-4-[2-amino-6(cyclopropylamino)-9H-purin-9yl]-2-cyclopentane-I-methanol (abacavir, ABC) and 2′,3′-dideoxyinosine (didanosine, ddI) as guanosine and adenine analogues,

respectively (Figure 3) [3]. These agents, in order to inhibit reverse transcription, have to be phosphorylated by cellular kinases to their triphosphate derivatives. All NRTIs follow the same mechanism of RT inhibition: once activated to their triphosphate form, they are incorporated by RT into the growing primer (Figure 4), competing with the natural dNTPs and terminating DNA synthesis due to their lack of the 3′-hydroxyl group (Figure 5). Therefore, once incorporated into dsDNA they prevent the incorporation of the incoming nucleotide. Importantly, while HIV-1 RT uses these NRTIs as substrates, the cellular DNA polymerases do not recognize them with the same affinity.

Under selective drug pressure, drug resistant viral mutants can gain a competitive advantage over wt virus and become the dominant quasispecies. HIV-1 resistance to NRTIs usually involves two general mechanisms: NRTI discrimination, that reduces the NRTI incorporation rate, and NRTI excision that unblocks NRTI-terminated primers. A simple example of discrimination is steric hindrance in which there is a selective alteration of the NRTI binding and/or incorporation rate such as in the case of the M184V mutation and 3TC [36, 37], where the valine substitution makes steric contacts with the sulfur of the oxathiolane ring of 3TC triphosphate, preventing its proper positioning for catalysis [38]. Even though the discrimination mechanism is less obvious for other NRTIs, in which structurally poorer compounds (e.g., the ones just lacking the 3′-OH group) should be differentially recognized, mutations in the nucleoside-binding site such as K65R, T69D, L74V, V75T, located in the β3-β4 loop of the p66 fingers subdomain, have

HIV-1 Reverse Transcriptase Still Remains a New Drug Target: Structure, Function, Classical Inhibitors, and New Inhibitors with Innovative Mechanisms of Actions

163

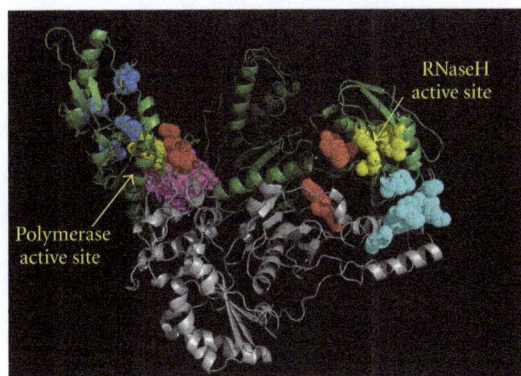

FIGURE 4: Amino acid residues involved in RTI binding. RT two subunits are in green (p66) and in gray (p51). The catalytic residues of the polymerase active site and the RNase H active site are colored in yellow. NRTIs and NtRTIs interact with residues close to the polymerase active site (blue). NNRTIs bind in a hydrophobic pocket next to the polymerase active site (magenta). RHRTIs such as DKAs, N-hydroxyimides, N-hydroxy quinazolinediones and naphthyridine derivatives bind in the RNase H active site (in yellow on the right). Vinylogous ureas bind to a hydrophobic pocket at the interface between the RNase H domain and the p51 subunit (cyan). Hydrazone derivatives have been proposed to bind two different sites (red). One located between the polymerase active site and the NNRTI-binding pocket (sharing a few residues with it) and the second one located between the RNase H and the connection domain. Anthraquinone derivatives have been proposed to bind to the first hydrazone pocket next to the NNRTI-binding site.

been reported to allow a better RT discrimination between NRTI triphosphates and natural dNTPs, since they are involved in the RT interaction with the incoming dNTP [39, 40]. Differently, M41L, D67N, D70R, L210W, T215F/Y, and K219Q mutations, located around the dNTP-binding pocket and also termed thymidine analogs mutations (TAMs), increase NRTI excision. In particular, D67N and K70R are the most important in the excision of 3'-end NRTI-terminated DNA while T215F/Y may increase the RT affinity for the excision substrate ATP so that the NRTI excision is reasonably efficient at ATP physiological concentrations [32, 40, 41]. Other TAMs such as M41L and L210W may stabilize the 215F/Y interaction with the dNTP-binding pocket [42], whereas the K219Q mutation may increase the RT processivity to compensate the higher rate of 3'-nucleotide removal [32, 34]. Recently, mutations in the connection and RNase H domains have also been shown to confer NRTI resistance [43–47]. In particular, connection mutations such as E312Q, G335C/D, N348I, A360I/V, V365I, and A376S have been shown to increase AZT resistance up to 500-fold in the context of TAMs by reducing RNase H activity [43]. This RNase H-dependent mechanism of NRTI resistance has been proposed to be due to an increase in NRTI excision determined by a reduction of RNase H activity [44]. In contrast, the connection mutation G333D, in the context of TAMs and M184V mutation, increases discrimination against 3TC-MP incorporation [48], suggesting an RNase H-independent mechanism of NRTI resistance probably due to

long-range interactions and conformational changes in the connection domain [49].

4.2. Nucleotide RT Inhibitors. NtRTIs, such as (R)-9-(2phosphonylmethoxypropyl)-adenine (tenofovir, PMPA) (Figure 6), are compounds that already have a phosphonate group resistant to hydrolysis [3]. Therefore, they only need two phosphorylation steps to be converted to their active diphosphate derivatives, abbreviating the intracellular activation pathway and allowing a more rapid and complete conversion to the active agent [50, 51]. Similarly to NRTIs, NtRTIs are phosphorylated to the corresponding diphosphates by cellular enzymes and serve as alternative substrates (competitive inhibitors); once incorporated into the growing viral DNA, they act as obligatory chain terminators [50]. NtRTIs such as tenofovir are taken as prodrugs to facilitate penetration of target cell membranes. Subsequently, endogenous chemolytic enzymes release the original nucleoside monophosphate analogue that exerts its action [51].

4.3. Nonnucleoside RT Inhibitors. NNRTIs are structurally and chemically dissimilar compounds that bind in noncompetitive manner to a hydrophobic RT pocket close to the polymerase active site (Figure 4), distorting the protein and inhibiting the chemical step of polymerization [3, 52]. In fact, NNRTIs binding to RT induces rotamer conformational changes in some residues (Y181 and Y188) and makes the thumb region more rigid, blocking DNA synthesis. Importantly, unlike NRTIs, NNRTIs do not require intracellular metabolism to exert their activity. More than thirty different classes of compounds could be considered to be NNRTIs [3]. The currently approved NNRTIs are 11-cyclopropyl-4-methyl-5H-dipyrido[3,2-b:2',3'-e][1,4]diazepin-6(11H)-one (nevirapine), (S)-6-chloro-4-(cyclopropylethynyl)-4-(trifluoromethyl)-1H-benzo[d][1,3]oxazin-2(4H)-one (efavirenz), N-(2-(4-(3-(isopropylamino)pyridin-2-yl) piperazine-1-carbonyl)indolin-5-yl)methanesulfonamide (delavirdine) and 4-((6-amino-5-bromo-2-((4-cyanophenyl)amino)pyrimidine-4-yl)oxy)-3,5-dimethylbenzonitrile (etravirine) and 4-((4-((4-(cyanomethyl)-2,6-dimethylphenyl)amino)pyrimidin-2-yl)amino)benzonitrile (rilpivirine) (Figure 7).

Crystallography, molecular modeling and docking studies have revealed that first generation NNRTIs assume a butterfly-like conformation [53–57]. The stabilization of the NNRTI binding in the allosteric site is accomplished through (i) stacking interactions between the NNRTIs aromatic rings and the side chains of Y181, Y188, W229, and Y318 residues in the RT lipophilic pocket; (ii) electrostatic forces (particularly significant for K101, K103, and E138 residues); (iii) van der Waals interactions with L100, V106, V179, Y181, G190, W229, L234, and Y318 residues; (iv) hydrogen bonds between NNRTI and the main chain (carbonyl/amino) peptide bonds of RT [53, 54, 58, 59]. Larger first-generation inhibitors, such as delavirdine, extend towards the flexible loop containing the P236 residue, while maintaining stacking interactions with the tyrosine residues 181 and 188 and hydrogen bonding with K103 [60]. Stacking interactions

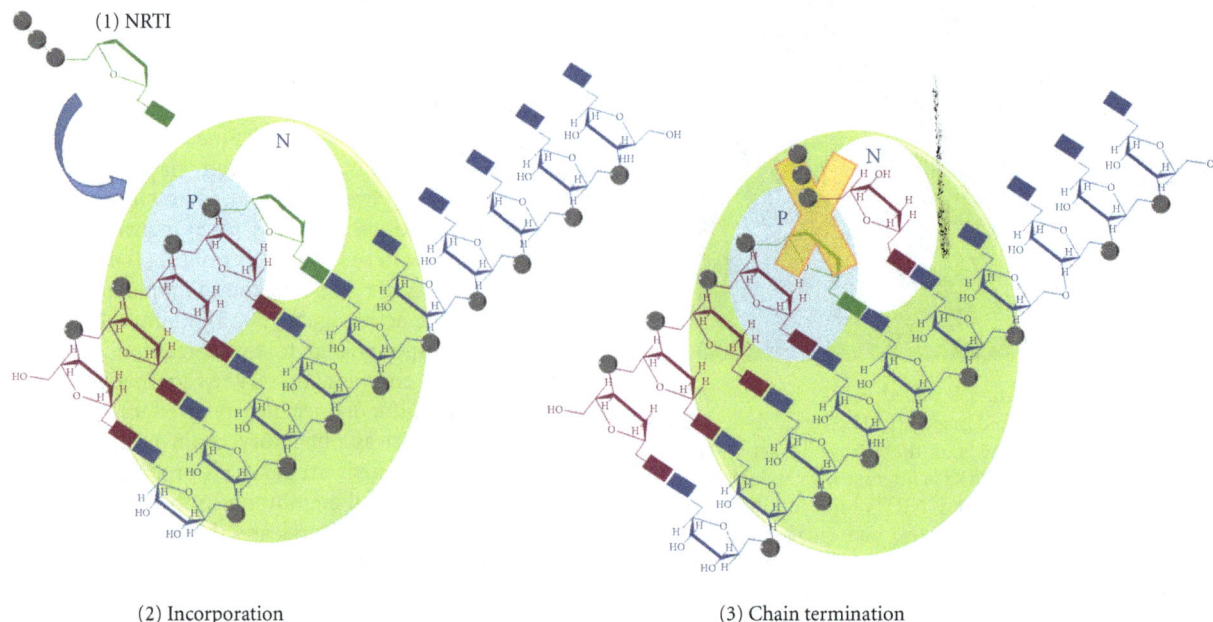

(1) NRTI

(2) Incorporation

(3) Chain termination

FIGURE 5: Mechanism of action of RT inhibitors acting as chain terminators. The RT is represented as a pale green circle with the priming binding site in cyan (P) and the nucleotide binding site in white (N). The RNA template is showed in blue and the (−)strand DNA in purple. The NRTI triphosphate (strong green) (1) competes for the binding with the natural dNTPs, it is incorporated into the growing DNA (2) and it blocks the further DNA elongation because it lacks the 3′-hydroxyl group (3).

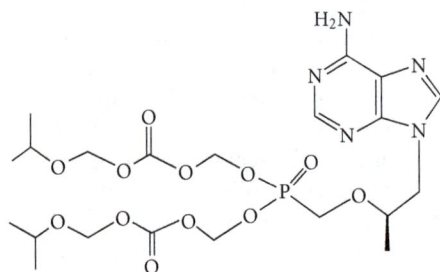

Tenofovir, PMPA

FIGURE 6: Chemical structure of approved NtRTI.

are less important in the case of efavirenz binding, while hydrogen bonds between the inhibitor and the protein backbone of K101 and K103 residues are critical [61].

First-generation NNRTIs, such as nevirapine and delavirdine, easily select resistant RTs that contain single amino acid mutations such as Y181C, K103N, and Y188C [62, 63], that change their key hydrophobic interactions at the NNRTI binding site. Second-generation NNRTIs, such as efavirenz and dapivirine, usually require two or more mutations in the HIV-1 RT before significantly decreasing their antiviral potency. In general, two or more HIV-1 RT mutations are clustered in the NNRTI pocket, suggesting a direct stereochemical mode of reduction of NNRTI binding, even though other mechanisms may also be present such as the one shown by V108I mutation that induces resistance by

perturbing the Y181 and Y188 residues [61] or the one proposed for K103N mutation that should stabilize the apo-RT conformation and, hence, create an energy barrier to NNRTIs binding, reducing their potency [61]. Interestingly, NRTI-resistant mutant virus strains keep full sensitivity to the inhibitory effects of NNRTIs, and vice versa. Recently, however, mutations in the connection and RNase H domains such as N384I, T369I, and E399D have been shown to confer resistance to both NRTIs and NNRTIs probably by altering the template:primer positioning [44, 47, 64].

5. New Nucleoside RT Inhibitors

The NRTIs therapeutic use is limited by several factors [65]. Firstly, drug-drug interactions with other NRTIs used in combination treatments such as the one observed between AZT and D4T, that share the same phosphorylation pathway and show a less than additive effect when used in combination [66], or between ddI and tenofovir which determine an increase in single drugs toxicity [65]. Secondly, drug-drug interactions with other molecules such as the one observed when ABC or tenofovir is administered with some protease inhibitors [65, 67], or when ABC is administered with ethanol [68]. Thirdly, several adverse events such as mitochondrial toxicity (linked to myopathy, cardiomyopathy, anemia, lipoatrophy), drug hypersensitivity reactions, and renal dysfunctions have been associated with NRTI treatment [65]. Fourthly, as described above, the selection of NRTI-resistant strains, which is still the main limitation in view

HIV-1 Reverse Transcriptase Still Remains a New Drug Target: Structure, Function, Classical Inhibitors, and New Inhibitors with Innovative Mechanisms of Actions

165

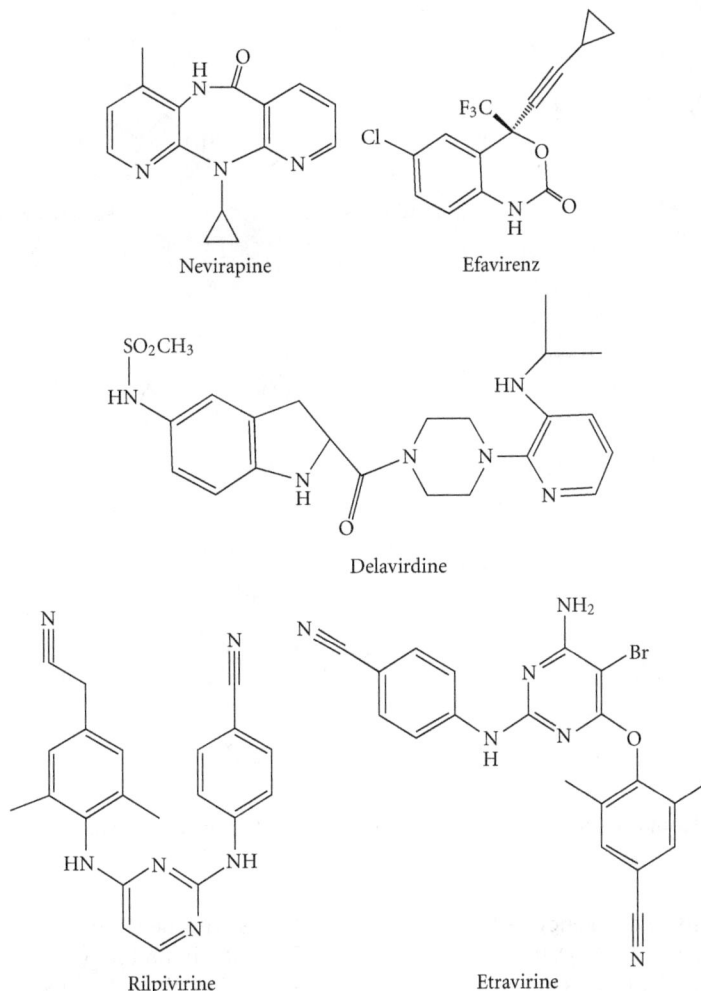

FIGURE 7: Chemical structures of approved NNRTIs.

of the need for life-long antiviral treatments. Particularly, it has been reported that almost 50% of the viremic patients actually harbor M184V RT mutant strains and that 6–16% of the patients have been infected with viruses resistant to at least one drug and, hence, have a poorer response to therapy and a lower barrier to select further drug-resistant strains [65, 69]. Given this scenario, the new NRTIs which are currently under investigation are sought to have a favorable resistance profile, reduced adverse effects, and/or a novel mechanism of action.

5.1. Nucleoside RT Inhibitors in Development Acting as Chain Terminators. (−)-2′-deoxy-3′-oxa-4′-thiocytidine (Apricitabine, ATC) (Figure 8) is a (−)enantiomer deoxycytidine analog with a favorable resistance profile. In fact, ATC shows only a 2-fold potency reduction on TAM strains, with or without the M184V mutation, and on K65R mutant strain, while it shows a 10-fold potency reduction on Q151M mutant strains [70–72]. ATC has a favorable toxic profile with little effects on mitochondrial DNA levels [73], while it shows negative drug-drug interactions when administered

with 3TC or FTC [74]. Overall, ATC seems to be a good candidate in NRTI-experienced patients including individuals who have experienced virological failure on 3TC and FTC containing regimens or harboring M184V mutant strains. In fact, ATC has successfully completed the primary endpoint of a phase IIb trial in drug-resistant HIV patients with the M184V mutation.

L-β-2′,3′-didehydro-2′,3′-dideoxy-5-fluorocytidine (Elvucitabine, L-d4TC) (Figure 8) is an L-cytidine analog under investigation in phase I/II clinical trials that is more potent than 3TC and that shows no mitochondrial toxicity [75] and an interesting protecting effect on the mitochondrial toxicity due to other NRTIs [76]. L-d4TC resistance profile shows that it selects for M14V RT mutants [77] and has a 10-fold potency reduction on K65R mutant strains [78].

1-β-D-2,6-diaminopurine dioxolane (Amdoxovir, DAPD) (Figure 8) is a prodrug under investigation in phase II clinical trials which is deaminated to 1-β-D-dioxolane guanosine (DXG) that, upon triphosphorylation, is the active drug. DAPD has a favorable resistance profile since it shows minimal resistance to TAM- and M184V-resistant strains

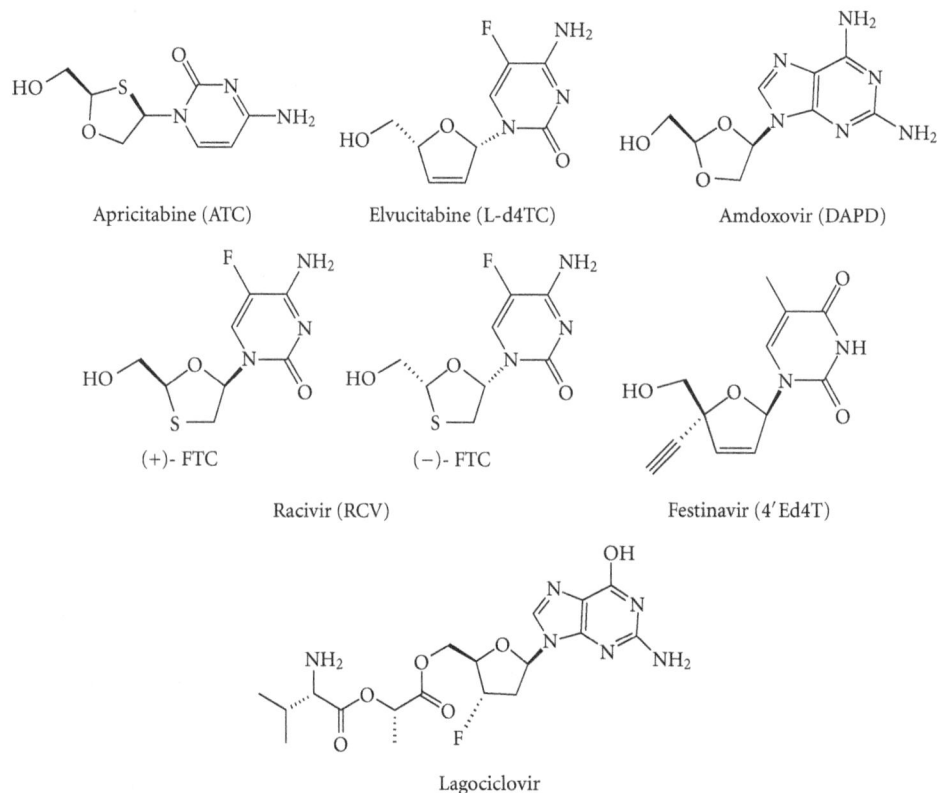

FIGURE 8: Chemical structures of new NRTIs acting as chain terminators.

[79, 80], while it shows a >10-fold potency reduction on K65R and Q151M strains [81]. While DAPD, *in vitro*, reduces the mitochondrial DNA content, DXG does not affect it [82].

$(\pm)$-β-2',3'-dideoxy-3'-thia-5-fluorocytosine (Racivir, RCV) (Figure 8) is a racemic mixture of (+) and (−)FTC currently under evaluation in phase II/III clinical trials as part of a combination therapy. While both molecules inhibit RT [83], (−)FTC is better phosphorylated than (+)FTC in cells [84], and, therefore, it shows a higher potency in virus inhibition [85]. The RCV resistance profile is interesting; in fact, (−)FTC selects for M184V-resistant strains, while (+)FTC selects for T215Y-resistant strains [86]. Since the simultaneous selection of these two amino acid mutations is incompatible, such racemic mixture orthogonal resistance profile determines a delay in the onset of the drug resistance selection [87]. The long-term mitochondrial toxicity, however, is still to be fully assessed since (+)FTC triphosphate is only 36-fold selective for RT versus DNA polymerase γ [88].

In addition, the chain terminator NRTIs Festinavir (4'-Ed4T) [89] and Lagociclovir [90] (Figure 8) are currently under development.

5.2. Nucleoside RT Inhibitors with Innovative Mode of Action.
The RT inhibition by NRTIs can also be achieved by mechanisms different from the classical chain termination due to the lack of a 3'-hydroxyl group. In particular, new classes of

inhibitors with new modes of action are the translocation-defective RT inhibitors (TDRTI), the delayed chain terminators RT inhibitors (DCTRTI), the lethal mutagenesis RT inhibitors (LMRTI), and the dinucleotide tetraphosphates $(N_{p4}N_s)$.

5.2.1. Translocation-Defective RT Inhibitors.
TDRTIs are NRTIs with modifications of the sugar moiety that block the RT translocation after the NRTI incorporation. 4'-ethynyl-2-fluoro-2'-deoxyadenosine (EFdA) (Figure 9) is the most potent derivative of a series of 4'-substituted nucleoside analogs which, differently from the other NRTIs, have a 3'-hydroxyl group [91]. EFdA is able to inhibit many drug-resistant strains several orders of magnitude more potently than the other approved NRTIs. For instance, it inhibits the M184V mutant strain with an EC_{50} value of 8 nM, while some other drug-resistant strains are even hypersensitive to EFdA [92]. Importantly, RT can use EFdA triphosphate (EFdA-TP) as substrate but, despite the presence of the 3'-hydroxyl group, the incorporated EFdA monophosphate (EFdA-MP) blocks further DNA synthesis since the enzyme is not able to efficiently translocate on a RNA:DNA or a DNA:DNA hybrid containing a 3'-terminal EFdA-MP [93] (Figure 10). In fact, on the one hand, the North (C2'-exo/C3'-endo) EFdA sugar ring conformation (which is the proper 3'-terminus position for in-line nucleophilic attack on the α-phosphate of the incoming dNTP) has been shown to be required for efficient binding at the primer-binding and

HIV-1 Reverse Transcriptase Still Remains a New Drug Target: Structure, Function, Classical Inhibitors, and New Inhibitors with Innovative Mechanisms of Actions

167

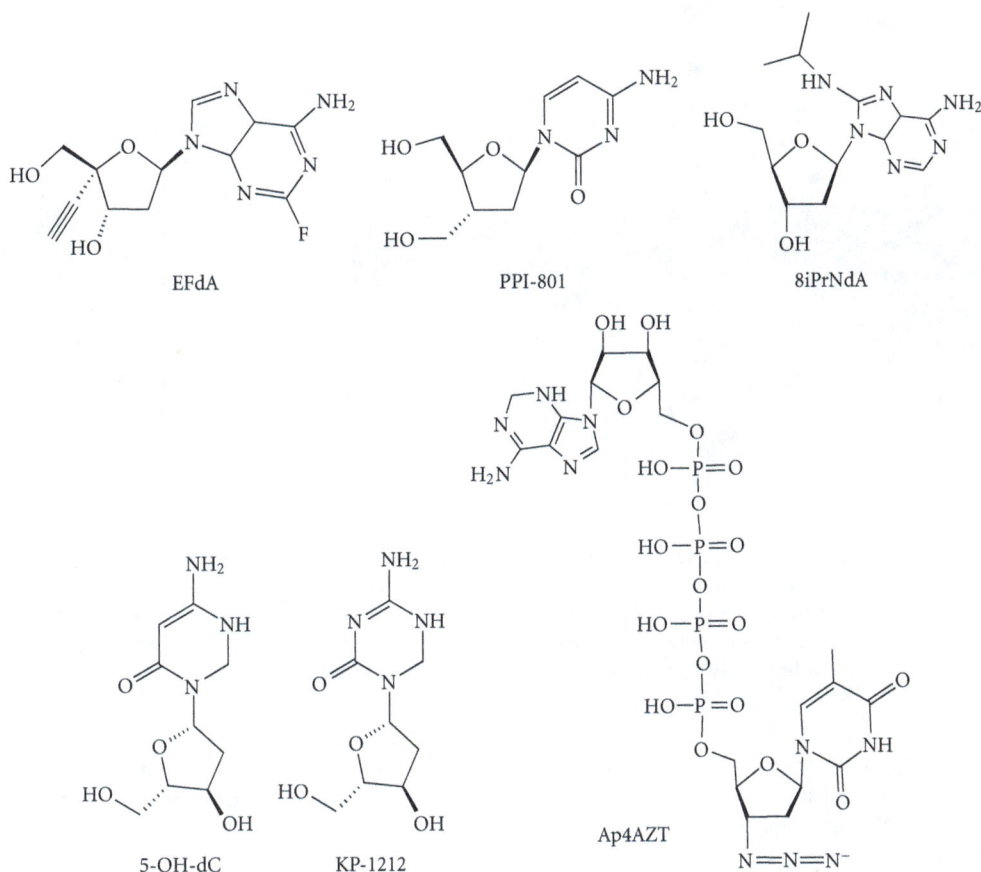

EFdA

PPI-801

8iPrNdA

5-OH-dC

KP-1212

Ap4AZT

FIGURE 9: Chemical structures of NRTIs with new mechanisms of action.

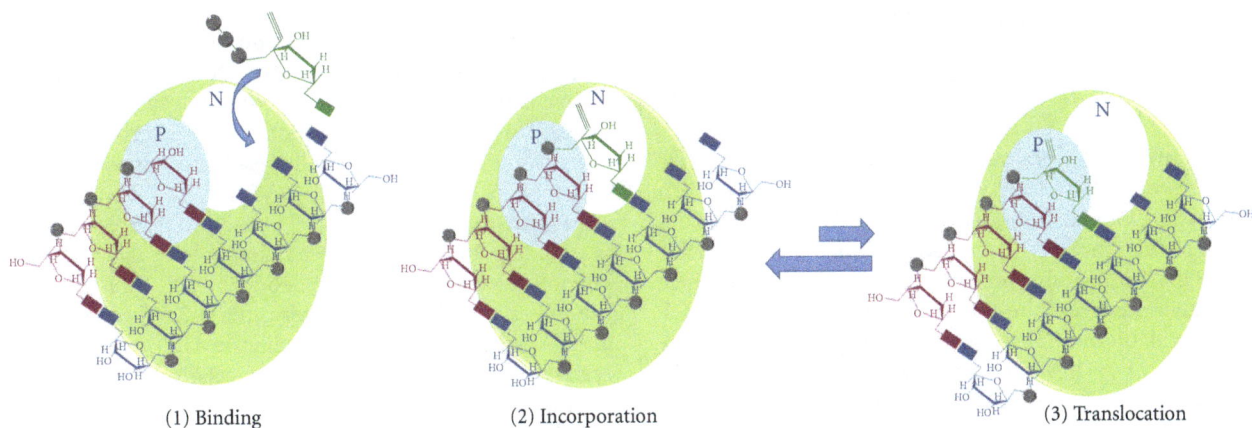

(1) Binding

(2) Incorporation

(3) Translocation

FIGURE 10: Mechanism of action of TDRTIs. The RT is represented as a pale green circle with the priming binding site in cyan (P) and the nucleotide binding site in white (N). The RNA template is shown in blue and the (−)strand DNA in purple. The TDRTI triphosphate (strong green) can be used as RT substrate (1) and is incorporated in the nucleic acid (2). The incorporated TDRTI blocks the further DNA synthesis since the enzyme is not able to efficiently translocate (3).

RT polymerase active sites suggesting that, once incorporated into the DNA, the EFdA 3′-hydroxyl group is not likely to prevent by itself additional nucleotides incorporation, and, thus, it does not contribute to the mechanism of chain termination [94]. On the other hand, molecular modeling studies suggested that the 4′-ethynyl of EFdA may fit into a hydrophobic pocket defined by residues A114, Y215, F160,

M184 and the aliphatic D185 chain [93]. Hence, it has been proposed that the presence of a 4′-ethynyl substitution on the ribose ring possibly hampers RT to translocate the 3′-EFdA-MP terminus DNA. Under these circumstances, RT is stabilized in a pretranslocation state which antagonizes the further nucleotide addition, since the dNTP-binding site is not accessible and the incorporation of the next

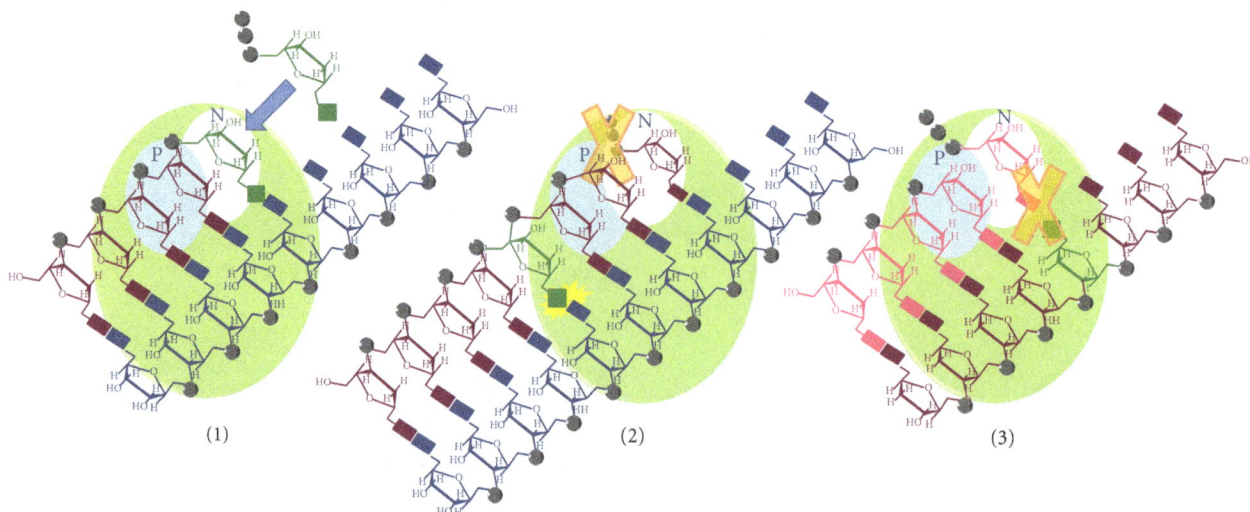

FIGURE 11: Mechanism of action of DCTRTIs. The RT is represented as a pale green circle with the priming binding site in cyan (P) and the nucleotide binding site in white (N). The RNA template is shown in blue and the (−)strand DNA in purple. DCTRTI triphosphate (strong green) is incorporated into the growing DNA chain (1). After further nucleotides addition, its presence blocks DNA elongation, probably through steric hindrance interference (yellow) between the RNA:DNA hybrid and the RT nucleic-acid-binding cleft (2). In addition, their incorporation can also block the synthesis of the (+)strand DNA affecting the base pairing (3).

complementary nucleotide is prevented [93]. Notably, in spite of the fact that the diminished translocation makes the 3′-EFdA-MP terminus DNA an excellent substrate for NRTI excision, the net excision process has been reported to be not very efficient, apparently because once the nucleotide is excised through pyrophosphorolysis to form EFdA-TP, the latter is rapidly reincorporated [93]. Moreover, it has been recently reported that EFdA is a poor substrate for DNA polymerase γ (it is incorporated 4,300-fold less than dATP), suggesting minimal mitochondrial toxicity [95].

5.2.2. Delayed Chain Terminators RT Inhibitors.
DCTRTIs are NRTIs that allow further incorporation of dNTPs into the growing DNA chain since they have a 3′-hydroxyl group. However, after further nucleotide addition, their presence blocks DNA elongation, probably through steric hindrance interference between the RNA:DNA hybrid and the RT nucleic acid binding cleft, close to the polymerase active site (Figure 11). They can also block the synthesis of the (+)strand DNA affecting the base pairing.

2′,3′-dideoxy-3′C-hydroxymethyl cytidine (PPI-801) (Figure 9) has been reported to allow the incorporation of one additional dNTP prior to mediating chain termination [65]. Interestingly, the incorporated PPI-801 is not accessible for nucleotide excision, and, therefore, this class of compounds is proposed to be attractive because it should be active also on NRTI-resistant strains with enhanced 3′-end nucleotide excision.

8-isopropyl-amino-2′-deoxyadenosine (8iPrNdA) (Figure 9) is a recently reported molecule belonging to a series of nucleoside analogs with a natural deoxyribose moiety and modifications at position 8 of the adenine base [96]. These modifications may induce a steric clash with helix αH in the thumb domain of the p66 subunit, causing delayed chain termination. In fact, once incorporated into the elongated DNA, 8iPrNdA stops the further DNA synthesis after the incorporation of three additional dNTPs [96]. Even though the potency and selectivity of 8iPrNdA are not very high, it is an interesting example of an NRTI with modifications on the adenine base and not on the sugar moiety.

5.2.3. Lethal Mutagenesis RT Inhibitors.
LMRTIs are NRTIs that allow further incorporation of dNTPs into the growing DNA chain. However, their incorporation causes a significant increase of nucleotide mismatches that determines a high mutation rate that eventually leads to viral replication suppression.

5-hydroxydeoxycytidine (5-OH-dC) (Figure 9) is a deoxycytidine analog that can efficiently base pair with both guanosine and adenosine nucleotides [97]. Viral growth in the presence of 5-OH-dC determines a 2.5-fold increase in G to A substitutions and a decline in viral infectivity over serial passages [97]. The fact that a relatively small increase in the HIV mutation frequency has a large effect on viral lethality substantiates the concept that the HIV mutation frequency is close to the error threshold for the viability of the quasispecies and that NRTIs that may significantly increase mutation frequency can act almost analogously to the cellular cytidine deaminase APOBEC3G [97].

5-aza-5,6-dihydro-2′-deoxycytidine (KP-1212) (Figure 9) is a deoxycytidine analog with a modified base and a natural sugar moiety that can also base pair with both guanosine and adenosine nucleotides [98]. The virus grown in the presence of KP-1212 accumulates a number of mutations that, eventually, stops its replication [98]. KP-1212 has been reported to interact also with DNA polymerase γ [99], suggesting a possible mitochondrial toxicity that, however, has not been observed in cells [98].

HIV-1 Reverse Transcriptase Still Remains a New Drug Target: Structure, Function, Classical Inhibitors, and New Inhibitors with Innovative Mechanisms of Actions

169

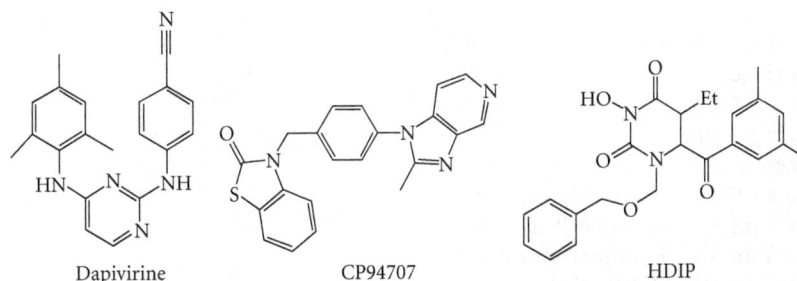

FIGURE 12: Chemical structures of new NNRTIs.

5.2.4. Dinucleotide Tetraphosphates. As described above, nucleotides excision is a major mechanism of NRTI resistance. During this mechanism RT catalyzes the pyrophosphorolysis of, for instance, a 3′-AZT-MP terminated DNA. In fact, in the presence of the PPi donor ATP, RT catalyzes the excision reaction which results in the production of a dinucleoside tetraphosphate (i.i. $A_{p4}AZT$) freeing the 3′-end for further DNA elongation. Notably, X-ray crystal studies have shown that the AMP part of the $A_{p4}AZT$ dinucleotide (Figure 9) binds differently to wt and drug-resistant mutant RTs [100]. These observations demonstrate that (i) RT can catalyze the reverse reaction and (ii) drug resistance mutations create a high-affinity ATP-binding site and open the possibility of designing drugs that can inhibit the enzyme mimicking the $N_{p4}N$ excision product that may be particularly active on NRTI-resistant strains. Up to now, a few $N_{p4}Ns$ have been synthesized that are able to inhibit wt and AZT-resistant RTs in the low micromolar range [101]. Notably, while the tetraphosphate linker, that avoids the intracellular phosphorylation step, is a potential advantage of these molecules, it is also an obstacle for their stability and cellular permeability. More studies dedicated to a further exploration of the ATP-binding site may lead to potent and innovative drugs.

6. New Nonnucleoside RT Inhibitors

The NNRTIs therapeutic use is limited mainly by the selection of NNRTI resistant virus, even though drug hypersensitivity and/or serious central nervous system dysfunctions are also toxicity issues for some NNRTIs. For this reason, there is still an active focus on the development of new NNRTIs, especially for compounds with high potency against K103N, Y181C, and Y188V mutant viruses. Besides the fact that more than 30 different conformational classes of NNRTIs have been reported to date [102, 103], the recent development of new NNRTIs has been focused on the identification of molecules that retain high conformation flexibility and positional adaptability in order to adjust the inhibitor conformation to the NNRTI-binding pocket, whose shape is different according to the presence of the diverse amino acid residues involved in NNRTI resistance. In fact, while first-generation NNRTIs, such as nevirapine, delavirdine, or efavirenz, bind to RT in "two-wing" (or "butterfly-like") conformation, the most recently developed NNRTIs show a "U" (or "horseshoe") conformation which gives an increased

plasticity to these derivatives [104, 105]. Success stories of such an approach are the latest approved NNRTIs, etravirine and rilpivirine (Figure 7), and another compound under clinical investigation in phase I/II clinical trials, dapivirine (Figure 12) [104, 105].

Another complementary strategy used to improve the NNRTIs performance is to design derivatives that make strong interactions with highly conserved amino acid residues in the NNRTI- binding pocket such as F227, W229, L234, and Y318 [105, 106]. In fact, these first three residues are part of the primer grip region that maintains the primer terminus in an appropriate orientation for the nucleophilic attack on the incoming dNTP. Specifically, the W229 residue is the prime candidate residue for drug design, and, in fact, among others, the above-mentioned rilpivirine has been reported to make strong interactions with the indole ring of W229.

Another reported interesting NNRTI is 3-(4-(2-methyl-1H-imidazo[4,5-c]pyridin-1-yl)benzyl)benzo[d]thiazol-2(3H)-one (CP94707) (Figure 12) that inhibits, even though not very potently, wt and mutant Y181C and Y188C RTs at the same concentrations and shows only a 2-fold reduction in potency of inhibition on K103N RT [107]. CP94707 makes little contact with Y181 and Y188 residues, while it makes aromatic ring stacking interactions with W229 amino acid [107]. In addition, CP94707 binding to RT results in rearrangement of the distally positioned Y115 side chain, 15 Å away, to a conformation that is incompatible with binding of dNTPs. Y115, in fact, can act as a gatekeeper residue that discriminates between deoxynucleotides and ribonucleotides. Therefore, it has been proposed that CP94707 may have a nonconventional mode of action [108].

An NNRTIs series of N-hydroxyimide derivatives, such as compound 1-((benzyloxy)methyl)-6-(3,5-dimethylbenzoyl)-5-ethyl-3-hydroxydihydropyrimidine-2,4(1H,3H)-dione (HDIP) (Figure 12), have been developed as dual RT and IN inhibitors (DRT-INI). In fact, they have been reported to inhibit both the RT-associated RDDP function and the IN activity [109, 110] and have been proposed to bind to the NNRTI-binding site and also chelate the magnesium ion in the IN active site [109, 110].

7. Nucleotide Competing RT Inhibitors

A series of indolopyridones, therefore belonging to the NN-RTIs, have been shown to inhibit RT interacting differently from the classic NNRTIs. In particular, 5-methyl-1-(4-nitrophenyl)-2-oxo-2,5-dihydro-1H pyrido[3,2-b]indole-3-carbonitrile (INDOPY-1) (Figure 13) (i) inhibits also HIV-2 RT [111], while the other NNRTIs are inactive against this enzyme; (ii) it is active against K103N, Y181C, and Y188C mutant RTs as potently as on wt RT, while it is 3.6-fold less active against the K103N/L100I double-mutant RT [112]; (iii) it is active on TAM viruses, while it is 3- to 8-fold less effective on M184V or Y115F mutant viruses, it is more than 100-fold less potent on the M184V/Y115F double-mutant virus, and it is slightly more effective on K65R mutant virus [111–113]. In addition, the INDOPY-1 analog 1-(4-nitrophenyl)-2-oxo-2,5-dihydro-1H-pyrido[3,2-b]indole-3-carbonitrile (VRX329747) (Figure 13) selected HIV-1 RT mutated at the amino acid residues M41L, A62V, S68N, G112S, V118I, and M184V, which are all located around the incoming nucleotide-binding site [112]. Further, binding and biochemical studies revealed that (i) the M184V mutation reduces the affinity to INDOPY-1, while the Y115F mutation facilitates the dNTP binding, and their combined effects enhance the ability of the enzyme to discriminate against the inhibitor [113]; (ii) RT complexed with INDOPY-1 is trapped in the posttranslocational state [113]; (iii) the INDOPY-1 has preference with respect to substrate primer identity since its binding to RT is higher on a DNA:DNA versus a RNA:DNA primer:template [114]; (iv) when assayed by steady-state kinetic analysis with homopolymeric template primers, INDOPY-1 inhibits RT-catalyzed DNA polymerization with a competitive [111] or mixed-type [112] mode with respect to dNTPs. Overall, these observations suggest that the binding site of the indolopyridones and nucleotide substrates can at least partially overlap and they are therefore proposed as Nucleotide competing RT inhibitors (NcRTIs).

4-dimethylamino-6-vinylpyrimidines (DAVPs) is another class of compounds that have been reported to compete with the incoming dNTP and therefore can be considered NcRTI [115, 116]. However, differently from INDOPY-1, DAVP1 (Figure 13) is 4000- and 5000-fold less potent on mutant K103N and Y181C RTs, respectively [115], and binds also to unligated RT (while INDOPY-1 binds only to the RT:template:primer complex) [116]. X-ray crystal studies have confirmed that DAVP1 binds to an RT site that is distinct from the NNRTI-binding pocket, and it is close to the RT polymerase catalytic site [117]. This site is located in a hinge region, at the interface between the p66 thumb and p66 palm subdomains, that comprises the amino acid residues M230 and G231 (participating to the primer grip region and helping in the correct positioning of the 3′-OH end of the DNA primer), G262, K263 and W266 (involved in the template primer recognition), M184 and D186 (the first is involved in DNA synthesis fidelity, while the second is part of the catalytically essential YXDD motif) [117]. Hence, the DAVP1 binding site is located in a region critical for the correct positioning of the 3′-OH primer for the in-line nucleophilic attack by the incoming dNTP and the subsequent chemical bond formation with its α-phosphate. Notably, the X-ray study also revealed that in the RT/DAVP-1 complex the RT conformation is analogous to the "closed"

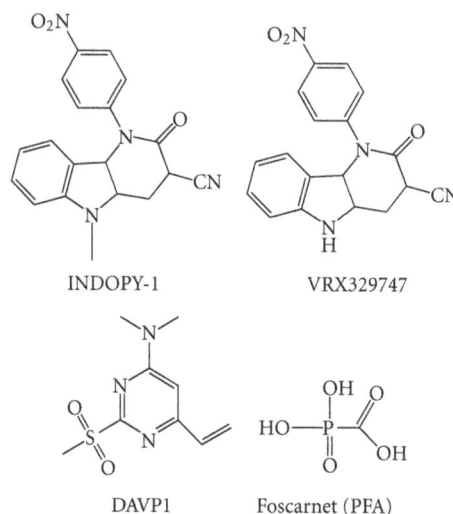

FIGURE 13: Chemical structures of NcRTIs.

conformation observed in unliganded RTs (with the p66 thumb subdomain folded into the DNA-binding cleft) and differs from that observed in RT/NNRTI complexes that has a hyperextended "open" conformation [117]. However, considering the proposed binding site, the reason for the loss of DAVP1 activity against K103N and Y181C mutant RTs remains unclear. While it has been hypothesized that DAVP1, owing to its small size, could travel between the NNRTI and nucleoside-binding pockets [117], more studies are needed to understand the DAVP1 mode of action.

8. PPi Analogs Inhibitors

Foscarnet (phosphonoformate, PFA) (Figure 13) is a PPi analogue that targets the DNA polymerase of herpes viruses as well as the RT of retroviruses [118]. Foscarnet is used intravenously to treat opportunistic viral infections, particularly CMV retinitis in patients with AIDS, but its pharmacokinetic profile is complicated by nephrotoxicity [119]. When assayed against HIV-1 RT, it competitively blocks pyrophosphorolysis and PPi exchange reactions, suggesting that foscarnet and PPi share overlapping binding sites [120]. It has been shown that foscarnet traps the RT pretranslocated complex preventing the binding of the next nucleotide, and, thus, the pretranslocated complex has been proposed as a target for drug discovery [121]. In vivo and in vitro foscarnet-resistant HIV-1 variants have been shown to carry mutations in the RT gene at several positions, including W88G/S, E89K/G, L92I, A114S, S156A, Q161L, and H208Y [122–125]. Notably, most of the mutations that reduce the susceptibility to PFA also confer hypersensitivity to AZT and it has been suggested that foscarnet analogs may inhibit the phosphorolytic rescue of NRTI-terminated primers and be used to prevent the excision-based mode of NRTI resistance [126].

HIV-1 Reverse Transcriptase Still Remains a New Drug Target: Structure, Function, Classical Inhibitors, and New Inhibitors with Innovative Mechanisms of Actions

171

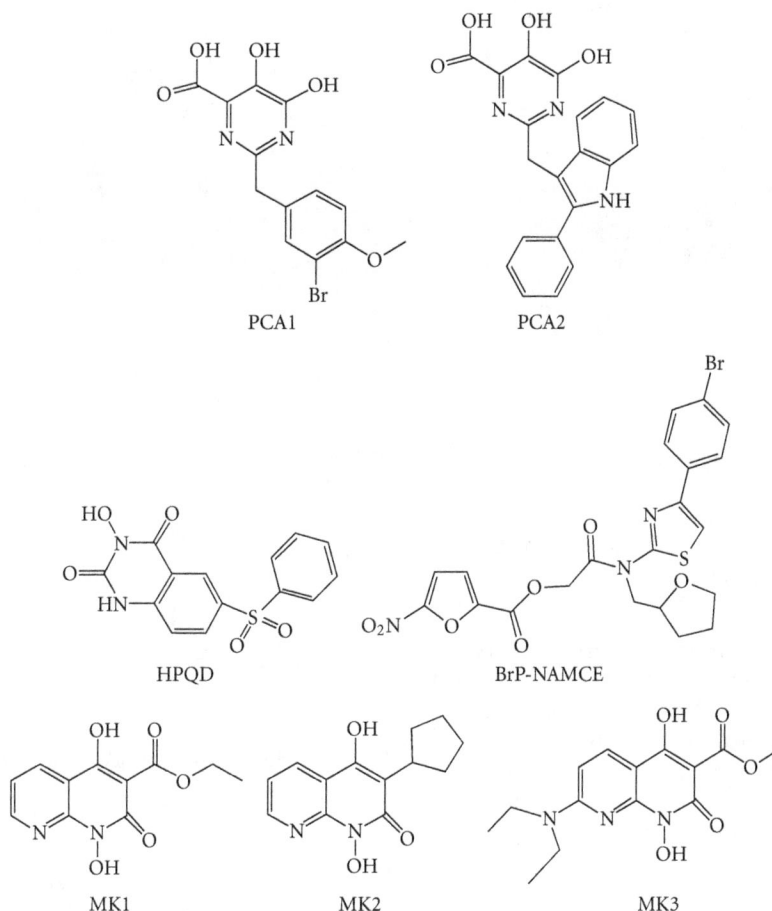

FIGURE 14: Chemical structures of metal chelating RHRTIs.

9. RNase H Inhibitors

Despite the fact that the RT-associated RNase H function is essential for the reverse transcription process as well as the RT-associated DNA polymerase function, no effective RNase H RTIs (RHRTIs) have been developed yet. In the last few years, however, a few classes of RHRTI that are specifically targeted to the RNase H active site (Figure 4) have been identified [19, 127]. Most of them are able to chelate the divalent magnesium ion within the RNase H active site, but they also exert a high cellular toxicity, possibly due to an unspecific metal chelation, since the RNase H active site is an open pocket and offers, at least so far, little elements for selective small-molecule optimization.

9.1. Metal Chelating RHRTI. Pyrimidinol carboxylic acids 2-(3-bromo-4-methoxybenzyl)-5,6-dihydroxypyrimidine-4-carboxylic acid (PCA1), 5,6-dihydroxy-2-((2-phenyl-1H-indol-3-yl)methyl)pyrimidine-4-carboxylic acid (PCA2) and N-hydroxy quinazolinedione inhibitors 3-hydroxy-6-(phenylsulfonyl)quinazoline-2,4(1H,3H)-dione) (HPQD) (Figure 14) were designed to coordinate the two metal ions in the active site of RNase H and showed no interactions with the polymerase metal-binding site [128]. However, so far they have not been further developed.

Similarly, Nitrofuran-2-carboxylic acids derivatives such as the 5-nitro-furan-2-carboxylic acid [[4-(4-bromophenyl)-thiazol-2-yl]-(tetrahydro-furan-2-ylmethyl)-carbamoyl]-methyl ester (BrP-NAMCE) (Figure 14) were identified to inhibit the RNase H function by chelating the magnesium ion [129], and other analogs were also reported [130], but more derivatization studies are needed in order to develop effective inhibitors.

Naphthyridine derivatives ethyl 1,4-dihydroxy-2-oxo-1,2-dihydro-1,8-naphthyridine-3-carboxylate (MK1), 3-cyclopentyl-1,4-dihydroxy-1,8-naphthyridin-2(1H)-one (MK2) and methyl 7-(diethylamino)-1,4-dihydroxy-2-oxo-1,2-dihydro-1,8-naphthyridine-3-carboxylate (MK3) (Figure 14) have been reported to bind to the RNase H active site by coordinating the two metal ions, engaging the conserved catalytic DDE motif [131]. Interestingly, they were reported to be sandwiched by a loop containing residues A538 and H539 residues on the one side and N474 on the opposite side. In addition, MK3 was also shown to bind to a site adjacent to the NNRTI including amino acid residues L100, V108, Y181, Y183, D186, L187, K223, F227, L228, W229, and L234 [131]. Unlike the binding to the RNase H active site, the binding to this alternate site appears to be predominantly mediated via the hydrophobic interactions with the diethylaminophenoxy group unique to MK3. The rilevance of the MK3 binding

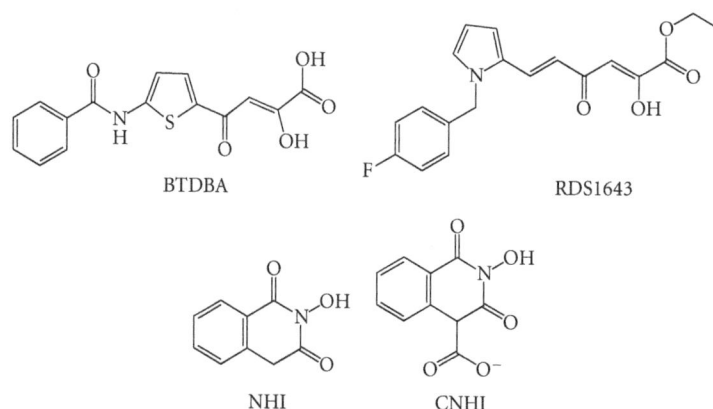

FIGURE 15: Chemical structures of dual RHRTI-INIs.

to this site is not clear; however, the site is similar to the binding site for DHBNH (see later).

9.2. Dual RHRTI and IN Inhibitors. The first recently discovered RHRTIs were the diketo acid (DKA) derivatives 4-[5-(benzoylamino)thien-2-yl]-2,4-dioxobutanoic acid (BTDBA) (Figure 15) [132] and 6-[1-(4-fluorophenyl)methyl-1*H*pyrrol-2-yl)]-2,4-dioxo-5-hexenoic acid ethyl ester (RDS-1643) (Figure 15) [133], that were independently developed against the HIV-1 IN. Due to similarities between RNase H and IN active sites, they were explored as RHRTIs and found to be active. Both of them are able to chelate Mg^{2+} in the RNase H catalytic site and are inactive on the DNA polymerase function [132, 133]. For this reason DKAs are currently under development as dual RNase H and INIs (DRH-INI) [19, 134–136].

Other derivatives that have also been developed as DRH-INIs are *N*-hydroxyimide. The prototype of these inhibitors was the 2-hydroxyisoquinoline-1,3(2*H*,4*H*)-diones (NHI) (Figure 15) [137, 138] that was shown, by crystal structures with the isolated RNase H domain, to bind to RT in a strictly metal dependent manner, confirming the metal-ion-mediated mode of action. More recently, other *N*-hydroxyimide derivatives were synthesized such as DRH-INIs [139, 140]. Interestingly, the methyl 2-Hydroxy-1,3-dioxo-1,2,3,4-tetrahydroisoquinoline-4-carboxylate analog (CNHI) (Figure 15) has also been shown to inhibit the replication of the double-mutant G140S/Q148H, which is the most resistant strain to the INI raltegravir [140], indicating that it is possible to design compounds with the same scaffold that may (i) inhibit both RNase H and IN and (ii) inhibit specifically one of the two enzymes. Further studies will be needed to dissect the specifics of the two active sites.

9.3. Nonmetal Chelator RHRTI. Unlike the above-mentioned compounds, vinylogous ureas compounds 2-amino-5,6,7,8-tetrahydro-4*H*-cyclohepta[*b*]thiophene-3-carboxamide (NSC727447) and *N*-[3-(aminocarbonyl)-4,5-dimethyl-2-thienyl]-2-furancarboxamide (NSC727448) (Figure 16) that inhibit the RNase H function are ineffective on the DNA polymerase function, but they do not chelate the magnesium

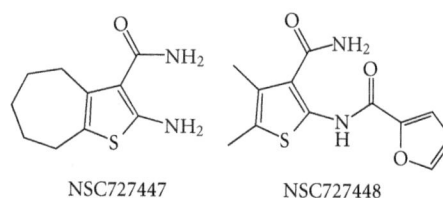

FIGURE 16: Chemical structures of nonmetal chelating RHRTIs.

ion [141]. These two derivatives were further developed into more potent analogs that, however, were devoid of antiviral activity in cell culture [142]. Molecular modeling studies showed that they bind to an hydrophobic pocket comprising residues V276, C280, K281, K275, R277, and R284 of the p51 thumb and residues G541 and H539 of the RNase H domain (Figure 4) [142]. Further studies are certainly warranted since this new pocket is highly attractive for RHRTIs development.

10. Dual RNase H and Polymerase Inhibitors

An interesting class of RHRTIs is the hydrazone derivatives, whose first reported analog was *N*-(4-*tert*-butylbenzoyl)-2-hydroxy-1-naphthaldehyde hydrazone (BBNH) (Figure 17). Unlike other NNRTIs or RHRTIs, BBNH inhibits both the polymerase and the RNase H activities of HIV-1 RT [143] and therefore can be classified as dual NNRTI (DNNRTI). In addition, BBNH inhibits both RT-associated RNase H and RDDP activities of K103N, Y181I, Y188H, and Y188L mutant RTs with potency similar to wt RT, while, when assayed on Y181C mutant RT, it inhibits only the RDDP function and is inactive on the RNase H function [144]. This information, together with the data on other hydrazone derivatives that chelate the metal ion cofactor in the RNase H site [145], led to propose that two BBNH molecules could bind RT in two different sites, the first one in the polymerase domain, possibly near the NNRTI-binding site, and the second one possibly located in the RNase H domain. Subsequently, another

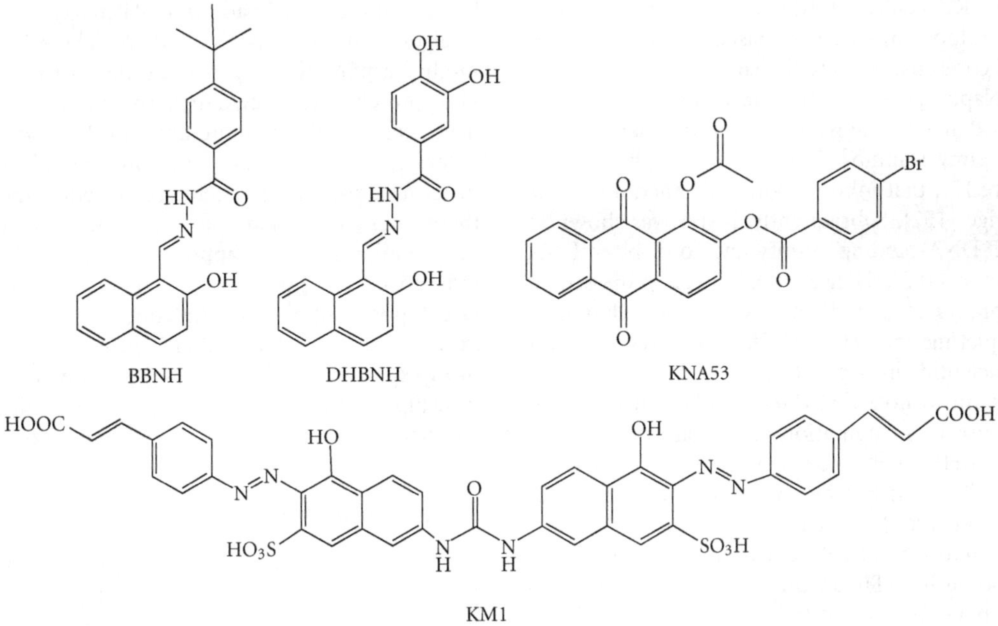

HIV-1 Reverse Transcriptase Still Remains a New Drug Target: Structure, Function, Classical Inhibitors, and New Inhibitors with Innovative Mechanisms of Actions

173

FIGURE 17: Chemical structures of dual RNase H and polymerase inhibitors.

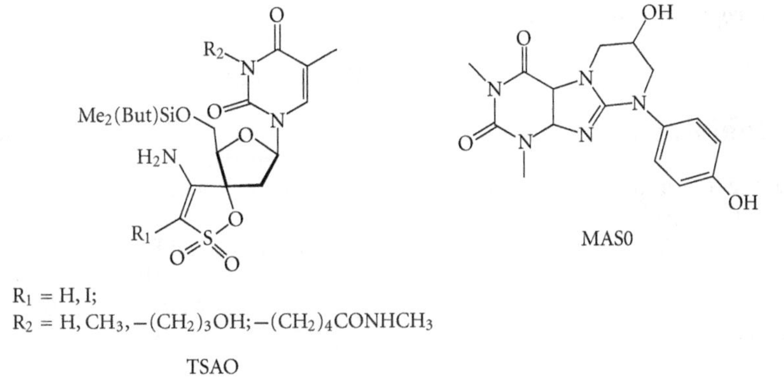

$R_1 = H, I;$
$R_2 = H, CH_3, -(CH_2)_3OH; -(CH_2)_4CONHCH_3$

TSAO

FIGURE 18: Chemical structures of DimRTIs.

derivative, (E)-3,4-dihydroxy-N'-((2-hydroxynaphthalen-1-yl)methylene)benzohydrazide (DHBNH) (Figure 17), has been reported to bind near the polymerase active site in a pocket different from the NNRTI-binding site and also >50 Å away from the RNase H active site (Figure 4) [146]. Hence, it was hypothesized that DHBNH may either perturb the trajectory of the template primer, so that RNase H cannot operate on its substrate, or that it may also bind to a second site, in or near the RNase H domain, that was not seen in the crystal. More recently, molecular docking studies on a series of hydrazone analogs proposed that they bind to a pocket that includes residues Y405, W406, Q500, and Y501 of p66 subunit, and, hence, they form hydrophobic interactions with RT and with base pairs in the groove of the RNA:DNA substrate [147]. In fact, residues D499 and A502, adjacent to Q500, which were perturbed by the hydrazone derivatives presence [147], are part of the primer grip of the RNase H domain and play a role in aligning the DNA:RNA substrate with the active site. Therefore, the hydrazones binding to

Q500 may disrupt the primer grip's role in the activity of RNase H.

A second class of DNNRTI is a series of emodin [148] and alizarine anthraquinone derivatives [149, 150] such as 1-acetoxy-9,10-dioxo-9,10-dihydroanthracen-2-yl 4-bromobenzoate (KNA-53) (Figure 17), that inhibits both RT-associated functions of wt and K103N RTs and only the RNase H function of Y181C RT. Mode of action studies and molecular dynamic simulation led to proposing that the anthraquinone derivatives bind to the site adjacent to the NNRTI pocket, which was originally reported [146] for the hydrazones derivatives (Figure 4) [149]. Accordingly, it has been suggested that the anthraquinone inhibition of the RNase H function may be due to a change in the RNA:DNA hybrid RT accommodation, induced by their binding, which results in a possible variation in the nucleic acid trajectory toward the RNase H catalytic site [149].

A third class of DNNRTI is the naphthalenesulfonic acid derivatives that were originally reported to have a selective activity on the RT-associated RDDP function [151] and were further developed by structure-based design, molecular similarity, and combinatorial medicinal chemistry to obtain compound 2-Naphthalenesulfonic acid (4-hydroxy-7-[[[[5-hydroxy-6-[(4 cinnamylphenyl)azo]-7-sulfo-2-naphthalen-yl]amino]-carbonyl]amino]-3-[(4-cinnamylphenyl)]azo (KM-1) (Figure 17), that inhibits both RT functions in the nanomolar range [152]. Subsequently, KM-1 was shown to weaken the RT DNA-binding affinity and to displace DNA from the enzyme [153]. Hence, it has been proposed to preclude the proper alignment of DNA at the polymerase active site, depleting the active DNA-bound RT complex required for nucleotide incorporation [153].

It is important to note that questions have been raised regarding the use of combinations between RHRTIs and NRTIs. In fact, RHRTIs have been proposed to lead to an increase in NRTIs resistance by mimicking the RNase H-dependent mechanism of NRTI resistance of some connection domain mutations [43]. Recently, however, studies on the effects of some RHRTIs on the HIV-1 susceptibility to AZT and 3TC have shown that none of the tested RHRTIs decreased NRTI susceptibility, while only one DNNRTI decreased AZT susceptibility by 5-fold [154]. More studies are needed to fully understand the interplay between RNase H inhibition and NRTIs susceptibility as well as its clinical relevance.

11. RT Dimerization Inhibitors

RT dimerization is an absolute requirement for all enzymatic activities, and, accordingly, the development of inhibitors targeting the dimerization of RT represents a promising alternative antiviral strategy [155]. Up to now only a series of small molecules have been found which are able to inhibit RT dimerization. Among them are the above-mentioned BBNH derivative [143, 145] and the [2′,5′-bis-O-(tert-butyldimethylsilyl)-beta-D-ribofuranose]-3′-spiro-5″-(4″-amino-1″,2″-oxathiole-2″,2″-dioxide) (TSAO) (Figure 18) derivatives [156], that make extensive contact with the $\beta7/\beta8$ loop of the p51 subunit, that forms the "floor" of the NNRTI binding pocket and fits in a groove-like structure that constitutes the template:primer binding site in the p66 subunit. More recently, a structure-based ligand study has identified compounds 7-hydroxy-9-(4-hydroxyphenyl)-1,3-dimethyl-1,6,7,8,9,10a-hexahydropyrimido [2,1-f]purine-2,4(3H,4aH)-dione MAS0 as potent dimerization RT inhibitors (DimRTIs) (Figure 18) [157].

12. Other Potential Targets in RT

The increase in knowledge regarding HIV life cycle and specifically the function of the HIV RT and its essential interactions with other proteins will reveal potential drug targets. Even though no inhibitors have been identified yet, to the best of our knowledge, the DNA synthesis initiation (with an RNA:RNA primer), the PPT hydrolysis, the strand transfer, and pyrophosphorolysis RT functions are all potential aspects of the RT activities that may be targeted by small molecules. In addition, RT makes contact with other viral proteins such as NC and IN. These binding surfaces might be potential targets since their disruption may alter viral protein efficiency. Furthermore, some cellular factors have been described to interact with RT (and with the RT:IN complex) during reverse transcription and may have a role in its function [158]. Therefore, a better understanding of these interactions may offer other new target sites. Finally, intracellular immunity approaches may also involve proteins that affect RT functions and may thus offer additional target possibilities [31]. In conclusion, although RT has been the very first targeted HIV protein and is probably the most explored one, it still presents uninvestigated (or under investigation) functions and aspects that still make it a new fascinating target for innovative drug development.

Acknowledgments

This work was supported by RAS Grant LR 7/2007 CRP-2_450 and by Fondazione Banco di Sardegna. F. Esposito was supported by RAS fellowships, cofinanced with funds from PO Sardinia FSE 2007–2013 and of LR 7/2007, project CRP2_683. A. Corona was supported by MIUR fellowship DM 198/2003.

References

[1] F. Barré-Sinoussi, J. C. Chermann, F. Rey et al., "Isolation of a T-lymphotropic retrovirus from a patient at risk for acquired immune deficiency syndrome (AIDS)," *Science*, vol. 220, no. 4599, pp. 868–871, 1983.

[2] S. Broder and R. C. Gallo, "A pathogenic retrovirus (HTLV-III) linked to AIDS," *New England Journal of Medicine*, vol. 311, no. 20, pp. 1292–1297, 1984.

[3] Y. Mehellou and E. De Clercq, "Twenty-six years of anti-HIV drug discovery: where do we stand and where do we go?" *Journal of Medicinal Chemistry*, vol. 53, no. 2, pp. 521–538, 2010.

[4] A. M. N. Tsibris and M. S. Hirsch, "Antiretroviral therapy in the clinic," *Journal of Virology*, vol. 84, no. 11, pp. 5458–5464, 2010.

[5] L. Ratner, W. Haseltine, and R. Patarca, "Complete nucleotide sequence of the AIDS virus, HTLV-III," *Nature*, vol. 313, no. 6000, pp. 277–284, 1985.

[6] S. H. Hughes, E. Arnold, and Z. Hostomsky, "RNase H of retroviral reverse transcriptases," in *Ribonucleases H*, R. J. Crouch and J. J. Toulmé, Eds., pp. 195–224, Les Editions Inserm, Paris, France, 1998.

[7] H. E. Huber and C. C. Richardson, "Processing of the primer for plus strand DNA synthesis by human immunodeficiency virus 1 reverse transcriptase," *Journal of Biological Chemistry*, vol. 265, no. 18, pp. 10565–10573, 1990.

[8] J. W. Rausch and S. F. J. Le Grice, "'Binding, bending and bonding': polypurine tract-primed initiation of plus-strand DNA synthesis in human immunodeficiency virus," *International Journal of Biochemistry and Cell Biology*, vol. 36, no. 9, pp. 1752–1766, 2004.

[9] V. P. Basu, M. Song, L. Gao, S. T. Rigby, M. N. Hanson, and R. A. Bambara, "Strand transfer events during HIV-1 reverse transcription," *Virus Research*, vol. 134, no. 1-2, pp. 19–38, 2008.

HIV-1 Reverse Transcriptase Still Remains a New Drug Target: Structure, Function, Classical Inhibitors, and New Inhibitors with Innovative Mechanisms of Actions

175

[10] G. Divita, K. Rittinger, C. Geourjon, G. Deleage, and R. S. Goody, "Dimerization kinetics of HIV-1 and HIV-2 reverse transcriptase: a two step process," *Journal of Molecular Biology*, vol. 245, no. 5, pp. 508–521, 1995.

[11] L. A. Kohlstaedt, J. Wang, J. M. Friedman, P. A. Rice, and T. A. Steitz, "Crystal structure at 3.5 A resolution of HIV-1 reverse transcriptase complexes with an inhibitor," *Science*, vol. 256, no. 5065, pp. 1783–1790, 1992.

[12] A. Jacobo-Molina, J. Ding, R. G. Nanni et al., "Crystal structure of human immunodeficiency virus type 1 reverse transcriptase complexed with double-stranded DNA at 3.0 A resolution shows bent DNA," *Proceedings of the National Academy of Sciences of the United States of America*, vol. 90, no. 13, pp. 6320–6324, 1993.

[13] S. Liu, E. A. Abbondanzieri, J. W. Rausch, S. F. J. Le Grice, and X. Zhuang, "Slide into action: dynamic shuttling of HIV reverse transcriptase on nucleic acid substrates," *Science*, vol. 322, no. 5904, pp. 1092–1097, 2008.

[14] E. A. Abbondanzieri, G. Bokinsky, J. W. Rausch, J. X. Zhang, S. F. J. Le Grice, and X. Zhuang, "Dynamic binding orientations direct activity of HIV reverse transcriptase," *Nature*, vol. 453, no. 7192, pp. 184–189, 2008.

[15] T. A. Stelz, "A mechanism for all polymerases," *Nature*, vol. 391, no. 6664, pp. 231–232, 1998.

[16] M. Ghosh, P. S. Jacques, D. W. Rodgers, M. Ottman, J. L. Darlix, and S. F. J. Le Grice, "Alterations to the primer grip of p66 HIV-1 reverse transcriptase and their consequences for template-primer utilization," *Biochemistry*, vol. 35, no. 26, pp. 8553–8562, 1996.

[17] S. G. Sarafianos, B. Marchand, K. Das et al., "Structure and function of HIV-1 reverse transcriptase: molecular mechanisms of polymerization and inhibition," *Journal of Molecular Biology*, vol. 385, no. 3, pp. 693–713, 2009.

[18] M. W. Kellinger and K. A. Johnson, "Nucleotide-dependent conformational change governs specificity and analog discrimination by HIV reverse transcriptase," *Proceedings of the National Academy of Sciences of the United States of America*, vol. 107, no. 17, pp. 7734–7739, 2010.

[19] E. Tramontano and R. Di Santo, "HIV-1 RT-associated Rnase H function inhibitors: recent advances in drug development," *Current Medicinal Chemistry*, vol. 17, no. 26, pp. 2837–2853, 2010.

[20] M. Nowotny, S. A. Gaidamakov, R. J. Crouch, and W. Yang, "Crystal structures of RNase H bound to an RNA/DNA hybrid: substrate specificity and metal-dependent catalysis," *Cell*, vol. 121, no. 7, pp. 1005–1016, 2005.

[21] E. Rosta, M. Nowotny, W. Yang, and G. Hummer, "Catalytic mechanism of RNA backbone cleavage by ribonuclease H from quantum mechanics/molecular mechanics simulations," *Journal of the American Chemical Society*, vol. 133, no. 23, pp. 8934–8941, 2011.

[22] V. Mizrahi, M. T. Usdin, A. Harington, and L. R. Dudding, "Site-directed mutagenesis of the conserved Asp-443 and Asp-498 carboxy-terminal residues of HIV-1 reverse transcriptase," *Nucleic Acids Research*, vol. 18, no. 18, pp. 5359–5363, 1990.

[23] V. Mizrahi, R. L. Brooksbank, and N. C. Nkabinde, "Mutagenesis of the conserved aspartic acid 443, glutamic acid 478, asparagine 494, and aspartic acid 498 residues in the ribonuclease H domain of p66/p51 human immunodeficiency virus type I reverse transcriptase. Expression and biochemical analysis," *Journal of Biological Chemistry*, vol. 269, no. 30, pp. 19245–19249, 1994.

[24] G. L. Beilhartz and M. Götte, "HIV-1 ribonuclease H: structure, catalytic mechanism and inhibitors," *Viruses*, vol. 2, no. 4, pp. 900–926, 2010.

[25] S. G. Sarafianos, K. Das, C. Tantillo et al., "Crystal structure of HIV-1 reverse transcriptase in complex with a polypurine tract RNA:DNA," *EMBO Journal*, vol. 20, no. 6, pp. 1449–1461, 2001.

[26] J. J. Champoux and S. J. Schultz, "Ribonuclease H: properties, substrate specificity and roles in retroviral reverse transcription," *FEBS Journal*, vol. 276, no. 6, pp. 1506–1516, 2009.

[27] E. S. Furfine and J. E. Reardon, "Reverse transcriptase. RNase H from the human immunodeficiency virus. Relationship of the DNA polymerase and RNA hydrolysis activities," *Journal of Biological Chemistry*, vol. 266, no. 1, pp. 406–412, 1991.

[28] A. Telesnitsky and S. P. Goff, "Reverse Transcriptase and the generation of retroviral DNA," in *Retroviruses*, J. M. Coffin, S. H. Hughes, and H. E. Varmus, Eds., pp. 121–160, Cold Spring Harbor Laboratory Press, New York, NY, USA, 1997.

[29] X. Ji, G. J. Klarmann, and B. D. Preston, "Effect of human immunodeficiency virus type 1 (HIV-1) nucleocapsid protein on HIV-1 reverse transcriptase activity in vitro," *Biochemistry*, vol. 35, no. 1, pp. 132–143, 1996.

[30] D. Grohmann, J. Godet, Y. Mély, J. L. Darlix, and T. Restle, "HIV-1 nucleocapsid traps reverse transcriptase on nucleic acid substrates," *Biochemistry*, vol. 47, no. 46, pp. 12230–12240, 2008.

[31] R. S. Aguiar and B. M. Peterlin, "APOBEC3 proteins and reverse transcription," *Virus Research*, vol. 134, no. 1-2, pp. 74–85, 2008.

[32] D. Arion, N. Kaushik, S. McCormick, G. Borkow, and M. A. Parniak, "Phenotypic mechanism of HIV-1 resistance to 3′-azido-3′-deoxythymidine (AZT): increased polymerization processivity and enhanced sensitivity to pyrophosphate of the mutant viral reverse transcriptase," *Biochemistry*, vol. 37, no. 45, pp. 15908–15917, 1998.

[33] P. R. Meyer, S. E. Matsuura, A. G. So, and W. A. Scott, "Unblocking of chain-terminated primer by HIV-1 reverse transcriptase through a nucleotide-dependent mechanism," *Proceedings of the National Academy of Sciences of the United States of America*, vol. 95, no. 23, pp. 13471–13476, 1998.

[34] P. R. Meyer, S. E. Matsuura, A. Mohsin Mian, A. G. So, and W. A. Scott, "A mechanism of AZT resistance: an increase in nucleotide-dependent primer unblocking by mutant HIV-1 reverse transcriptase," *Molecular Cell*, vol. 4, no. 1, pp. 35–43, 1999.

[35] D. M. Simpson and M. Tagliati, "Nucleoside analogue-associated peripheral neuropathy in human immunodeficiency virus infection," *Journal of Acquired Immune Deficiency Syndromes and Human Retrovirology*, vol. 9, no. 2, pp. 153–161, 1995.

[36] R. F. Schinazi, R. M. Lloyd, M. H. Nguyen et al., "Characterization of human immunodeficiency viruses resistant to oxathiolane-cytosine nucleosides," *Antimicrobial Agents and Chemotherapy*, vol. 37, no. 4, pp. 875–881, 1993.

[37] R. Schuurman, M. Nijhuis, R. Van Leeuwen et al., "Rapid changes in human immunodeficiency virus type 1 RNA load and appearance of drug-resistant virus populations in persons treated with lamivudine (3TC)," *Journal of Infectious Diseases*, vol. 171, no. 6, pp. 1411–1419, 1995.

[38] S. G. Sarafianos, K. Das, A. D. Clark et al., "Lamivudine (3TC) resistance in HIV-1 reverse transcriptase involves steric hindrance with β-branched amino acids," *Proceedings*

of the National Academy of Sciences of the United States of America, vol. 96, no. 18, pp. 10027–10032, 1999.

[39] H. Huang, R. Chopra, G. L. Verdine, and S. C. Harrison, "Structure of a covalently trapped catalytic complex of HIV-1 reverse transcriptase: implications for drug resistance," *Science*, vol. 282, no. 5394, pp. 1669–1675, 1998.

[40] P. L. Boyer, S. G. Sarafianos, E. Arnold, and S. H. Hughes, "Selective excision of AZTMP by drug-resistant human immunodeficiency virus reverse transcriptase," *Journal of Virology*, vol. 75, no. 10, pp. 4832–4842, 2001.

[41] S. Dharmasena, Z. Pongracz, E. Arnold, S. G. Sarafianos, and M. A. Parniak, "3′-azido-3′-deoxythymidine-(5′)-tetraphospho-(5′)-adenosine, the product of ATP-mediated excision of chain-terminating AZTMP, is a potent chain-terminating substrate for HIV-1 reverse transcriptase," *Biochemistry*, vol. 46, no. 3, pp. 828–836, 2007.

[42] N. Yahi, C. Tamalet, C. Tourres, N. Tivoli, and J. Fantini, "Mutation L210W of HIV-1 reverse transcriptase in patients receiving combination therapy: incidence, association with other mutations, and effects on the structure of mutated reverse transcriptase," *Journal of Biomedical Science*, vol. 7, no. 6, pp. 507–513, 2000.

[43] G. N. Nikolenko, K. A. Delviks-Frankenberry, S. Palmer et al., "Mutations in the connection domain of HIV-1 reverse transcriptase increase 3′-azido-3′-deoxythymidine resistance," *Proceedings of the National Academy of Sciences of the United States of America*, vol. 104, no. 1, pp. 317–322, 2007.

[44] S. H. Yap, C. W. Sheen, J. Fahey et al., "N348I in the connection domain of HIV-1 reverse transcriptase confers zidovudine and nevirapine resistance," *PLoS Medicine*, vol. 4, no. 12, Article ID e335, 2007.

[45] K. A. Delviks-Frankenberry, G. N. Nikolenko, R. Barr, and V. K. Pathak, "Mutations in human immunodeficiency virus type 1 RNase H primer grip enhance 3′-azido-3′-deoxythymidine resistance," *Journal of Virology*, vol. 81, no. 13, pp. 6837–6845, 2007.

[46] J. H. Brehm, D. Koontz, J. D. Meteer, V. Pathak, N. Sluis-Cremer, and J. W. Mellors, "Selection of mutations in the connection and RNase H domains of human immunodeficiency virus type 1 reverse transcriptase that increase resistance to 3′-azido-3′-dideoxythymidine," *Journal of Virology*, vol. 81, no. 15, pp. 7852–7859, 2007.

[47] A. Hachiya, E. N. Kodama, S. G. Sarafianos et al., "Amino acid mutation N348I in the connection subdomain of human immunodeficiency virus type 1 reverse transcriptase confers multiclass resistance to nucleoside and nonnucleoside reverse transcriptase inhibitors," *Journal of Virology*, vol. 82, no. 7, pp. 3261–3270, 2008.

[48] S. Zelina, C. W. Sheen, J. Radzio, J. W. Mellors, and N. Sluis-Cremer, "Mechanisms by which the G333D mutation in human immunodeficiency virus type 1 reverse transcriptase facilitates dual resistance to zidovudine and lamivudine," *Antimicrobial Agents and Chemotherapy*, vol. 52, no. 1, pp. 157–163, 2008.

[49] K. A. Delviks-Frankenberry, G. N. Nikolenko, and V. K. Pathak, "The "connection" between HIV drug resistance and RNase H," *Viruses*, vol. 2, no. 7, pp. 1476–1503, 2010.

[50] J. Balzarini, L. Naesens, S. Aquaro et al., "Intracellular metabolism of CycloSaligenyl 3′azido-2′,3′-dideoxythymidine monophosphate, a prodrug of 3′-azido-2′,3′-dideoxythymidine (zidovudine)," *Molecular Pharmacology*, vol. 56, no. 6, pp. 1354–1361, 1999.

[51] K. E. Squires, "An introduction to nucleoside and nucleotide analogues," *Antiviral Therapy*, vol. 6, no. 3, pp. 1–14, 2001.

[52] K. Das, S. E. Martinez, J. D. Bauman, and E. Arnold, "HIV-1 reverse transcriptase complex with DNA and nevirapine reveals non-nucleoside inhibition mechanism," *Nature Structural & Molecular Biology*, vol. 19, pp. 253–259, 2012.

[53] P. W. Mui, S. P. Jacober, K. D. Hargrave, and J. Adams, "Crystal structure of nevirapine, a non-nucleoside inhibitor of HIV-1 reverse transcriptase, and computational alignment with a structurally diverse inhibitor," *Journal of Medicinal Chemistry*, vol. 35, no. 1, pp. 201–202, 1992.

[54] W. Schäfer, W. G. Friebe, H. Leinert et al., "Non-nucleoside inhibitors of HIV-1 reverse transcriptase: molecular modeling and X-ray structure investigations," *Journal of Medicinal Chemistry*, vol. 36, no. 6, pp. 726–732, 1993.

[55] J. Ding, K. Das, C. Tantillo et al., "Structure of HIV-1 reverse transcriptase in a complex with the non-nucleoside inhibitor α-APA R 95845 at 2.8 A resolution," *Structure*, vol. 3, no. 4, pp. 365–379, 1995.

[56] P. P. Mager, "Hybrid canonical-correlation neural-network approach applied to nonnucleoside HIV-1 reverse transcriptase inhibitors (HEPT derivatives)," *Current Medicinal Chemistry*, vol. 10, no. 17, pp. 1643–1659, 2003.

[57] N. Sluis-Cremer, N. A. Temiz, and I. Bahar, "Conformational changes in HIV-1 reverse transcriptase induced by nonnucleoside reverse transcriptase inhibitor binding," *Current HIV Research*, vol. 2, no. 4, pp. 323–332, 2004.

[58] E. De Clercq, "Perspectives of non-nucleoside reverse transcriptase inhibitors (NNRTIs) in the therapy of HIV-1 infection," *Farmaco*, vol. 54, no. 1-2, pp. 26–45, 1999.

[59] J. Balzarini, "Current status of the non-nucleoside reverse transcriptase inhibitors of human immunodeficiency virus type 1," *Current Topics in Medicinal Chemistry*, vol. 4, no. 9, pp. 921–944, 2004.

[60] R. M. Esnouf, J. Ren, A. L. Hopkins et al., "Unique features in the structure of the complex between HIV-1 reverse transcriptase and the bis(heteroaryl)piperazine (BHAP) U-90152 explain resistance mutations for this nonnucleoside inhibitor," *Proceedings of the National Academy of Sciences of the United States of America*, vol. 94, no. 8, pp. 3984–3989, 1997.

[61] J. Ren and D. K. Stammers, "Structural basis for drug resistance mechanisms for non-nucleoside inhibitors of HIV reverse transcriptase," *Virus Research*, vol. 134, no. 1-2, pp. 157–170, 2008.

[62] J. W. Mellors, G. E. Dutschman, G. J. Im, E. Tramontano, S. R. Winkler, and Y. C. Cheng, "In vitro selection and molecular characterization of human immunodeficiency virus-1 resistant to non-nucleoside inhibitors of reverse transcriptase," *Molecular Pharmacology*, vol. 41, no. 3, pp. 446–451, 1992.

[63] J. W. Mellors, G. J. Im, E. Tramontano et al., "A single conservative amino acid substitution in the reverse transcriptase of human immunodeficiency virus-1 confers resistance to (+)-(5S)-4,5,6,7-tetrahydro-5-methyl-6-(3-methyl-2-butenyl)imidazo[4,5,1-jk][1,4]benzodiazepin-2(1H)-thione (TIBO R82150)," *Molecular Pharmacology*, vol. 43, no. 1, pp. 11–16, 1993.

[64] G. N. Nikolenko, K. A. Delviks-Frankenberry, and V. K. Pathak, "A novel molecular mechanism of dual resistance to nucleoside and nonnucleoside reverse transcriptase inhibitors," *Journal of Virology*, vol. 84, no. 10, pp. 5238–5249, 2010.

HIV-1 Reverse Transcriptase Still Remains a New Drug Target: Structure, Function, Classical Inhibitors, and New Inhibitors with Innovative Mechanisms of Actions

177

[65] T. Cihlar and A. S. Ray, "Nucleoside and nucleotide HIV reverse transcriptase inhibitors: 25 years after zidovudine," *Antiviral Research*, vol. 85, no. 1, pp. 39–58, 2010.

[66] H. T. Ho and M. J. Hitchcock, "Cellular pharmacology of 2′,3′-dideoxy-2′,3′-didehydrothymidine, a nucleoside analog active against human immunodeficiency virus," *Antimicrobial Agents and Chemotherapy*, vol. 33, no. 6, pp. 844–849, 1989.

[67] L. J. Waters, G. Moyle, S. Bonora et al., "Abacavir plasma pharmacokinetics in the absence and presence of atazanavir/ritonavir or lopinavir/ritonavir and vice versa in HIV-infected patients," *Antiviral Therapy*, vol. 12, no. 5, pp. 825–830, 2007.

[68] J. A. Mcdowell, G. E. Chittick, C. P. Stevens, K. D. Edwards, and D. S. Stein, "Pharmacokinetic interaction of abacavir (1592U89) and ethanol in human immunodeficiency virus-infected adults," *Antimicrobial Agents and Chemotherapy*, vol. 44, no. 6, pp. 1686–1690, 2000.

[69] J. A. Johnson, J. F. Li, X. Wei et al., "Minority HIV-1 drug resistance mutations are present in antiretroviral treatment-naive populations and associate with reduced treatment efficacy," *PLoS Medicine*, vol. 5, no. 7, Article ID e158, 2008.

[70] R. C. Bethell, Y. S. Lie, and N. T. Parkin, "In vitro activity of SPD754, a new deoxycytidine nucleoside reverse transcriptase inhibitor (NRTI), against 215 HIV-1 isolates resistant to other NRTIs," *Antiviral Chemistry and Chemotherapy*, vol. 16, no. 5, pp. 295–302, 2005.

[71] Z. Gu, B. Allard, J. M. De Muys et al., "In vitro antiretroviral activity and in vitro toxicity profile of SPD754, a new deoxycytidine nucleoside reverse transcriptase inhibitor for treatment of human immunodeficiency virus infection," *Antimicrobial Agents and Chemotherapy*, vol. 50, no. 2, pp. 625–631, 2006.

[72] S. Cox and J. Southby, "Apricitabine—a novel nucleoside reverse transcriptase inhibitor for the treatment of HIV infection that is refractory to existing drugs," *Expert Opinion on Investigational Drugs*, vol. 18, no. 2, pp. 199–209, 2009.

[73] M. P. de Baar, E. R. de Rooij, K. G. M. Smolders, H. B. van Schijndel, E. C. Timmermans, and R. Bethell, "Effects of apricitabine and other nucleoside reverse transcriptase inhibitors on replication of mitochondrial DNA in HepG2 cells," *Antiviral Research*, vol. 76, no. 1, pp. 68–74, 2007.

[74] R. Bethell, J. De Muys, J. Lippens et al., "In vitro interactions between apricitabine and other deoxycytidine analogues," *Antimicrobial Agents and Chemotherapy*, vol. 51, no. 8, pp. 2948–2953, 2007.

[75] T. S. Lin, M. Z. Luo, M. C. Liu et al., "Design and synthesis of 2′,3′-dideoxy-2′,3′-didehydro-β-L-cytidine (β- L-d4C) and 2′,3′-dideoxy-2′-3′-didehydro-β-L-5-fluorocytidine (β-L-Fd4C), two exceptionally potent inhibitors of human hepatitis B virus (HBV) and potent inhibitors of human immunodeficiency virus (HIV) in vitro," *Journal of Medicinal Chemistry*, vol. 39, no. 9, pp. 1757–1759, 1996.

[76] G. E. Dutschman, E. G. Bridges, S. H. Liu et al., "Metabolism of 2′,3′-dideoxy-2′,3′-didehydro-β-L(-)-5-fluorocytidine and its activity in combination with clinically approved anti-human immunodeficiency virus β-D(+) nucleoside analogs in vitro," *Antimicrobial Agents and Chemotherapy*, vol. 42, no. 7, pp. 1799–1804, 1998.

[77] J. L. Hammond, U. M. Parikh, D. L. Koontz et al., "In vitro selection and analysis of human immunodeficiency virus type 1 resistant to derivatives of β-2′,3′-didehydro-2′,3′-dideoxy-5-fluorocytidine," *Antimicrobial Agents and Chemotherapy*, vol. 49, no. 9, pp. 3930–3932, 2005.

[78] U. M. Parikh, D. L. Koontz, C. K. Chu, R. F. Schinazi, and J. W. Mellors, "In vitro activity of structurally diverse nucleoside analogs against human immunodeficiency virus type 1 with the K65R mutation in reverse transcriptase," *Antimicrobial Agents and Chemotherapy*, vol. 49, no. 3, pp. 1139–1144, 2005.

[79] Z. Gu, M. A. Wainberg, N. Nguyen-Ba et al., "Mechanism of action and in vitro activity of 1′,3′-dioxolanylpurine nucleoside analogues against sensitive and drug-resistant human immunodeficiency virus type 1 variants," *Antimicrobial Agents and Chemotherapy*, vol. 43, no. 10, pp. 2376–2382, 1999.

[80] J. P. Mewshaw, F. T. Myrick, D. A. C. S. Wakefield et al., "Dioxolane guanosine, the active form of the prodrug diaminopurine dioxolane, is a potent inhibitor of drug-resistant HIV-1 isolates from patients for whom standard nucleoside therapy fails," *Journal of Acquired Immune Deficiency Syndromes*, vol. 29, no. 1, pp. 11–20, 2002.

[81] H. Z. Bazmi, J. L. Hammond, S. C. H. Cavalcanti, C. K. Chu, R. F. Schinazi, and J. W. Mellors, "In vitro selection of mutations in the human immunodeficiency virus type 1 reverse transcriptase that decrease susceptibility to (-)-β-D-dioxolane- guanosine and suppress resistance to 3′-azido-3′-deoxythymidine," *Antimicrobial Agents and Chemotherapy*, vol. 44, no. 7, pp. 1783–1788, 2000.

[82] P. A. Furman, J. Jeffrey, L. L. Kiefer et al., "Mechanism of action of 1-β-D-2,6-diaminopurine dioxolane, a prodrug of the human immunodeficiency virus type 1 inhibitor 1-β-D-dioxolane guanosine," *Antimicrobial Agents and Chemotherapy*, vol. 45, no. 1, pp. 158–165, 2001.

[83] J. Y. Feng and K. S. Anderson, "Mechanistic studies comparing the incorporation of (+) and (-) isomers of 3TCTP by HIV-1 reverse transcriptase," *Biochemistry*, vol. 38, no. 1, pp. 55–63, 1999.

[84] D. S. Shewach, D. C. Liotta, and R. F. Schinazi, "Affinity of the antiviral enantiomers of oxathiolane cytosine nucleosides for human 2′-deoxycytidine kinase," *Biochemical Pharmacology*, vol. 45, no. 7, pp. 1540–1543, 1993.

[85] R. F. Schinazi, A. McMillan, D. Cannon et al., "Selective inhibition of human immunodeficiency viruses by racemates and enantiomers of cis-5-fluoro-1-[2-(hydroxymethyl)-1,3-oxathiolan-5-yl]cytosine," *Antimicrobial Agents and Chemotherapy*, vol. 36, no. 11, pp. 2423–2431, 1992.

[86] R. F. Schinazi, A. McMillan, R. L. Lloyd, S. Schlueter-Wirtz, D. C. Liotta, and C. K. Chu, "Molecular properties of HIV-1 resistant to (+)-enantiomers and racemates of oxathiolane cytosine nucleosides and their potential for the treatment of HIV and HBV infections," *Antiviral Research*, vol. 34, article A42, 1997.

[87] C. Herzmann, K. Arasteh, R. L. Murphy et al., "Safety, pharmacokinetics, and efficacy of (+/-)-β-2′,3′- dideoxy-5-fluoro-3′-thiacytidine with efavirenz and stavudine in antiretroviral-naive human immunodeficiency virus-infected patients," *Antimicrobial Agents and Chemotherapy*, vol. 49, no. 7, pp. 2828–2833, 2005.

[88] J. Y. Feng, E. Murakami, S. M. Zorca et al., "Relationship between antiviral activity and host toxicity: comparison of the incorporation efficiencies of 2′,3′-Dideoxy-5-Fluoro-3′ -Thiacytidine-triphosphate analogs by human immunodeficiency virus type 1 reverse transcriptase and human mitochondrial DNA polymerase," *Antimicrobial Agents and Chemotherapy*, vol. 48, no. 4, pp. 1300–1306, 2004.

[89] G. E. Dutschman, S. P. Grill, E. A. Gullen et al., "Novel 4′-substituted stavudine analog with improved anti-human

immunodeficiency virus activity and decreased cytotoxicity," *Antimicrobial Agents and Chemotherapy*, vol. 48, no. 5, pp. 1640–1646, 2004.

[90] P. Herdewijn, J. Balzarini, M. Baba et al., "Synthesis and anti-HIV activity of different sugar-modified pyrimidine and purine nucleosides," *Journal of Medicinal Chemistry*, vol. 31, no. 10, pp. 2040–2048, 1988.

[91] H. Hayakawa, S. Kohgo, K. Kitano et al., "Potential of 4′-C-substituted nucleosides for the treatment of HIV-1," *Antiviral Chemistry and Chemotherapy*, vol. 15, no. 4, pp. 169–187, 2004.

[92] A. Kawamoto, E. Kodama, S. G. Sarafianos et al., "2′-Deoxy-4′-C-ethynyl-2-halo-adenosines active against drug-resistant human immunodeficiency virus type 1 variants," *International Journal of Biochemistry and Cell Biology*, vol. 40, no. 11, pp. 2410–2420, 2008.

[93] E. Michailidis, B. Marchand, E. N. Kodama et al., "Mechanism of inhibition of HIV-1 reverse transcriptase by 4′-ethynyl-2-fluoro-2′-deoxyadenosine triphosphate, a translocation-defective reverse transcriptase inhibitor," *Journal of Biological Chemistry*, vol. 284, no. 51, pp. 35681–35691, 2009.

[94] K. A. Kirby, K. Singh, E. Michailidis et al., "The sugar ring conformation of 4′-ethynyl-2-fluoro-2′-deoxyadenosine and its recognition by the polymerase active site of HIV reverse transcriptase," *Cellular and Molecular Biology*, vol. 57, no. 1, pp. 40–46, 2011.

[95] C. D. Sohl, K. Singh, R. Kasiviswanathan et al., "The mechanism of interaction of human mitochondrial DNA γ with the novel nucleoside reverse transcriptase inhibitor 4′-Ethynyl-2-Fluoro-2′-deoxyadenosine indicates a low potential for host toxicity," *Antimicrobial Agents and Chemotherapy*, vol. 56, pp. 1630–1634, 2012.

[96] V. Vivet?Boudou, C. Isel, M. Sleiman et al., "8-modified-2′-deoxyadenosine analogues induce delayed polymerization arrest during HIV-1 reverse transcription," *PLoS ONE*, vol. 6, no. 11, Article ID e27456, 2011.

[97] L. A. Loeb, J. M. Essigmann, F. Kazazi, J. Zhang, K. D. Rose, and J. I. Mullins, "Lethal mutagenesis of HIV with mutagenic nucleoside analogs," *Proceedings of the National Academy of Sciences of the United States of America*, vol. 96, no. 4, pp. 1492–1497, 1999.

[98] K. S. Harris, W. Brabant, S. Styrchak, A. Gall, and R. Daifuku, "KP-1212/1461, a nucleoside designed for the treatment of HIV by viral mutagenesis," *Antiviral Research*, vol. 67, no. 1, pp. 1–9, 2005.

[99] R. A. Smith, L. A. Loeb, and B. D. Preston, "Lethal mutagenesis of HIV," *Virus Research*, vol. 107, no. 2, pp. 215–228, 2005.

[100] X. Tu, K. Das, Q. Han et al., "Structural basis of HIV-1 resistance to AZT by excision," *Nature Structural and Molecular Biology*, vol. 17, no. 10, pp. 1202–1209, 2010.

[101] P. R. Meyer, A. J. Smith, S. E. Matsuura, and W. A. Scott, "Chain-terminating dinucleoside tetraphosphates are substrates for DNA polymerization by human immunodeficiency virus type 1 reverse transcriptase with increased activity against thymidine analogue-resistant mutants," *Antimicrobial Agents and Chemotherapy*, vol. 50, no. 11, pp. 3607–3614, 2006.

[102] N. Sluis-Cremer and G. Tachedjian, "Mechanisms of inhibition of HIV replication by non-nucleoside reverse transcriptase inhibitors," *Virus Research*, vol. 134, no. 1-2, pp. 147–156, 2008.

[103] P. Zhan, X. Chen, D. Li, Z. Fang, E. De Clercq, and X. Liu, "HIV-1 NNRTIs: structural diversity, pharmacophore similarity, and implications for drug design," *Medicinal Research Reviews*. In press.

[104] K. Das, P. J. Lewi, S. H. Hughes, and E. Arnold, "Crystallography and the design of anti-AIDS drugs: conformational flexibility and positional adaptability are important in the design of non-nucleoside HIV-1 reverse transcriptase inhibitors," *Progress in Biophysics and Molecular Biology*, vol. 88, no. 2, pp. 209–231, 2005.

[105] P. Zhan, X. Liu, Z. Li, C. Pannecouque, and E. De Clercq, "Design strategies of novel NNRTIs to overcome drug resistance," *Current Medicinal Chemistry*, vol. 16, no. 29, pp. 3903–3917, 2009.

[106] P. A. J. Janssen, P. J. Lewi, E. Arnold et al., "In search of a novel anti-HIV drug: multidisciplinary coordination in the discovery of 4-[[4-[[4-[(1E)-2-cyanoethenyl]-2,6-dimethylphenyl]amino]-2- pyrimidinyl]amino]benzonitrile (R278474, rilpivirine)," *Journal of Medicinal Chemistry*, vol. 48, no. 6, pp. 1901–1909, 2005.

[107] J. D. Pata, W. G. Stirtan, S. W. Goldstein, and T. A. Steitz, "Structure of HIV-1 reverse transcriptase bound to an inhibitor active against mutant reverse transcriptases resistant to other nonnucleoside inhibitors," *Proceedings of the National Academy of Sciences of the United States of America*, vol. 101, no. 29, pp. 10548–10553, 2004.

[108] J. Ren and D. K. Stammers, "HIV reverse transcriptase structures: designing new inhibitors and understanding mechanisms of drug resistance," *Trends in Pharmacological Sciences*, vol. 26, no. 1, pp. 4–7, 2005.

[109] J. Tang, K. Maddali, C. D. Dreis et al., "N-3 hydroxylation of pyrimidine-2,4-diones yields dual inhibitors of HIV reverse transcriptase and integrase," *ACS Medicinal Chemistry Letters*, vol. 2, no. 1, pp. 63–67, 2011.

[110] J. Tang, K. Maddali, C. D. Dreis et al., "6-Benzoyl-3-hydroxypyrimidine-2,4-diones as dual inhibitors of HIV reverse transcriptase and integrase," *Bioorganic and Medicinal Chemistry Letters*, vol. 21, no. 8, pp. 2400–2402, 2011.

[111] D. Jochmans, J. Deval, B. Kesteleyn et al., "Indolopyridones inhibit human immunodeficiency virus reverse transcriptase with a novel mechanism of action," *Journal of Virology*, vol. 80, no. 24, pp. 12283–12292, 2006.

[112] Z. Zhang, M. Walker, W. Xu et al., "Novel nonnucleoside inhibitors that select nucleoside inhibitor resistance mutations in human immunodeficiency virus type 1 reverse transcriptase," *Antimicrobial Agents and Chemotherapy*, vol. 50, no. 8, pp. 2772–2781, 2006.

[113] M. Ehteshami, B. J. Scarth, E. P. Tchesnokov et al., "Mutations M184V and Y115F in HIV-1 reverse transcriptase discriminate against "nucleotide-competing reverse transcriptase inhibitors"," *Journal of Biological Chemistry*, vol. 283, no. 44, pp. 29904–29911, 2008.

[114] A. Auger, G. L. Beilhartz, S. Zhu et al., "Impact of primer-induced conformational dynamics of HIV-1 reverse transcriptase on polymerase translocation and inhibition," *The Journal of Biological Chemistry*, vol. 286, pp. 29575–29583, 2011.

[115] G. Maga, M. Radi, S. Zanoli et al., "Discovery of non-nucleoside inhibitors of HIV-1 reverse transcriptase competing with the nucleotide substrate," *Angewandte Chemie*, vol. 46, no. 11, pp. 1810–1813, 2007.

[116] M. Radi, C. Falciani, L. Contemori et al., "A multidisciplinary approach for the identification of novel HIV-1 non-nucleoside reverse transcriptase inhibitors: S-DABOCs

HIV-1 Reverse Transcriptase Still Remains a New Drug Target: Structure, Function, Classical Inhibitors, and New Inhibitors with Innovative Mechanisms of Actions

179

and DAVPs," *ChemMedChem*, vol. 3, no. 4, pp. 573–593, 2008.

[117] S. Freisz, G. Bec, M. Radi et al., "Crystal structure of HIV-1 reverse transcriptase bound to a non-nucleoside inhibitor with a novel mechanism of action," *Angewandte Chemie*, vol. 49, no. 10, pp. 1805–1808, 2010.

[118] B. Oberg, "Antiviral effects of phosphonoformate (PFA, foscarnet sodium)," *Pharmacology and Therapeutics*, vol. 40, no. 2, pp. 213–285, 1989.

[119] R. R. Razonable, "Antiviral drugs for viruses other than human immunodeficiency virus," *Mayo Clinic Proceedings*, vol. 86, pp. 1009–1026, 2011.

[120] D. Derse, K. F. Bastow, and Y. Cheng, "Characterization of the DNA polymerases induced by a group of herpes simplex virus type I variants selected for growth in the presence of phosphonoformic acid," *Journal of Biological Chemistry*, vol. 257, no. 17, pp. 10251–10260, 1982.

[121] B. Marchand, E. P. Tchesnokov, and M. Götte, "The pyrophosphate analogue foscarnet traps the pre-translocational state of HIV-1 reverse transcriptase in a Brownian ratchet model of polymerase translocation," *Journal of Biological Chemistry*, vol. 282, no. 5, pp. 3337–3346, 2007.

[122] J. W. Mellors, H. Z. Bazmi, R. F. Schinazi et al., "Novel mutations in reverse transcriptase of human immunodeficiency virus type 1 reduce susceptibility to foscarnet in laboratory and clinical isolates," *Antimicrobial Agents and Chemotherapy*, vol. 39, no. 5, pp. 1087–1092, 1995.

[123] G. Tachedjian, D. J. Hooker, A. D. Gurusinghe et al., "Characterisation of foscarnet-resistant strains of human immunodeficiency virus type 1," *Virology*, vol. 212, no. 1, pp. 58–68, 1995.

[124] G. J. Im, E. Tramontano, C. J. Gonzalez, and Y. C. Cheng, "Identification of the amino acid in the human immunodeficiency virus type 1 reverse transcriptase involved in the pyrophosphate binding of antiviral nucleoside triphosphate analogs and phosphonoformate. Implications for multiple drug resistance," *Biochemical Pharmacology*, vol. 46, no. 12, pp. 2307–2313, 1993.

[125] E. Tramontano, G. Piras, J. W. Mellors, M. Putzolu, H. Z. Bazmi, and P. La Colla, "Biochemical characterization of HIV-1 reverse transcriptases encoding mutations at amino acid residues 161 and 208 involved in resistance to phosphonoformate," *Biochemical Pharmacology*, vol. 56, no. 12, pp. 1583–1589, 1998.

[126] C. Cruchaga, E. Ansó, A. Rouzaut, and J. J. Martínez-Irujo, "Selective excision of chain-terminating nucleotides by HIV-1 reverse transcriptase with phosphonoformate as substrate," *Journal of Biological Chemistry*, vol. 281, no. 38, pp. 27744–27752, 2006.

[127] E. Tramontano, "HIV-1 RNase H: recent progress in an exciting, yet little explored, drug target," *Mini-Reviews in Medicinal Chemistry*, vol. 6, no. 6, pp. 727–737, 2006.

[128] E. B. Lansdon, Q. Liu, S. A. Leavitt et al., "Structural and binding analysis of pyrimidinol carboxylic acid and N-hydroxy quinazolinedione HIV-1 RNase H inhibitors," *Antimicrobial Agents and Chemotherapy*, vol. 55, no. 6, pp. 2905–2915, 2011.

[129] H. Fuji, E. Urano, Y. Futahashi et al., "Derivatives of 5-nitro-furan-2-carboxylic acid carbamoylmethyl ester inhibit RNase H activity associated with HIV-1 reverse transcriptase," *Journal of Medicinal Chemistry*, vol. 52, no. 5, pp. 1380–1387, 2009.

[130] H. Yanagita, E. Urano, K. Matsumoto et al., "Structural and biochemical study on the inhibitory activity of derivatives of 5-nitro-furan-2-carboxylic acid for RNase H function of HIV-1 reverse transcriptase," *Bioorganic and Medicinal Chemistry*, vol. 19, no. 2, pp. 816–825, 2011.

[131] H. P. Su, Y. Yan, G. S. Prasad et al., "Structural basis for the inhibition of RNase H activity of HIV-1 reverse transcriptase by RNase H active site-directed inhibitors," *Journal of Virology*, vol. 84, no. 15, pp. 7625–7633, 2010.

[132] C. A. Shaw-Reid, V. Munshi, P. Graham et al., "Inhibition of HIV-1 ribonuclease H by a novel diketo acid, 4-[5-(benzoylamino)thien-2-yl]-2,4-dioxobutanoic acid," *Journal of Biological Chemistry*, vol. 278, no. 5, pp. 2777–2780, 2003.

[133] E. Tramontano, F. Esposito, R. Badas, R. Di Santo, R. Costi, and P. La Colla, "6-[1-(4-Fluorophenyl)methyl-1H-pyrrol-2-yl)]-2,4-dioxo-5-hexenoic acid ethyl ester a novel diketo acid derivative which selectively inhibits the HIV-1 viral replication in cell culture and the ribonuclease H activity in vitro," *Antiviral Research*, vol. 65, no. 2, pp. 117–124, 2005.

[134] R. Di Santo, R. Costi, M. Artico, E. Tramontano, P. La Colla, and A. Pani, "HIV-1 integrase inhibitors that block HIV-1 replication in infected cells. Planning synthetic derivatives from natural products," *Pure and Applied Chemistry*, vol. 75, no. 2-3, pp. 195–206, 2003.

[135] R. Costi, R. Di Santo, M. Artico et al., "6-Aryl-2,4-dioxo-5-hexenoic acids, novel integrase inhibitors active against HIV-1 multiplication in cell-based assays," *Bioorganic and Medicinal Chemistry Letters*, vol. 14, no. 7, pp. 1745–1749, 2004.

[136] R. Costi, R. Di Santo, M. Artico et al., "2,6-Bis(3,4,5-trihydroxybenzylydene) derivatives of cyclohexanone: novel potent HIV-1 integrase inhibitors that prevent HIV-1 multiplication in cell-based assays," *Bioorganic and Medicinal Chemistry*, vol. 12, no. 1, pp. 199–215, 2004.

[137] K. Klumpp, J. Q. Hang, S. Rajendran et al., "Two-metal ion mechanism of RNA cleavage by HIV RNase H and mechanism-based design of selective HIV RNase H inhibitors," *Nucleic Acids Research*, vol. 31, no. 23, pp. 6852–6859, 2003.

[138] J. Q. Hang, S. Rajendran, Y. Yang et al., "Activity of the isolated HIV RNase H domain and specific inhibition by N-hydroxyimides," *Biochemical and Biophysical Research Communications*, vol. 317, no. 2, pp. 321–329, 2004.

[139] M. Billamboz, F. Bailly, M. L. Barreca et al., "Design, synthesis, and biological evaluation of a series of 2-hydroxyisoquinoline-1,3(2H,4H)-diones as dual inhibitors of human immunodeficiency virus type 1 integrase and the reverse transcriptase RNase H domain," *Journal of Medicinal Chemistry*, vol. 51, no. 24, pp. 7717–7730, 2008.

[140] M. Billamboz, F. Bailly, C. Lion et al., "Magnesium chelating 2-hydroxyisoquinoline-1,3(2H, 4H)-diones, as inhibitors of HIV-1 integrase and/or the HIV-1 reverse transcriptase ribonuclease H domain: discovery of a novel selective inhibitor of the ribonuclease H function," *Journal of Medicinal Chemistry*, vol. 54, no. 6, pp. 1812–1824, 2011.

[141] M. Wendeler, H. F. Lee, A. Bermingham et al., "Vinylogous ureas as a novel class of inhibitors of reverse transcriptase-associated ribonuclease H activity," *ACS Chemical Biology*, vol. 3, no. 10, pp. 635–644, 2008.

[142] S. Chung, M. Wendeler, J. W. Rausch et al., "Structure-activity analysis of vinylogous urea inhibitors of human immunodeficiency virus-encoded ribonuclease H," *Antimicrobial Agents and Chemotherapy*, vol. 54, no. 9, pp. 3913–3921, 2010.

[143] G. Borkow, R. S. Fletcher, J. Barnard et al., "Inhibition of the ribonuclease H and DNA polymerase activities of HIV-1

reverse transcriptase by N-(4-tert-butylbenzoyl)-2-hydroxy-1- naphthaldehyde hydrazone," *Biochemistry*, vol. 36, no. 11, pp. 3179–3185, 1997.

[144] D. Arion, N. Sluis-Cremer, K. L. Min, M. E. Abram, R. S. Fletcher, and M. A. Parniak, "Mutational analysis of tyr-501 of HIV-1 reverse transcriptase: effects on ribonuclease H activity and inhibition of this activity by N-acylhydrazones," *Journal of Biological Chemistry*, vol. 277, no. 2, pp. 1370–1374, 2002.

[145] N. Sluis-Cremer, D. Arion, and M. A. Parniak, "Destabilization of the HIV-1 reverse transcriptase dimer upon interaction with N-acyl hydrazone inhibitors," *Molecular Pharmacology*, vol. 62, no. 2, pp. 398–405, 2002.

[146] D. M. Himmel, S. G. Sarafianos, S. Dharmasena et al., "HIV-1 reverse transcriptase structure with RNase H inhibitor dihydroxy benzoyl naphthyl hydrazone bound at a novel site," *ACS Chemical Biology*, vol. 1, no. 11, pp. 702–712, 2006.

[147] A. K. Felts, K. La Barge, J. D. Bauman et al., "Identification of alternative binding sites for inhibitors of HIV-1 ribonuclease H through comparative analysis of virtual enrichment studies," *Journal of Chemical Information and Modeling*, vol. 51, no. 5, pp. 1986–1998, 2011.

[148] K. Tatyana, E. Francesca, Z. Luca et al., "Inhibition of HIV-1 ribonuclease H activity by novel frangula-emodine derivatives," *Medicinal Chemistry*, vol. 5, no. 5, pp. 398–410, 2009.

[149] F. Esposito, T. Kharlamova, S. Distinto et al., "Alizarine derivatives as new dual inhibitors of the HIV-1 reverse transcriptase-associated DNA polymerase and RNase H activities effective also on the RNase H activity of non-nucleoside resistant reverse transcriptases," *FEBS Journal*, vol. 278, no. 9, pp. 1444–1457, 2011.

[150] E. Tramontano, T. Kharlamova, and F. Esposito, "Effect of new quinizarin derivatives on both HCV NS5B RNA polymerase and HIV-1 reverse transcriptase associated ribonuclease H activities," *Journal of Chemotherapy*, vol. 23, pp. 273–276, 2011.

[151] P. Mohan, S. Loya, O. Avidan et al., "Synthesis of naphthalenesulfonic acid small molecules as selective inhibitors of the DNA polymerase and ribonuclease H activities of HIV-1 reverse transcriptase," *Journal of Medicinal Chemistry*, vol. 37, no. 16, pp. 2513–2519, 1994.

[152] A. G. Skillman, K. W. Maurer, D. C. Roe et al., "A novel mechanism for inhibition of HIV-1 reverse transcriptase," *Bioorganic Chemistry*, vol. 30, no. 6, pp. 443–458, 2002.

[153] L. Z. Wang, G. L. Kenyon, and K. A. Johnson, "Novel mechanism of inhibition of HIV-1 reverse transcriptase by a new non-nucleoside analog, KM-1," *Journal of Biological Chemistry*, vol. 279, no. 37, pp. 38424–38432, 2004.

[154] C. A. Davis, M. A. Parniak, and S. H. Hughes, "The effects of RNase H inhibitors and nevirapine on the susceptibility of HIV-1 to AZT and 3TC," *Virology*, vol. 419, pp. 64–71, 2011.

[155] S. Srivastava, N. Sluis-Cremer, and G. Tachedjian, "Dimerization of human immunodeficiency virus type 1 reverse transcriptase as an antiviral target," *Current Pharmaceutical Design*, vol. 12, no. 15, pp. 1879–1894, 2006.

[156] M. J. Camarasa, S. Velázquez, A. San-Félix, M. J. Pérez-Pérez, and F. Gago, "Dimerization inhibitors of HIV-1 reverse transcriptase, protease and integrase: a single mode of inhibition for the three HIV enzymes?" *Antiviral Research*, vol. 71, no. 2-3, pp. 260–267, 2006.

[157] D. Grohmann, V. Corradi, M. Elbasyouny et al., "Small molecule inhibitors targeting HIV-1 reverse transcriptase dimerization," *ChemBioChem*, vol. 9, no. 6, pp. 916–922, 2008.

[158] K. Warren, D. Warrilow, L. Meredith, and D. Harrich, "Reverse transcriptase and cellular factors: regulators of HIV-1 reverse transcriptase," *Viruses*, vol. 1, pp. 873–894, 2009.

18

GAPDH Pseudogenes and the Quantification of Feline Genomic DNA Equivalents

A. Katrin Helfer-Hungerbuehler, Stefan Widmer, and Regina Hofmann-Lehmann

Clinical Laboratory, Vetsuisse Faculty, University of Zurich, Winterthurerstrasse 260, 8057 Zurich, Switzerland

Correspondence should be addressed to A. Katrin Helfer-Hungerbuehler; khungerbuehler@vetclinics.uzh.ch

Academic Editor: Emanuel Strehler

Quantitative real-time PCR (qPCR) is broadly used to detect and quantify nucleic acid targets. In order to determine cell copy number and genome equivalents, a suitable reference gene that is present in a defined number in the genome is needed, preferably as a single copy gene. For most organisms, a variable number of glyceraldehyde-3-phosphate dehydrogenase (GAPDH) pseudogenes have been reported. However, it has been suggested that a single-copy of the GAPDH pseudogene is present in the feline genome and that a GAPDH assay can therefore be used to quantify feline genomic DNA (gDNA). The aim of this study was to determine whether one or more GAPDH pseudogenes are present in the feline genome and to provide a suitable alternative qPCR system for the quantification of feline cell copy number and genome equivalents. Bioinformatics and sequencing results revealed that not just one but several closely related GAPDH-like sequences were present in the cat genome. We thus identified, developed, optimized, and validated an alternative reference gene assay using feline albumin (fALB). Our data emphasize the need for an alternative reference gene, apart from the GAPDH pseudogene, for the normalization of gDNA levels. We recommend using the fALB qPCR assay for future studies.

1. Introduction

Fluorescence-based quantitative real-time PCR (qPCR) is a highly sensitive method for the detection and quantification of nucleic acids. Due to its conceptual simplicity, sensitivity, specificity, and speed, qPCR applications can be found in a variety of fields, including medicine and the life sciences [1, 2]. In clinical diagnostics, qPCR is broadly used for the detection and quantification of bacterial and viral loads, gene dosage determination, cancer diagnostics, and applications in forensic medicine [3–7].

To assess the cell number present in a PCR reaction, the coanalysis of suitable reference genes is crucial. Such reference genes should be single-copy number genes and should not frequently undergo genetic alterations, to allow the accurate normalization of genomic DNA (gDNA) samples. In addition, internal control genes are also used to investigate abnormalities in gene number, and amplified oncogenes have been shown to have diagnostic, prognostic, and therapeutic relevance. Thus, TaqMan PCR-based gene quantification assays are also used to identify allelic imbalances (germ-line deletions or amplifications), for example, in individuals suffering from breast cancer, cutaneous melanoma, or nervous system tumors [8, 9].

Glyceraldehyde-3-phosphate dehydrogenase (GAPDH) represents a universally expressed reference gene that has many biological roles in addition to its function in glycolysis. A GAPDH assay has been developed for the accurate normalization of feline messenger RNA (mRNA) expression [10], and this assay was recently validated and compared to other potential feline mRNA reference gene assays [11]. Subsequently, the GAPDH assay was also applied as quality control to test for the integrity of the gDNA and the absence of PCR inhibitors [10, 12, 13]. One report suggested that a single copy of the GAPDH pseudogene is present in the feline genome and that the feline GAPDH assay can therefore be used to quantify cell number in feline samples [14]. However, no information on the exact position or the sequence of this GAPDH pseudogene was provided [14]. In contrast, a variable number of GAPDH pseudogenes has been reported for other organisms [15, 16].

Pseudogenes for many different genes have been found in all animal genomes studied so far [17]. Due to the high sequence similarity of the pseudogenes to their "parent" gene, pseudogenes often interfere with PCR or hybridization experiments that are intended to detect the genes only [18, 19]. Specifically, processed pseudogenes typically lack introns and are therefore well known to hamper data interpretation in mRNA transcription analysis [18, 20]. Promising clinical applications of RT-PCR assays for the purpose of early diagnosis and relapse monitoring of micrometastatic tumor cells have suffered from false-positive results due to their interference with corresponding pseudogenes in the past [19]. Thus, the aim of this study was to (i) investigate the number and sequence(s) of the GAPDH pseudogene(s) in the feline genome, with a specific focus on the GAPDH assay region that is frequently used to normalize genome equivalents [14] and (ii) provide a suitable alternative qPCR system for the normalization of the amount of input gDNA and the determination of cell number in feline samples.

2. Materials and Methods

2.1. Sample Description. All cats included in this study were in experimental studies officially approved by the Veterinary Office of the Swiss Canton of Zurich (TVB 30/2003, 59/2005, and 99/2007). The cats were kept in groups under optimal ethological conditions in a barrier facility, as previously described [21]. All cats were euthanized for reasons unrelated to the present study.

Tissue samples were collected upon necropsy from two feline leukemia virus-infected cats (cat B8, female 3.8 years old, and cat 15, neutered male, 1.3 years old). The cats underwent histopathological examination, and samples from the rectum, colon, spleen, and jejunum were collected. Tissues for histology were processed as described [22], histologically examined, and verified to be free of pathological abnormalities. The samples for molecular analysis were snap frozen in liquid nitrogen following collection and stored at −80°C until nucleic acid extraction.

2.2. Nucleic Acid Extractions. Tissues (approximately 25 mg) were homogenized prior to extraction in 180 μL ATL buffer (Qiagen, Hombrechtikon, Switzerland) for genomic DNA isolation. The samples were processed using the DNeasy Blood & Tissue Kit (Qiagen) following the manufacturer's recommendations. For all nucleic acid extractions, negative controls consisting of 100 μL of phosphate-buffered saline were prepared with each batch to monitor cross-contamination.

2.3. GAPDH Pseudogene Sequencing. For the analysis of the GAPDH-like sequences of the domestic cat, gDNA was obtained from tissue samples from two cats (cat B8, rectum and colon, and cat 15, spleen and jejunum).

Primers were designed using Primer Express software (Version 3, Applied Biosystems, Rotkreuz, Switzerland) based on the GAPDH mRNA sequence [23] to encompass the previously published TaqMan sequence [10]. PCR amplification of the GAPDH pseudogenes using the forward primer binding the exon 2 (GCGCCTGGTCACCAGGGCTGC; pb position: 39–59) and the reverse primer located on the exon 4 (GACTCCACAACATACTCAGCACCAGCATCAC; bp position: 287–257) resulted in an approximately 249 bp fragment. To ensure a high fidelity of amplification, the PCR was performed using Phusion polymerase and HF buffer (Finnzymes, Ipswich, UK), 500 nM of each primer, and 2 μL of extracted gDNA in a final volume of 20 μL. The cycling conditions were as follows: initial denaturation for 2 min at 98°C, followed by 40 cycles of 20 sec at 98°C, 30 sec at 70°C, and 20 sec at 72°C prior to a final elongation at 72°C for 10 min. The PCR products were separated by 2.5% agarose gel electrophoresis, excised from the gel, purified using the GenElute Gel Extraction Kit (Sigma-Aldrich, Buchs, Switzerland), and cloned into the pCRII-TOPO TA cloning vector (Invitrogen, Basel, Switzerland) according to the manufacturer's instructions. A total of 15 of the obtained clones were sequenced by Microsynth (Balgach, Switzerland).

The sequences were analyzed using Clone Manager Professional software version 7.01 (Scientific & Educational Software, Cary, NC, USA).

2.4. Alternative qPCR Assay Design. In other species, albumin (ALB) is known to be a single-copy gene and therefore it has been used as an internal control gene [8, 9, 24, 25]. In order to determine whether feline ALB (fALB) is also present as a single-copy gene, sequence and gene organization information of the feline genome were retrieved from GenBank (no. NM_001009961) and Ensembl (ENSFCAG00000011854). Subsequently, a fALB qPCR assay was designed. The TaqMan hydrolysis probe and primers for the fALB assay were designed using Primer Express software (Version 2, Applied Biosystems; Table 1). The primer pair (Microsynth, Table 1) was tested to ensure that it amplified a product of the appropriate length using 5 μL gDNA in a total volume of 25 μL per reaction with an ABI Prism 7700 sequence detection system (Applied Biosystems) and the TaqMan Fast Universal PCR Master Mix (Applied Biosystems). The PCR products were analyzed by gel electrophoresis in a 3% agarose gel stained with ethidium bromide, and the bands were visualized using the ChemiGenius 2 Bio Imaging System (Syngene, Cambridge, UK).

The qPCR reactions were performed as described [11] from gDNA extracted from tissues (diluted 1:10). The thermocycling conditions consisted of initial denaturation at 95°C for 20 sec followed by 45 cycles of 95°C for 3 sec and 60°C for 45 sec. The gDNA was amplified and quantified using a Rotor-Gene 6000 real-time rotary analyzer (Corbett, Mortlake, VIC, Australia) and an ABI Prism 7700 sequence detection system (Applied Biosystems).

2.5. Production of a Standard for Absolute fALB Quantification. Cat gDNA was used to generate an fALB standard template for absolute quantification. A 150 bp sequence, consisting of the fALB TaqMan sequence, was amplified from 2 μL target gDNA using the fALB primers (Table 1). The PCR

TABLE 1: Details of the TaqMan real-time PCR assays.

Gene	Oligo	Sequence (5′ → 3′)	Amplicon size (bp)	Final conc. (nM)
GAPDH[1]	Forward	GCCATCAATGACCCCTTCAT		480
	Reverse	GCCGTGGAATTTGCCGT	82	480
	Probe	CTCAACTACATGGTCTACATGTTCCAGTATGATTCCA[2]		160
ALB[3]	Forward	GATGGCTGATTGCTGTGAGA		500
	Reverse	CCCAGGAACCTCTGTTCATT	150	500
	Probe	ATCCCGGCTTCGGTCAGCTG[2,4]		200

[1][10]; [2]5′FAM/3′TAMRA; [3]present study; [4]HPLC purification.

reaction contained 2.5 units Taq DNA polymerase (Sigma), final concentrations of 250 nM of each primer, 200 μM dNTPs (Sigma), 2 μL 10 × PCR buffer (Sigma), and 2 μL of template in a final volume of 20 μL. The amplification was performed using a Biometra TPersonal thermal cycler (Biolabo, Châtel-St-Denis, Switzerland). The cycling conditions consisted of an initial denaturation at 94°C for 2 min followed by 40 cycles of 94°C for 30 sec, 55°C for 30 sec, and 72°C for 15 sec, with a final extension at 72°C for 2 min. The resulting amplicon was gel-purified and cloned into the TOPO TA cloning vector (Invitrogen). The inserts in selected clones were verified by sequencing using an ABI PRISM 310 genetic analyzer (Applied Biosystems), as described [26]. The fALB reference plasmid was linearized by restriction digestion using the enzyme SalI (Roche, Rotkreuz, Switzerland) and gel-purified (Gen Elute PCR Clean-Up Kit, Sigma-Aldrich), and the copy number was determined spectrophotometrically (GeneQuant, Pharmacia Biotech) and by agarose gel electrophoresis (ChemiGenius 2 Bio Imaging System). Tenfold serial dilutions of the standard templates in 30 μg/mL carrier salmon sperm DNA (Invitrogen) were aliquoted and frozen at −20°C, as described [27].

2.6. Efficiency, Analytical Sensitivity, Linear Range, and Precision of the fALB qPCR Assay. The newly designed fALB qPCR assay was evaluated according to the MIQE guidelines [28] using the Rotor-Gene 6000 real-time rotary analyzer (Corbett). The efficiency of the assay was calculated as previously described [29], using the following equation: $E = (10^{(-1/slope)}) - 1$. The analytical sensitivity of the system was defined by an endpoint dilution experiment using the tenfold serially diluted standard template and ten replicates per dilution, as described [27]. The linear range of amplification of the fALB qPCR assay was determined by ten-fold serial dilution of the linearized standard template. For the precision analysis, a dilution of a standard template containing 10^5 copies of fALB per reaction was chosen and assayed 10 and 13 times for the intrarun and interrun precision, respectively, and the mean value, standard deviation, and coefficients of variation were calculated.

2.7. Nucleotide Sequence Accession Numbers. The sequence of the partial GAPDH pseudogene of the cat B8 sample (clone 14) was submitted to GenBank under accession number JX523658.

3. Results

3.1. Investigation of the GAPDH Pseudogene Sequence. An approach combining sequencing and bioinformatics was chosen to investigate the presence of potential GAPDH pseudogenes in the domestic cat genome. The potential target sequence of the GAPDH qPCR assay was amplified and cloned, and 15 clones were sequenced: 10 from cat B8 and 5 from cat 15. The sequencing of these 15 clones yielded 11 distinct sequences similar to that of the feline GAPDH mRNA (GenBank: NM_001009307). Eight different sequences were found within the target region of the primers and/or the hydrolysis probe of the GAPDH TaqMan assay (Figure 1). Overall, the sequence identity ranged from 82% (cat 15, clone 20: cat 15_20, Figure 1) to 96% (cat B8, clone 9: cat B8_9) in comparison to the reference sequence (GenBank: NM_001009307).

Sequence and gene organization information were retrieved from the Genome Annotation Resource Fields *Felis catus* v12.2 website (GARFIELD) http://lgd.abcc.ncifcrf.gov/ [30], and five sequences termed "similar to GAPDH" were found in the feline genome. One of these sequences on chromosome F2 exhibited 100% similarity to clone 14 of cat B8 (cat B8_14, Figure 1). In addition, the GAPDH assay sequence was searched using BLAST against the "AANG WGS Contigs" database containing the *Felis catus* whole-genome shotgun sequencing project (*Felis catus*-6.2; 14× coverage; [GenBank: AANG02000000]). Over 60 different contigs containing sequence similar to the GAPDH assay sequence were detected. The sequence with the best fit had three mismatches with the sequence of the GAPDH assay: two mismatches with the forward primer and one mismatch with the reverse primer.

3.2. Efficiency, Analytical Sensitivity, Linear Range, and Precision of the fALB qPCR Assay. In order to confirm the absence of pseudogenes for feline ALB, a bioinformatics approach was chosen. Sequence and gene organization information was retrieved from the GARFIELD website [30] and the second draft assembly, *Felis catus*-6.2 (GenBank: AANG02000000). Only one sequence termed "albumin" indicative of the albumin gene, was detected; it is located on chromosome B1.

Subsequently, a fALB real-time PCR assay was designed to amplify a sequence located within the exon IV of the feline albumin. The amplification of feline gDNA using the newly designed primers yielded PCR products with the expected

```
                                                      GAPDH.138r
NM_001009307 1 TTTTAACTCTGGCAAAGTGGACATTGTCGCCATCAATGACCCCTTCATTGACCTCAACTA 60
AF097177     1 ............................................................ 60
Cat B8_8     1 ......C.....G.......T...A.....G..C.......................... 60
Cat 15_1     1 .......A...........T.C....AA.......-........................ 59
Cat 15_20    1 .....T.....T.........T......A.....G..G........C.........G... 60
Cat 15_14    1 ....G.....A.T......TG--...A.................C............... 58
Cat B8_9     1 ..................................G......................... 60
Cat B8_10    1 ..................G.......A................................. 60
Cat B8_5     1 ..................G.......A................................. 60
Cat 15_9     1 ...C..............G.......A................................. 60
Cat 15_17    1 ..................G.......A................................. 60
Cat B8_14    1 ..................G.......A................................. 60
Cat B8_4     1 ......T.....T.....G.......A................................. 60
```

Intron in gDNA (83 bp)
GTGAGTGCTGCCCCCACAGCTGGCGTAGGGAGCAGACCCTGGCTGAAGTGCAGCCGCTTGATGGCCCTCACTTGTCCCTCCAG

```
                    GAPDH.77p                  GAPDH.57f
NM_001009307 61 CATGGTCTACATGTTCCAGTATGATTCCACCCACGGCAAATTCCACGGCACAGTCAAGGC 120
AF097177     61 ............................................................ 120
Cat B8_8     61 ...A.....................A.............AA...G............. 120
Cat 15_1     60 ..CA.............C...........T.........T..T...CA.A....... 119
Cat 15_20    61 ...T.....T.......C.C....A...GTT.T........T.............A. 120
Cat 15_14    59 ..................C.....A.T............. 118
Cat B8_9     61 ................A.................A.......... 120
Cat B8_10    61 ...........T..C............T.........A............ 120
Cat B8_5     61 ..................A....-..........----..A.. 115
Cat 15_9     61 ..................A....-.........G.----..A.. 115
Cat 15_17    61 ..................A....-.........G.----..A.. 115
Cat B8_14    61 ...........T......A....-..........----..A.. 115
Cat B8_4     61 ..................A....-..........----..A.. 115
```

Intron in gDNA (122 bp)
GTAAGTGTAGAAGACGGAACAGGGTGGAATTTGCTTTGAGGGAGTTACTAGGATGGGCTGACAACCTTGGGTGGTACATGGTA
CCCCATGTCCCCGAGCTTTCGACTTGTCTCCCTTTATAG

```
NM_001009307 121 TGAGAACGGGAAACTTGTCATCAATGGAAAGCCCATCACCATCTTCCAGGA-GCGAGATC 179
AF097177     121 ......................T.............. 166
Cat B8_8     121 .....TAA.........G.....G.......T.............-..A..... 179
Cat 15_1     120 .....T.............G.....................-..A..... 178
Cat 15_20    121 ..C....A.A...T....T........G..AT...C.T....G........A....A... 180
Cat 15_14    119 ...................G................T..G........A.-..... 177
Cat B8_9     121 .....................................A.-.T...... 179
Cat B8_10    121 ................A.......T.......CA......-..... 179
Cat B8_5     116 ......T.A.....G......C................-..A.... 174
Cat 15_9     116 ......T.A.....G......C................-..A.... 174
Cat 15_17    116 ......T.A.....G......C................-..A.... 174
Cat B8_14    116 ......T.A.....G......C................-..A.... 174
Cat B8_4     116 ......T.A.....G......C................-..A.... 174
```

```
NM_001009307 180 CCGCCAACATCAAATGGG 197
AF097177     166                    166
Cat B8_8     180 ..A...........A..197
Cat 15_1     179 ................196
Cat 15_20    181 .T....G.........198          ↓ Splice sites
Cat 15_14    178 T.A.............195
Cat B8_9     180 T...............197
Cat B8_10    180 .........C......197
Cat B8_5     175 ................192
Cat 15_9     175 ................192
Cat 15_17    175 ................192
Cat B8_14    175 ................192
Cat B8_4     175 ................192
```

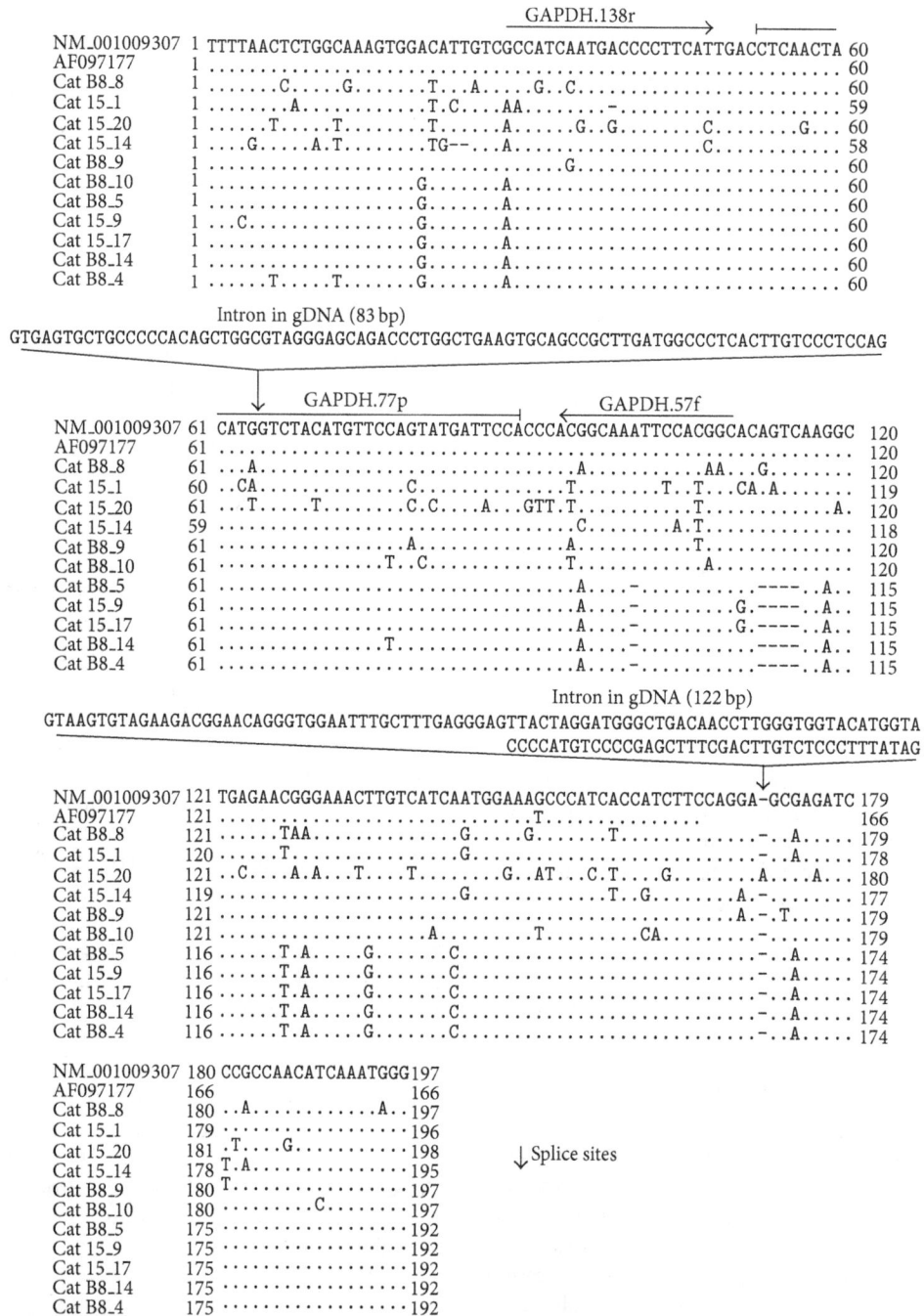

FIGURE 1: Comparison of feline GAPDH and pseudo-GAPDH nucleotide sequences. A comparison of the nucleotide sequences of feline GAPDH mRNA (GenBank: NM_001009307, position 60 to 256, and GenBank: AF097177, position 63 to 228) and gDNA sequences similar to GAPDH retrieved from cat B8 and cat 15. The number listed after the animal identifier (B8 and 15) represents the clone identity. Sequences found in multiple clones are shown only once. Nucleotides that differ from those in the two reference strains are indicated. Primer sequences for the qPCR reported previously [10] were located at positions 29 to 48 and 109 to 93, and the hydrolysis probe spans positions 53 to 89, as indicated. Arrows point to the splicing sites and the intron sequence from the gDNA (Ensembl: ENSFCAG00000006874) is depicted (Intron in gDNA). The exon sequences of the GAPDH gDNA (Ensembl: ENSFCAG00000006874) on which the GAPDH assay is located (exon 2 and 3) are 100% identical to the GAPDH mRNA sequence depicted in this figure (GenBank: NM_001009307).

size of 150 bp, as determined by agarose gel electrophoresis. The primers were then used in combination with the newly designed probe, and the amplification efficiency of the fALB qPCR was determined to be 99.5% to 100% using 10-fold serial dilutions of the standard template (Supplementary Material available online at http://dx.doi.org/10.1155/2013/587680 and data not shown). The highest dilution that still resulted in a positive signal in the qPCR assay contained an average of 1 copy of the standard in a $5\,\mu L$ reaction; in an endpoint dilution experiment, 6 of the 10 replicates of this dilution were positive. The fALB qPCR assay was linear over eight orders of magnitude, from 10^1 to 10^8 copies (Supplementary Material). The qPCR assay displayed a good precision; the coefficient of variation for the absolute number using 10^5 copies/reaction was 5.51% for the intrarun precision analysis and 6.39% for the interrun analysis.

4. Discussion

In this report, we describe the detection of several GAPDH-like sequences that are characteristic for processed GAPDH pseudogenes in the domestic cat genome. Based on the assumption that there is only one copy of the GAPDH pseudogene in the domestic cat genome [14], the GAPDH qPCR assay that was previously designed to amplify GAPDH mRNA/complementary DNA (cDNA) [10] was regularly used to determine the cell number of input gDNA. However, in our experience, the analysis of gDNA samples resulted in a lower amplification efficiency compared to cDNA (unpublished observations); we hypothesized that this might occur due to mismatches between the primers and/or hydrolysis probe and the gDNA sample. Thus, we performed a sequence analysis of the binding region of the primers and the hydrolysis probe of the GAPDH assay. The sequencing of 15 different clones comprising the GAPDH assay sequence revealed 11 different GAPDH-like sequences; however, none exhibited 100% similarity to the GAPDH mRNA sequence (GenBank: NM_001009307). It is possible that there are additional GAPDH pseudogenes present in the feline genome that were not detected in the present study because the sequencing was restricted to 15 clones. The use of an alternative sequencing technique (deep-sequencing rather than cloning followed by Sanger dideoxy sequencing) may provide a broader picture of the GAPDH-like sequences present in the cat genome. Furthermore, the primer binding sites chosen within the GAPDH sequence may have additionally restricted the number of recognized GAPDH pseudogenes. Thus, the number of GAPDH pseudogenes or GAPDH-like sequences in the feline genome may have been underestimated in our study. However, our sequencing results readily demonstrate that more than one feline GAPDH pseudogene is present in the genomic DNA of feline cells, and the GAPDH pseudogene sequences differed from the GAPDH mRNA sequence to some extent.

This finding was supported by the sequence information retrieved from the GARFIELD website [30] and the second draft assembly, *Felis catus*-6.2 (GenBank: AANG02000000), in which no sequence with 100% similarity to the GAPDH mRNA sequence was found, but a multitude of closely related sequences were present. Moreover, our data are in agreement with studies investigating GAPDH pseudogenes in other species. In humans, between 56 and 62 GAPDH pseudogenes have been detected [15, 17]. Of note, the copy number of pseudogenes may vary within a population, as has been documented for the ATP-binding cassette transporter pseudogene within the Chinese population [31]. The number of recognized processed GAPDH pseudogenes may be as high as 120 in dogs and is over 300 in murine rodents [15–17]. For glycolytic genes, including GAPDH, it has been shown that there is a positive correlation between the level of gene expression and the abundance of processed pseudogenes in mice; GAPDH was found to be particularly highly expressed and to generate the highest number of pseudogenes among all glycolytic genes [17]. The authors explained the over-abundance of GAPDH by the fact that GAPDH has many additional functions other than those related to glycolysis [17].

An alternative assay for the quantification of gDNA and cell number in feline samples was designed, validated, and implemented. In other species, different genes have been used for the quantification of cells, including the chemokine receptor CCR5 [32] and ALB [9]. ALB is known to be a single-copy gene in other species, such as humans and rhesus macaques, and has been used in both species as an internal control gene [8, 9, 24, 25]. According to the GARFIELD website [30], ALB is also present as a single-copy gene in the feline genome. ALB has been used for the normalization of retroviral provirus loads [24, 25, 33, 34]. We chose to develop a feline albumin assay because one of our group's main scientific interests is in retroviral infections. The newly implemented fALB assay was shown to be highly sensitive and efficient for gDNA, with a wide range of linearity. Our database searches resulted in only one hit termed "ALB." Thus, to the best of our knowledge, fALB is a single-copy number gene.

5. Conclusions

This study investigated the number of GAPDH pseudogenes in the domestic cat. Our results indicate that several closely related GAPDH-like sequences are indeed present in the cat genome. The GAPDH assay may still be used for quality control to test for the integrity of gDNA and the absence of PCR inhibitors. However, it appears to be a suboptimal choice for the quantification of gDNA equivalents. A newly designed assay using the fALB reference gene for the normalization of gDNA was validated and implemented. We recommend using this highly sensitive fALB qPCR assay for the normalization of input genomic DNA equivalents in future studies.

Abbreviations

qPCR: Quantitative real-time PCR
gDNA: Genomic DNA
GAPDH: Glyceraldehyde-3-phosphate dehydrogenase
fALB: Feline albumin
mRNA: Messenger RNA

cDNA: Complementary DNA
CT: Cycle threshold.

Conflict of Interests

The authors declare that they have no conflict of interests.

Authors' Contribution

A. Katrin Helfer-Hungerbuehler performed the experiments, analyzed the data, and drafted the paper. Stefan Widmer performed the sequencing experiments as a partial fulfillment of the requirement of a term paper. RHL conceived and supervised the follow-up study and edited the paper.

Acknowledgments

The authors would particularly like to thank Dr. V. Cattori and Dr. M. L. Meli for excellent assistance and helpful discussions. They are grateful to the animal caretakers for expert technical aid with the cats and to the technicians and doctoral students of the Clinical Laboratory, particularly T. Meili Prodan and B. Weibel, and the Institute for Veterinary Pathology, Vetsuisse Faculty, for their excellent laboratory assistance. The molecular biology work was performed using the logistics of the Center for Clinical Studies, Vetsuisse Faculty, University of Zurich. This study was financially supported by a Research Grant from the Swiss National Science Foundation (310030_135586).

References

[1] M. Kubista, J. M. Andrade, M. Bengtsson et al., "The real-time polymerase chain reaction," *Molecular Aspects of Medicine*, vol. 27, no. 2-3, pp. 95–125, 2006.

[2] S. A. Bustin, "Absolute quantification of mRNA using real-time reverse transcription polymerase chain reaction assays," *Journal of Molecular Endocrinology*, vol. 25, no. 2, pp. 169–193, 2000.

[3] P. S. Bernard and C. T. Wittwer, "Real-time PCR technology for cancer diagnostics," *Clinical Chemistry*, vol. 48, no. 8, pp. 1178–1185, 2002.

[4] I. M. Mackay, K. E. Arden, and A. Nitsche, "Real-time PCR in virology," *Nucleic Acids Research*, vol. 30, no. 6, pp. 1292–1305, 2002.

[5] I. M. Mackay, "Real-time PCR in the microbiology laboratory," *Clinical Microbiology and Infection*, vol. 10, no. 3, pp. 190–212, 2004.

[6] S. A. Bustin and R. Mueller, "Real-time reverse transcription PCR (qRT-PCR) and its potential use in clinical diagnosis," *Clinical Science*, vol. 109, no. 4, pp. 365–379, 2005.

[7] L. I. Moreno, C. M. Tate, E. L. Knott et al., "Determination of an effective housekeeping gene for the quantification of mRNA for forensic applications," *Journal of Forensic Sciences*, vol. 57, no. 4, pp. 1051–1058, 2012.

[8] I. Laurendeau, M. Bahuau, N. Vodovar et al., "TaqMan PCR-based gene dosage assay for predictive testing in individuals from a cancer family with INK4 locus haploinsufficiency," *Clinical Chemistry*, vol. 45, no. 7, pp. 982–986, 1999.

[9] I. Bieche, M. H. Champème, D. Vidaud, R. Lidereau, and M. Vidaud, "Novel approach to quantitative polymerase chain reaction using real-time detection: application to the detection of gene amplification in breast cancer," *International Journal of Cancer*, vol. 78, no. 5, pp. 661–666, 1998.

[10] C. M. Leutenegger, C. N. Mislin, B. Sigrist, M. U. Ehrengruber, R. Hofmann-Lehmann, and H. Lutz, "Quantitative real-time PCR for the measurement of feline cytokine mRNA," *Veterinary Immunology and Immunopathology*, vol. 71, no. 3-4, pp. 291–305, 1999.

[11] Y. Kessler, A. K. Helfer-Hungerbuehler, V. Cattori et al., "Quantitative TaqMan® real-time PCR assays for gene expression normalisation in feline tissues," *BMC Molecular Biology*, vol. 10, article no. 106, 2009.

[12] G. A. Wolf-Jackel, V. Cattori, C. P. Geret et al., "Quantification of the humoral immune response and hemoplasma blood and tissue loads in cats coinfected with "Candidatus Mycoplasma haemominutum" and feline leukemia virus," *Microbial Pathogenesis*, vol. 53, no. 2, pp. 74–80, 2012.

[13] M. Novacco, F. S. Boretti, G. A. Wolf-Jackel et al., "Chronic, "Candidatus Mycoplasma turicensis" Infection," *Veterinary Research*, vol. 42, no. 1, article 59, 2011.

[14] S. Molia, B. B. Chomel, R. W. Kasten et al., "Prevalence of Bartonella infection in wild African lions (Panthera leo) and cheetahs (Acinonyx jubatus)," *Veterinary Microbiology*, vol. 100, no. 1-2, pp. 31–41, 2004.

[15] Y. J. Liu, D. Zheng, S. Balasubramanian et al., "Comprehensive analysis of the pseudogenes of glycolytic enzymes in vertebrates: the anomalously high number of GAPDH pseudogenes highlights a recent burst of retrotrans-positional activity," *BMC Genomics*, vol. 10, article 480, 2009.

[16] S. Riad-El Sabrouty, J. M. Blanchard, L. Marty, P. Jeanteur, and M. Piechaczyk, "The Muridae glyceraldehyde-3-phosphate dehydrogenase family," *Journal of Molecular Evolution*, vol. 29, no. 3, pp. 212–222, 1989.

[17] L. McDonell and G. Drouin, "The abundance of processed pseudogenes derived from glycolytic genes is correlated with their expression level," *Genome*, vol. 55, no. 2, pp. 147–151, 2012.

[18] Z. Zhang, N. Carriero, and M. Gerstein, "Comparative analysis of processed pseudogenes in the mouse and human genomes," *Trends in Genetics*, vol. 20, no. 2, pp. 62–67, 2004.

[19] P. Ruud, O. Fodstad, and E. Hovig, "Identification of a novel cytokeratin 19 pseudogene that may interfere with reverse transcriptase-polymerase chain reaction assays used to detect micrometastatic tumor cells," *International Journal of Cancer*, vol. 80, no. 1, pp. 119–125, 1999.

[20] B. Garbay, E. Boue-Grabot, and M. Garret, "Processed pseudogenes interfere with reverse transcriptase-polymerase chain reaction controls," *Analytical Biochemistry*, vol. 237, no. 1, pp. 157–159, 1996.

[21] C. P. Geret, B. Riond, V. Cattori, M. L. Meli, R. Hofmann-Lehmann, and H. Lutz, "Housing and care of laboratory cats: from requirements to practice," *Schweizer Archiv fur Tierheilkunde*, vol. 153, no. 4, pp. 157–164, 2011.

[22] A. K. Helfer-Hungerbuehler, V. Cattori, F. S. Boretti et al., "Dominance of highly divergent feline leukemia virus A progeny variants in a cat with recurrent viremia and fatal lymphoma," *Retrovirology*, vol. 7, no. 1, article 14, 2010.

[23] M. Kullberg, M. A. Nilsson, U. Arnason, E. H. Harley, and A. Janke, "Housekeeping genes for phylogenetic analysis of eutherian relationships," *Molecular Biology and Evolution*, vol. 23, no. 8, pp. 1493–1503, 2006.

[24] A. Dehée, R. Césaire, N. Désiré et al., "Quantitation of HTLV-I proviral load by a TaqMan real-time PCR assay," *Journal of Virological Methods*, vol. 102, no. 1-2, pp. 37–51, 2002.

[25] H. K. Chung, T. Unangst, J. Treece, D. Weiss, and P. Markham, "Development of real-time PCR assays for quantitation of simian betaretrovirus serotype-1, -2, -3, and -5 viral DNA in Asian monkeys," *Journal of Virological Methods*, vol. 152, no. 1-2, pp. 91–97, 2008.

[26] B. Willi, F. S. Boretti, V. Cattori et al., "Identification, molecular characterization, and experimental transmission of a new hemoplasma isolate from a cat with hemolytic anemia in Switzerland," *Journal of Clinical Microbiology*, vol. 43, no. 6, pp. 2581–2585, 2005.

[27] V. Cattori and R. Hofmann-Lehmann, "Absolute quantitation of feline leukemia virus proviral DNA and viral RNA loads by TaqMan real-time PCR and RT-PCR," *Methods in Molecular Biology*, vol. 429, pp. 73–87, 2008.

[28] S. A. Bustin, V. Benes, J. A. Garson et al., "The MIQE guidelines: minimum information for publication of quantitative real-time PCR experiments," *Clinical Chemistry*, vol. 55, no. 4, pp. 611–622, 2009.

[29] D. Klein, P. Janda, R. Steinborn, M. Müller, B. Salmons, and W. H. Günzburg, "Proviral load determination of different feline immunodeficiency virus isolates using real-time polymerase chain reaction: influence of mismatches on quantification," *Electrophoresis*, vol. 20, no. 2, pp. 291–299, 1999.

[30] J. U. Pontius and S. J. O'Brien, "Genome annotation resource fields—GARFIELD: a genome browser for Felis catus," *Journal of Heredity*, vol. 98, no. 5, pp. 386–389, 2007.

[31] M. K. Kringen, C. Stormo, R. M. Grimholt, J. P. Berg, and A. P. Piehler, "Copy number variations of the ATP-binding cassette transporter ABCC6 gene and its pseudogenes," *BMC Research Notes*, vol. 5, no. 1, article 425, 2012.

[32] J. A. Thomas, T. D. Gagliardi, W. G. Alvord, M. Lubomirski, W. J. Bosche, and R. J. Gorelick, "Human immunodeficiency virus type 1 nucleocapsid zinc-finger mutations cause defects in reverse transcription and integration," *Virology*, vol. 353, no. 1, pp. 41–51, 2006.

[33] N. Désiré, A. Dehée, V. Schneider et al., "Quantification of human immunodeficiency virus type 1 proviral load by a TaqMan real-time PCR assay," *Journal of Clinical Microbiology*, vol. 39, no. 4, pp. 1303–1310, 2001.

[34] A. Waters, A. L. A. Oliveira, S. Coughlan et al., "Multiplex real-time PCR for the detection and quantitation of HTLV-1 and HTLV-2 proviral load: addressing the issue of indeterminate HTLV results," *Journal of Clinical Virology*, vol. 52, no. 1, pp. 38–44, 2011.

19

Genotoxicity Studies Performed in the Ecuadorian Population

César Paz-y-Miño, Nadia Cumbal, and María Eugenia Sánchez

Instituto de Investigaciones Biomédicas, Facultad de Ciencias de la Salud, Universidad de las Américas,
Ave. de los Granados y Colimes Quito, 1712842, Ecuador

Correspondence should be addressed to César Paz-y-Miño, cpazymino@udla.edu.ec

Academic Editor: Mark Berneburg

Genotoxicity studies in Ecuador have been carried out during the past two decades. The focuses of the research were mainly the area of environmental issues, where the populations have been accidentally exposed to contaminants and the area of occupational exposure of individuals at the workplace. This paper includes studies carried out in the population of the Amazon region, a zone known for its rich biodiversity as well as for the ecological damage caused by oil spills and chemical sprayings whose consequences continue to be controversial. Additionally, we show the results of studies comprised of individuals occupationally exposed to toxic agents in two very different settings: flower plantation workers exposed to pesticide mixtures and X-ray exposure of hospital workers. The results from these studies confirm that genotoxicity studies can help evaluate current conditions and prevent further damage in the populations exposed to contaminants. As such, they are evidence of the need for biomonitoring employers at risk, stricter law enforcement regarding the use of pesticides, and increasingly conscientious oil extraction activities.

1. Introduction

Genotoxicity is a collective term that refers to any process that affects the structural integrity of DNA [1]. This multidisciplinary field of research aims to detect compounds capable of causing DNA damage in hopes of understanding the biological consequences of genotoxic agents and their involvement in the alteration of the molecular mechanisms of the genetic material [2]. These consequences can eventually lead to carcinogenic processes [3]. Over the past century, industrialization and globalization of the western hemisphere lead to the high volume production of different chemicals and complex preparations that are still currently released into the environment [4]. Living organisms are increasingly being exposed to genotoxic agents whose growing presence in the biosphere can substantially harm the population [5]. Activities such as fuel extraction and glyphosate spraying in the Amazon region of Ecuador are the two most controversial environmental health issues in the nation and are still considered as latent threats whose consequences continue to be studied [6, 7]. Agriculture in Ecuador is the second most important productive activity that contributes to national income [8]. However, the

lack of regulation regarding pesticide use and occupational safety pose a significant threat to the workers' health [9]. Additionally, various studies have focused on individuals exposed to radiation in the workplace, such as medical radiation workers constantly exposed to ionizing radiation that has well-known DNA-damaging effects [10–12]. The present paper intends to show a summary on the work carried out in Ecuador for the past two decades in the field of genotoxicity. All the cytogenetic studies have been performed on blood lymphocytes and the results obtained refer only to somatic mutations. The studies have included cytogenetic findings, such as the ones presented in Table 1 as well as molecular results which are shown in Table 2.

2. Glyphosate Genotoxicity Studies

The northeastern Ecuadorian border underwent the aerial spraying of an herbicide mix during the period of 2002–2007 and was supported by the Colombian government [13]. The Roundup mix presumably contained high doses of glyphosate plus a surfactant known as polyethoxylated tallowamine (POEA) and the adjuvant Cosmoflux 411F [14].

TABLE 1: Cytogenetic findings in genotoxicity studies.

Glyphosate	Paz-y-Miño et al. [6]	Comet assay: 35.5 μm DNA migration for exposed, 25.94 μm for controls.
	Paz-y-Miño et al. [13]	All the studied population showed low or no chromosomal fragility.
Other pesticides	Paz-y-Miño et al. [9, 10, 93]	Chromosomal aberrations: 20.59% in exposed and 2.73% in controls.
	Paz-y-Miño et al. [55]	Comet assay: 31.58 μm DNA migration for exposed, 25.94 μm for controls. Chromosomal aberrations: 5.48% in exposed and 0.45% in controls.
Hydrocarbons	Paz-y-Miño et al. [7]	Chromosomal aberrations: 20% in exposed and 1-2% in controls. 12% type A DNA damage and 1% type E DNA damage in exposed group while 81% type A and 0% type E in controls.
	Paz-y-Miño et al. [88]	48.8% type A DNA damage and 0.1% type E DNA damage in exposed group while 67.9% type A and 0% type E in controls.
Radiation	Paz-y-Miño et al. [102]	Chromosomal aberrations: 29% in exposed, 26.0% in the followup, and 3.5% in controls.
	Paz-y-Miño et al. [112]	12.6% metaphases with telomeric associations in the exposed smoker group; 6.0% TA in exposed nonsmokers; 9.0% in unexposed smokers and 0.1% TA in control group.
	Paz-y-Miño et al. [9, 10, 93]	Comet assay: 26.55 μm DNA migration; mean chromosomal aberrations without gaps: 5.39% ($r = 0.50$, $P < 0.05$). Mean chromosomal aberrations including gaps: 12.08% ($r = 0.78$, $P < 0.01$).
	Muñoz et al. [128]	Comet assay: 29.08 μm DNA migration for exposed group, 25.91 μm for controls. Chromosomal aberrations: 50% in exposed, 26.0% in the followup, and 4% in controls.

TABLE 2: Molecular findings in genotoxicity studies.

Glyphosate	Paz-y-Miño et al. [13]	Regarding the GSTP1 Ile105Val polymorphism, the frequency of the Val allele was higher in exposed individuals (0.48) than control individuals (0.28). The Val/Val variant represented a 4.88-fold risk of acquiring detoxification problems, whereas the combination of the Ile/Val and Val/Val alleles was associated with a 2.6-fold risk of presenting a GSTP1 gene dysfunction. As for the GPX-1 Pro198Leu polymorphism, the Leu allele had a higher frequency in exposed individuals (0.41), unlike control individuals (0.32). The Leu/Leu variant was associated with an 8.5-fold risk of having problems in the function of the GPX-1 gene.
Other pesticides	Paz-y-Miño et al. [55]	The level of damage was not significantly influenced by genetic polymorphisms of the CYP 1A1 gene in the studied population.
Hydrocarbons	Paz-y-Miño et al. [7]	As far as the MSH2 gene is concerned, there is a relation between polymorphisms of the exon 13 and the DNA damage evaluated in the individuals exposed to hydrocarbons ($P < 0.001$), the study of the CYP 1A1 gene found no relation between its polymorphisms and having greater susceptibility to DNA damage.

Glyphosate is an effective organophosphorous herbicide used worldwide [15, 16] known to cause variable levels of toxicity in different organisms, such as the alteration of metabolic pathways, cytotoxicity in humans, metamorphosis alterations in amphibians, and abnormal development of sea urchin eggs [17–22]. Such reports add to the already numerous concerns over the compound's many effects in the environment. The cytogenetic study of blood lymphocytes from individuals that lived in the area that endured the sprayings showed the absence of chromosomal aberrations two years after the last spraying in Ecuadorian soil [13]. Also, since glyphosate has been known to cause oxidative stress in microorganisms, plant and animal species [23–29], the study analyzed three gene polymorphisms (GSTP1 Ile105Val, GPX-1 Pro198Leu, and XRCC1 Arg399Gln) that have been previously associated to the alteration of antioxidant activity, DNA detoxicating processes, and protective functions [30–33]. The GSTP1 gene encodes for glutathione S-transferase pi, an enzyme that is involved in the protection against exogenous and endogenous oxidative damage [34]. As a member of the gluthathione-S-transferase superfamily of enzymes, GSTpi participates in the conjugation of xenobiotics, such as herbicides, insecticides, and other environmental carcinogens, to form glutathione and facilitate their excretion [35, 36]. Specifically, the GSTP1 Ile105Val polymorphism has been associated to higher levels of DNA damage in pesticide-exposed populations [37]. In the Ecuadorian population studied, the prevalence of the GSTP1 Val/Val genotype associated with enzyme dysfunction was observed in the exposed individuals. On the other hand, the GPX-1 gene encodes for one of the most important detoxifying enzymes: gluthathione peroxidase. This enzyme protects mammalian cells, especially human erythrocytes, against oxidative damage [38]. Studies have shown that the loss of gluthathione peroxidase activity can generate tissue damage [39] and offer more sensitivity towards toxic xenobiotics, such as paraquat and adriamycin [38]. Although the *GPX-1* Pro198Leu polymorphism has been exhaustively studied in relation to cancer, the study in Ecuador identified the prevalence of the Leu allele of the *GPX-1* gene in glyphosate-exposed individuals that suggests a higher risk of DNA damage and increased sensitivity to herbicides. The third gene that was part of the study, *XRCC1*, is involved in the mechanisms of DNA single-strand breaks (SSBs) and base-excision repair (BER) that could modify the individual susceptibility to the genotoxic effect of xenobiotics [40, 41]. Although other studied populations have found an association between *XRCC1* gene genotypes and an increased risk of DNA damage due to pesticide exposure, similar results were not found for the Ecuadorian population studied [42]. Aside from the cytogenetic and molecular analysis carried out, the social conditions of the population were surveyed and psychological assessment was offered to all the individuals. The results of these two activities suggested the negative effect of the fumigations on the individuals' mental health, social conditions, and quality of life [13].

A previous study took place two years before the aforementioned study at the Ecuadorian border. It involved individuals living within 200 m to 3 Km from the areas under continuous and sporadic spraying [14, 43]. The comet assay technique, described by Singh et al., 1988 [44], was carried out on blood samples from exposed individuals and corresponding controls to show the occurrence of DNA fragmentation. DNA damage was classified into five categories and the mean of DNA migration was recorded. The results showed that the exposed group displayed significantly higher mean DNA migration than the control group. Similarly, there was a higher degree of DNA damage in the exposed group in comparison to the control group. These results suggest a negative effect of the glyphosate formulation since none of the studied individuals had been previously or simultaneously exposed to other toxic compounds, such as pesticides or tobacco [6]. The northern strip on the Ecuadorian border has not gone unnoticed in the controversy regarding aerial sprayings and their consequences. Nevertheless, the genotoxic and overall toxic potential of glyphosate remains under study and in vitro findings [45–49] have reached a variety of results [29]. The two studies carried out in the Ecuadorian border suggest a significant and immediate risk arising from the use of this chemical and prompt to continue its investigation.

3. Other Pesticide Genotoxicity Studies

Despite the known risks of the use of some pesticides due to their potential health consequences [25], many of those catalogued as extremely toxic continue being used in certain agricultural zones and flower plantations in Ecuador [9]. Pesticides are widely used all over the world in agriculture to protect crops and in public health to control diseases [50, 51]. The risk of developing malignancies such as cancer in occupationally exposed populations is of great concern and has drawn attention to workers in various activities, from the manufacturing workers to the pesticide applicators [52]. Studies available in scientific literature have focused

their methodology on cytogenetic endpoints to evaluate the potential genotoxicity of pesticides, including chromosomal aberrations (CAs), micronuclei (MN), and sister chromatid exchanges (SCEs) [52, 53]. The analysis of chromosomal aberrations such as breaks, dicentric chromosomes, and rings was part of the methodology used in a leading pesticide exposure study carried out in flower plantation workers in Quito. These workers were exposed to 27 different pesticides, some of which have been previously labeled as highly toxic [9]. Also, the level of erythrocyte acetylcholinesterase was measures in every individual as a marker to evaluate the exposure to organophosphate pesticides [54]. The study found an overall CA percentage of 20.59% in the exposed group and 2.73% of CA in the control group. Additionally, the exposed group showed a higher proportion of chromatid-type aberrations and numeric alterations. This does not only reflect genomic instability but also comprises outstanding evidence of damage supported by the abnormal low levels of acetylcholinesterase seen in the exposed group [55].

Pesticide genotoxicity was also studied in individuals working as pesticide applicators in the zone of Cayambe, northeast from Quito. The workers were exposed to 46 pesticides of different degrees of toxicity and at different concentrations and mixtures during work at the plantation [56]. The methodology involved chromosomal aberration test matched with alkaline comet assay [55]. Additionally, the samples were analyzed at a molecular level focusing on the *CYP1A1* gene, a gene that has been extensively studied in relation to occupationally exposure to pesticides [56]. Because the gene is involved in the human xenobiotic metabolism, its alteration presumably increases the risk of developing lung, colorectal, prostate, and breast cancer [57–61]. In accordance to the first study carried out in workers from Quito, the results of the CA analysis in this study showed the significantly high presence of chromosomal damage in the exposed group as compared to the control group. Furthermore, the comet assay test offered results that supported the CA analysis by showing that the DNA migration of the exposed group was certainly higher than that of the control group. On the other hand, the study presented no correlation between the cytogenetic findings and genotyping of the *CYP1A1* MspI and Ile/Val gene polymorphisms. Though CA and comet assay showed interesting results, the gene was not linked to pesticide exposure in the studied population, as opposed to other populations [62, 63].

As an important element of the agricultural production, pesticides have become a necessary tool for crop management in developing countries [64, 65]. Nonetheless, the lack of adequate legislation and enforcement of existing pesticide laws and regulations places agricultural workers, their families, and nearby populations in great risk of developing cancer and other diseases [66]. Our study shows evidence of genotoxic damage in individuals occupationally exposed to pesticides in Ecuador. These are results that demand the establishment of effective exposure biomarkers that could be used for biomonitoring the threatened workers in order to prevent the future development of illness [5].

4. Hydrocarbons Genotoxicity Studies

The Ecuadorian Amazon is one of the ecologically richest regions in the world and it is also sparsely populated. The oil extraction activity in Ecuador began in 1972, it became economically fundamental immediately and continues to be the principal source of national income [67]. Unfortunately, along the process, millions of gallons of oil and toxic residues have been discarded directly onto the environment causing health and environmental issues [68–71]. Indeed, more than 30 billion gallons of toxic wastes and crude oil had been discharged into the land and waterways of the Ecuadorian Amazon up until 1993 [72]. Crude oil is a complex mixture of many chemical compounds. It contains a variety of hydrocarbons of diverse toxicological power such as benzene, toluene, xylene and polynuclear aromatic hydrocarbons [73]. High concentrations of benzene can cause neurotoxin symptoms that cause injuries to the bone marrow and, less frequently, pancytopenia [74]. Similarly, benzene is known to cause leukemia and the development of hematological tumors [75]. The exposure to carcinogen compounds used in the oil industry increases the development of cancer in men, women, and children. In men, an increase of lung, esophagus, rectum, skin, and kidney cancer has been noticed. In women, researchers have seen an increase of cervical, lymphatic ganglion, and bladder cancer. In children, an increase of hematopoietic cancer has been shown among other types of cancer [75–86]. Studies carried out in the Ecuadorian Amazon Basin were found to be compatible with international studies. A relationship between cancer incidence and living in proximity to oil fields has been established [87]. An initial study was carried out in the province of Orellana including 23 women living not more than 10 Km away from a crude oil extraction zone, with the corresponding control group. In order to assess genotoxicity, the comet assay test was used to measure DNA damage by classifying the nucleus morphology into five categories. The sampling zones closest to the extraction wells showed a greater evidence of DNA damage than those that are farther away which suggests a distance-damage relationship. This relationship is also supported by the increased occurrence of type A nuclei (no damage) as the distance from the wells increases [88]. Another study was carried out with a significantly bigger sample size from the nearby zone of San Carlos and matching controls from both San Carlos and the country's capital Quito. Comet assay showed that the affected group has a high occurrence of type B nuclei fragmentation, as opposed to the prevalence of type A cells in the control groups from San Carlos and Quito. Additionally, the analysis of chromosomal aberrations showed that 20% of the exposed individuals presented chromosomal breaks and gaps while only 2% of the control individuals had such aberrations [88]. At a molecular level, we incorporated the analysis of the polymorphisms of the genes CYP1A1 (MspI and Ile/Val) and MSH2 (gIVS12-6T>C), both related to the development of cancer [58, 59, 89, 90]. The results of this last part of the study showed that the CYP1A1 gene polymorphisms were not related to either group, as it has been reported previously in the Caucasian population [58].

However, the study showed a significant difference of the MSH2 gene polymorphism between groups which suggests a higher susceptibility to DNA damage in the exposed group [7]. The chemical complexity of petroleum causes that, once a spill occurs, the constituents disseminate into different extents between the oil phase and the air, soil and water phases of the environment. Physical, chemical, and biological processes age the spilled product resulting in additional changes in composition and complexity [91]. Taking in to consideration the toxicity of these fractions, the risk at petroleum extraction sites is an issue that must be addressed by making informed decisions. By comparing the affected group with the control individuals living in the same town though far from the extraction sites, the study has been able to evidence of the genotoxicity in the exposed population living in nearby petroleum extraction wells. This suggests that the contaminating material resulting from this activity has created an altered environment that exposes the population to chemical fractions considered as dangerous and may also cause genotoxic effects.

5. Radiation Genotoxicity Studies

Environmental mutagens can be broadly classified as radiation and chemicals [92]. Ionizing radiation is capable of extracting electrons of the radiated material due to its high energy. This is only a start point for other ionizing reactions that produce more unstable molecules that eventually cause mutations in DNA [93].

Ever since X-rays were shown to induce mutation in Drosophila over 70 years ago, the established idea has been that the genotoxic effects of ionizing radiation, such as mutations and carcinogenesis, are caused by the direct damage of the cell nucleus [94, 95]. Diagnostic radiology is a field of physical medicine that uses X-rays in order to obtain functional and anatomical information on the human body [96]. Because of the benefits of this diagnostic tool that allows real-time visualization, it is frequently used by the medical professionals [96].

Ionizing radiation is capable of acting on the living cell causing several effects that result from the excitation of atoms and molecules that ultimately cause structural changes. At a molecular level, DNA is possibly affected by water ionization that forms free radicals and promotes the oxidation of several compounds and hydrogen oxide [97]. Even small doses of this radiation could cause great damage because a simple electron excitation can break up to 20 hydrogen bonds [98–100]. The damage resulting from radiation exposure can be seen in the form of chromosomal aberrations in the cell nucleus that are associated with an elevating risk of developing cancer [101]. A first study focused on 10 individuals exposed to radiation in the workplace. They were exposed to 1.84 mSv/year and received a follow-up cytogenetic study after a year from the first blood sampling. The chromosomal aberration results found by the cytogenetic analysis showed interesting results in both instances [102]. First, they showed that complex chromosome alterations, such as rings and dicentrics, are present in low percentages contrary to the occurrence of simple alterations (gaps, breaks, and acentrics). This is due to the fact that 72 h cultures were used. By then, cells have gone through a second and third mitotic divisions. Therefore, primary alterations have been kept and turned into secondary aberrations in the growing generations; meanwhile, early cells with complex alterations have already died [103, 104]. Two individuals exposed to higher doses of radiation (4.54 and 1.07 mSv, resp.) did show complex alterations in the second sampling. This is an unusual finding likely to be caused by the individuals' sporadic exposure to higher amounts of radiation. A significant increase of chromosomal aberrations was observed by comparing CA during the first and the second sampling. Out of these, there was a higher number of lesions at the chromatid level (mostly gaps) possibly due to the proper action of DNA repair mechanisms at low doses of exposure over long periods of time [102, 105]. Other studies have also found an increase of CA in individuals exposed to similar doses of radiation, but have not addressed the importance of periodic biomonitoring [106]. Though numerical aberrations were not the focus of the study, the exposed individuals showed an increased frequency of hyperploidies and hypoploidies possibly due to the imbalance in the cell cycle caused by the exposure to toxic agents [107]. This data evidences the importance of periodic control of the occupational exposure and of monitoring the dosage-time levels of exposure at the workplace [102].

A second study focused on telomeric associations in individuals exposed both to X-rays and smoke in order to determine the existence of these associations as chromosomal markers of exposure to these carcinogenic agents. The phenomenon of telomeric association is an intermediate step in the progression towards chromosomal instability that also comprises a risk of developing cancer [108, 109]. Cytogenetic monitoring is currently accepted as an evaluation tool for exposed populations at risk as it has been used in studies regarding ionizing radiation [5]. Cytogenetic analysis of cigarette smokers has shown the occurrence of chromosomal aberrations in populations from Colombia and India [110, 111]. In this study, mitotic indices determined in all groups (smokers exposed to radiation, nonsmokers exposed to radiation, smokers unexposed, and unexposed nonsmokers) showed no correlations between the exposure to both carcinogens and the mitotic indices and cell proliferation. Nonetheless, the three different exposed groups showed high frequencies of telomeric associations [112]. These results were particularly surprising for cigarette smokers since no cytogenetic biomarker of exposure for cigarette had been demonstrated to be consistent so far. Also, the group of smokers unexposed to radiation showed higher frequencies of TA than the nonsmoking X-ray-exposed group, a result that was possibly due to the many carcinogens present in cigarette smoke [110, 111, 113]. Though it has been reported that both agents have a synergistic effect [113], our study did not find such a tendency. However, the group exposed to both agents did show the highest frequency of telomeric associations. The study suggested that telomeric associations can assess the genomic instability phenomena in populations

exposed to mutagens. Although, telomere length has been reported as a biomarker for age, stress and cancer, telomere biology and the molecular pathways that protect telomeres continue to be studied in order to determining the outcome of radiation exposure [114].

In another study, the inclusion of gaps as chromosomal aberration was investigated. Gaps are defined as the unstained regions of a chromosome that contain zones of lesser width than that of a chromatid [115, 116]. A gap is observed as an empty space because the DNA thread is so thin that it becomes practically invisible to the usual technique [113]. Since genotoxic agents such as ionizing radiation are capable of inducing chromosomal uncoiling events and affecting DNA condensation, gaps can certainly be the product of exposure to genotoxic agents [117]. Nonetheless, some authors had considered gaps to be structures that lack biological significance [118]. The study involved individuals exposed to X-rays and the unexposed control group. The findings of the CA analysis and comet assay were compared, including and excluding gaps. These two complementary techniques do not detect the same kind of lesions. On one hand, chromosomal aberrations are originated from double-strand breaks; on the other hand, comet assay can detect single-strand breaks, double-strand breaks, and alkali labile sites (when using the alkaline version) [119] and has proven to be a useful way of assessing X-ray damage to lymphocytes [120]. The correlation between the two methods including gaps as CA was positive. Gaps measured damage in the DNA since there was a stronger correlation between the results of both applied techniques when gaps were included as a CA. Although there is an increasing interest in studying the more complex chromosomal aberrations such as dicentrics and translocations, current studies still include gaps as part of the genotoxicity studies [121–124]. These findings suggested a revision of the biological importance of gaps in population occupational biomonitoring [124]. Furthermore, another study involved a group of radiologists and technicians exposed to X-rays at the workplace, excluding those with family and personal history of cancer and smoke exposure. The mean dose of ionizing radiation for the affected group was 0.99 mSv and the chromosomal aberrations observed involved gaps, breaks, dicentric, rings, and double minutes. The cytogenetic analysis showed that CAs were present in 50% of the individuals in the exposed group and in 2% of the control individuals. However, these results were not statistically significant. On the other hand, the comet assay did show a highly significant difference of migration in the exposed group as compared to the control group, possibly because the comet assay shows a wider set of damage consequences [93, 125]. Similarl to a study carried out in Iran that found no specific relation with the characteristics of the occupational setting and the duration of exposure [126], this study found no correlation between the results of both tests and the duration and dose of exposure due to a lack of significant variation between individual doses. Such results may also support the idea of hormesis taking place as a way of adapting to the workplace after several years, though the idea remains to be controversial [127]. Even though there were no significant results regarding CA, relative risk calculation showed that exposed individuals had a risk 20 times higher of showing aberrations than the control group [128]. These aberrations may lead to the alteration of cell control mechanisms such as apoptosis and tumor suppressor genes, besides the loss of genetic material due to cell death as a result of changes in division and repairing mechanisms [129–131]. Although this study did not show an association between CA and exposure time or dose level, other studies have agreed on the fact that long-term exposure to low radiation levels are the cause for higher percentages of CA [132, 133].

Cytogenetic findings are of great importance because they are associated with the mechanisms of carcinogenesis. The interaction with physical agents, such as ionizing radiation, produces a variety of primary lesions [134] whose prevalence can determine cancer risk [135]. Because of the importance of biomonitoring occupationally exposed populations, proper research guidelines have been established [136]. mFISH assays are currently being put in use in order to carry out a more detailed analysis of simple and complex aberrations that could model the effects of radiation on lymphocytes [107]. Nonetheless, earlier cytogenetic techniques are still held as the golden standard for biomonitoring populations. Environmental and occupational health issues are increasingly being studied because of its importance in public health. In Ecuador, going through a preliminary cytogenetic testing to evaluate the genotoxic effects of different agents is a rather voluntary decision and the toxic qualities of certain widely used chemicals, such as herbicides and pesticides, are not of common knowledge. According to the results obtained in this set of studies, adequate biomonitoring laws should be enforced. In the case of glyphosate, research has helped to consider changes regarding the targeted areas, duration of sprayings and chemical composition of herbicides. Though plantations and industries do offer protective equipment, this gear is not always used by the small farm owner. Reports on the use of highly toxic pesticides conclude that the lack of regulation of pesticide use benefits the informal distribution of these hazardous compounds, not only in flower plantations but increasingly in small farms. Lastly, occupational health risks must be studied in all professions that face any level of exposure to physical or chemical agents suspected to cause illness. Radiologists and other professionals exposed to radiation should have access to cytogenetics testing and follow-up studies that can report on any unusual results in order to prevent diseases as part of occupational health and safety laws.

References

[1] J. Bohne and T. Cathomen, "Genotoxicity in gene therapy: an account of vector integration and designer nucleases," *Current Opinion in Molecular Therapeutics*, vol. 10, no. 3, pp. 214–223, 2008.

[2] M. Uhl, M. J. Plewa, B. J. Majer, and S. Knasmüller, "Basic principles of genetic toxicology with an emphasis on plant bioassays," in *Bioassays in Plant Cells for Improvement of Ecosystem and Human Health*, J. Maluszynska and M. Plewa, Eds., pp. 11–30, Katowice, Poland, 2003.

[3] G. H. Westphalen, L. M. Menezes, D. Prá et al., "In vivo determination of genotoxicity induced by metals from orthodontic appliances using micronucleus and comet assays," *Genetics and Molecular Research*, vol. 7, no. 4, pp. 1259–1266, 2008.

[4] F. Henkler and A. Luch, "Adverse health effects of environmental chemical agents through non-genotoxic mechanisms," *Journal of Epidemiology and Community Health*, vol. 65, no. 1, pp. 1–3, 2011.

[5] T. Adamus, I. Mikulenková, L. Dobiáš, J. Havránková, and T. Pek, "Cytogenetic methods and biomonitoring of occupational exposure to genotoxic factors," *Journal of Applied Biomedicine*, vol. 4, no. 4, pp. 197–203, 2006.

[6] C. Paz-y-Miño, M. E. Sánchez, M. Arévalo et al., "Evaluation of DNA damage in an Ecuadorian population exposed to glyphosate," *Genetics and Molecular Biology*, vol. 30, no. 2, pp. 456–460, 2007.

[7] C. Paz-y-Miño, A. López-Cortés, M. Arévalo, and M. E. Sánchez, "Monitoring of DNA damage in individuals exposed to petroleum hydrocarbons in Ecuador," *Annals of the New York Academy of Sciences*, vol. 1140, pp. 121–128, 2008.

[8] A. Joaquín, S. Vallejo, and R. Trejos, *Más que Alimentos en la Mesa: La Real Contribución de la Agricultura a la Economía del Ecuador*, vol. 11, Instituto Interamericano de Cooperación para la Agricultura (IICA), Quito, Ecuador, 2005.

[9] C. Paz-y-Miño, G. Bustamante, M. E. Sáchez, and P. E. Leone, "Cytogenetic monitoring in a population occupationally exposed to pesticides in ecuador," *Environmental Health Perspectives*, vol. 110, no. 11, pp. 1077–1080, 2002.

[10] C. Paz-y-Miño, M. V. Dávalos, M. E. Sánchez, M. Arévalo, and P. E. Leone, "Should gaps be included in chromosomal aberration analysis?: evidence based on the comet assay," *Mutation Research*, vol. 516, no. 1-2, pp. 57–61, 2002.

[11] M. Stoia, S. Oancea, and D. C. Obreja, "Comparative study of genotoxic effects in workers exposed to inorganic lead and low dose irradiation using micronucleus test," *Romanian Journal of Legal Medicine*, vol. 17, no. 4, pp. 287–294, 2009.

[12] V. Garaj-Vrhovac and N. Kopjar, "The alkaline Comet assay as biomarker in assessment of DNA damage in medical personnel occupationally exposed to ionizing radiation," *Mutagenesis*, vol. 18, no. 3, pp. 265–271, 2003.

[13] C. Paz-y-Miño, M. J. Muñoz, A. Maldonado et al., "Baseline determination in social, health, and genetic areas in communities affected by glyphosate aerial spraying on the northeastern Ecuadorian border," *Reviews on Environmental Health*, vol. 26, no. 1, pp. 45–51, 2011.

[14] Ministerio de Relaciones Exteriores (MREE), *Misión de Verificación: Impactos en el Ecuador de las Fumigaciones Realizadas en el Departamento del Putumayo dentro del Plan Colombia*, Ministerio de Relaciones Exteriores del Ecuador, Quito, Ecuador, 2002.

[15] S. O. Duke and S. B. Powles, "Glyphosate: a once-in-a-century herbicide," *Pest Management Science*, vol. 64, no. 4, pp. 319–325, 2008.

[16] J. F. Acquavella, B. H. Alexander, J. S. Mandel, C. Gustin, B. Baker, and P. Chapman, "Glyphosate biomonitoring for farmers and their families: results from the farm family exposure study," *Environmental Health Perspectives*, vol. 112, no. 3, pp. 321–326, 2004.

[17] N. Benachour and G. E. Séralini, "Glyphosate formulations induce apoptosis and necrosis in human umbilical, embryonic, and placental cells," *Chemical Research in Toxicology*, vol. 22, no. 1, pp. 97–105, 2009.

[18] A. Martínez, I. Reyes, and N. Reyes, "Cytotoxicity of the herbicide glyphosate in human peripheral blood mononuclear cells," *Biomedica*, vol. 27, no. 4, pp. 594–604, 2007.

[19] C. M. Howe, M. Berrill, B. D. Pauli, C. C. Helbing, K. Werry, and N. Veldhoen, "Toxicity of glyphosate-based pesticides to four North American frog species," *Environmental Toxicology and Chemistry*, vol. 23, no. 8, pp. 1928–1938, 2004.

[20] S. K. Dinehart, L. M. Smith, S. T. McMurry, T. A. Anderson, P. N. Smith, and D. A. Haukos, "Toxicity of a glufosinate- and several glyphosate-based herbicides to juvenile amphibians from the Southern High Plains, USA," *Science of the Total Environment*, vol. 407, no. 3, pp. 1065–1071, 2009.

[21] J. Marc, O. Mulner-Lorillon, S. Boulben, D. Hureau, G. Durand, and R. Bellé, "Pesticide roundup provokes cell division dysfunction at the level of CDK1/cyclin B activation," *Chemical Research in Toxicology*, vol. 15, no. 3, pp. 326–331, 2002.

[22] R. Bellé, R. Le Bouffant, J. Morales, B. Cosson, P. Cormier, and O. Mulner-Lorillon, "Sea urchin embryo, DNA-damaged cell cycle checkpoint and the mechanisms initiating cancer development," *Journal de la Societe de Biologie*, vol. 201, no. 3, pp. 317–327, 2007.

[23] D. M. Romero, M. C. Ríos de Molina, and Á. B. Juárez, "Oxidative stress induced by a commercial glyphosate formulation in a tolerant strain of Chlorella kessleri," *Ecotoxicology and Environmental Safety*, vol. 74, no. 4, pp. 741–747, 2011.

[24] N. S. El-Shenawy, "Oxidative stress responses of rats exposed to Roundup and its active ingredient glyphosate," *Environmental Toxicology and Pharmacology*, vol. 28, no. 3, pp. 379–385, 2009.

[25] M. Mladinic, S. Berend, A. L. Vrdoljak, N. Kopjar, B. Radic, and D. Zeljezic, "Evaluation of genome damage and its relation to oxidative stress induced by glyphosate in human lymphocytes in vitro," *Environmental and Molecular Mutagenesis*, vol. 50, no. 9, pp. 800–807, 2009.

[26] O. V. Lushchak, O. I. Kubrak, J. M. Storey, K. B. Storey, and V. I. Lushchak, "Low toxic herbicide Roundup induces mild oxidative stress in goldfish tissues," *Chemosphere*, vol. 76, no. 7, pp. 932–937, 2009.

[27] K. A. Modesto and C. B. R. Martinez, "Roundup® causes oxidative stress in liver and inhibits acetylcholinesterase in muscle and brain of the fish Prochilodus lineatus," *Chemosphere*, vol. 78, no. 3, pp. 294–299, 2010.

[28] N. Ahsan, D. G. Lee, K. W. Lee et al., "Glyphosate-induced oxidative stress in rice leaves revealed by proteomic approach," *Plant Physiology and Biochemistry*, vol. 46, no. 12, pp. 1062–1070, 2008.

[29] L. Goldman, *Childhood Pesticide Poisoning*, United Nations Environment Program, Geneva, Switzerland, 2004, http://www.who.int/ceh/publications/pestpoisoning.pdf.

[30] I. Meiers, J. H. Shanks, and D. G. Bostwick, "Glutathione S-transferase pi (GSTP1) hypermethylation in prostate cancer: review 2007," *Pathology*, vol. 39, no. 3, pp. 299–304, 2007.

[31] H. W. Lo, L. Stephenson, X. Cao, M. Milas, R. Pollock, and F. Ali-Osman, "Identification and functional characterization of the human Glutathione S-transferase P1 gene as a novel transcriptional target of the p53 tumor suppressor gene," *Molecular Cancer Research*, vol. 6, no. 5, pp. 843–850, 2008.

[32] A. M. Moyer, O. E. Salavaggione, T. Y. Wu et al., "Glutathione S-transferase P1: gene sequence variation and functional genomic studies," *Cancer Research*, vol. 68, no. 12, pp. 4791–4801, 2008.

[33] R. H. Wong, C. L. Du, J. D. Wang, C. C. Chan, J. C. J. Luo, and T. J. Cheng, "XRCC1 and CYP2E1 polymorphisms as susceptibility factors of plasma mutant p53 protein and anti-p53 antibody expression in vinyl chloride monomer-exposed polyvinyl chloride workers," *Cancer Epidemiology Biomarkers and Prevention*, vol. 11, no. 5, pp. 475–482, 2002.

[34] C. Jerónimo, G. Varzim, R. Henrique et al., "I105V polymorphism and promoter methylation of the GSTP1 gene in prostate adenocarcinoma," *Cancer Epidemiology Biomarkers and Prevention*, vol. 11, no. 5, pp. 445–450, 2002.

[35] L. Kadouri, Z. Kote-Jarai, A. Hubert et al., "Glutathione-S-transferase M1, T1 and P1 polymorphisms, and breast cancer risk, in BRCA1/2 mutation carriers," *British Journal of Cancer*, vol. 98, no. 12, pp. 2006–2010, 2008.

[36] C. Martínez, E. García-Martín, H. Alonso-Navarro et al., "Glutathione-S-transferase P1 polymorphism and risk for essential tremor," *European Journal of Neurology*, vol. 15, no. 3, pp. 234–238, 2008.

[37] Y. J. Liu, P. L. Huang, Y. F. Chang et al., "GSTP1 genetic polymorphism is associated with a higher risk of DNA damage in pesticide-exposed fruit growers," *Cancer Epidemiology Biomarkers and Prevention*, vol. 15, no. 4, pp. 659–666, 2006.

[38] J. M. Matés, "Effects of antioxidant enzymes in the molecular control of reactive oxygen species toxicology," *Toxicology*, vol. 153, no. 1–3, pp. 83–104, 2000.

[39] J. R. Arthur, "The glutathione peroxidases," *Cellular and Molecular Life Sciences*, vol. 57, no. 13-14, pp. 1825–1835, 2000.

[40] P. Rohr, J. da Silva, B. Erdtmann et al., "BER gene polymorphisms (OGG1 Ser326Cys and XRCC1 Arg194Trp) and modulation of DNA damage due to pesticides exposure," *Environmental and Molecular Mutagenesis*, vol. 52, no. 1, pp. 20–27, 2011.

[41] M. C. Stern, D. M. Umbach, C. H. Van Gils, R. M. Lunn, and J. A. Taylor, "DNA repair gene XRCC1 polymorphisms, smoking, and bladder cancer risk," *Cancer Epidemiology Biomarkers and Prevention*, vol. 10, no. 2, pp. 125–131, 2001.

[42] R. H. Wong, S. Y. Chang, S. W. Ho et al., "Polymorphisms in metabolic GSTP1 and DNA-repair XRCC1 genes with an increased risk of DNA damage in pesticide-exposed fruit growers," *Mutation Research*, vol. 654, no. 2, pp. 168–175, 2008.

[43] Acción Ecológica, *Frontera: Daños Genéticos Por las Fumigaciones del Plan Colombia*, Acción Ecológica, Quito, Ecuador, 2004.

[44] N. P. Singh, M. T. McCoy, R. R. Tice, and E. L. Schneider, "A simple technique for quantitation of low levels of DNA damage in individual cells," *Experimental Cell Research*, vol. 175, no. 1, pp. 184–191, 1988.

[45] D. D. Evans and M. J. Batty, "Effects of high dietary concentrations of glyphosate (Roundup®) on a species of bird, marsupial and rodent indigenous to Australia," *Environmental Toxicology and Chemistry*, vol. 5, no. 4, pp. 399–401, 1986.

[46] G. M. Williams, R. Kroes, and I. C. Munro, "Safety evaluation and risk assessment of the herbicide Roundup and its active ingredient, glyphosate, for humans," *Regulatory Toxicology and Pharmacology*, vol. 31, no. 2, pp. 117–165, 2000.

[47] D. A. Goldstein, J. F. Acquavella, R. M. Mannion, and D. R. Farmer, "An analysis of glyphosate data from the California Environmental Protection Agency pesticide illness surveillance program," *Journal of Toxicology*, vol. 40, no. 7, pp. 885–892, 2002.

[48] M. H. Bernal, K. R. Solomon, and G. Carrasquilla, "Toxicity of formulated glyphosate (Glyphos) and cosmo-flux to larval and juvenile colombian frogs 2. field and laboratory microcosm acute toxicity," *Journal of Toxicology and Environmental Health—Part A*, vol. 72, no. 15-16, pp. 966–973, 2009.

[49] K. R. Solomon, A. Anadón, G. Carrasquilla, A. L. Cerdeira, J. Marshall, and L.-H. Sanin, "Coca and poppy eradication in Colombia: environmental and human health assessment of aerially applied glyphosate," *Reviews of Environmental Contamination and Toxicology*, vol. 190, pp. 43–125, 2007.

[50] N. Sailaja, M. Chandrasekhar, P. V. Rekhadevi et al., "Genotoxic evaluation of workers employed in pesticide production," *Mutation Research*, vol. 609, no. 1, pp. 74–80, 2006.

[51] S. Bull, K. Fletcher, A. R. Boobis, and J. M. Battershill, "Evidence for genotoxicity of pesticides in pesticide applicators: a review," *Mutagenesis*, vol. 21, no. 2, pp. 93–103, 2006.

[52] C. Bolognesi, "Genotoxicity of pesticides: a review of human biomonitoring studies," *Mutation Research*, vol. 543, no. 3, pp. 251–272, 2003.

[53] P. Grover, K. Danadevi, M. Mahboob, R. Rozati, B. S. Banu, and M. F. Rahman, "Evaluation of genetic damage in workers employed in pesticide production utilizing the Comet assay," *Mutagenesis*, vol. 18, no. 2, pp. 201–205, 2003.

[54] V. Ng, D. Koh, A. Wee, and S. E. Chia, "Salivary acetylcholinesterase as a biomarker for organophosphate exposure," *Occupational Medicine*, vol. 59, no. 2, pp. 120–122, 2009.

[55] C. Paz-y-Miño, M. Arévalo, M. E. Sanchez, and P. E. Leone, "Chromosome and DNA damage analysis in individuals occupationally exposed to pesticides with relation to genetic polymorphism for CYP 1A1 gene in Ecuador," *Mutation Research*, vol. 562, no. 1-2, pp. 77–89, 2004.

[56] V. Kumar, C. S. Yadav, S. Singh et al., "CYP 1A1 polymorphism and organochlorine pesticides levels in the etiology of prostate cancer," *Chemosphere*, vol. 81, no. 4, pp. 464–468, 2010.

[57] A. M. Tsatsakis, A. Zafiropoulos, M. N. Tzatzarakis, G. N. Tzanakakis, and A. Kafatos, "Relation of PON1 and CYP1A1 genetic polymorphisms to clinical findings in a cross-sectional study of a Greek rural population professionally exposed to pesticides," *Toxicology Letters*, vol. 186, no. 1, pp. 66–72, 2009.

[58] C. San Jose, A. Cabanillas, J. Benitez, J. A. Carrillo, M. Jimenez, and G. Gervasini, "CYP1A1 gene polymorphisms increase lung cancer risk in a high-incidence region of Spain: a case control study," *BMC Cancer*, vol. 10, article 463, 2010.

[59] E. Taioli, L. Gaspari, S. Benhamou et al., "Polymorphisms in CYP1A1, GSTM1, GSTT1 and lung cancer below the age of 45 years," *International Journal of Epidemiology*, vol. 32, no. 1, pp. 60–63, 2003.

[60] J. Little, L. Sharp, L. F. Masson et al., "Colorectal cancer and genetic polymorphisms of CYP1A1, GSTM1 and GSTT1: a case-control study in the Grampian region of Scotland," *International Journal of Cancer*, vol. 119, no. 9, pp. 2155–2164, 2006.

[61] T. N. Sergentanis and K. P. Economopoulos, "Four polymorphisms in cytochrome P450 1A1 (CYP1A1) gene and breast cancer risk: a meta-analysis," *Breast Cancer Research and Treatment*, vol. 122, no. 2, pp. 459–469, 2010.

[62] V. M. Basham, P. D. P. Pharoah, C. S. Healey et al., "Polymorphisms in CYP1A1 and smoking: no association with breast cancer risk," *Carcinogenesis*, vol. 22, no. 11, pp. 1797–1800, 2001.

[63] K. Kvitko, J. C. B. Nunes, T. A. Weimer, F. M. Salzano, and M. H. Hutz, "Cytochrome P4501A1 polymorphisms in South American Indians," *Human Biology*, vol. 72, no. 6, pp. 1039–1043, 2000.

[64] A. V. Ngowi, *Health Impact of Exposure to Pesticides in Agriculture in Tanzania*, University of Tampere, Tampere, Finland, 2002.

[65] R. Naravaneni and K. Jamil, "Determination of AChE levels and genotoxic effects in farmers occupationally exposed to pesticides," *Human and Experimental Toxicology*, vol. 26, no. 9, pp. 723–731, 2007.

[66] IARC, *Evaluation of Carcinogenic Risks to Humans: Occupational Exposures in Insecticide Application and Some Pesticides*, vol. 53, The International Agency for Research on Cancer, Lyon, France, 1991, http://monographs.iarc.fr/ENG/Monographs/vol53/mono53.pdf.

[67] Banco Central del Ecuador (BCE), *Cifras Económicas del Ecuador Abril 2009*, Banco Central del Ecuador, Quito, Ecuador, 2009, http://www.bce.fin.ec/documentos/Estadisticas/SectorReal/Previsiones/IndCoyuntura/CifrasEconomicas/cie200904.pdf.

[68] P. R. Epstein and J. Selber, *A Life Cycle Analysis of Its Health and Environmental Impacts*, The Center for Health and the Global Environment, Boston, Mass, USA, 2002, http://chge.med.harvard.edu/publications/documents/oil-fullreport.pdf.

[69] J. Kimmerling, *Amazon Crude*, Brickfron Graphics, New York, NY, USA, 1993.

[70] M. Neri, D. Ugolini, S. Bonassi et al., "Children's exposure to environmental pollutants and biomarkers of genetic damage: II. Results of a comprehensive literature search and meta-analysis," *Mutation Research*, vol. 612, no. 1, pp. 14–39, 2006.

[71] K. L. Platt, S. Aderhold, K. Kulpe, and M. Fickler, "Unexpected DNA damage caused by polycyclic aromatic hydrocarbons under standard laboratory conditions," *Mutation Research*, vol. 650, no. 2, pp. 96–103, 2008.

[72] C. Jochnick, R. Normand, and S. Zaidi, "Rights violations in the Ecuadorian Amazon: the human consequences of oil development," *Health & Human Rights*, vol. 1, no. 1, pp. 82–100, 1994.

[73] IARC, *Evaluation of the Carcinogenic Risk of Chemicals to Man: Occupational Exposures to Petroleum Refining; Crude Oil and Major Petroleum Fuels*, vol. 45, The International Agency for Research on Cancer, Lyon , France, 1989, http://monographs.iarc.fr/ENG/Monographs/vol45/mono45.pdf.

[74] I. Rahman, K. Narasimhan, S. Aziz, and W. Owens, "Gasoline ingestion: a rare cause of pancytopenia," *American Journal of the Medical Sciences*, vol. 338, no. 5, pp. 433–434, 2009.

[75] R. B. Hayes, Y. Songnian, M. Dosemeci, and M. Linet, "Benzene and lymphohematopoietic malignancies in humans," *American Journal of Industrial Medicine*, vol. 40, no. 2, pp. 117–126, 2001.

[76] T. A. McDonald, *Public Health Goal for Benzene in Drinking Water*, Office of Environmental Health Hazard Assessment, Sacramento, Calif, USA, 2001, http://oehha.ca.gov/water/phg/pdf/BenzeneFinPHG.pdf.

[77] W. J. Blot, L. A. Brinton, J. F. Fraumeni, and B. J. Stone, "Cancer mortality in U.S. counties with petroleum industries," *Science*, vol. 198, no. 4312, pp. 51–53, 1977.

[78] R. G. Olin, A. Ahlbom, and I. Lindberg-Navier, "Occupational factors associated with astrocytomas: a case-control study," *American Journal of Industrial Medicine*, vol. 11, no. 6, pp. 615–625, 1987.

[79] R. A. Lyons, S. P. Monaghan, M. Heaven, B. N. C. Littlepage, T. J. Vincent, and G. J. Draper, "Incidence of leukaemia and lymphoma in young people in the vicinity of the petrochemical plant at Baglan Bay, South Wales, 1974 to 1991," *Occupational and Environmental Medicine*, vol. 52, no. 4, pp. 225–228, 1995.

[80] J. Kaldor, J. A. Harris, and E. Glazer, "Statistical association between cancer incidence and major-cause mortality, and estimated residential exposure to air emissions from petroleum and chemical plants," *Environmental Health Perspectives*, vol. 54, pp. 319–332, 1983.

[81] M. Gérin, J. Siemiatycki, M. Désy, and D. Krewski, "Associations between several sites of cancer and occupational exposure to benzene, toluene, xylene, and styrene: results of a case-control study in Montreal," *American Journal of Industrial Medicine*, vol. 34, no. 2, pp. 144–156, 1998.

[82] J. D. Everall and P. M. Dowd, "Influence of environmental factors excluding ultra violet radiation on the incidence of skin cancer," *Bulletin du Cancer*, vol. 65, no. 3, pp. 241–247, 1978.

[83] P. Boffetta, N. Jourenkova, and P. Gustavsson, "Cancer risk from occupational and environmental exposure to polycyclic aromatic hydrocarbons," *Cancer Causes and Control*, vol. 8, no. 3, pp. 444–472, 1997.

[84] M. S. Gottlieb, C. L. Shear, and D. B. Seale, "Lung cancer mortality and residential proximity to industry," *Environmental Health Perspectives*, vol. 45, pp. 157–164, 1982.

[85] C. Y. Yang, M. F. Cheng, J. F. Chiu, and S. S. Tsai, "Female lung cancer and petrochemical air pollution in Taiwan," *Archives of Environmental Health*, vol. 54, no. 3, pp. 180–185, 1999.

[86] B. J. Pan, Y. J. Hong, G. C. Chang, M. T. Wang, F. F. Cinkotai, and Y. C. Ko, "Excess cancer mortality among children and adolescents in residential districts polluted by petrochemical manufacturing plants in Taiwan," *Journal of Toxicology and Environmental Health*, vol. 43, no. 1, pp. 117–129, 1994.

[87] A. K. Hurtig and M. San Sebastián, "Geographical differences in cancer incidence in the Amazon basin of Ecuador in relation to residence near oil fields," *International Journal of Epidemiology*, vol. 31, no. 5, pp. 1021–1027, 2002.

[88] C. Paz-y-Miño, B. Castro, A. López-Cortés et al., "Impacto genético en comunidades amazónicas del Ecuador localizadas en zonas petroleras," *Revista Ecuatoriana de Medicina y Ciencias Biológicas*, vol. 1, no. 1-2, pp. 7–19, 2010.

[89] G. Thodi, F. Fostira, R. Sandaltzopoulos et al., "Screening of the DNA mismatch repair genes MLH1, MSH2 and MSH6 in a Greek cohort of Lynch syndrome suspected families," *BMC Cancer*, vol. 10, article 544, 2010.

[90] D. A. Lawes, T. Pearson, S. SenGupta, and P. B. Boulos, "The role of *MLH1, MSH2* and *MSH6* in the development of multiple colorectal cancers," *British Journal of Cancer*, vol. 93, no. 4, pp. 472–477, 2005.

[91] Environment Agency, *Principles for Evaluating the Human Health Risks from Petroleum Hydrocarbons in Soils: A Consultation Paper*, Environment Agency, Bristol, UK, 2003, http://www.environment-agency.gov.uk/static/documents/Research/petroleum_hydrocarbons1.pdf.

[92] I. Kovalchuk, O. Kovalchuk, and B. Hohn, "Biomonitoring the genotoxicity of environmental factors with transgenic plants," *Trends in Plant Science*, vol. 6, no. 7, pp. 306–310, 2001.

[93] C. Paz-y-Miño, A. Creus, O. Cabré, and P. E. Leone, *Genética, Toxicología y Carcinogenesis*, PUCE, Quito, Ecuador, 2002.

[94] T. K. Hei, R. Persaud, H. Zhou, and M. Suzuki, "Genotoxicity in the eyes of bystander cells," *Mutation Research*, vol. 568, no. 1, pp. 111–120, 2004.

[95] A. V. Carrano, "Chromosome aberrations and radiation-induced cell death. I. Transmission and survival parameters of aberrations," *Mutation Research*, vol. 17, no. 3, pp. 341–353, 1973.

[96] E. De Souza and J. P. D. M. Soares, "Occupational and technichal correlations of interventional radiology," *Jornal Vascular Brasileiro*, vol. 7, no. 4, pp. 341–350, 2008.

[97] J. A. V. Butler, "Effects of ultra-violet light on nucleic acid and nucleoproteins and other biological systems," *Experientia*, vol. 11, no. 8, pp. 289–293, 1955.

[98] D. R. Boreham, "Cellular defense mechanisms against the biological effects of ionizing radiation," in *Proceedings of the 10th International Congress of International Radiation Protection Association (IRPA '00)*, Hiroshima, Japan, May 2000, http://w3.tue.nl/fileadmin/sbd/Documenten/IRPA_refresher_courses/Cellular_Defense_Mechanisms_Against_the_Biological_Effects_of_Ionizing_Radiation.pdf.

[99] J. Chung, H. Ward, K. Teschke, P. A. Ratner, and Y. Chow, *A Retrospective Cohort Study of Cancer Risks among Nurses in British Columbia: Potential Exposure to Ionizing Radiation Report*, British Columbia: Research Secretariat of the Workers' Compensation Board of British Columbia, 2005, http://www.cher.ubc.ca/PDFs/Ionizing_Radiation_2005.pdf.

[100] G. Obe, P. Pfeiffer, J. R. K. Savage et al., "Chromosomal aberrations: formation, identification and distribution," *Mutation Research*, vol. 504, no. 1-2, pp. 17–36, 2002.

[101] V. Garaj-Vrhovac and D. Zeljezic, "Comet assay in the assessment of the human genome damage induced by γ-radiation in vitro," *Radiology and Oncology*, vol. 38, no. 1, pp. 43–47, 2004.

[102] C. Paz-y-Mino, P. E. Leone, M. Chavez et al., "Follow up study of chromosome aberrations in lymphocytes in hospital workers occupationally exposed to low levels of ionizing radiation," *Mutation Research*, vol. 335, no. 3, pp. 245–251, 1995.

[103] A. V. Carrano, "Chromosome aberrations and radiation-induced cell death. II. Predicted and observed cell survival," *Mutation Research*, vol. 17, no. 3, pp. 355–366, 1973.

[104] A. P. Krishnaja and N. K. Sharma, "Transmission of γ-ray-induced unstable chromosomal aberrations through successive mitotic divisions in human lymphocytes in vitro," *Mutagenesis*, vol. 19, no. 4, pp. 299–305, 2004.

[105] P. K. Gadhia, M. Gadhia, S. Georje, K. R. Vinod, and M. Pithawala, "Induction of chromosomal aberrations in mitotic chromosomes of fish *Boleophthalmus dussumieri* after exposure in vivo to antineoplastics Bleomycin, Mitomycin-C and Doxorubicin," *Indian Journal of Science and Technology*, vol. 1, no. 1, pp. 1–6, 2008.

[106] R. S. Cardoso, S. Takahashi-Hyodo, P. Peitl Jr., T. Ghilardi-Neto, and E. T. Sakamoto-Hojo, "Evaluation of chromosomal aberrations, micronuclei, and sister chromatid exchanges in hospital workers chronically exposed to ionizing radiation," *Teratogenesis Carcinogenesis and Mutagenesis*, vol. 21, no. 6, pp. 431–439, 2001.

[107] L. Hlatky, R. K. Sachs, M. Vazquez, and M. N. Cornforth, "Radiation-induced chromosome aberrations: insights gained from biophysical modeling," *BioEssays*, vol. 24, no. 8, pp. 714–723, 2002.

[108] R. Gertler, R. Rosenberg, D. Stricker et al., "Telomere length and human telomerase reverse transcriptase expression as

markers for progression and prognosis of colorectal carcinoma," *Journal of Clinical Oncology*, vol. 22, no. 10, pp. 1807–1814, 2004.

[109] K. I. Nakamura, E. Furugori, Y. Esaki et al., "Correlation of telomere lengths in normal and cancers tissue in the large bowel," *Cancer Letters*, vol. 158, no. 2, pp. 179–184, 2000.

[110] V. Balachandar, B. L. Kumar, K. Suresh, and K. Sasikala, "Evaluation of chromosome aberrations in subjects exposed to environmental tobacco smoke in Tamilnadu, India," *Bulletin of Environmental Contamination and Toxicology*, vol. 81, no. 3, pp. 270–276, 2008.

[111] M. S. Sierra-Torres, Y. Y. Arboleda-Moreno, L. S. Hoyos, and C. H. Sierra-Torres, "Chromosome aberrations among cigarette smokers in Colombia," *Mutation Research*, vol. 562, no. 1-2, pp. 67–75, 2004.

[112] C. Paz-y-Miño, J. C. Pérez, V. Dávalos, M. E. Sánchez, and P. E. Leone, "Telomeric associations in cigarette smokers exposed to low levels of X-rays," *Mutation Research*, vol. 490, no. 1, pp. 77–80, 2001.

[113] K. B. S. Kumar, R. Ankathil, and K. S. Devi, "Chromosomal aberrations induced by methyl parathion in human peripheral lymphocytes of alcoholics and smokers," *Human and Experimental Toxicology*, vol. 12, no. 4, pp. 285–288, 1993.

[114] S. D. Bouffler, M. A. Blasco, R. Cox, and P. J. Smith, "Telomeric sequences, radiation sensitivity and genomic instability," *International Journal of Radiation Biology*, vol. 77, no. 10, pp. 995–1005, 2001.

[115] Y. Saitoh, Y. Harata, F. Mizuhashi, M. Nakajima, and N. Miwa, "Biological safety of neutral-pH hydrogen-enriched electrolyzed water upon mutagenicity, genotoxicity and subchronic oral toxicity," *Toxicology and Industrial Health*, vol. 26, no. 4, pp. 203–216, 2010.

[116] U. Von Recklinghausen, C. Johannes, L. Riedel, and G. Obe, "Aberration patterns and cell cycle progression following exposure of lymphocytes to the alkylating agent Trenimon," *Chromosome Alterations*, pp. 315–324, 2007.

[117] L. C. Sánchez-Peña, B. E. Reyes, L. López-Carrillo et al., "Organophosphorous pesticide exposure alters sperm chromatin structure in Mexican agricultural workers," *Toxicology and Applied Pharmacology*, vol. 196, no. 1, pp. 108–113, 2004.

[118] J. Friedman, F. Shabtai, L. S. Levy, and M. Djaldetti, "Chromium chloride induces chromosomal aberrations in human lymphocytes via indirect action," *Mutation Research*, vol. 191, no. 3-4, pp. 207–210, 1987.

[119] P. L. Olive and J. P. Banáth, "The comet assay: a method to measure DNA damage in individual cells," *Nature Protocols*, vol. 1, no. 1, pp. 23–29, 2006.

[120] D. Milković, V. Garaj-Vrhovac, M. Ranogajec-Komor et al., "Primary DNA damage assessed with the comet assay and comparison to the absorbed dose of diagnostic X-rays in children," *International Journal of Toxicology*, vol. 28, no. 5, pp. 405–416, 2009.

[121] L. Jiunn-Wang, H. Ching-I, M. Isao, and C. Yng-Tay, "Chloroacetaldehyde indices chromosome aberrations and micronucleus formation but not 2-chloroethanol," *Journal of Health Science*, vol. 57, no. 3, pp. 300–303, 2011.

[122] L. C. Silva-Pereira, P. C. S. Cardoso, D. S. Leite et al., "Cytotoxicity and genotoxicity of low doses of mercury chloride and methylmercury chloride on human lymphocytes in vitro," *Brazilian Journal of Medical and Biological Research*, vol. 38, no. 6, pp. 901–907, 2005.

[123] T. Kyoya, Y. Obara, and A. Nakata, "Chromosomal aberrations in Japanese grass voles in and around an illegal dumpsite at the Aomori-Iwate prefectural boundary," *Zoological Science*, vol. 25, no. 3, pp. 307–312, 2008.

[124] K. Guleria and V. Sambyal, "Spectrum of chromosomal aberrations in peripheral blood lymphocytes of gastrointestinal tract (GIT) and breast cancer patients," *International Journal of Human Genetics*, vol. 10, no. 1–3, pp. 147–158, 2010.

[125] E. Horváthová, D. Slameňová, L. Hlinčíková, T. K. Mandal, A. Gábelová, and A. R. Collins, "The nature and origin of DNA single-strand breaks determined with the comet assay," *Mutation Research*, vol. 409, no. 3, pp. 163–171, 1998.

[126] H. Samavat and H. Mozdarani, "Chromosomal aberrations in Iranian radiation workers due to chronic exposure of X-irradiation," *International Journal of Low Radiation*, vol. 1, no. 2, pp. 216–222, 2004.

[127] J. M. Kauffman, "Radiation Hormesis: demonstrated, deconstructed, denied, dismissed, and some implications for public policy," *Journal of Scientific Exploration*, vol. 17, no. 3, pp. 389–407, 2003.

[128] M. J. Muñoz, A. López-Cortés, I Sarmiento, C. Herrera, M. E. Sánchez, and C. Paz-y-Miño, "Genetic biomonitoring of individuals exposed to ionizing radiation and the relationship with cáncer," *Oncología*, vol. 18, no. 1, pp. 75–82, 2008.

[129] N. Bayo, "Reacción celular ante la radiación," *Radiobiología*, vol. 1, no. 1, pp. 9–11, 2001.

[130] B. Leffon, B. Perez-Candahía, J. Loueiro, J. Mendez, and E. Pásaro, "Papel de los polimorfismos para enzimas de reparación en el daño del ADN inducido por estierno y estireno-7, 8-óxido," *Reviews in Toxicology*, no. 21, pp. 92–97, 2004.

[131] E. L. Goode, C. M. Ulrich, and J. D. Potter, "Polymorphisms in DNA repair genes and associations with cancer risk," *Cancer Epidemiology Biomarkers and Prevention*, vol. 11, no. 12, pp. 1513–1530, 2002.

[132] M. Díaz-Valecillos, J. Fernández, A. Rojas, J. Valecillos, and J. Cañizales, "Chromosome alterations in workers exposed to ionizing radiation," *Investigacion Clinica*, vol. 45, no. 3, pp. 197–211, 2004.

[133] F. Zakeri and T. Hirobe, "A cytogenetic approach to the effects of low levels of ionizing radiations on occupationally exposed individuals," *European Journal of Radiology*, vol. 73, no. 1, pp. 191–195, 2010.

[134] H. E. Jiliang, C. Weilin, J. Lifen, and J. Haiyan, "Comet assay and cytokinesis-blocked micronucleus test for monitoring the genotoxic effects of X-ray radiation in humans," *Chinese Medical Journal*, vol. 113, no. 10, pp. 911–914, 2000.

[135] L. Hagmar, S. Bonassi, U. Strömberg et al., "Chromosomal aberrations in lymphocytes predict human cancer: a report from the European study group on cytogenetic biomarkers and health (ESCH)," *Cancer Research*, vol. 58, no. 18, pp. 4117–4121, 1998.

[136] R. J. Albertini, D. Anderson, G. R. Douglas et al., "IPCS guidelines for the monitoring of genotoxic effects of carcinogens in humans," *Mutation Research*, vol. 463, no. 2, pp. 111–172, 2000.

20

Protease-Mediated Maturation of HIV: Inhibitors of Protease and the Maturation Process

Catherine S. Adamson

Biomedical Sciences Research Complex, School of Medicine, University of St. Andrews, North Haugh, St. Andrews, Fife KY16 9ST, UK

Correspondence should be addressed to Catherine S. Adamson, csa21@st-andrews.ac.uk

Academic Editor: Abdul Waheed

Protease-mediated maturation of HIV-1 virus particles is essential for virus infectivity. Maturation occurs concomitant with immature virus particle release and is mediated by the viral protease (PR), which sequentially cleaves the Gag and Gag-Pol polyproteins into mature protein domains. Maturation triggers a second assembly event that generates a condensed conical capsid core. The capsid core organizes the viral RNA genome and viral proteins to facilitate viral replication in the next round of infection. The fundamental role of proteolytic maturation in the generation of mature infectious particles has made it an attractive target for therapeutic intervention. Development of small molecules that target the PR active site has been highly successful and nine protease inhibitors (PIs) have been approved for clinical use. This paper provides an overview of their development and clinical use together with a discussion of problems associated with drug resistance. The second-half of the paper discusses a novel class of antiretroviral drug termed maturation inhibitors, which target cleavage sites in Gag not PR itself. The paper focuses on bevirimat (BVM) the first-in-class maturation inhibitor: its mechanism of action and the implications of naturally occurring polymorphisms that confer reduced susceptibility to BVM in phase II clinical trials.

1. Introduction

Human Immunodeficiency Virus Type 1 (HIV-1) is the causative agent of the worldwide Acquired Immunodeficiency Syndrome (AIDS) epidemic. Approximately 34 million people were estimated to be living with HIV at the end of 2010. The number of people infected is a consequence of continued large numbers of new HIV-1 infections together with a reduction in AIDS-related deaths due to a significant expansion in access to antiretroviral drug therapy [1]. In the absence of an effective vaccine or cure, antiviral drugs are currently the only treatment option available to HIV-infected patients. Therapeutic regimes commonly termed HAART (highly active antiretroviral therapy) suppress viral replication but do not eradicate the virus; therefore, treatment must be administered on a lifelong basis [2, 3]. HAART consists of the simultaneous use of a combination of three or four different antiretroviral drugs.

This combinational approach is required due to the ease with which HIV-1 can acquire drug resistance to a single drug administered as monotherapy [3, 4]. Drug resistance arises due to the high degree of HIV-1 genetic diversity within the virus population (quasi-species) infecting an individual patient. This genetic diversity is created as a consequence of a rapid rate of viral replication combined with the error prone nature of the viral reverse transcriptase (RT), which copies the viral RNA genome into a double-stranded DNA copy and the frequent recombination events that occur during genome replication [3, 5, 6]. HAART is possible due to the successful development and clinical use of more than 20 antiretroviral drugs, which belong to six different mechanistic classes. These drugs primarily target the viral enzymes: RT inhibitors (which fall into two classes based on their mode of action: the nucleoside-analog RT inhibitors (NRTIs) and nonnucleoside-analog RT inhibitors (NNRTIs)), protease (PR) inhibitors (PIs), and an integrase (IN) inhibitor [7–10]. Most clinical treatment regimens use a combination of either a PI or NNRTI with two NRTIs, though since its approval for clinical use in 2007 the first IN inhibitor (insentress) has increasingly been used in therapy regimens. The remaining two mechanistic drug classes each contain one approved drug and target the viral entry process by either blocking viral fusion by targeting the viral gp41 envelope protein or acting as an antagonist against the host cell coreceptor CCR5 [11]. The viral entry inhibitors are in general reserved for salvage therapy. Salvage therapy is required upon treatment failure primarily due to the emergence of drug resistance and to be effective should ideally include at least one new drug targeting a novel site of action. Until a cure for HIV infection is achieved, the continued threat of drug resistance makes the identification and development of a continuous pipeline of new drugs with a novel mechanism of action an ongoing requirement [12]. In this paper we discuss protease-mediated maturation of HIV-1 particles and the strategies to target this step in HIV-1 replication for therapeutic intervention.

2. Proteolytic Maturation and Its Role in HIV-1 Replication

Proteolytic maturation is essential for the production of infectious HIV-1 virus particles and has been extensively reviewed [16–18]. Particle assembly is driven by the Gag (Pr55^{Gag}) polyprotein, which is transported to the cellular plasma membrane where it undergoes higher-order Gag-Gag multimerization. A second polyprotein Gag-Pol (Pr160$^{Gag-Pol}$) is also incorporated into the assembling particle through Gag-Gag interactions. Gag-Pol is expressed via a −1 ribosomal frameshift during approximately 5–10% of Gag translation events. The Pol domain encodes the viral PR, RT, and IN proteins. Gag-Gag multimerization forces membrane curvature and assembly is completed upon budding of the particle from the plasma membrane. Initially, the newly formed particles have a noninfectious immature morphology. However, concomitant with virus budding, PR is activated to facilitate particle maturation. The exact mechanism of PR activation is not clearly understood, but it is known to require Gag-Pol dimerization. Once PR is liberated from the polyprotein through autocatalysis, it cleaves Gag and Gag-Pol into their respective proteins. Cleavage of the Pol domain results in the enzymatic proteins PR, RT, and IN. Cleavage of Gag results in four protein domains: matrix (MA or p17), capsid (CA or p24), nucleocapsid (NC or p7), p6, and two spacer peptides SP1 (p2) and SP2 (p1) (Figure 1(a)). Gag cleavage follows a sequential cascade that is kinetically controlled by the differential rate of processing at each of the five cleavage sites in Gag. The first cleavage creates an N-terminal fragment that contains the MA-CA-SP1 domains and a C-terminal fragment that contains the NC-SP2-p6 domains. Subsequent cleavage events occur at the MA-CA and SP1-p6 sites and finally the CA-SP1 and NC-SP2 sites are cleaved.

The physical consequence of Gag cleavage is a morphological rearrangement of the non-infectious immature particle to a mature infectious particle containing a conical core, which is generated by a second assembly event upon release of the CA domain (Figure 1(b)). The conical CA core contains the RT and IN enzymes along with the dimeric viral RNA genome in complex with NC and is essential for virus replication upon infection of a new cell. Therefore, correct core formation is essential for the production of infectious particles and this has been shown to be dependent on accurate proteolytic processing of Gag as mutations that disrupt the cleavage of individual sites or alter the order in which sites are cleaved result in aberrant particles that have significantly reduced infectivity. The fundamental role of proteolytic maturation in the generation of infectious particles makes inhibiting this process an attractive target for therapeutic intervention. In this paper we discuss how this has been approached by (i) the successful development and clinical use of PIs which target the PR enzyme itself and (ii) research to develop a novel class of antiretroviral drug termed maturation inhibitors which target the Gag cleavage sites that act as the substrate for PR.

3. Protease Inhibitors

3.1. Introduction. Protease inhibitors (PIs) target and inhibit the enzymatic activity of the HIV-1 PR. PIs inhibit PR activity to the extent that is sufficient to prevent cleavage events in Gag and Gag-Pol that result in the production of non-infectious virus particles. The development of PIs in the mid 1990s was a critical step forward in the successful treatment of HIV-1 patients. This is because their development provided a second mechanistic class of antiretroviral drug, which made HAART combination therapy possible. PIs have remained a key component of HIV-1 patient treatment regimens right up to the current day. To date, nine PIs have been approved for clinical use, they are saquinavir, ritonavir, indinavir, nelfinavir, fosamprenavir, lopinavir, atazanavir, tipranavir, and darunavir [8] (Table 1).

3.2. Protease Inhibitor Design. Design of PIs has been primarily driven by structural knowledge of PR (Figure 2), its substrate, and the chemical reaction of peptide bond cleavage [16]. Like other retroviruses, HIV-1 PR, is related to the cellular aspartyl PR family, which include pepsin and renin. This family of proteases are typified by an active site that uses two apposed catalytic aspartic acid (Asp) residues, each within a conserved Asp-Thr-Gly motif. To function, the cellular PRs form a pseudodimer utilizing two Asp residues from within the same molecule to create an active site. In contrast, retroviral PRs only contain one Asp-Thr-Gly motif and must therefore form a true dimer. Indeed X-ray crystallography has shown that the HIV-1 PR exists as a dimer consisting of two identical monomers [19–21]. The crystal structure of the dimer reveals that four-stranded β sheets derived from both ends of each monomer hold the dimer together. A long substrate-binding cleft is created between the monomers and the active site is situated near its centre with the two Asp residues located at its base. Two β-hairpin flaps originating from each monomer cover the

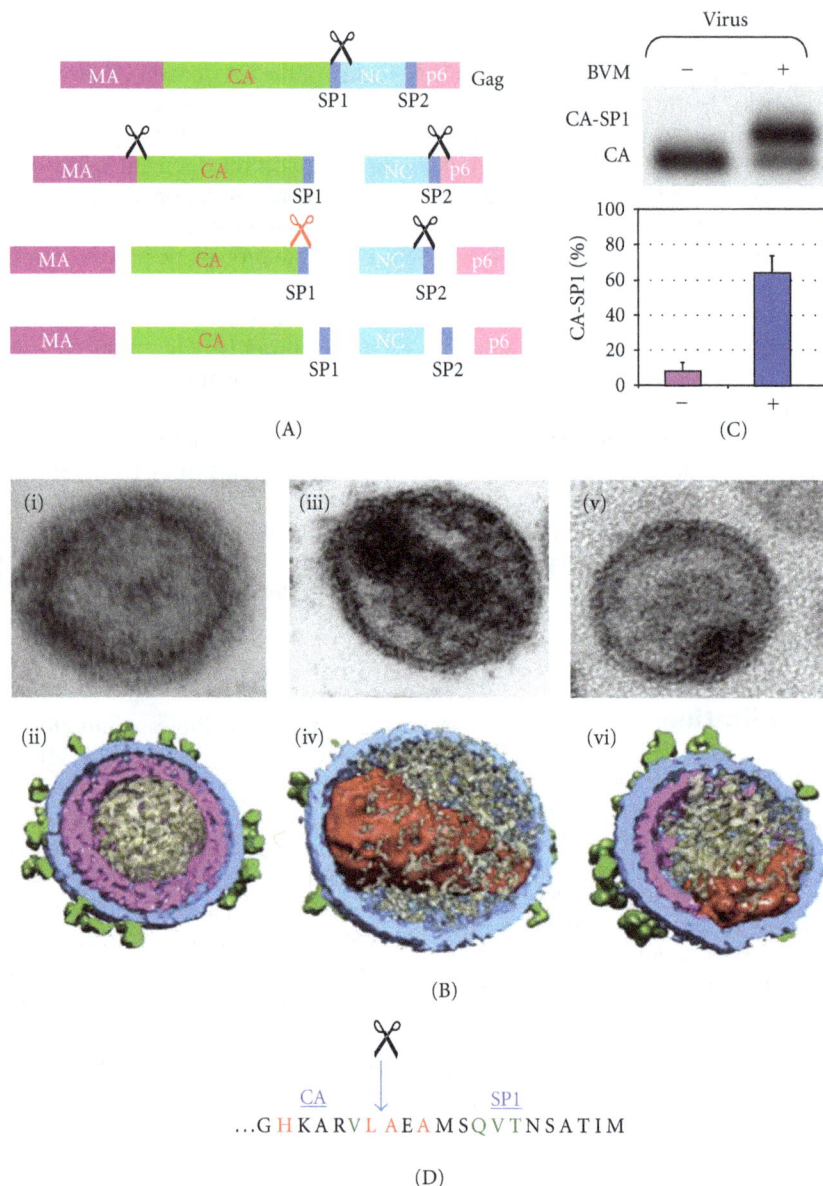

(A)

(B)

(C)

(D)

FIGURE 1: Proteolytic maturation of HIV-1 and its inhibition by bevirimat (BVM). (A) Gag processing cascade, illustrating the order in which the Gag precursor is cleaved by the viral protease. Each cleavage site is indicated by a scissor symbol, the red scissor symbol depicts the cleavage event blocked by BVM. (B) Virion morphology visualized by transmission electron microscopy (i, iii, v) and cryoelectron tomography models generated by segmented surface rendering. The glycoprotein spikes are coloured green, the membrane and MA layer in blue, Gag related shells in magenta, core structures in red, and other internal density in beige (ii, iv, vi). Immature particles (i and ii), mature (iii and iv), and BVM-treated (v and vi). (C) Biochemical data demonstrating accumulation of the uncleaved CA-SP1 precursor in virus particles in the presence of 1 μg/mL BVM. (D) Amino acid sequence at the CA-SP1 junction region; amino acids highlighted in green indicate the highly polymorphic residues to which reduced susceptibility to BVM in clinical trials has been mapped and amino acids highlighted in red indicate those that at which BVM resistance arises *in vitro*. Adapted with permission from Elsevier and the American Society for Microbiology [12, 13].

active site and are thought to function by stabilizing the substrate within the binding cleft.

The substrate-binding cleft interacts with multiple different substrate cleavage site sequences in Gag and Gag-Pol. The sequence of these sites are at least seven amino acids long and termed P4-P3′, with P1 and P1′ directly flanking the cleavage site [16]. There is no clear consensus amino acid recognition sequence; however, general patterns have been recognised and most substrate sites have a branched amino acid at the P2 site, a hydrophobic residue at P1, and an aromatic or proline at P1′. Instead of amino acid sequence, the topology of the cleavage site is primarily important for their recognition and interaction with PR [22]. Each of the substrate recognition sites has a super-imposable structure, known as the substrate

TABLE 1: FDA approved protease inhibitors. Key protease resistance mutations sourced from the 2011 data review of HIV drug resistance by the international AIDS society USA [15].

Protease inhibitor	Year of FDA approval	Key resistance mutations
Saquinavir	1995	G48V, L90M
Ritonavir	1996	Used for boosting
Indinavir	1996	M46I/L, V82A/F/T, I84V
Nelfinavir	1997	D30N, L90M
Fosamprenavir	1999	I50V, I84V
Lopinavir	2000	V32I, I47V/A, L76V, V82A/F/T/S
Atazanavir	2003	I50L, I84V, N88S
Tipranavir	2005	I47V, Q58E, T74P, V82L/T, N83, I84V
Darunavir	2006	I47V, I50V, I54M/L, V76V, I84V

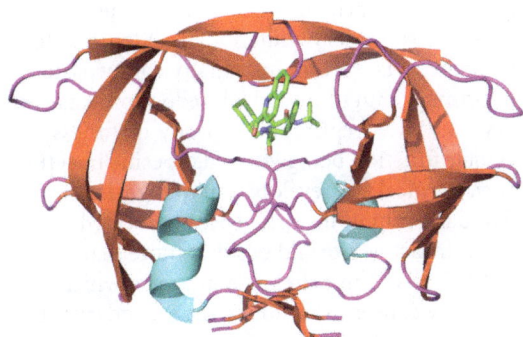

FIGURE 2: Three-dimensional structure of the HIV protease dimer in complex with the protease inhibitor saquinavir bound at the active site. Adapted by Jerry Alexandrators with permission from Annual Reviews [14].

envelope, which fits within the PR substrate-binding cleft. The divergent amino acid sequences of substrate recognition sites do however result in subtle structural differences, which are caused by different side chain protrusions from the substrate envelope. These side chains extend into pockets or subsites in the substrate-binding cleft. Each subsite is named for the corresponding substrate side chain, for example, the S1 subsite corresponds to the P1 side chain. These differences are thought to alter the rate at which cleavage occurs at individual sites in Gag facilitating the regulated proteolytic processing cascade of Gag that is essential for correct particle formation.

The catalytic mechanism of substrate cleavage requires the Asp residues to coordinate a water molecule that is used to hydrolyze the target peptide (scissile) bond [23]. During the reaction, a transition state intermediate is formed which has been mimicked in the design of most PIs, which are peptidomimetic transition-state analogues. The principle of this design strategy is that the normal peptide linkage [–NH–CO–] is replaced by a hydroxyethylene group [–CH$_2$–CH(OH)–], which cannot be cleaved by simple hydrolysis. Saquinavir was the first PI to be approved for clinical use and its design is based on this principle. The following PIs ritonavir, indinavir, nelfinavir, amprenavir, lopinaivr, atazanavir, fosamprenavir, and darunavir also all contain a central core motif of a hydroxyethylene scaffold.

The exception is tipranavir, which has a coumarin scaffold and is therefore the only clinically approved PI, which is not a peptidomimetic [8]. Knowledge of the catalytic mechanism and a strategy to generate a transition state analogue was coupled with the ability to cocrystallize candidate inhibitors in complex with PR. This facilitated structure-based drug design that enabled consecutive rounds of lead optimization to develop inhibitors, which competitively bind the active site with affinities to purified PR in the low nanomolar to low picomolar range. The rational design strategy was also used to develop inhibitors that aim to combat problems encountered in the clinic, including poor bioavailability, aberrant side effects, and drug resistance.

3.3. Clinical Application and Resistance. Clinical use of PIs began in 1995 with the FDA approval of saquinavir [8]. Saquinavir's approval was closely followed by ritonavir and indinavir in 1996 and nelfinavir in 1997 [8]. In vitro studies demonstrated that all of these "first generation" PIs inhibit HIV-1 replication in the nanomolar range in a selection of cell types relevant to HIV-1 infection [24–27]. Initial clinical trials with these drugs were conducted as monotherapy and encouragingly demonstrated declines in HIV-1 RNA levels although the antiviral effect was not sustained for long periods of time due to the rapid acquisition of drug resistance [28–33]. Improved and more sustained reductions in viral RNA levels along with increased CD4 cell counts were obtained when a PI was included in triple therapy combinations with two NRTIs [34–40]. Importantly, these triple-drug regimens (HAART) significantly reduced disease progression and mortality in HIV/AIDS patients [8, 38]. Therefore, the development of PIs facilitated a pivotal step forward in the clinical management of HIV/AIDS and dramatically improved the clinical outcome of the disease.

Despite the successes, antiviral suppression was not always durable and these early clinical trials highlighted a number of key problems associated with PIs. As indicated above, drug resistance was problematic from the outset and the complex mechanisms of resistance will be discussed in more detail below. Acquisition of drug resistance was compounded by problems with adverse side effects (abnormal lipid and glucose metabolism) and low bioavailability (typical of peptide-like molecules), which led to suboptimal drug concentrations, high pill burdens, and difficulties with patient adherence to treatment regimens [8]. A notable observation to help overcome the pharmacological problems was that ritonavir acts as a potent inhibitor of the cytochrome P450 3A4 metabolic pathway [41]. As a consequence it has been demonstrated that coadministering a low nontherapeutic dose of ritonavir with other PIs leads to dramatically improved bioavailability, half-life, and potency of these PIs [41]. Ritonavir boosting has become a standard procedure when using most PIs in the clinic.

The next generation of PIs aimed to improve upon the problems highlighted above. The first was amprenavir, which was approved for clinical use in 1999, next came lopinavir which was approved in 2000, followed by atazanavir in 2003, tipranavir in 2005, and lastly darunavir in 2006 [8]. In 2003 amprenavir was subsequently reformulated as the prodrug fosamprenavir, which improved drug plasma concentrations and afforded a lower pill burden [42]. Reduction in pill burden was also achieved by the coformulation of lopinavir with a low dose of ritonavir and further progress in the simplification of drug regimens came with atazanavir, which was the first PI with a once daily dosing regimen. Drug potency has also been improved, in vitro studies have shown atazanavir and darunavir to be particularly potent with IC50 values of between 1 and 5 nM however, tipranavir is the least potent because of its novel nonpeptidomimetic chemical structure [43–47]. Clinical trials demonstrated that the next generation PIs performed well with superior virological efficacy when tested against a placebo or another comparator PI in a background of two NRTIs [8, 48–52]. Finally many of these PIs acquire different drug resistance mutation profiles from the earlier PIs and/or have a higher genetic barrier to resistance.

Resistance has been encountered for all nine PIs and has been extensively reviewed [8, 53–55]. A current summary of the key mutations acquired by each of them is provided in a data review of HIV-1 drug resistance by the International AIDS Society-USA [15]. The genetic barrier for acquisition of PI resistance is relatively high, that is, it requires two or more amino acid changes to confer significant resistance. This is because PI drug resistance is a stepwise pathway that results in complex interdependent combinations of multiple mutations. All of these interdependent changes are required to act in synergy to confer drug resistance whilst simultaneously maintaining the fitness of the virus.

The mutations that arise first are referred to as primary or major mutations and they are usually located in the PR substrate-binding cleft or its immediate vicinity. Examples of primary mutations include D30N, G48V, I50L/V, V82A/F/L/S/T, I84V, and L90M [15]. These primary mutations are principally responsible for acquisition of drug resistance by causing conformational changes in and around the active site that prevent inhibitor binding [53]. More specifically, PIs bound to the substrate-binding cleft occupy a similar space as the substrate envelope, but atoms of the PI protrude from this space and interact with residues in PR. Therefore, it has been proposed that drug resistance mutations arise at PR residues involved in these points of contact to inhibit PI binding [56]. In addition to the

direct mechanism of resistance described above resistance-conferring mutations may also result in conformational changes to PR beyond the active site and nonactive site mutations can also contribute to drug resistance [53]. Recently rare amino acid insertions, particularly between residues 32 and 42 have been observed to occur more frequently and in correlation with the introduction of atazanavir, lopinavir, amprenavir, and tipranavir into the clinic. The insertions have been proposed to be associated with PI resistance by imposing minor structural changes to the PR flap and substrate-binding cleft, although they always appear in combination with other well-described PI resistance mutations [57].

Primary mutations are accompanied by secondary or minor mutations, which can be preexisting polymorphisms or acquired after primary mutations. The function of many of these secondary mutations is often not to confer drug resistance *per se* but instead to compensate for the effect of primary mutations, which reduce protease catalytic efficiency and virus replication capacity or fitness [58–62]. Despite their function being less drug specific in action, they are however critical for development of high-level resistance. The secondary mutations are generally located at residues distal from the active site and occur at more than 20 residues of PR [15]. Unlike the primary mutations, which generally occur at highly conserved residues, the secondary mutations, are often polymorphic in PI treatment-naïve patient isolates, a well-documented example is the L63P substitution [58], and thus favour the selection of primary mutations in the presence of drug. Despite the presence of multiple secondary mutations almost all clinical strains of HIV-1 with high-level PI drug resistance display some degree of fitness loss [58, 61, 63, 64].

Resistance to PIs is a compromise between resistance and PR enzyme function. The mutations in PR described above primarily have an impact on inhibitor binding while still allowing the enzyme to recognise and cleave its Gag and Gag-Pol substrates to some degree. In addition to the changes in PR itself, amino acid changes in the Gag substrate have also been described [54]. These mutations are primarily located at or near to Gag cleavage sites and more specifically the sites in the NC-SP2-p6 region of Gag. Key mutations observed at the NC-SP2 cleavage site are A431V and I437V, which are commonly found in association with the PR primary mutation V82A and key mutations observed at the SP2-p6 cleavage site are L449F and P453L, which are commonly found in association with the PR primary mutations I50V and I84V [65–72]. *In vitro* selection experiments have also shown that mutations at the NC-SP2 cleavage site (A431V, K436E, and/or I437V/T) can also be selected in the presence of PIs without any accompanying resistance mutations in PR [73]. Mutations in Gag located at positions distal to cleavage sites have also been documented [74–76]. The impact of mutations in Gag has been attributed to (i) acting as compensatory mutations that improve fitness defects imposed by PI resistance-conferring mutations in PR and (ii) directly contributing to PI resistance [65–67, 73, 77]. The mechanism by which Gag cleavage-site mutations compensate for a loss in viral fitness is by improving the interaction between the substrate and the mutant enzyme and hence increasing the ability of the mutant PR to cleave [78]. Noncleavage site mutations are thought to improve fitness by causing more broad conformational changes in Gag making cleavage sites more accessible to PR [74–76]. The mechanism by which Gag cleavage-site mutations directly contribute to PI resistance is however not clearly understood [54].

Despite the complex interdependent combinations of multiple mutations in both PR and its Gag substrate that are required to attain high-level PI drug resistance, many of the PIs have a distinctive primary mutation that can be considered a signatory of drug resistance to that particular PI. For example the D30N mutation is a signatory of nelfinavir resistance, I50L is a signatory of atazanavir resistance, the I50V mutation is a signatory of amprenavir and darunavir resistance, and the G48V mutation is a signatory of saquinavir resistance [15]. Unfortunately, however, many mutations confer drug resistance to multiple PIs leading to broad cross-resistance amongst most PIs [15]. For example, the I84V mutation is the most important as it affects all eight PIs in clinical use and acts as a key mutation for five of them (atazanavir, darunavir, fosamprenavir, indinavir, and tipranavir). Mutations at residue 82 affect all of the PIs except darunavir. The I54V substitution acts as a key mutation for darunavir but it also affects all the other PIs with the exception of nelfinavir and the L90M mutation affects PIs with the exception of darunavir and tipranavir. Cross-resistance is likely due to the fact that although chemically different, most of the PIs were designed using the same basic principle and have similar structures and interactions with the PR substrate-binding cleft. Extensive cross-resistance has serious clinical consequences that threatens the usefulness of PIs and drives an ongoing need for new PIs with improved resistance profiles.

3.4. Conclusion. The introduction of PIs into the clinic more than 15 years ago heralded the era of HAART and resulted in a significant reduction in morbidity and mortality among HIV-infected patients. Due to their clinical potency, PIs are still commonly used in treatment regimens, although only three (lopinavir, atazanavir and darunavir) of the nine approved PIs are in widespread use. Despite the clinical benefits, the usefulness of first generation PIs was particularly hampered by toxic side effects and low bioavailability, which resulted in high pill burdens and low patient adherence. A significant advance in resolving these issues was the introduction of low-dose ritonavir boosting, which increases plasma PI levels by inhibiting the cytochrome P450 metabolic pathway. Ritonavir-boosting is itself; however, associated with toxicity; therefore, alternative boosting compounds with improved properties are being developed.

PI drug resistance is a major cause of therapy failure despite the relatively high genetic barrier to resistance. Unfortunately, PR has proven to be a highly flexible and adaptable drug target due to diverse mutational profiles and the complex interplay between PR and its Gag substrate.

Extensive cross-resistance to PIs has also been a key problem that has limited the overall usefulness of the drug class despite the development of new inhibitors such as darunavir with favourable resistance profiles. Therefore, there is a need to develop further novel inhibitors with improved resistance profiles to address these ongoing issues [79]. One strategy to develop such new PIs is to build on the design of existing inhibitors that target the PR active site by introducing novel modifications to established PI chemical entities. One such example is the novel inhibitor GS-8374, which is a modification of a darunavir-like analogue [80]. GS-8374 has been shown to be highly potent with a resistance profile superior to all clinically approved PIs including the parent molecule darunavir [80]. A second strategy is to identify molecules with novel chemical scaffolds, for example PPL-100 is a nonpeptidomimetic inhibitor that incorporates a new lysine-based scaffold and binds the flap region of PR via a novel mechanism [81]. PPL-100 has been shown to have a favourable resistance profile against known PI resistant HIV-1 isolates and its *in vitro* selection pattern results in two previously undocumented mutations T80I and P81S together with two previously reported compensatory mutations K45R and M46I [81, 82]. Allosteric inhibitors that bind a site other than the PR active site via a noncompetitive mechanism of action have also been identified and shown to be effective against both wildtype and PI resistant purified PR [83]. A further novel strategy, discussed below, is to design inhibitors that prevent proteolytic maturation by targeting the Gag substrate rather than the PR enzyme itself.

4. Maturation Inhibitors

4.1. Introduction. PIs directly target the PR enzyme; however, an alternative approach to inhibiting HIV-1 proteolytic maturation is to identify small molecules that bind its Gag substrate and specifically block individual cleavage events. Such a strategy would be successful because accurate proteolytic processing of Gag is essential for the production of infectious particles as mutations that disrupt the cleavage of individual sites or alter the order in which sites are cleaved result in aberrant particles that have significantly reduced infectivity. Molecules with this mechanism of action have been termed maturation inhibitors and the first-in-class is 3-O-(3′,3′-dimethylsuccinyl)betulinic acid (DSB), also known as PA-457, MPC-4326, or bevirimat (BVM).

4.2. Mechanism of Action. BVM specifically inhibits CA-SP1 cleavage, which occurs late in the Gag proteolytic cleavage cascade [84, 85]. This has been demonstrated by a number of key observations: (i) biochemical studies have demonstrated an accumulation of the uncleaved CA-SP1 intermediate in both cell and virus-associated protein fractions from HIV-1 expressing cells treated with BVM [84–86] (Figure 1(c)); (ii) viruses such as HIV-2 and SIV which have a divergent sequence at the CA-SP1 junction are not sensitive to BVM [87]; (iii) the majority of BVM drug-resistance conferring mutations map to the CA-SP1 junction or within SP1 [84–86, 88–95]. A second molecule PF-46396 has been identified that also inhibits CA-SP1 cleavage [96]. Interestingly, although PF-46396 has a similar mechanism of action as BVM, it belongs to a distinct chemical class as it is a pyridone-based compound not a betulinic acid derivative like BVM [96].

The consequence of BVM blocking SP1 cleavage from the C-terminus of CA is the formation of noninfectious particles with an aberrant morphology [84] (Figure 1(b)). Three-dimensional (3D) imaging of BVM-treated particles by cryoelectron tomography showed that they contain an incomplete protein shell, which has a hexagonal honeycomb lattice in the CA layer that is similar in structure to the Gag lattice of immature virus particles [13]. This partial shell is consistent with the aberrant electron dense crescent inside the viral membrane observed in BVM-treated particles by conventional thin-sectioning electron microscopy [84]. Both imaging techniques also showed most BVM-treated particles to contain an acentric mass, which represents an abnormal core-like structure [13, 84]. The general morphological features of BVM-treated particles are shared by particles generated by the CA5 mutant, which has two amino acid substitutions that completely block CA-SP1 cleavage [84, 97]. However, these particles have a thinner CA layer with no visible evidence of honeycomb lattice organization [13]. The presence of structural organization in the BVM-treated but not the CA5 CA layer suggests that BVM binding stabilizes the immature lattice as well as blocking CA-SP1 cleavage and that both modes of action may potentially contribute to the generation of non-infectious particles [13].

The assembly state of Gag is a determinant of BVMs activity. BVM does not inhibit CA-SP1 processing in the context of monomeric Gag in solution [84], but instead requires Gag assembly for its activity [84, 98, 99]. Therefore, it can be hypothesized that BVM binds to a pocket formed during Gag-Gag multimerization. Conversely, Gag processing disrupts the putative binding site because BVM has been shown to bind immature but not mature HIV-1 particles [99]. The BVM binding site has been mapped to the CA-SP1 junction within immature virus particles using photoaffinity BVM analogues and mass spectroscopy [100]. This provides the first direct evidence that the BVM binding site spans the CA-SP1 junction and is consistent with previous biochemical and genetic data that have implicated this region of Gag in BVM binding. Indeed, BVM binding is disrupted in a selection of BVM-drug resistant mutations with amino acid substitutions that map to the CA-SP1 junction [100, 101]. Positioning of BVM across the CA-SP1 junction supports a mechanism of action whereby binding blocks access of the viral PR to the CA-SP1 cleavage site. A second related hypothesis is that BVM binding alters the conformation, exposure, or flexibility of this region such that PR cleaves it less efficiently. The binding study [100] also identified a second BVM binding site in the major homology region (MHR) of CA, a region of Gag known to function in virus assembly [17]. The significance of a potential second BVM binding site has yet to be established but may provide an explanation for the observation that at high concentrations BVM inhibits virus particle assembly [102].

The structure of the BVM binding site remains unknown because this region of Gag has been disordered in X-ray crystallographic studies [103, 104]. The disorder has been attributed to a structural flexibility, which permits higher-order Gag-Gag multimerization during virus particle assembly [105–110]. It is, however, generally accepted that the CA-SP1 region of Gag adopts a α-helical conformation. The evidence for a helical structure is based on (i) secondary structure computer modelling predictions [111], (ii) genetic data demonstrating that mutation of key residues predicted to be helix breakers results in a disruption of virus particle assembly [106, 111] and (iii) biophysical and NMR techniques that have shown the CA-SP1 region to have a propensity to adopt a helical conformation under certain environmental conditions [110, 112, 113]. Although the interactions formed by the proposed CA-SP1 junction helices in the Gag lattice are not known, a cryoelectron tomography study of immature particles led to the hypothesis that the CA-SP1 region exists as a six-helix bundle that lies directly below the hexagonal honeycomb CA lattice [114]. Because BVM activity is known to require higher-order Gag-Gag multimerization, it has been suggested that the BVM binding pocket might involve more than one helix and hence bound BVM may occupy a cleft formed between helices [100]. The considerable technical challenges of obtaining high-resolution structural information of the CA-SP1 junction in the context of higher-order multimerized Gag make the prospect of rational drug design using inhibitor cocomplexes not currently possible. However, further understanding of the interactions involved is important for the development of second-generation maturation inhibitors. Such new molecules are now required as clinical development of BVM was suspended in 2010 due to problems with intrinsic BVM drug resistance in HIV-1 infected patients during phase II clinical trials.

4.3. Clinical Development and Resistance. BVM was considered an attractive candidate for clinical development because of its potent in vitro activity with a mean IC50 value of 10 nM and its novel mechanism of action, which makes it equally effective against viruses that have acquired resistance to key antiretroviral drugs in clinical use [84]. Additional attributes including promising pharmacological and safety studies in animal models and phase I clinical trials [115] led to the testing of BVM in HIV-1 infected patients. Initial success in these phase II clinical trials demonstrated significant BVM dose-dependent viral load reductions [115]. However, further studies quickly showed that approximately 50% of BVM-treated patients did not effectively respond to the drug and exhibited viral load reductions of less than 0.5 log [93]. Failure to respond was not due to suboptimal BVM plasma concentrations but has been attributed to virological parameters instead.

Examination of patient-derived virus revealed amino acid assignment at SP1 residues 6, 7, and 8 (Gag positions 369, 370, and 371) is associated with response to BVM [92, 93, 95] (Figure 1(d)). This trio of residues map to the C-terminal half of SP1, which is relatively nonconserved but commonly encodes a QVT (glutamine-valine-threonine) motif in clade B HIV-1 isolates [95]. Patients most likely to respond to BVM are infected with virus encoding the QVT motif, while patients infected with virus encoding polymorphisms at SP1 residues 6–8 are less likely to respond [93]. Studies to investigate the contribution of individual substitutions at SP1 residues 6–8 have shown that mutations at SP1 residue 7 and 8 (e.g., SP1-V7A, -V7M, -T8Δ, -T8N) all confer varying degrees of reduced susceptibility to BVM [88, 92, 94, 95]. Most notably, a critical role for BVM resistance has been attributed to the SP1-V7A polymorphism as it confers full resistance to BVM [88, 92, 94, 95]. BVM susceptibility was not however reduced by mutations at SP1 residue 6 (e.g., SP1-Q6H, Q6A) or the SP1-T8A polymorphism [88, 92, 94, 95]. Therefore, any contribution of these substitutions to reduced BVM susceptibility maybe dependent on the synergistic effects of a combinations of different polymorphisms, i.e. the context of the wider Gag background. Indeed, one study identified five patient-derived virus samples with significantly reduced BVM susceptibility *in vitro* but still encoded the QVT motif [92]. In two of these isolates, BVM resistance has been demonstrated to be conferred by a polymorphism in CA (CA-V230I) situated at the P2 position of the CA-SP1 cleavage site [92] (Figure 1(d)). In the other three isolates, the determinants of reduced BVM susceptibility were not resolved [92], indicating that in some instances the factors conferring BVM susceptibility are likely to be more complex than the parameters that have been established to date.

The CA-V230I and SP1-V7A substitutions have also been acquired in in vitro BVM drug-resistance selection experiments [88, 90, 91]. In vitro studies have also identified a panel of other BVM-resistance mutations (CA-H226Y, CA-L231M, CA-L231F, SP1-A1V, SP1-A3V, and SP1-A3T) [84–86, 90] (Figure 1(d)). Unlike, the clinically important innate polymorphisms discussed above, these *in vitro* selected BVM-resistance mutations map to residues in the vicinity of the CA-SP1 cleavage site that are highly conserved throughout HIV-1 isolates [86]. As a likely consequence, these mutations have not been observed in most patient-derived virus samples either with [93, 95] or without BVM treatment [92, 95]. However, it should be noted that the most frequently acquired mutation SP1-A1V has been shown not to impose a significant defect on virus replication *in vitro* [86, 89, 90] and replicates efficiently in SCID-hu Thy/Liv mice [116]. Therefore, it remains a hypothetical possibility that the SP1-A1V mutation could be acquired over time in patients that initially respond well to BVM treatment.

Initial failure to select the key BVM-resistance conferring polymorphisms *in vitro* has been attributed to the experimental conditions utilized [91]; however, later experiments did result in selection of some of the key polymorphic mutations albeit at low frequency [88, 90]. Nevertheless, a recent study used a more sophisticated *in vitro* method of serial passage of quasi-species containing recombinant HIV-1 and deep sequencing that more accurately mimicked *in vitro* the selection of BVM-resistance observed *in vivo* [91]. In hindsight use of this *in vitro* selection method or more extensive testing of the spectrum of activity

across a diverse panel of clinical isolates may have more accurately predicted the clinical response to BVM and either led to discontinuation of BVM development at an earlier stage thereby avoiding costly clinical studies or alternatively steered BVM's clinical development to include a genotyping test to screen for preexisting key polymorphisms to enable prior identification of patients most likely to effectively respond to BVM treatment [91].

The clinically important polymorphisms preexist in the HIV-1 population without prior BVM treatment. This intrinsic resistance has caused problems for BVM's clinical development, which was consequently discontinued in 2010. Genotypic analysis has demonstrated a high prevalence of polymorphisms at the QVT motif and their frequency is dependent on the genetic clade of HIV-1 [94, 95, 117]. In clade B viruses, which are predominant in the US and Europe, polymorphism frequency at the QVT motif has been reported to occur at a rate of ~30–60% [91, 95, 117]. This genotypic analysis matches BVM susceptibility rates in the *in vitro* phenotypic and clinical trial studies discussed above [92, 93, 95]. In nonclade B viruses, QVT polymorphism rates are much higher with rates of >90% [95, 117]. Typically polymorphisms occur most frequently at SP1 residue 7, followed by residue 8, and then residue 6 [91, 94, 95]. The critical SP1-V7A polymorphism has been shown to be largely predominant and occurs at a frequency of ~16% in clade B viruses and ~65–70% in clade C viruses, which are mostly found in Southern Africa [94, 95]. The high frequency of the SP1-V7A polymorphism combined with its known capacity to confer full resistance to BVM therefore poses the biggest threat to the potential effectiveness and clinical development of BVM.

The prevalence of the key polymorphisms in relation to HAART and the presence of PI resistance mutations has been investigated due to the complex interplay between PR and its Gag substrate. Being a new class of antiretroviral drug BVM was most likely in the first instance to be used as salvage therapy for patients harbouring multidrug resistant HIV-1 isolates. Studies have shown no association between the prevalence of key QVT polymorphisms and HAART treatment experience but in the absence of BVM [91, 92, 117]. One study also reported no association between prevalence of QVT polymorphisms and PI resistance-conferring mutations [92]; however, two other studies with bigger sample sizes demonstrated a higher frequency of BVM resistance mutations in PI resistant patient isolates [117, 118]. The effect of PI resistance on acquisition of BVM resistance *in vitro* has also been investigated [89, 90]. These two studies made different conclusions about the impact of the PI mutations on the temporal acquisition of BVM-resistance conferring mutations, with one study reporting a delay in the emergence of BVM-resistance [89]. The reported differences may be dependent on the type of PI mutations or the study systems used. Interestingly, the other study [90] demonstrated that the PR background influenced the type and diversity of BVM resistance conferring mutations. Viruses with a wildtype PR predominantly acquired the SP1-A1V mutation, whereas viruses with a PI resistance PR acquired a significantly higher prevalence of mutations at the QVT motif (SP1-V7A, V7N, and SP1-T8N), at the polymorphic CA-230 residue (CA-V230I) and also a previously unreported mutation SP1-S5N. The PR genetic background was also found to effect BVM susceptibility and virus replication capacity [90]. While these studies have not fully resolved the complex interplay between PR, the Gag substrate and susceptibility to BVM they clearly demonstrate that this parameter should be considered in future development of maturation inhibitors.

4.4. Conclusion. Maturation inhibitors are a novel mechanistic class of antiretroviral drug that target PR cleavage sites in Gag. BVM is the first-in-class maturation inhibitor, which specifically inhibits cleavage of SP1 from the C-terminus of CA. A number of other small molecules that target Gag have also been identified. PF-46396 is a second maturation inhibitor, which also inhibits CA-SP1 cleavage but is chemically distinct from BVM. There are also a small number of molecules that target CA and inhibit assembly of the immature particle and/or the CA core [12]. BVM is however the only molecule that targets Gag, which has been tested in clinical trials. BVM was considered a good candidate for clinical development because of its *in vitro* potency, novel mechanism of action, and good safety profile in animal models and phase I clinical trials. Although initial results of BVM efficacy in HIV-1 infected patients were encouraging, it was quickly established that approximately 50% of patients do not effectively respond to the drug. Failure to respond is due to virological parameters, more specifically, intrinsic polymorphisms primarily located at SP1 residues 6, 7, and 8. These polymorphisms have a high prevalence, particularly in non-clade B HIV-1 isolates. The existence of BVM-resistance conferring polymorphisms in BVM-treatment naïve patients severely limits the clinical usefulness of BVM and consequently clinical development of BVM was suspended in 2010.

Halted clinical development of BVM necessitates the need for a second-generation maturation inhibitor to overcome the problem of intrinsic drug resistance encountered by BVM. BVM targets an as yet undefined drug-binding pocket, which is hypothesized to be created upon higher-order multimerization of Gag during virus particle assembly. The significant technical challenge of obtaining high-resolution structural information of this hypothetical drug target makes rational structure-based drug design unfeasible at the current time. However, the need to develop improved maturation inhibitors has highlighted a need to further our understanding of the CA-SP1 region of Gag and its role in HIV-1 particle assembly. BVM, PF-46396, and their analogues can be utilized as tools to further explore drug-binding requirements to inform future strategies to improve drug resistance profiles. Development of BVM has provided evidence that small molecules to inhibit HIV-1 replication can target Gag cleavage sites. Four other cleavage sites are present in Gag and a genetic study predicted that a small molecule that blocks MA-CA cleavage maybe a particularly potent inhibitor of HIV-1 replication [119]. However, the intrinsic flexibility in Gag cleavage sites and wide variation

in substrate sequence recognition by HIV PR may represent insurmountable problems for the future development of maturation inhibitors.

References

[1] "UNAIDS World AIDS Day Report," 2011.

[2] L. F. Chen, J. Hoy, and S. R. Lewin, "Ten years of highly active antiretroviral therapy for HIV infection," *Medical Journal of Australia*, vol. 186, no. 3, pp. 146–151, 2007.

[3] V. Simon, D. D. Ho, and Q. Abdool Karim, "HIV/AIDS epidemiology, pathogenesis, prevention, and treatment," *Lancet*, vol. 368, no. 9534, pp. 489–504, 2006.

[4] Z. Temesgen, F. Cainelli, E. M. Poeschla, S. A. Vlahakis, and S. Vento, "Approach to salvage antiretroviral therapy in heavily antiretroviral-experienced HIV-positive adults," *Lancet Infectious Diseases*, vol. 6, no. 8, pp. 496–507, 2006.

[5] W.-S. Hu, T. Rhodes, Q. Dang, and V. Pathak, "Retroviral recombination: review of genetic analyses," *Frontiers in Bioscience*, vol. 8, pp. d143–d155, 2003.

[6] E. S. Svarovskaia, S. R. Cheslock, W. H. Zhang, W. S. Hu, and V. K. Pathak, "Retroviral mutation rates and reverse transcriptase fidelity," *Frontiers in Bioscience*, vol. 8, pp. d117–d134, 2003.

[7] T. Cihlar and A. S. Ray, "Nucleoside and nucleotide HIV reverse transcriptase inhibitors: 25 years after zidovudine," *Antiviral Research*, vol. 85, no. 1, pp. 39–58, 2010.

[8] A. M. J. Wensing, N. M. van Maarseveen, and M. Nijhuis, "Fifteen years of HIV protease inhibitors: raising the barrier to resistance," *Antiviral Research*, vol. 85, no. 1, pp. 59–74, 2010.

[9] M. P. de Béthune, "Non-nucleoside reverse transcriptase inhibitors (NNRTIs), their discovery, development, and use in the treatment of HIV-1 infection: a review of the last 20 years (1989–2009)," *Antiviral Research*, vol. 85, no. 1, pp. 75–90, 2010.

[10] D. J. McColl and X. Chen, "Strand transfer inhibitors of HIV-1 integrase: bringing IN a new era of antiretroviral therapy," *Antiviral Research*, vol. 85, no. 1, pp. 101–118, 2010.

[11] J. C. Tilton and R. W. Doms, "Entry inhibitors in the treatment of HIV-1 infection," *Antiviral Research*, vol. 85, no. 1, pp. 91–100, 2010.

[12] C. S. Adamson and E. O. Freed, "Novel approaches to inhibiting HIV-1 replication," *Antiviral Research*, vol. 85, no. 1, pp. 119–141, 2010.

[13] P. W. Keller, C. S. Adamson, J. Bernard Heymann, E. O. Freed, and A. C. Steven, "HIV-1 maturation inhibitor bevirimat stabilizes the immature gag lattice," *Journal of Virology*, vol. 85, no. 4, pp. 1420–1428, 2011.

[14] A. Wlodawer and J. W. Erickson, "Structure-based inhibitors of HIV-1 protease," *Annual Review of Biochemistry*, vol. 62, pp. 543–585, 1993.

[15] V. A. Johnson, V. Calvez, H. F. Günthard et al., "2011 update of the drug resistance mutations in HIV-1," *Topics in Antiviral Medicine*, vol. 19, no. 4, pp. 156–164, 2011.

[16] R. Swanstrom and J. W. Willis, "Synthesis, assembly and processing of viral proteins," in *Retroviruses*, J. M. Coffin, S. H. Hughes, and H. E. Varmus, Eds., Cold Spring Harbor Laboratory Press, 1997.

[17] C. S. Adamson and E. O. Freed, "HIV-1 assembly, release and maturation," in *Advances in Pharmacolgy, HIV-1: Molecular*

Biology and Pathogenesis: Viral Mechansims, K.-T. Jeang, Ed., Elsevier, 2007.

[18] B. K. Ganser-Pornillos, M. Yeager, and W. I. Sundquist, "The structural biology of HIV assembly," *Current Opinion in Structural Biology*, vol. 18, no. 2, pp. 203–217, 2008.

[19] M. A. Navia, P. M. D. Fitzgerald, B. M. McKeever et al., "Three-dimensional structure of aspartyl protease from human immunodeficiency virus HIV-1," *Nature*, vol. 337, no. 6208, pp. 615–620, 1989.

[20] R. Lapatto, T. Blundell, A. Hemmings et al., "X-ray analysis of HIV-1 proteinase at 2.7 Å resolution confirms structural homology among retroviral enzymes," *Nature*, vol. 342, no. 6247, pp. 299–302, 1989.

[21] A. Wlodawer, M. Miller, M. Jaskolski et al., "Conserved folding in retroviral proteases: crystal structure of a synthetic HIV-1 protease," *Science*, vol. 245, no. 4918, pp. 616–621, 1989.

[22] M. Prabu-Jeyabalan, E. Nalivaika, and C. A. Schiffer, "Substrate shape determines specificity of recognition for HIV-1 protease: analysis of crystal structures of six substrate complexes," *Structure*, vol. 10, no. 3, pp. 369–381, 2002.

[23] J. Anderson, "Viral protease inhibitors," *Handbook of Experimental Pharmacology*, vol. 189, pp. 85–110, 2009.

[24] J. C. Craig, I. B. Duncan, D. Hockley, C. Grief, N. A. Roberts, and J. S. Mills, "Antiviral properties of Ro 31-8959, an inhibitor of human immunodeficiency virus (HIV) proteinase," *Antiviral Research*, vol. 16, no. 4, pp. 295–305, 1991.

[25] D. J. Kempf, K. C. Marsh, J. F. Denissen et al., "ABT-538 is a potent inhibitor of human immunodeficiency virus protease and has high oral bioavailability in humans," *Proceedings of the National Academy of Sciences of the United States of America*, vol. 92, no. 7, pp. 2484–2488, 1995.

[26] J. P. Vacca, B. D. Dorsey, W. A. Schleif et al., "L-735,524: an orally bioavailable human immunodeficiency virus type 1 protease inhibitor," *Proceedings of the National Academy of Sciences of the United States of America*, vol. 91, no. 9, pp. 4096–4100, 1994.

[27] A. K. Patick, H. Mo, M. Markowitz et al., "Antiviral and resistance studies of AG1343, an orally bioavailable inhibitor of human immunodeficiency virus protease," *Antimicrobial Agents and Chemotherapy*, vol. 40, no. 2, pp. 292–297, 1996.

[28] V. S. Kitchen, C. Skinner, K. Ariyoshi et al., "Safety and activity of saquinavir in HIV infection," *Lancet*, vol. 345, no. 8955, pp. 952–955, 1995.

[29] H. Jacobsen, M. Haenggi, M. Ott et al., "Reduced sensitivity of saquinavir: an update on genotyping from phase I/II trials," *Antiviral Research*, vol. 29, no. 1, pp. 95–97, 1996.

[30] S. A. Danner, A. Carr, J. M. Leonard et al., "A short-term study of the safety, pharmacokinetics, and efficacy of ritonavir, an inhibitor of HIV-1 protease," *New England Journal of Medicine*, vol. 333, no. 23, pp. 1528–1533, 1995.

[31] M. Markowitz, M. Saag, W. G. Powderly et al., "A preliminary study of ritonavir, an inhibitor of HIV-1 protease, to treat HIV-1 infection," *New England Journal of Medicine*, vol. 333, no. 23, pp. 1534–1539, 1995.

[32] D. S. Stein, D. G. Fish, J. A. Bilello, S. L. Preston, G. L. Martineau, and G. L. Drusano, "A 24-week open-label phase I/II evaluation of the HIV protease inhibitor MK-639 (indinavir)," *AIDS*, vol. 10, no. 5, pp. 485–492, 1996.

[33] M. Markowitz, M. Conant, A. Hurley et al., "A preliminary evaluation of nelfinavir mesylate, an inhibitor of human immunodeficiency virus (HIV)-1 protease, to treat HIV infection," *Journal of Infectious Diseases*, vol. 177, no. 6, pp. 1533–1540, 1998.

[34] A. C. Collier, R. W. Coombs, D. A. Schoenfeld et al., "Treatment of human immunodeficiency virus infection with saquinavir, zidovudine, and zalcitabine," *New England Journal of Medicine*, vol. 334, no. 16, pp. 1011–1017, 1996.

[35] D. W. Notermans, S. Jurriaans, F. De Wolf et al., "Decrease of HIV-1 RNA levels in lymphoid tissue and peripheral blood during treatment with ritonavir, lamivudine and zidovudine," *AIDS*, vol. 12, no. 2, pp. 167–173, 1998.

[36] D. Mathez et al., "Reductions in viral load and increases in T lymphocyte numbers in treatment-naive patients with advanced HIV-1 infection treated with ritonavir, zidovudine and zalcitabine triple therapy," *Antiviral Therapy*, vol. 2, no. 3, pp. 175–183, 1997.

[37] R. M. Gulick, J. W. Mellors, D. Havlir et al., "Treatment with indinavir, zidovudine, and lamivudine in adults with human immunodeficiency virus infection and prior antiretroviral therapy," *New England Journal of Medicine*, vol. 337, no. 11, pp. 734–739, 1997.

[38] S. M. Hammer, K. E. Squires, M. D. Hughes et al., "A controlled trial of two nucleoside analogues plus indinavir in persons with human immunodeficiency virus infection and CD4 cell counts of 200 per cubic millimeter or less," *New England Journal of Medicine*, vol. 337, no. 11, pp. 725–733, 1997.

[39] M. Gartland, "AVANTI 3: a randomized, double-blind trial to compare the efficacy and safety of lamivudine plus zidovudine versus lamivudine plus zidovudine plus nelfinavir in HIV-1-infected antiretroviral-naive patients," *Antiviral Therapy*, vol. 6, no. 2, pp. 127–134, 2001.

[40] M. S. Saag, P. Tebas, M. Sension et al., "Randomized, double-blind comparison of two nelfinavir doses plus nucleosides in HIV-infected patients (Agouron study 511)," *AIDS*, vol. 15, no. 15, pp. 1971–1978, 2001.

[41] R. P. G. Van Heeswijk, A. I. Veldkamp, J. W. Mulder et al., "Combination of protease inhibitors for the treatment of HIV-1-infected patients: a review of pharmacokinetics and clinical experience," *Antiviral Therapy*, vol. 6, no. 4, pp. 201–229, 2001.

[42] C. Falcoz, J. M. Jenkins, C. Bye et al., "Pharmacokinetics of GW433908, a prodrug of amprenavir, in healthy male volunteers," *Journal of Clinical Pharmacology*, vol. 42, no. 8, pp. 887–898, 2002.

[43] M. H. St. Clair, J. Millard, J. Rooney et al., "In vitro antiviral activity of 141W94 (VX-478) in combination with other antiretroviral agents," *Antiviral Research*, vol. 29, no. 1, pp. 53–56, 1996.

[44] H. L. Sham, D. J. Kempf, A. Molla et al., "ABT-378, a highly potent inhibitor of the human immunodeficiency virus protease," *Antimicrobial Agents and Chemotherapy*, vol. 42, no. 12, pp. 3218–3224, 1998.

[45] B. S. Robinson, K. A. Riccardi, Y. F. Gong et al., "BMS-232632, a highly potent human immunodeficiency virus protease inhibitor that can be used in combination with other available antiretroviral agents," *Antimicrobial Agents and Chemotherapy*, vol. 44, no. 8, pp. 2093–2099, 2000.

[46] S. M. Poppe, D. E. Slade, K. T. Chong et al., "Antiviral activity of the dihydropyrone PNU-140690, a new nonpeptidic human immunodeficiency virus protease inhibitor,"

Antimicrobial Agents and Chemotherapy, vol. 41, no. 5, pp. 1058–1063, 1997.

[47] Y. Koh, H. Nakata, K. Maeda et al., "Novel bis-tetrahydrofuranylurethane-containing nonpeptidic protease inhibitor (PI) UIC-94017 (TMC114) with potent activity against multi-PI-resistant human immunodeficiency virus in vitro," *Antimicrobial Agents and Chemotherapy*, vol. 47, no. 10, pp. 3123–3129, 2003.

[48] A. Rodriguez-French, J. Boghossian, G. E. Gray et al., "The NEAT Study: a 48-week open-label study to compare the antiviral efficacy and safety of GW433908 versus nelfinavir in antiretroviral therapy-naive HIV-1-infected patients," *Journal of Acquired Immune Deficiency Syndromes*, vol. 35, no. 1, pp. 22–32, 2004.

[49] S. Walmsley, B. Bernstein, M. King et al., "Lopinavir-ritonavir versus nelfinavir for the initial treatment of HIV infection," *New England Journal of Medicine*, vol. 346, no. 26, pp. 2039–2046, 2002.

[50] R. L. Murphy, I. Sanne, P. Cahn et al., "Dose-ranging, randomized, clinical trial of atazanavir with lamivudine and stavudine in antiretroviral-naive subjects: 48-week results," *AIDS*, vol. 17, no. 18, pp. 2603–2614, 2003.

[51] C. B. Hicks, P. Cahn, D. A. Cooper et al., "Durable efficacy of tipranavir-ritonavir in combination with an optimised background regimen of antiretroviral drugs for treatment-experienced HIV-1-infected patients at 48 weeks in the Randomized Evaluation of Strategic Intervention in multi-drug reSistant patients with Tipranavir (RESIST) studies: an analysis of combined data from two randomised open-label trials," *Lancet*, vol. 368, no. 9534, pp. 466–475, 2006.

[52] A. M. Mills, M. Nelson, D. Jayaweera et al., "Once-daily darunavir/ritonavir vs. lopinavir/ritonavir in treatment-naive, HIV-1-infected patients: 96-week analysis," *AIDS*, vol. 23, no. 13, pp. 1679–1688, 2009.

[53] A. Ali, R. M. Bandaranayake, Y. Cai et al., "Molecular basis for drug resistance in HIV-1 protease," *Viruses*, vol. 2, no. 11, pp. 2509–2535, 2010.

[54] F. Clavel and F. Mammano, "Role of gag in HIV resistance to protease inhibitors," *Viruses*, vol. 2, no. 7, pp. 1411–1426, 2010.

[55] J. L. Martinez-Cajas and M. A. Wainberg, "Protease inhibitor resistance in HIV-infected patients: molecular and clinical perspectives," *Antiviral Research*, vol. 76, no. 3, pp. 203–221, 2007.

[56] N. M. King, M. Prabu-Jeyabalan, E. A. Nalivaika, and C. A. Schiffer, "Combating susceptibility to drug resistance: lessons from HIV-1 protease," *Chemistry and Biology*, vol. 11, no. 10, pp. 1333–1338, 2004.

[57] M. Kožíšek, K. G. Šašková, P. Řezáčová et al., "Ninety-nine is not enough: molecular characterization of inhibitor-resistant human immunodeficiency virus type 1 protease mutants with insertions in the flap region," *Journal of Virology*, vol. 82, no. 12, pp. 5869–5878, 2008.

[58] J. Martinez-Picado, A. V. Savara, L. Sutton, and R. T. D'Aquila, "Replicative fitness of protease inhibitor-resistant mutants of human immunodeficiency virus type 1," *Journal of Virology*, vol. 73, no. 5, pp. 3744–3752, 1999.

[59] M. Nijhuis, R. Schuurman, D. De Jong et al., "Increased fitness of drug resistant HIV-1 protease as a result of acquisition of compensatory mutations during suboptimal therapy," *AIDS*, vol. 13, no. 17, pp. 2349–2359, 1999.

[60] G. Croteau, L. Doyon, D. Thibeault, G. Mckercher, L. Pilote, and D. Lamarre, "Impaired fitness of human immunodeficiency virus type 1 variants with high-level resistance to protease inhibitors," *Journal of Virology*, vol. 71, no. 2, pp. 1089–1096, 1997.

[61] F. Mammano, V. Trouplin, V. Zennou, and F. Clavel, "Retracing the evolutionary pathways of human immunodeficiency virus type 1 resistance to protease inhibitors: virus fitness in the absence and in the presence of drug," *Journal of Virology*, vol. 74, no. 18, pp. 8524–8531, 2000.

[62] L. Menéndez-Arias, M. A. Martínez, M. E. Quiñones-Mateu, and J. Martinez-Picado, "Fitness variations and their impact on the evolution of antiretroviral drug resistance," *Current Drug Targets-Infectious Disorders*, vol. 3, no. 4, pp. 355–371, 2003.

[63] J. D. Barbour, T. Wrin, R. M. Grant et al., "Evolution of phenotypic drug susceptibility and viral replication capacity during long-term virologic failure of protease inhibitor therapy in human immunodeficiency virus-infected adults," *Journal of Virology*, vol. 76, no. 21, pp. 11104–11112, 2002.

[64] G. Bleiber, M. Munoz, A. Ciuffi, P. Meylan, and A. Telenti, "Individual contributions of mutant protease and reverse transcriptase to viral infectivity, replication, and protein maturation of antiretroviral drug-resistant human immunodeficiency virus type 1," *Journal of Virology*, vol. 75, no. 7, pp. 3291–3300, 2001.

[65] F. Mammano, C. Petit, and F. Clavel, "Resistance-associated loss of viral fitness in human immunodeficiency virus type 1: phenotypic analysis of protease and gag coevolution in protease inhibitor-treated patients," *Journal of Virology*, vol. 72, no. 9, pp. 7632–7637, 1998.

[66] Y. M. Zhang, H. Imamichi, T. Imamichi et al., "Drug resistance during Indinavir therapy is caused by mutations in the protease gene and in its gag substrate cleavage sites," *Journal of Virology*, vol. 71, no. 9, pp. 6662–6670, 1997.

[67] L. Doyon, G. Croteau, D. Thibeault, F. Poulin, L. Pilote, and D. Lamarre, "Second locus involved in human immunodeficiency virus type 1 resistance to protease inhibitors," *Journal of Virology*, vol. 70, no. 6, pp. 3763–3769, 1996.

[68] F. Bally, R. Martinez, S. Peters, P. Sudre, and A. Telenti, "Polymorphism of HIV type 1 Gag p7/p1 and p1/p6 cleavage sites: clinical significance and implications for resistance to protease inhibitors," *AIDS Research and Human Retroviruses*, vol. 16, no. 13, pp. 1209–1213, 2000.

[69] M. F. Maguire, R. Guinea, P. Griffin et al., "Changes in human immunodeficiency virus type 1 Gag at positions L449 and P453 are linked to I50V protease mutants in vivo and cause reduction of sensitivity to amprenavir and improved viral fitness in vitro," *Journal of Virology*, vol. 76, no. 15, pp. 7398–7406, 2002.

[70] H. C. F. Côté, Z. L. Brumme, and P. R. Harrigan, "Human immunodeficiency virus type 1 protease cleavage site mutations associated with protease inhibitor cross-resistance selected by indinavir, ritonavir, and/or saquinavir," *Journal of Virology*, vol. 75, no. 2, pp. 589–594, 2001.

[71] I. Malet, B. Roquebert, C. Dalban et al., "Association of Gag cleavage sites to protease mutations and to virological response in HIV-1 treated patients," *Journal of Infection*, vol. 54, no. 4, pp. 367–374, 2007.

[72] J. Verheyen, E. Litau, T. Sing et al., "Compensatory mutations at the HIV cleavage sites p7/p1 and p1/p6-gag in therapy-naive and therapy-experienced patients," *Antiviral Therapy*, vol. 11, no. 7, pp. 879–887, 2006.

[73] M. Nijhuis, N. M. Van Maarseveen, S. Lastere et al., "A novel substrate-based HIV-1 protease inhibitor drug resistance mechanism," *PLoS Medicine*, vol. 4, no. 1, article e36, 2007.

[74] C. M. Parry, A. Kohli, C. J. Boinett, G. J. Towers, A. L. McCormick, and D. Pillay, "Gag determinants of fitness and drug susceptibility in protease inhibitor-resistant human immunodeficiency virus type 1," *Journal of Virology*, vol. 83, no. 18, pp. 9094–9101, 2009.

[75] H. Gatanaga, Y. Suzuki, H. Tsang et al., "Amino acid substitutions in Gag protein at non-cleavage sites are indispensable for the development of a high multitude of HIV-1 resistance against protease inhibitors," *Journal of Biological Chemistry*, vol. 277, no. 8, pp. 5952–5961, 2002.

[76] L. Myint, M. Matsuda, Z. Matsuda et al., "Gag non-cleavage site mutations contribute to full recovery of viral fitness in protease inhibitor-resistant human immunodeficiency virus type 1," *Antimicrobial Agents and Chemotherapy*, vol. 48, no. 2, pp. 444–452, 2004.

[77] E. Dam, R. Quercia, B. Glass et al., "Gag mutations strongly contribute to HIV-1 resistance to protease inhibitors in highly drug-experienced patients besides compensating for fitness loss," *PLoS Pathogens*, vol. 5, no. 3, Article ID e1000345, 2009.

[78] M. Prabu-Jeyabalan, E. A. Nalivaika, N. M. King, and C. A. Schiffer, "Structural basis for coevolution of a human immunodeficiency virus type 1 nucleocapsid-p1 cleavage site with a V82A drug-resistant mutation in viral protease," *Journal of Virology*, vol. 78, no. 22, pp. 12446–12454, 2004.

[79] S. V. Gulnik and M. Eissenstat, "Approaches to the design of HIV protease inhibitors with improved resistance profiles," *Current Opinion in HIV and AIDS*, vol. 3, no. 6, pp. 633–641, 2008.

[80] C. Callebaut, K. Stray, L. Tsai et al., "In vitro characterization of GS-8374, a novel phosphonate-containing inhibitor of HIV-1 protease with a favorable resistance profile," *Antimicrobial Agents and Chemotherapy*, vol. 55, no. 4, pp. 1366–1376, 2011.

[81] M. N. L. Nalam, A. Peeters, T. H. M. Jonckers, I. Dierynck, and C. A. Schiffer, "Crystal structure of lysine sulfonamide inhibitor reveals the displacement of the conserved flap water molecule in human immunodeficiency virus type 1 protease," *Journal of Virology*, vol. 81, no. 17, pp. 9512–9518, 2007.

[82] S. Dandache, G. Sévigny, J. Yelle et al., "In vitro antiviral activity and cross-resistance profile of PL-100, a novel protease inhibitor of human immunodeficiency virus type 1," *Antimicrobial Agents and Chemotherapy*, vol. 51, no. 11, pp. 4036–4043, 2007.

[83] M. W. Chang, M. J. Giffin, R. Muller et al., "Identification of broad-based HIV-1 protease inhibitors from combinatorial libraries," *Biochemical Journal*, vol. 429, no. 3, pp. 527–532, 2010.

[84] F. Li, R. Goila-Gaur, K. Salzwedel et al., "PA-457: a potent HIV inhibitor that disrupts core condensation by targeting a late step in Gag processing," *Proceedings of the National Academy of Sciences of the United States of America*, vol. 100, no. 23, pp. 13555–13560, 2003.

[85] J. Zhou, X. Yuan, D. Dismuke et al., "Small-molecule inhibition of human immunodeficiency virus type 1 replication by specific targeting of the final step of virion maturation," *Journal of Virology*, vol. 78, no. 2, pp. 922–929, 2004.

[86] C. S. Adamson, S. D. Ablan, I. Boeras et al., "In vitro resistance to the human immunodeficiency virus type 1 maturation inhibitor PA-457 (Beviriniat)," *Journal of Virology*, vol. 80, no. 22, pp. 10957–10971, 2006.

[87] J. Zhou, C. H. Chen, and C. Aiken, "The sequence of the CA-SP1 junction accounts for the differential sensitivity of HIV-1 and SIV to the small molecule maturation inhibitor 3-O-3′, 3′-dimethylsuccinyl-betulinic acid," *Retrovirology*, vol. 1, article 15, 2004.

[88] C. S. Adamson, M. Sakalian, K. Salzwedel, and E. O. Freed, "Polymorphisms in Gag spacer peptide 1 confer varying levels of resistance to the HIV-1maturation inhibitor bevirimat," *Retrovirology*, vol. 7, article 36, 2010.

[89] C. S. Adamson, K. Waki, S. D. Ablan, K. Salzwedel, and E. O. Freed, "Impact of human immunodeficiency virus type 1 resistance to protease inhibitors on evolution of resistance to the maturation inhibitor bevirimat (PA-457)," *Journal of Virology*, vol. 83, no. 10, pp. 4884–4894, 2009.

[90] A. Fun, "HIV-1 protease inhibitor mutations affect the development of HIV-1 resistance to the maturation inhibitor bevirimat," *Retrovirology*, vol. 8, article 70, 2011.

[91] D. J. H. F. Knapp, P. R. Harrigan, A. F. Y. Poon, Z. L. Brumme, M. Brockman, and P. K. Cheung, "In vitro selection of clinically relevant bevirimat resistance mutations revealed by "deep" sequencing of serially passaged, quasispecies-containing recombinant HIV-1," *Journal of Clinical Microbiology*, vol. 49, no. 1, pp. 201–208, 2011.

[92] N. A. Margot, C. S. Gibbs, and M. D. Miller, "Phenotypic susceptibility to bevirimat in isolates from HIV-1-infected patients without prior exposure to bevirimat," *Antimicrobial Agents and Chemotherapy*, vol. 54, no. 6, pp. 2345–2353, 2010.

[93] S. McCallister, "HIV-1 Gag polymorphisms determine treatment respose to bevirimat (PA-457)," *Antiviral Therapy*, vol. 13, p. A10, 2008.

[94] W. Lu, K. Salzwedel, D. Wang et al., "A single polymorphism in HIV-1 subtype C SP1 is sufficient to confer natural resistance to the maturation inhibitor bevirimat," *Antimicrobial Agents and Chemotherapy*, vol. 55, no. 7, pp. 3324–3329, 2011.

[95] K. Van Baelen, K. Salzwedel, E. Rondelez et al., "Susceptibility of human immunodeficiency virus type 1 to the maturation inhibitor bevirimat is modulated by baseline polymorphisms in Gag spacer peptide," *Antimicrobial Agents and Chemotherapy*, vol. 53, no. 5, pp. 2185–2188, 2009.

[96] W. S. Blair, J. Cao, J. Fok-Seang et al., "New small-molecule inhibitor class targeting human immunodeficiency virus type 1 virion maturation," *Antimicrobial Agents and Chemotherapy*, vol. 53, no. 12, pp. 5080–5087, 2009.

[97] K. Wiegers, G. Rutter, H. Kottler, U. Tessmer, H. Hohenberg, and H. G. Kräusslich, "Sequential steps in human immunodeficiency virus particle maturation revealed by alterations of individual Gag polyprotein cleavage sites," *Journal of Virology*, vol. 72, no. 4, pp. 2846–2854, 1998.

[98] M. Sakalian, C. P. McMurtrey, F. J. Deeg et al., "3-O-(3′, 3′-dimethysuccinyl) betulinic acid inhibits maturation of the human immunodeficiency virus type 1 gag precursor assembled in vitro," *Journal of Virology*, vol. 80, no. 12, pp. 5716–5722, 2006.

[99] J. Zhou, L. Huang, D. L. Hachey, C. H. Chen, and C. Aiken, "Inhibition of HIV-1 maturation via drug association with

the viral Gag protein in immature HIV-1 particles," *Journal of Biological Chemistry*, vol. 280, no. 51, pp. 42149–42155, 2005.

[100] A. T. Nguyen, C. L. Feasley, K. W. Jackson et al., "The prototype HIV-1 maturation inhibitor, bevirimat, binds to the CA-SP1 cleavage site in immature Gag particles," *Retrovirology*, vol. 8, article 101, 2011.

[101] J. Zhou, H. C. Chin, and C. Aiken, "Human immunodeficiency virus type 1 resistance to the small molecule maturation inhibitor 3-O-(3′, 3′-dimethylsuccinyl)-betulinic acid is conferred by a variety of single amino acid substitutions at the CA-SP1 cleavage site in Gag," *Journal of Virology*, vol. 80, no. 24, pp. 12095–12101, 2006.

[102] S. DaFonseca, A. Blommaert, P. Coric, S. H. Saw, S. Bouaziz, and P. Boulanger, "The 3-O-(3′, 3′-dimethylsuccinyl) derivative of betulinic acid (DSB) inhibits the assembly of virus-like particles in HIV-1 Gag precursor-expressing cells," *Antiviral Therapy*, vol. 12, no. 8, pp. 1185–1203, 2007.

[103] T. R. Gamble, S. Yoo, F. F. Vajdos et al., "Structure of the carboxyl-terminal dimerization domain of the HIV-1 capsid protein," *Science*, vol. 278, no. 5339, pp. 849–853, 1997.

[104] D. K. Worthylake, H. Wang, S. Yoo, W. I. Sundquist, and C. P. Hill, "Structures of the HIV-1 capsid protein dimerization domain at 2.6 Å resolution," *Acta Crystallographica Section D*, vol. 55, no. 1, pp. 85–92, 1999.

[105] X. Guo, A. Roldan, J. Hu, M. A. Wainberg, and C. Liang, "Mutation of the SP1 sequence impairs both multimerization and membrane-binding activities of human immunodeficiency virus type 1 Gag," *Journal of Virology*, vol. 79, no. 3, pp. 1803–1812, 2005.

[106] C. Liang, J. Hu, R. S. Russell, A. Roldan, L. Kleiman, and M. A. Wainberg, "Characterization of a putative α-helix across the capsid-SP1 boundary that is critical for the multimerization of human immunodeficiency virus type 1 Gag," *Journal of Virology*, vol. 76, no. 22, pp. 11729–11737, 2002.

[107] C. Liang, J. Hu, J. B. Whitney, L. Kleiman, and M. A. Wainberg, "A structurally disordered region at the C terminus of capsid plays essential roles in multimerization and membrane binding of the Gag protein of human immunodeficiency virus type 1," *Journal of Virology*, vol. 77, no. 3, pp. 1772–1783, 2003.

[108] Y. Morikawa, D. J. Hockley, M. V. Nermut, and I. M. Jones, "Roles of matrix, p2, and N-terminal myristoylation in human immunodeficiency virus type 1 Gag assembly," *Journal of Virology*, vol. 74, no. 1, pp. 16–23, 2000.

[109] A. Ono, D. Demirov, and E. O. Freed, "Relationship between human immunodeficiency virus type 1 Gag multimerization and membrane binding," *Journal of Virology*, vol. 74, no. 11, pp. 5142–5150, 2000.

[110] S. A. K. Datta, L. G. Temeselew, R. M. Crist et al., "On the role of the SP1 domain in HIV-1 particle assembly: a molecular switch?" *Journal of Virology*, vol. 85, no. 9, pp. 4111–4121, 2011.

[111] M. A. Accola, S. Höglund, and H. G. Göttlinger, "A putative α-helical structure which overlaps the capsid-p2 boundary in the human immunodeficiency virus type 1 Gag precursor is crucial for viral particle assembly," *Journal of Virology*, vol. 72, no. 3, pp. 2072–2078, 1998.

[112] J. L. Newman, E. W. Butcher, D. T. Patel, Y. Mikhaylenko, and M. F. Summers, "Flexibility in the P2 domain of the HIV-1 Gag polyprotein," *Protein Science*, vol. 13, no. 8, pp. 2101–2107, 2004.

[113] N. Morellet, S. Druillennec, C. Lenoir, S. Bouaziz, and B. P. Roques, "Helical structure determined by NMR of the HIV-1 (345-392)Gag sequence, surrounding p2: implications for particle assembly and RNA packaging," *Protein Science*, vol. 14, no. 2, pp. 375–386, 2005.

[114] E. R. Wright, J. B. Schooler, H. J. Ding et al., "Electron cryotomography of immature HIV-1 virions reveals the structure of the CA and SP1 Gag shells," *EMBO Journal*, vol. 26, no. 8, pp. 2218–2226, 2007.

[115] P. F. Smith, A. Ogundele, A. Forrest et al., "Phase I and II study of the safety, virologic effect, and pharmacokinetics/pharmacodynamics of single-dose 3-O-(3′,3′-dimethylsuccinyl)betulinic acid (bevirimat) against human immunodeficiency virus Infection," *Antimicrobial Agents and Chemotherapy*, vol. 51, no. 10, pp. 3574–3581, 2007.

[116] C. A. Stoddart, P. Joshi, B. Sloan et al., "Potent activity of the HIV-1 maturation inhibitor bevirimat in SCID-hu Thy/Liv mice," *PLoS ONE*, vol. 2, no. 11, Article ID e1251, 2007.

[117] E. Seclén, M. D. M. González, A. Corral, C. De Mendoza, V. Soriano, and E. Poveda, "High prevalence of natural polymorphisms in Gag (CA-SP1) associated with reduced response to Bevirimat, an HIV-1 maturation inhibitor," *AIDS*, vol. 24, no. 3, pp. 467–469, 2010.

[118] J. Verheyen, C. Verhofstede, E. Knops et al., "High prevalence of bevirimat resistance mutations in protease inhibitor-resistant HIV isolates," *AIDS*, vol. 24, no. 5, pp. 669–673, 2010.

[119] S. K. Lee, J. Harris, and R. Swanstrom, "A strongly transdominant mutation in the human immunodeficiency virus type 1 gag gene defines an achilles heel in the virus life cycle," *Journal of Virology*, vol. 83, no. 17, pp. 8536–8543, 2009.

Permissions

The contributors of this book come from diverse backgrounds, making this book a truly international effort. This book will bring forth new frontiers with its revolutionizing research information and detailed analysis of the nascent developments around the world.

We would like to thank all the contributing authors for lending their expertise to make the book truly unique. They have played a crucial role in the development of this book. Without their invaluable contributions this book wouldn't have been possible. They have made vital efforts to compile up to date information on the varied aspects of this subject to make this book a valuable addition to the collection of many professionals and students.

This book was conceptualized with the vision of imparting up-to-date information and advanced data in this field. To ensure the same, a matchless editorial board was set up. Every individual on the board went through rigorous rounds of assessment to prove their worth. After which they invested a large part of their time researching and compiling the most relevant data for our readers. Conferences and sessions were held from time to time between the editorial board and the contributing authors to present the data in the most comprehensible form. The editorial team has worked tirelessly to provide valuable and valid information to help people across the globe.

Every chapter published in this book has been scrutinized by our experts. Their significance has been extensively debated. The topics covered herein carry significant findings which will fuel the growth of the discipline. They may even be implemented as practical applications or may be referred to as a beginning point for another development. Chapters in this book were first published by Hindawi Publishing Corporation; hereby published with permission under the Creative Commons Attribution License or equivalent.

The editorial board has been involved in producing this book since its inception. They have spent rigorous hours researching and exploring the diverse topics which have resulted in the successful publishing of this book. They have passed on their knowledge of decades through this book. To expedite this challenging task, the publisher supported the team at every step. A small team of assistant editors was also appointed to further simplify the editing procedure and attain best results for the readers.

Our editorial team has been hand-picked from every corner of the world. Their multi-ethnicity adds dynamic inputs to the discussions which result in innovative outcomes. These outcomes are then further discussed with the researchers and contributors who give their valuable feedback and opinion regarding the same. The feedback is then collaborated with the researches and they are edited in a comprehensive manner to aid the understanding of the subject.

Apart from the editorial board, the designing team has also invested a significant amount of their time in understanding the subject and creating the most relevant covers. They scrutinized every image to scout for the most suitable representation of the subject and create an appropriate cover for the book.

The publishing team has been involved in this book since its early stages. They were actively engaged in every process, be it collecting the data, connecting with the contributors or procuring relevant information. The team has been an ardent support to the editorial, designing and production team. Their endless efforts to recruit the best for this project, has resulted in the accomplishment of this book. They are a veteran in the field of academics and their pool of knowledge is as vast as their experience in printing. Their expertise and guidance has proved useful at every step. Their uncompromising quality standards have made this book an exceptional effort. Their encouragement from time to time has been an inspiration for everyone.

The publisher and the editorial board hope that this book will prove to be a valuable piece of knowledge for researchers, students, practitioners and scholars across the globe.

List of Contributors

Simon F. Scrace and Eric O'Neill
Department of Oncology, The Gray Institute, University of Oxford, Roosevelt Drive, Oxford OX3 7DQ, UK

Clayton J. Hattlmann, Jenna N. Kelly and Stephen D. Barr
Department of Microbiology and Immunology, Center for Human Immunology, The University of Western Ontario, London, ON, Canada N6A 5C1

Dominic P. Del Re and Junichi Sadoshima
Cardiovascular Research Institute, Department of Cell Biology and Molecular Medicine, UMDNJ-New Jersey Medical School, 185 South Orange Avenue, MSB G-609, Newark, NJ 07103-2714, USA

Claudia Dittfeld
AWG Tumor Genetics of the Medical Faculty, Martin-Luther-University Halle-Wittenberg, 06108 Halle, Germany
OncoRay, National Center for Radiation Research in Oncology, Medical Faculty Carl Gustav Carus, University of Technology, 06108 Halle, Dresden, Germany

Katrin Steinmann and Reinhard H. Dammann
AWG Tumor Genetics of the Medical Faculty, Martin-Luther-University Halle-Wittenberg, 06108 Halle, Germany
Institute for Genetics, Justus-Liebig University Giessen, 35392 Giessen, Germany

Antje M. Richter
Institute for Genetics, Justus-Liebig University Giessen, 35392 Giessen, Germany

Antje Klagge-Ulonska
AWG Tumor Genetics of the Medical Faculty, Martin-Luther-University Halle-Wittenberg, 06108 Halle, Germany

Nezar Noor Al-hebshi
Molecular Research Laboratory, Faculty of Medical Sciences, University of Science and Technology, Sana'a, Yemen
Faculty of Dentistry, Jazan University, P.O. Box 114, Jazan, Saudi Arabia

Amat-alrahman Ahmed Shamsan
Molecular Research Laboratory, Faculty of Medical Sciences, University of Science and Technology, Sana'a, Yemen

Mohammed Sultan Al-ak'hali
Department of Periodontology, Faculty of Dentistry, University of Sana'a, Sana'a, Yemen

Mark A. Wainberg and Bluma G. Brenner
Jewish General Hospital AIDS Centre, McGill University, 3755 Cote-Ste-Catherine Road, Montreal, QC, Canada H3T 1E2

Susannah Kassler, Howard Donninger and Geoffrey J. Clark
J.G. Brown Cancer Center, University of Louisville, 417 CTR Building, 505 S. Hancock Street, Louisville, KY 40202, USA

Michael J. Birrer
Massachusetts General Hospital Cancer Center, Massachusetts General Hospital, Harvard Medical School, Boston, MA 02114, USA

Rebecca Louise Harris, Carmen Wilma van den Berg and Derrick John Bowen
Institute of Molecular and Experimental Medicine, Cardiff University School of Medicine, Heath Park, Cardiff CF14 4XN, UK

Francesca Fausti and Sabrina Strano
Molecular Chemoprevention Group, Molecular Medicine Area, Regina Elena Cancer Institute, Via Elio Chianesi 53, 00143 Rome, Italy

Silvia Di Agostino, Andrea Sacconi and Giovanni Blandino
Translational Oncogenomic Unit, Molecular Medicine Area, Regina Elena Cancer Institute, Via Elio Chianesi 53, 00143 Rome, Italy

Marilyn Gordon, Mohamed El-Kalla and Shairaz Baksh
Department of Pediatrics, Faculty of Medicine and Dentistry, University of Alberta, 3-055 Katz Group Centre for Pharmacy and Health Research, 113 Street 87 Avenue, Edmonton, AB, Canada T6G 2E1
Women and Children's Health Research Institute, University of Alberta, 4-081 Edmonton Clinic Health Academy, 11405-87 Avenue, Edmonton, AB, Canada T6G 1C9

Tracy L. Hartman and Robert W. Buckheit Jr.
Anti-Infective Research Department, ImQuest BioSciences, Inc., 7340 Executive Way, Suite R, Frederick, MD 21704, USA

Jason Hammonds, Jaang-Jiun Wang and Paul Spearman
Department of Pediatrics, Emory University and Children's Healthcare of Atlanta, 2015 Uppergate Drive, Atlanta, GA 30322, USA

Jeremy Luban
Department of Microbiology and Molecular Medicine, University of Geneva, 1211 Geneva, Switzerland

Jennifer Law and Shairaz Baksh
Department of Pediatrics, Faculty of Medicine and Dentistry, University of Alberta, 3055 Katz Group Centre for Pharmacy and Health Research, 113 Street 87 Avenue, Edmonton, AB, Canada T6G 2E1

Victor C. Yu
Department of Pharmacy, Faculty of Science, National University of Singapore, 18 Science Drive 4, Singapore 117543

Aparna Laskar and Somnath Chatterjee
Infectious Diseases and Immunology Division, CSIR-Indian Institute of Chemical Biology, West Bengal, Kolkata 700032, India

Euan J. Rodger
Department of Pathology, Dunedin School of Medicine, University of Otago, P.O. Box 913, Dunedin 9054, New Zealand

Aniruddha Chatterjee
Department of Pathology, Dunedin School of Medicine, University of Otago, P.O. Box 913, Dunedin 9054, New Zealand
National Research Centre for Growth and Development, University of Auckland, Auckland 1142, New Zealand

Francesca Esposito, Angela Corona and Enzo Tramontano
Department of Life and Environmental Sciences, University of Cagliari, Cittadella Universitaria di Monserrato, SS 554, 09042 Monserrato, Italy

A. Katrin Helfer-Hungerbuehler, Stefan Widmer and Regina Hofmann-Lehmann
Clinical Laboratory, Vetsuisse Faculty, University of Zurich, Winterthurerstrasse 260, 8057 Zurich, Switzerland

César Paz-y-Miño, Nadia Cumbal and María Eugenia Sánchez
Instituto de Investigaciones Biomédicas, Facultad de Ciencias de la Salud, Universidad de las Américas, Ave. de los Granados y Colimes Quito, 1712842, Ecuador

Catherine S. Adamson
Biomedical Sciences Research Complex, School of Medicine, University of St. Andrews, North Haugh, St. Andrews, Fife, KY16 9ST, UK

www.ingramcontent.com/pod-product-compliance
Lightning Source LLC
Chambersburg PA
CBHW080639200326
41458CB00013B/4683